CONTROL APPLICATIONS IN POST-HARVEST AND PROCESSING TECHNOLOGY

(CAPPT'95)

A Postprint volume from the 1st IFAC/CIGR/EURAGENG/ISHS Workshop, Ostend, Belgium, 1 - 2 June 1995

Edited by

J. DE BAERDEMAEKER and J. VANDEWALLE
K.U. Leuven, Heverlee, Belgium

Published for the

INTERNATIONAL FEDERATION OF AUTOMATIC CONTROL

by

PERGAMON
An Imprint of Elsevier Science

UK Elsevier Science Ltd, The Boulevard, Langford Lane, Kidlington, Oxford, OX5 1GB, UK

USA Elsevier Science Inc., 660 White Plains Road, Tarrytown, New York 10591-5153, USA

JAPAN Elsevier Science Japan, Tsunashima Building Annex, 3-20-12 Yushima, Bunkyo-ku, Tokyo 113, Japan

First edition 1995

Library of Congress Cataloging in Publication Data

A catalogue record for this book is available from the Library of Congress

British Library Cataloguing in Publication Data

A catalogue record for this book is available from the British Library

ISBN 0-08-042598 4

This volume was reproduced by means of the photo-offset process using the manuscripts supplied by the authors of the different papers. The manuscripts have been typed using different typewriters and typefaces. The lay-out, figures and tables of some papers did not agree completely with the standard requirements: consequently the reproduction does not display complete uniformity. To ensure rapid publication this discrepancy could not be changed: nor could the English be checked completely. Therefore, the readers are asked to excuse any deficiencies of this publication which may be due to the above mentioned reasons.

The Editors

Printed in Great Britain

1st IFAC WORKSHOP ON CONTROL APPLICATIONS IN POST-HARVEST AND PROCESSING TECHNOLOGY

Organised by
BIRA
partner of the
BELGIAN FEDERATION OF AUTOMATIC CONTROL (IBRA-BIRA)
Belgian IFAC N.M.O.

Sponsored by
IFAC - International Federation of Automatic Control
CIGR, International Commission of Agricultural Engineering
EURAGENG, European Society of Agricultural Engineering
ISHS, International Society for Horticultural Science

◆

CONTENTS

KEYNOTE LECTURE

MATHEMATICAL MODELLING AND OPTIMIZATION

DECISION SUPPORT AND PLANNING

MEASUREMENT OF PROPERTIES OF PRODUCE

COMPUTER CONTROLLED PROCESS ENVIRONMENT

INTELLIGENT TECHNIQUES IN POST-HARVEST OPERATIONS

OBSERVATION FOR CONTROL

QUALITY CONTROL

Progress in Process Operation by **Goal Oriented Advanced Control**

G. van Straten and A.J.B. van Boxtel
Wageningen Agricultural University
Department of Agricultural Engineering and Physics
Bomenweg 4, 6703 HD Wageningen, The Netherlands

Abstract-Larger competition, increasing demand for quality, and the necessity for lower inculpation of resources set the scene for production, storage, and processing of agricultural products and derivates in the future. There is no doubt that automatic control will play a significant role in achieving these objectives in the chain from producer to customer. Advances in sensor technology, microelectronics, and control theory presently offer solutions far beyond the classical PID-controller. Yet, major gains will not be obtained if the ultimate goal is lost out of sight. This means that control solutions, how ever advanced they may be, will have to be placed in the frame of overall system operation in order to be profitable.

An analysis of the situation in post-harvest processing shows that the bottle-neck in achieving economically attractive solutions lies in the necessity to have suitable models describing the product's behaviour in function of the environmental variables. Once such models are available the concepts of optimal predictive control provides an attractive framework for profitable process operation and control. Optimal solutions can be computed either off-line, and implemented with - possibly advanced - low-level feedback compensators, or implemented directly on-line using a receding horizon approach. Successful implemenation of such goal oriented operation requires close cooperation between process engineers and control engineers.

The philosophy is illustrated by examples from ongoing research projects on advanced control methods and dynamic optimal operation in the field of production, post-harvest technology and food processing.

Key words: process control, process operation, dynamic optimization, post-harvest processing

1 Introduction

Larger competition, increasing demand for quality, and the necessity for lower inculpation of resources set the scene for production, storage, and processing of agricultural products and derivates in the future. There is no doubt that automatic control will play a significant role in achieving these objectives in the chain from producer to customer. Advances in sensor technology, microelectronics, and control theory presently offer solutions far beyond the classical on-off or PID-controller. Yet, major gains will not be obtained if the ultimate goal is lost out of sight. This means that control solutions, how ever advanced they may be, will have to be placed in the frame of overall system operation in order to be profitable. The major theme of this paper is to argue that profitability is best served by amalgamating process operation and process control. This requires a cooperation and cross-fertilization between the process engineer, traditionally responsible for process design and overall operation, and the control engineer, traditionally responsible for controller design and implementation.

The main concern during post-harvest operations is the storage and modification of agricultural produce. Product properties are formed during production, and modified during conditioning, storage, and further processing. A distinction can be made between mechanical handling and operations, and process operations. Examples of mechanical handling are harvesting, chopping, grading, animal food supply, manure collection, automatic milking, egg collection, feather picking, etc. Packing of the final product is also an important operation where automation is wide spread. In most of these mechanical applications, the emphasis is on automation by finite state machines (PLC's). Feed-back control is less common here and less necessary. Although computer vision and various forms of robotics do need feed-back, and find increasing application, we will restrict the discussion in this paper to process operations.

A common characteristic of post-harvest processing is that product modifications depend upon the immediate environment: temperature, moisture, radiation, chemical composition, flow conditions. A rough sketch of the present practice is as follows. By experience, trial and error or (rarely) computation, the process engineer knows what conditions will lead to the desired product properties. So, a set of environmental requirements, fixed in time or possibly as time trajectories, are available. Next the realization of these trajectories is a task of controllers, usually automatic controllers. Consequently, the task of control is seen as disjoined from the task of establishing and adjusting operation conditions. The latter is the principle field of process engineers, whereas the first is the field of control engineers.

The principle point of this presentation is our

conjecture, shared by many others, that the largest improvements in profitability can be made by 'closing the outer loop', i.e. by imposing the trajectories needed to achieve the best economical result. The second point of this paper is that this does not mean that the inner loop control problem always can be treated separately. There are two basic assumptions underlying the principle of hierarchical separation. The first is, that patterns of operating conditions or set-points specified are indeed realizable by the controller, and the second is that the assessment of operational conditions provides sufficient clues to formulate the design specifications of controllers at the basic level, but is otherwise independent of the controller design. In many cases, however, it is not obvious that these assumptions are sufficiently valid in order to achieve maximum profitability.

To illustrate this point, let us take the cool storage of potatoes. Usually, there is a desire to maintain both temperature and moisture content. However, cooling and ventilation heavily interact, and strict set-points cannot always be maintained under the prevailing environmental conditions. Knowing, on the other hand, that it is not so much the temperature itself, but rather a temperature integral that should be maintained, one could imagine a control that exploits the natural outside conditions by leaving room to moisture and temperature fluctuations. So, the emphasis should shift from maintaining pre-set conditions to steering them in a fashion that warrants product quality, and at the same time saves on energy costs. It is clear that this would require a re-thinking of the controller specifications, while at the same time allowing for a kind of feed-back adjustment in the operating conditions. In the ideal case, the total system should encompass these aspects in one go.

We argue that the separation between control in the strict sense and operation in many cases does not satisfy the high demands of today. Having said that, one should also look for possible solutions. We feel that such integrated solutions come closer thanks to the recent developments in the area of model based economic optimal control.

The structure of the remainder of this paper is as follows. We first explore the demands of post harvesting operations and processing. Next, we try to analyze where a few decennia of developments in the control field have brought us, stressing the conditions under which various advanced control schemes are applicable. Then the present solutions to the operational problem will briefly be reviewed. We then return to the question of profitability, by sketching an outline of the situation that we feel should be strived for in order to integrate control and operation to achieve the best economical result: *goal oriented advanced control*. We try to stress that this ideal demands close cooperation between process engineers and control engineers, and perhaps some additional educational training of both groups. In the end, some still limited but hopefully illustrative examples will be given.

2 Goals in Post-harvest and Processing operations

Post harvest operations and processing form just one part of the chain from producer to consumer. The aim of this section is to briefly discuss the objectives for control and operations. Although strictly speaking not part of the post-harvest stage, the discussion begins with the production stage, because of its effect upon the remainder of the chain. The next step is conditioning, to prepare the raw product for storage or later processing. Often, a storage is needed before the product can be transferred to the client or to further processing. In each of these steps the ultimate and derived goals will be discussed in order to position the potential options for control and operation.

2.1 Production

The ultimate goal of producers is to make profit. They will therefore attempt to outweigh the costs in investments, labour and operation to the benefits obtained when selling the product. An important aspect of the ability to obtain a good price is the quality of the product, i.e. the degree of fitness of the product for the intended use. A factor sometimes overlooked is that extreme good quality obtained at the expense of higher costs is only justifiable if the market is prepared to express the additional value in money. Otherwise, there is quality give away. Apart from quality, other aspects are important as well, e.g. timeliness (think of flowers for mother's day), product diversification, product market life cycle, and expected competition. In addition, growing awareness of consumers for environmental and energy issues, and believed or real health aspects, has lead to a tendency that market prices are not just determined by the intrinsic quality of the product itself, but also by the method of production used.

Climate conditioning is a dominant subject for application of control in many protected agricultural production processes. The traditional goal for low-level control is automation of short-term steering for disturbance rejection - e.g. reducing the influence of weather conditions. On the higher level, the objective is to maintain or manipulate environmental conditions in order to guarantee the 'best possible' conditions for plant growth or animal breeding. Since these conditions may be variable over time, the low level controllers also should provide for good tracking properties. The control system now becomes a tool to steer the production. The pathways required can be stored in advanced computer control systems, e.g. in greenhouse climate computers. Thus, the control system becomes a information system and a management tool.

A critical appraisal of the situation in plant or animal breeding in view of the decomposition assumptions shows the following. The climate control problem often involves simultaneous control of humidity and temperature, sometimes also control of gases such as CO_2. This is not an easy problem, because these loops are highly interactive. Moreover, there may be an uneven spatial distribution, which may be manageable and therefore should be taken into consideration (De Moor and Berckmans, 1995). The separation problem between set-point optimization and control is not easy either. Because time-scales of the physical processes involved are far shorter than those of the plant or animal a hierarchical decomposition seems possible. However, due to fact that the plant or animal itself exerts a non-negligible influence upon the environment, which is variable in time as production proceeds, the separation is less obvious. In glasshouses, moreover, the strongly varying irradiation input, which is not a disturbance but an essential production factor, cannot be 'regulated away' but rather should be exploited. This is not compatible with maintaining preset set-points trajectories, and adaptations may be profitable. Both the computation of the set-point trajectories and their adaptation to actual conditions require proper models relating plant growth and development to environmental conditions. In view of the ultimate goal, i.e. to make profit, it is quite remarkable to see that most controller systems for plant production or animal breeding do not make explicit reference to crop quantity and quality on the one hand, and operation costs (feed costs, nutrient costs, CO_2 costs, energy costs, etc.) on the other. Van Straten and Challa (1995) give an outline of the problems and their potential solution in greenhouse crop cultivation.

2.2 Conditioning

Many agricultural raw products require conditioning after harvesting in order to make them more suitable for later use. One can think of mechanical operations, like chopping. Other examples are drying of hay or grain and processing of seeds (pillage). The ultimate goal of these operations is to enhance the value by making the product less vulnerable to decay, to improve on conservation properties, and to preserve certain characteristics, like food value. Taking the example of drying, these goals can be translated in derived goals: to subject the product to a drying temperature regime that leads to a pre-specified moisture content at the lowest possible cost. 'Cost' is partly expressed as direct costs (energy), and partly as indirect costs (time of operation). Time is important in case of occupation of equipment, but also when no equipment is used, like in drying on the field, because of the increased risk for losses due to rain when time proceeds.

Reviewing the assumptions in this area it is clear that again it is necessary to carefully investigate the consequences of interaction between humidity and temperature, and the effect the product has on these quantities. Mass and heat transfer are largely determined by air flow rate, and so is the spatial distribution within the product. Moreover, the time scales of product transformations and environmental physics are in the same order of magnitude. Consequently, decomposition probably is not straight forward, and interference of controller objectives and operational objectives can be expected.

2.3 Storage

The natural limitations of production often cause an imbalance between the moment of production of agricultural produce and the demand from the market. Storage is then the common solution. Sometimes, storage is needed to provoke desired quality changes, e.g. ripening of banana's. Options are storage of the raw preconditioned product, or pre-processing and storage of half-products or storage of the final product. Because of the spatial demand, and the necessity to maintain proper environmental conditions, storage is expensive. Proper logistics sometimes can prevent the need for extensive storage (e.g. the ripening of banana's during ship transport).

Since agricultural products are of biological origin, usually considerable changes can take place in the product during storage. Plant material respires, for example. Also, weight losses due to evaporation occur. Gases, like ethylene, may be liberated, which may effect the freshness of other products stored in the same warehouse. The humidity and temperature conditions in the store are also instrumental to the risk of putrefaction, for instance condensation of water on fruits greatly enhances the chances for fungi to grow.

The most eye catching control aspect of storage is the generation of a controlled environment. The ultimate goal again is to make the best profit, or to loose as little as possible. Again there is a dichotomy:

the determination of the 'best possible' patterns or pathways in order to maintain or improve the quality of the product, and the realization of these pathways by proper control in the presence of changing disturbance conditions. The generation of pathways can be very complex, like e.g. in potato storage, where there is first a drying period, a wound healing period, a cooling down period, a period of long term storage and finally a reconditioning period. Experience has lead to a number of rules of thumb, and some of them have given rise to decision support systems in the form of expert systems.

Although cost effectiveness should be strived for, a major concern in storage is the avoidance of risk. Steering the climate control system to its economical limits may increase the vulnerability to failures or unexpected events, with the risk of loosing the whole lot. Consequently, risk assessment will have to be an important factor in any operation and control system for storage. Again, the remarkable thing is that only limited quantitative information is available to design the operational and control system to explicitly take costs, benefits and risks into account.

2.4 Processing

Some products can be sold directly on the market, like fruits and vegetables. Others require further processing. Examples here are all sorts of meat processing, including sterilisation, cooking and baking, and use of raw materials for fermentation and other bio-processes, as for cheese making and other diary operations, and in processes like wine production and beer brewing. In as much as processing is involved, control plays a similarly important role as in chemical processing. Due to the high proportion of batch or batch like processing in the food industry, there is more then proportionate attention to proper operation pathways and schemes.

The ultimate goal here is again to make profit, that is, the value added due to the operation should more than compensate for the total expenses. The derived goals are therefore the establishing of proper operational trajectories. In many cases these are found by a long tradition of trial and error, without explicit use of quantitative models. Like in the case of storage, process operators have a tendency to avoid risks. The assumption is that desired pathways can, in fact, be realized by low-level controllers.

2.5 Issues of profitability

A rather general picture of the present situation in the majority of post-harvest operations is given in Figure 1. The product is processed in a controlled

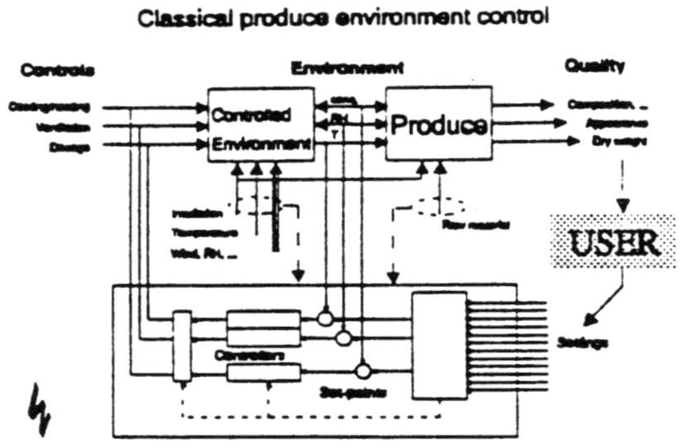

Figure 1. *Classical produce environment control*

environment. The quantity and quality of the product are dictated by the time history of the environmental variables. The environmental variables are controlled by control loops which influence the control inputs of the system. This part is closed loop, with the task to reject disturbances and to track set-point trajectories. However, the changes in the product are essentially open loop. The process operator must have recipes or blue prints which specify the desired environment in order to achieve the desired product. At best, he can make some adjustments on the basis of off-line observations on the product.

As shown before, the ultimate goal should be profitability. The three key elements that play a dominant role in the profitability issue are *quality*, *risk*, and *costs*. Whereas in a given operational set-up the expected costs can be computed, risk and quality are much more difficult to handle. Risk avoidance in effect means to stay away from operational constraints within a safety margin. It is obvious that conservative risk behaviour prevents the exploitation of the margin to the operational constraint. It is not easy to deal with risk in an objective way. Perhaps the most fruitful way of thinking is to treat the additional costs due to constraint avoidance as an insurance premium.

Quality is a central theme in post-harvest technology. At the same time, quality is difficult to define objectively. Quality can be defined as the degree of fitness for use. There are two difficulties associated to the quality concept in the frame of operation and control.

First, it is necessary to translate fitness for use into measurable quantities. Elements here are

- outer appearance: shape, colour, firmness, lack of bruises, texture;
- inner properties, either desirable or undesirable: water content, intracellular water, starch content, ion concentrations, sucrose content; pesticide

residues
- consumer properties: taste, odour

The second difficulty is to find out how the operational conditions influence the measurable quantities.

The above analysis leads to three areas of interest in the frame of enhanced profitability. The first, which will be discussed in the section 3, is to improve on the performance of controllers in the environmental control inner loop. The second deals with the (open loop) relation between environment and product quality and quantity dynamics. This part calls for dynamic product models. Finally, and ultimately, it seems logical to look for ways to also close the outer loop, i.e. to automate the process as a whole, using economic criteria. This will be discussed in section 5.

3 Developments in control

Control is concerned with manipulating a system in order to let it behave according to a certain objective. If the objective is specified in the form of a mathematical goal function, optimal control could be achieved if (a) the future external inputs to the system are given (b) the response of the systems behaviour to both disturbance and control inputs can be predicted exactly. Obviously, in practice, nor perfect knowledge of future inputs, nor a perfect model of the behaviour is achievable. The answer to these uncertainties is feed-back, i.e. making corrections in response to deviations in actual observed behaviour to desired behaviour. Although the overwhelming majority of practical applications applies feed-back, in recent years there is a revival of predictive methods, which, as we will see, if properly combined with the idea of feed-back, may offer interesting opportunities.

3.1 Feed-back control

3.1.1 On-off and time-proportional control

The most simple form of feed-back control is on-off control. It finds wide-spread application in many fields, in particular in air conditioning equipment. The main reasons are the intuitive appeal and easy understanding, the low costs, as actuators do not need to be variable, and the simple controller rules, which are easy to implement in hard-ware logic. A variant is formed by time-proportional controllers, which maintain the advantages but allow for refined control. An important prerequisite for the applicability of on-off control is that the system is well dimensioned, and that the requirements in terms of precision are moderate.

In passing it is interesting to remark that if the model and the goal function are both linear in the control, an optimal open loop control path is given by bang-singular-bang control. The bang-bang solutions are obviously on-off solutions, except that the switching rules may be different from what is achieved by the usual heuristic design.

3.1.2 PID control

The prototype of classical feed-back control is the celebrated Proportional-Integral-Derivative (PID) controller. The success of PID control is largely due to its good compensation properties on processes that behave in a first or second order like fashion. As many processes involving flow of material or heat have behaviours that can be approximated quite well with sigmoid type step responses, PID control has found widespread use in the process industries. A rough estimate is that some 80-90% of all control problems can satisfactorily be solved by the use of PID controllers. The fact that PID controllers can be tuned on the spot without formal reference to a process model also has largely contributed to its success.

Yet, PID controllers are not a panacea to all problems. In cases of large dead times, inverse response processes (non-minimum phase), and inherently multivariable loops with mutual interactions PID control does not provide satisfactory solutions.

An important limitation of PID controllers in the context of the present discussion on goal oriented control is the often overlooked fact that PID controllers control at all costs. This may not be the most desirable property from an economic point of view.

3.1.3 Cascade control

If the process consists of subsystems with widely varying time constants, an improvement on PID control of the slow subsystem is possible, provided a additional measurement signal on the fast process is made available. This is called cascade control, where the outer controller provides the set-point for the fast inner loop (Stephanopoulos, 1984). An example is the control of a ventilator flow rate controller by a humidity controller in a store house, or the control of the pipe temperature controller in greenhouse temperature control.

3.1.4 Multiple PID loops

PID control is essentially single input-single output control. Often, various process variables need to be controlled. Provided there are sufficient degrees of freedom, a possibility is to use multiple PID loops. In practice, it is not always obvious how to select proper control inputs, and how to arrange the pairings. The most desirable situation is achieved when loops can be decoupled, because otherwise the loops will interact and may cause oscillatory or even instable behaviour. Recently, the issue of practical controllability has regained attention, especially in the field of chemical process control. The relative gain array, Niederlinskii index and singular value decomposition are a few techniques to achieve acceptable multiple loops, and to study the controller integrity, i.e. the system behaviour if one of the loops fails (Luyben, 1990, Maciejowski, 1989).

3.1.5 LQ optimal control

Linear quadratic optimal control has been proposed to counteract two disadvantages of PID control: the dealing with multiple inputs and outputs, and the balancing of output performance versus control energy. LQ control theory is based upon the availability of a linear or linearized model of the process

$$\dot{x} = Ax + Bu$$
$$y = Cx + Du$$

and tries to optimize a goal function which must be in the special form of quadratic weighting of state deviations and control inputs:

$$J = \int_0^\infty (x^T P x + u^T Q u) dt$$

where x, u, y are the states, inputs and outputs, respectively. Provided that the model is available, this design leads to a closed loop control law in the form of multiple state feed-back:

$$u = -K_c x$$

where K_c is the feed-back gain matrix. Given the model the feed-back gain matrix can be computed in advance by solving an algebraic Ricatti equation. Thus, the actual implementation does not require an explicit on-line model.

Various modifications are known. For instance, an integral action can be introduced in order to abate offsets due to load variations. This involves the definition of auxiliary state variables (see e.g. Kwakernaak and Sivan, 1972). Also, if the states cannot all be measured, one could use a model to reconstruct the state from the available outputs. This is sometimes referred to as inferential control (Stephanopoulos, 1984). If the model is linear and time-invariant, the resulting observer can be used to compute the output feed-back, and no on-line model evaluation is necessary.

If the model and measurement have noise, the problem is called a linear quadratic Gaussian problem (LQG). This problem can be solved by the separation principle, sometimes called 'certainty equivalence' principle, which states that the optimal solution can be found by first estimating the state by an optimal state estimator (Kalman filter), and then solve the state feed-back, using the estimated state as if it were exact (e.g. Maciejowski, 1989). In this case, the actual controller implementation contains the Kalman filter. The controller itself is thus a dynamic system, with the same state dimension as the original plant:

$$\dot{z} = (A - BK_c)z - K_f(Cz - y)$$
$$y = -K_c z$$

where z is the internal controller state. The Kalman gain K_f and the controller gain K_c matrices appearing in the controller configuration can be computed in advance, and will only depend upon the original plant model (A,B,C; D assumed zero for simplicity), and the specifications of the penalty matrices P and Q.

The exposure above shows that the optimality of LQ(G) design depends upon three premises: there should be a sufficiently exact linear or linearized model, the noise should be Gaussian, and the goal should be expressed in quadratic terms. In practice, these premises are not always easy to meet. In particular the model assumption may be problematic. Many biologically based systems are not well understood, resulting in approximate models where the parameters may vary due to unmodelled sub-processes. Even seemingly simple physical systems, like the climate system in a greenhouse or a storage can have complex time-varying behaviour as a consequence of changes in the crop or the product. So, even if a linearization is reasonable, the system dynamics are often uncertain. Consequently, there is a need to look at the robustness of the control versus variations in the plant transfer function matrix.

There are two principally different approaches to the problem of robustness: robust design methods, or adaptive methods. The decision which approach to take is largely depending upon the question whether the system is essentially linear over the desired range of operating conditions, or not. In the latter case, adaptive control can be an attractive alternative to robust design methods (Åström and Wittenmark, 1989).

3.1.6 Robust designs

Loop transfer recovery

Although both the optimal state-feedback regulator and the Kalman filter in the LQG approach have good robustness and performance properties, their combination has not. This can be counteracted by redesigning the Kalman filter such that the good full state feed-back properties with respect to robustness are 'recovered'. This technique is known as loop transfer recovery (LQG/LTR, e.g. Maciejowski, 1989). The design generally results in a controller with a higher order than the LQG design.

H_∞ optimal control

The key issue of H-infinity controller design is to look for a stabilizing feed-back controller which minimizes a worst-case criterion function (infinite norm rather than the 2-norm in the quadratic case), usually formulated as a user weighted balance between the sensitivity function and the closed-loop transfer function. Although the specifications are given in frequency domain terms, the design result may still be formulated as a observer/state-feed-back combination. So, in the end, the design yields the same structure as LQG, but with different numbers. A good treaty of H_∞-optimal design is given in Doyle et al. (1992).

It is important to point out some basic features of all multivariable problems. Tight control of one particular state usually cannot go together with tight control of the other states. Also, the penalizing of steering energy may have the effect of seemingly less well behaviour to the judgement of the casual passer by. However, since a criterion is optimized, these designs give the best possible control, provided the criterion function really expresses the wishes. These wishes are based on some prior knowledge about plant uncertainty, disturbance frequency range, and expected frequency range of the command signal on the one hand, and some appraisal of the desired performance on the other. In general, the designer has quite some freedom to readjust the weights in the optimization procedure. Yet, it is not obvious how the design criteria are linked to the ultimate economic goal. It seems that in cases where the functioning of the controller has economic impact it would be desirable to restrict the designer's freedom by the ultimate - economic - goals of the controller. We are not aware of approaches where this is done.

3.1.7 Adaptive control

The other way of dealing with model uncertainty and variability is to try to steadily re-adjust the controller parameters (direct adaptive control) or to steadily recompute the controller parameters via a design procedure based upon an updated model (indirect adaptive control). Åström and Wittenmark (1989) give an excellent treatise on adaptive control, and discuss self-oscillating adaptive systems, gain scheduling (clearly showing the incentive to counteract the changes in dynamics of a non-linear plant if it is pushed into an other operating point), auto-tuning, model-reference adaptive systems and self-tuning regulators.

A key problem of adaptive controllers is that they can only successfully update the system model if there is information about the misfit. This deviation information, however, is hardly available over periods where the controller is successful. So, much of the adaptive control design theory is dealing with this problem.

3.1.8 Fuzzy control

A particular class of controllers that also tries answer the model uncertainty problem is formed by the class of fuzzy controllers. The key feature of fuzzy controllers is that they try to circumvent the need for an explicit quantitative model of the process. (Note that we do not refer here to the class of predictive controllers that make use of an explicit fuzzy model of the plant). Fuzzy controllers are usually set up by specifying rules derived from experience of human operators. Therefore, they may be viewed as idealized replacements of human controllers. The input-output mapping of a fuzzy controller is non-linear. If only the output error signals are used, the mapping is static, in case also the rate of change is used, as is generally the case, the mapping may be dynamic, as in PID control. Once the controller is designed on the basis of experience, experimentation, or simulation using any kind of suitable process model, the actual controller is just the set of membership functions and rules, and does not refer to a systems model.

Unlike common believe, fuzzy controllers do need sufficiently quantifiable input information. As they are derived from experience, they have the tendency to solidify present policies, and are therefore conservative. Nevertheless, fuzzy control has been applied in difficult processes to unify the performance of various human operators. Fuzzy controllers can also be used in cases where there is some common sense feeling on how a process could be controlled. The actual design, however, is difficult, and it is not clear how the results can be made optimal in an economic sense without giving up the advantage of not having the need for an explicit model.

3.2 Predictive control

The basic idea of predictive control is simple and attractive. Having a quantitative model for the process, one can at any time compute the control action needed to steer the process to its desired state according to some performance criterion. If there are no disturbances (or if future disturbances are known), and no model errors, this would amount to open-loop control. In practice, of course, these conditions are never really met, and consequently some form of feed-back is needed. It depends upon the specific situation how this is achieved, and some predictive control designs are actually equivalent to feed-back designs.

The different situations can be categorized as follows:

a) some disturbances measurable: feed-forward compensation, combined with feed-back design for the remaining disturbances, (note: feed-forward compensation does not imply knowledge of *future* disturbances)

b) linear model expressed as Markov parameters (impulse response) estimated from step response measurements: Dynamic Matrix Control (DMC). The solution is a fixed output feed-back controller (Luyben, 1990)

c) linear model in the form of CARIMA (controller autoregressive integrated moving average) model, and a quadratic goal function: Generalized Predictive Control (GPC) (Clarke et al., 1987). This approach leads to closed loop solutions which can be computed in advance.

d) non-linear model and a goal function of any shape, with small disturbances: open loop optimal control with feed-back compensator (Athens, 1971). The open loop solution must be computed off-line, which yields optimal state and control trajectories. On-line, corrections are applied using a linear (usually LQG designed) compensator.

e) non-linear model and a goal function of any shape, with solution of the optimal solution over some horizon, application of the first computed control action only, and repetition of the procedure as a new measurement comes in: Receding Horizon Optimal Control. The repetition ensures feed-back. (Mayne and Polak, 1993, Michalska and Mayne, 1993).

One of the attractive features of predictive control methods are that they can deal with state constraints. The latter two approaches are feed-back modifications of the general optimal control theory for any goal function with any exact model. This is sometimes referred to as 'dynamic optimal control'. The problem can be solved using the Hamiltonian approach, and using Pontryagin's optimal principle in case of controller constraints and a goal function that is linear in the controls (Lewis, 1986, Bryson and Ho, 1975). It is also conceivable to calculate a solution by suitable control parametrization and solving for the parameters by some global optimization algorithm, like random search or genetic algorithm.

4 Product quality models

The outline in section 3 shows that a large arsenal of controller design methods is available, which can solve the problem of controlling environmental variables by suppressing disturbances and maintaining fair tracking properties. Like in other fields dealing with product quality, in post-harvest processing the ultimate aim is, however, not the accurate control of environmental variables, but rather the control of product quality and quantity. Of course, the same techniques could be applied to this part of the problem, provided a sufficiently accurate model was available to describe the changes in product quality and quantity in function of the environmental variables.

Product quality models, therefore, are crucial if one wishes to achieve better operational results. At the same time, such models are not yet available for every product of interest. In the diary field, several relationships have been documented (e.g. Kessler, 1988). Another example is the inactivation of lactobacillus during drying (Lievense, 1991). A number of attempts are known in the field of drying, where models have been proposed to describe the drying process of rice (Toyoda, 1988), maize (Courtois et al., 1991), corn and other grain particles (e.g. O'Callaghan et al., 1971). Other situations where something is known are weight loss and respiration of potatoes and vegetables during storage, ethylene production of fruits, etc.

Also non-mechanistic models are sometimes available, e.g. in the form of fuzzy models, or fuzzy-neural models. As long as these models are quantitative, they can be viewed as non-linear dynamic or static descriptors of the product behaviour. In the frame of the general optimal predictive control such models could be used just like any other model.

It should be emphasized that the development of product models should be guided by the application in mind. Experience has shown that it is essential that this effort is done in close cooperation between the process engineer and the control engineer. Models developed for research purposes are seldom suitable for direct use, and normally should be reduced, which requires both process knowlege as well as knowledge of the requirements of control system design.

5 Perspective: optimal operation and control

From the discussion before it should now be clear that the achievement of profitability is closely related to closing the outer loop, i.e. by taking the product into account. In the ideal case, one would have a complete model of the process as a whole, some economic optimization function, and an optimization tool to compute the required controls. This is shown in Figure 2. Several aspects need to be considered before practical realization is possible.

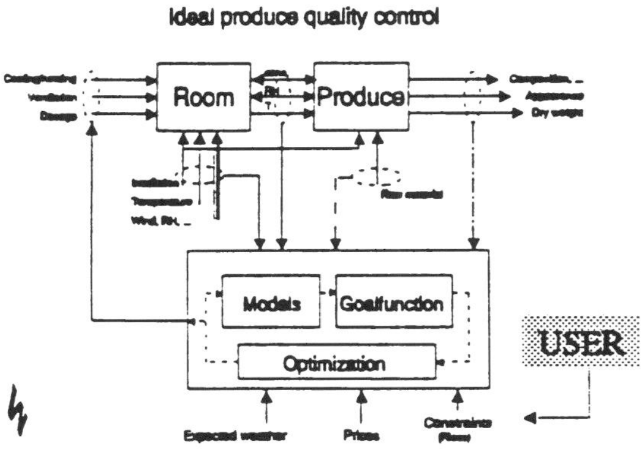

Figure 2. *Ideal scheme of economic goal oriented operation and control.*

With respect to the outer loop there are two main situations. In the first, some or all of the product quality features can be measured. Examples are the use of respirometric measurements, on-line weighing, on-line colour measurement, and so on. In that case, feed-back designs are possible. However, when we look at the feed-back design procedure (Figure 3) we see that the goals entering the design should take the form of controller specifications, to assure stability, good performance and robustness. This entails specifications like rise time, overshoot, band width, cross-over frequencies, and phase and gain margins, according to the procedures described in section 3. When we look at the controller thus designed from a user perspective, we see that the user should generate a command signal, which the controller will then try to realize on the basis of the design specifications. The user, however, has its own goals, and it is by no means trivial that the choices the designer has made when designing the system are consistent with what the user wants. Moreover, the user, on his part must have an idea about the consequences of the reference trajectory which are outside the actual control system.

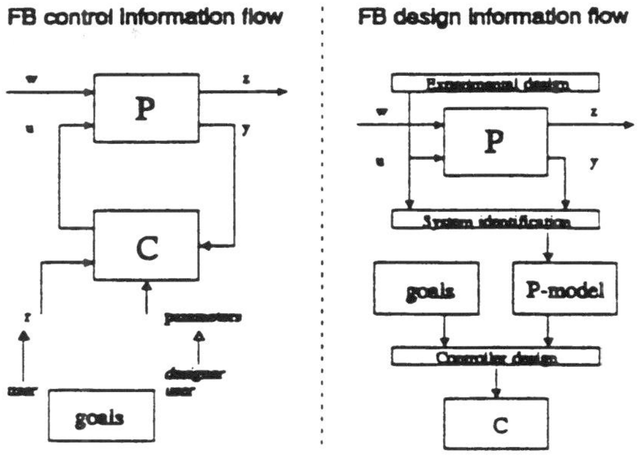

Figure 3. *Information flow during design (right) and in actual operation of the controller (left).*

We feel that these questions are easier to address in the frame of predictive controllers, because they allow, at least theoretically, designs that optimize economic criteria, while satisfying possible constraints.

The second situation occurs when the product properties cannot be measured on-line, or only occasionally. In that case, there is essentially no other choice than open loop predictive control methods to generate economically optimal trajectories.

The ideal of economical optimal control requires the following:

In the open loop part:

- Product quality and quantity model
- Physical environment model
- A goal function
- An optimization procedure

In the feed-back part:

- Sensors for product quality/quantity. These may be indirect, e.g. software sensors, speaking plant/fruit concept (De Baerdemaeker and Hashimoto, 1994)
- On-line physical sensors
- Suitable control implementation, maintaining the integrity of the overall optimal goal.

In the feed-forward part:

- Sensors to measure external disturbances
- Sometimes: a disturbance forecasting mechanism.

Prior to discussing some aspects of the actual implementation of optimal control schemes, we will first give two case studies to illustrate the potential of optimal path generation by optimization.

6 Examples of the goal oriented approach

6.1 Biscuit baking

At first sight baking of biscuits seems a rather easy control problem. The oven should be kept at a prescribed temperature and in industrial ovens, where biscuits are being transported through succeeding baking sections, the temperature of each section must be maintained at the set-point value. PID controllers can be applied, but also the on-off controller, as used in an ordinary kitchen oven, is very successful. In the controller design the dynamics from energy input to oven temperature should be taken into account.

Is baking really a simple temperature control problem? Let's examine the objectives of a baker preparing a product like biscuits.

- The first objective is to make a tasty product with a desired colour and which satisfies texture criteria. To realize this the prepared dough must rise and dried, while reactions are required to realize the colour and to enhance the flavour.
- The second objective is to realize an operation where a minimum of energy is being used and the oven is utilized efficiently.
- Third, the operation must be within safe bounds: self-ignition of the dough must be avoided.
- Finally, the operation is subject to constraints of minimum and maximum values of the available equipment.

From this summary of objectives it is clear that baking is not merely a temperature control problem but rather an operation problem that should satisfy a combination of objectives. The optimal operation strategy might possibly involve time dependent values of the input variables. The calculation of the trajectories of the input variables is a general optimal control problem which is non-linear and not quadratic. An essential ingredient of the dynamic optimization calculations is information about product formation. Furthermore suitable procedures for finding the solution of the control problem are needed.

In order to cope with the high degree of complexity of product formation the realization of the objectives is separated into several sub-problems in this case study. As a first step to full comprehension, operation strategies are worked out to achieve a major aspect of quality at the final baking time, i.e. the final mean water concentration.

Model background

Biscuits are porous and consequently heat transport is a result of conduction in the solid state and convection in the pores. The enthalpy transferred from the heating air to the dough is used to increase the temperature of the biscuit and also for evaporation of water from the dough into the pores. Mass transport is primarily by diffusion, but forced convection plays a part too, because the evaporation of water in the relatively small pores causes a vapour pressure rise in the biscuit.

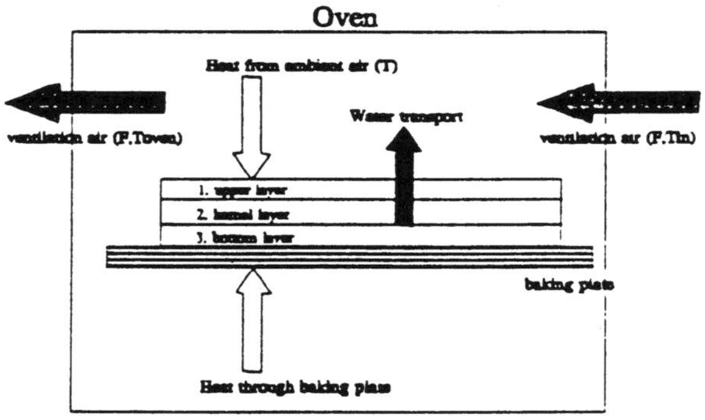

Figure 4 *Schematic representation of a drying biscuit*

Figure 4 shows a single biscuit in a gas fired oven. The biscuit is placed on a baking plate. Water transport is only possible to the air, not through the baking plate. The biscuit is about 3-5 mm high and has a diameter of about 5 cm. Therefore the main route for water and heat transport is along the vertical axis. The dough is heated up by air and the baking plate and as a result the temperature decreases from the edge to the centre of the biscuit. There is also a gradient in the water concentration in the biscuit. To handle the gradient in water concentration and temperature the biscuit has been divided into three coupled sections in this preliminary study.

Simulation

Prior to optimization the behaviour of the biscuit under traditional fixed baking strategy is examined by simulation. The incoming air is kept constant at 177 °C (450 K), and the baking plate at 147 °C (420 K). A characteristic simulation result is given in Figure 5. Three stages can be distinguished. Stage 1 is the heating phase where the temperature of the dough is being increased. In this period also water evaporation in the upper and bottom layer takes place which results in a decreasing water concentration in these

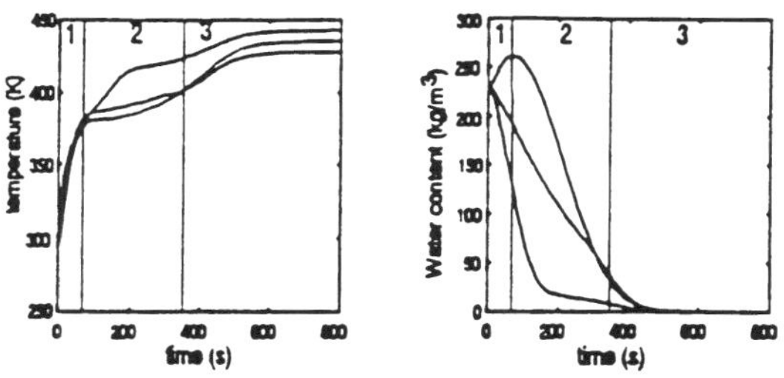

Figure 5 *Temperature and water concentration profiles in the biscuit during baking. —— upper layer, - - - kernel layer, - - - bottom layer*

layers. The temperature in the kernel layer is below that of the upper and bottom layer and therefore a part of the water vapour condensates in the kernel layer and thus the water concentration in this layer increases. In the second stage the mean evaporation rate is nearly constant, as can be seen from the almost linear decrease in water content, while the temperature changes in each layer are quite small. Stage 3 is the stage of decreasing drying rate characterised by a further increase in temperature.

Optimization

An optimization strategy was computed using the following objective function:

$$J_k = W(\rho_{mean} - \rho_{ref})^2 + \int_{t_0}^{t_f} E\,dt$$

where ρ is the water content averaged over the three layers of the biscuit (expressed in kg per m^3 dough), and E the energy costs, derived from the required ventilation air flow rate and the increase of ambient temperature to the required oven temperature. The units of the objective function are Dfl/m^3 dry product. W is a weight factor to translate the deviations from the desired reference water concentration into the units of J_k.

The control variables to be optimized are the ventilation flow rate and the temperature of the air into the oven. A typical result of the calculations is given in Figure 6.

During the baking time at first the energy input is mainly used to increase the temperature of the dough and after about 100-150 seconds for evaporation. Towards the end of the baking time evaporation becomes more difficult and the evaporation rate drops. At this point it suffices to use a low ventilation rate, which is profitable from the point of view of energy consumption. In this period the inlet temperature is below the temperature of the oven. This means that heat is being transferred from biscuit to oven. Thus, the drying strategy uses energy that has been stored in the mass of the biscuit in a previous stage, for drying at the end of the baking time.

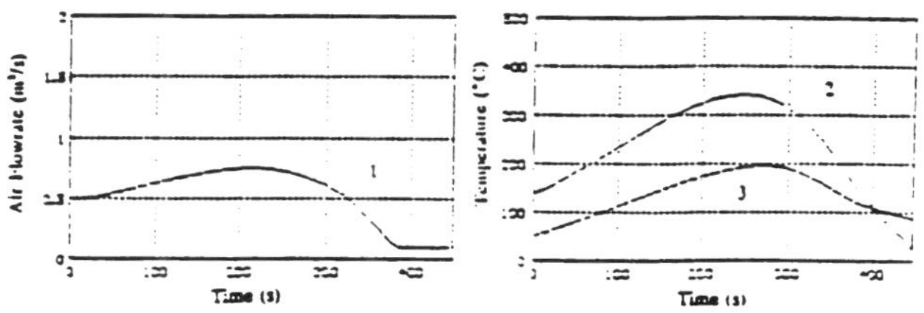

Figure 6 *Optimal trajectories. 1: air flow rate. 2:air inlet temperature. 3 temperature in oven.*

Remarks

Typical problems to be solved in this first step of the case study were:

- how to deal with the moisture and temperature gradient in the biscuits
- the occurrence of local minima in the optimization procedure
- interpretation of the background of optimal trajectories

These problems were solved by cooperation between control and process engineers.

6.2 Alfalfa drying using solar energy

A large number of agricultural products must be dried in order to ensure quality preservation during storage. Depending upon the climatic conditions the natural drying in the field may have to be complemented with artificial drying. Within charges of products the drying characteristics of the components may differ significantly, for instance because of particle size distribution, as in grain, or because the product is a composition of several components, e.g. stems and leaves. Drying of such products by a straightforward method results in differences in final water content between particles or components. Furthermore, the internal stress generated by moisture gradients within particles may lead to cracking which means loss of quality.

It becomes more and more attractive to utilize natural energy sources (e.g. solar and wind) in agriculture to reduce the consumption of fossil energy sources. A limitation in the application of these sources is the variation of the energy yield over the day time and variable weather between succeeding days.

To advance product qualities after drying in combination with the use of natural energy sources new operation methods have to be develloped. This case study demonstrates a possible procedure for the drying of alfalfa using solar energy.

Background

Due to the differences in structure the drying rate of the alfalfa stems differs from that of the leaves (Figure 7). As indicated in the figure the equilibrium water concentration of the stems is below that of the leaves. However, the product of the surface for mass transfer (A_s and A_l) and mass transfer coefficients (k_s and k_l) is about 5 times higher for the leaves than for the stems. This results in a significantly higher drying rate of the leaves.

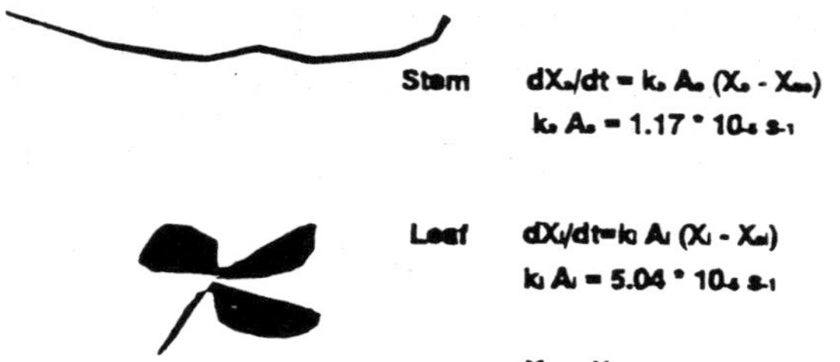

Figure 7 *Drying characteristics of stems (s) and leaves (l) differ. X: water concentrations, k: mass transfer coefficients, A: surfaces, X_e: equilibrium water concentrations (sorption isotherms), t: time*

If both stems and leaves are being dried at the same batch the water concentration in the leaves is always below that of the stems during the drying time. As soon as the stems approach the requested final water content the leaves will be overdried making them mechanically vulnerable. When the alfalfa is taken out of the dryer system most of the leaves crack and fall down. The damage to the leaves can be avoided if at the end of the drying time the water concentration of leaves can be forced to approach that of the stems.

Optimization

In Figure 8 a schematic overview of the system that has been optimized in this case study is given. The fan is used to transport the ambient air through the system. If solar radiation is available then the air is heated up in the solar energy system. If necessary, additional heating of the air takes place.

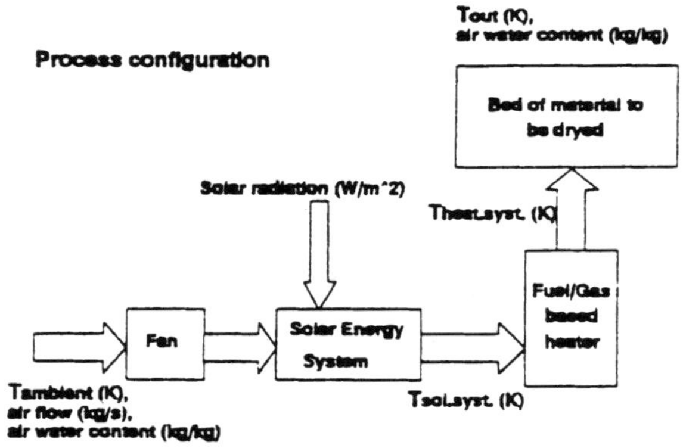

Figure 8 *Schematic overview of system to be optimized*

The solar energy input, ambient temperature and air humidity vary over a day and between months. Climatic data, as given in Figure 9 has been used as input data to optimize the drying system.

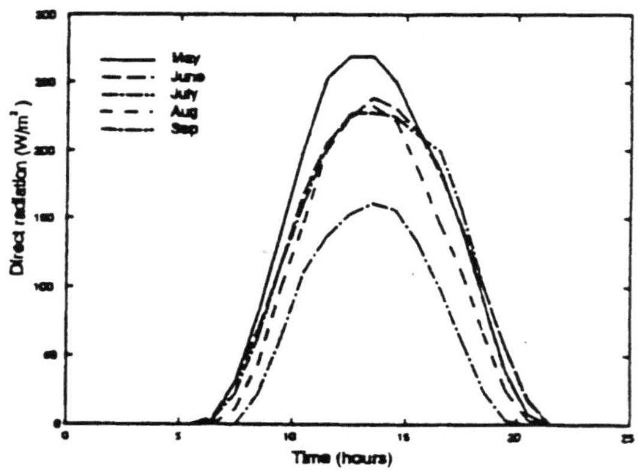

Figure 9 *Example of mean climatic data used.*

Several objective functions have been investigated in this case study. Some results are presented elsewhere (Farkas et al., 1995), but one of the more remarkable results was obtained by using an objective function aiming at minimizing energy costs, under the condition of prescribed water concents at the end of the drying time:

$$J_k = \int_{t_0}^{t_f} (C_{fan} + C_{heater})\, dt$$

under constraints:

$$x_l(t_f) = 0.4 \quad x_s(t_f) = 0.6$$

The result is shown in Figure 10. The curves of the water concentration show that a drying strategy is applied where at first the leaves are overdried. However, during the last 10 hours the temperature in the dryer is pushed to a minimal value. This results in rewetting of the leaves while in the mean time drying

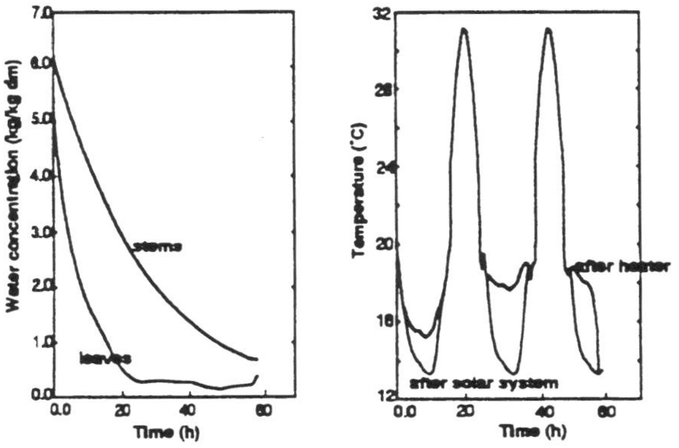

Figure 10 *Calculated operation strategy*

of the stems continues. This can be done thanks to the differences in equilibrium water contents of stem and leaves (X_{es} and X_{el}; sorption isotherms) at the prevailing operation conditions.

Remarks

- To develop the strategy a model was provided by drying experts, while a computer program for dynamic optimization was developed by a control engineer.
- In this study optimal operations were explored by using models from the process engineer and long year climatic data. However, the real-time weather conditions will deviate from that of the data used and will therefore call for adjustment of the input trajectories. It is clearly the task of the control engineer to generate solutions to exploit deviations between the averaged climatic data and actual real-time weather conditions.
- Succeeding charges of the product may have deviating drying curves. The development of methods to adapt to these differences is also a challenge for the control engineer.
- The results of this preliminary case study showed that potential energy savings are worthwhile (up to 30% as compared to a fixed strategy) and that there are possibilities to manipulate the water concentrations of different components.

7 Implementation aspects of Optimal Operation and Control

The actual implementation of the ideal scheme of optimal operation and control can be made on the basis of two considerations:

- whether or not the time scales of the physical part and the product part are wide apart
- whether the disturbances should be suppressed or rather exploited.

If the product changes are much slower than the environmental dynamics, the problem can be treated as a kind of (multiple) cascade control problem. In this case the product quality and quantity are controlled by controlling the inner environmental control loops, i.e. by providing set-points to the inner loop, computed from outer loop control. This is shown in Figure 11. In practice, often a further simplification is used by assuming tight, i.e. ideal, control. This greatly simplifies the computations of the outer loop steering or control, because the inner loop then reduces to algebraic equations (quasi-steady state, see e.g. Ioslovich et al., 1995 in greenhouse control). It should

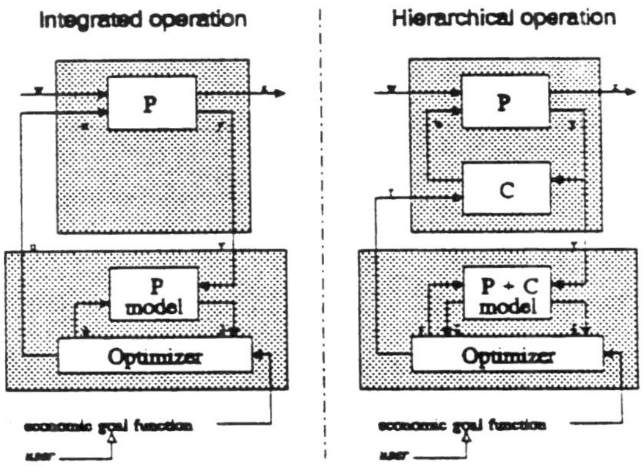

Figure 11. *Integrated (direct) versus hierarchical (indirect) operation and control.*

be noted, however, that this approach rests on the idea that desired trajectories are indeed realizable (i.e. not subject to severe interaction), and that they can be realized very fast and accurate. If there are indications that these conditions are not fulfilled, an improved but still sub-optimal approach is to design the outer loop by taking the physical process *plus* the inner loop controller as the actual system to be optimized. From the point of view of profitability, it is quite possible that even a sloppy design of the inner loop controllers can give almost optimal solutions, provided the model used in the optimization consists of the plant model plus inner controller.

In cases where there is no large difference in time scales, for instance in baking, and perhaps also in drying, the hierarchical approach above is still possible, but from the point of view of profitability, optimal predictive control procedures that directly compute the control inputs theoretically give the only solutions that are truly optimal. The hierarchical set-up, however, has certain advantages from the point of view of resilience to failure, because if the outer loop fails, the inner loops can be taken 'on hand' and still provide control. This is particularly so if hardware controllers are already present, and less so if the control is realized as centralized computer control. Whatever the situation, in any particular application where the choice is made to use the hierarchical set-up it would be necessary to see how large the sub-optimality loss is compared to direct optimal control.

The aspect of disturbance rejection or exploitation has not fully been investigated to date. If disturbances should be rejected, the classical feed-back designs are suitable. Also, a predictive controller, supplemented with a (linear) compensator will do the job. However, if disturbances should be exploited, it would be desirable to have a forecast of the disturbances to be expected, over the time horizon where they may affect the controlled environment. In greenhouse control, Tap et al. (19xx) have achieved some success by using the lazy man's prediction, i.e. assuming that the external weather remains the same as the actual measured weather. It is likely, however, that this is only true if the characteristic time scale of the physical process is in the order of magnitude of the prediction horizon. In cases of solar heat storage, for example, this may not be the case. Another solution would be to determine an average pattern, as in the drying case study above, and then treat the deviations as disturbances to be rejected. The economic aspects of these solution still remain to be investigated.

8 Conclusion

A first step to larger profitability in post-harvest processing is to improve our knowledge about the behaviour of the product quality in function of the operational variables.

As gradually more models become available, economically optimal pathways can be computed using the principles of optimal control on a goal function formulated in direct economical terms. Optimal solutions can be computed either off-line, and implemented with - possibly advanced - low-level feedback compensators, or implemented directly on-line using a receding horizon predictive control approach.

The realization of economically optimal operation and control requires the close cooperation between control engineers and process engineers.

Acknowledgements

The case study of biscuit baking was realized in cooperation with ir. U. de Vries of TNO-Nutrition Zeist the Netherland and the drying optimization is result of a cooperation with prof. I. Farkas of the Godollo Agricultural University in Hungary. The discussed case studies were derived from the MSc-thesis of F.G.M.P. Weijts and L.J.S. Lukasse.

9 References

Åström, K.J. and B. Wittenmark, Adaptive Control, Addison-Wesley, 1989.

Athens, M., The role and use of the stochastic linear-quadratic-Gaussian problem in control system design, IEEE Transactions on Automatic Control, 1971, AC16 (6), 529-552.

Bryson, A.E. and Ho, Y-C, Applied Optimal Control, Hemisphere Publishing Corp., New York, 1975.

Clarke, D.W., Mohtadi, C., Tuffs, P.S., Generalized Predictive Control, Automatica, 1987, 23, (2), Part I. The basic algorithm, 137-148, Part II. Extensions and Interpretations, 149-160.

Courtois, F., Lebert, A., Duquenoy, A., Lasseran, J.C. and Bimbinet, J.J., Modelling of drying in order to improve processing quality of maize, Drying Technology, 9(4), 927-945, 1991.

De Baerdemaeker, J. and Hashimoto, Y., Speaking fruit approach to the intelligent control of the storage system, Proc. of the 12th CIGR World Congress, Vol 1, 190-197, 1994.

De Moor, M. and Berkmans, D., Building a grey box model to model the energy and mass transfer in an imperfectly mixed fluid by using experimental data, 1st IMACS/IFAC Symposium on Mathematical modelling and simulation in agriculture & bio-industries, Brussels (B), 1995, Vol. III, VI.-A.4-1:10.

Doyle, J.C, Francis, B.A. and Tannenbaum, A.R., Feedback Control Theory, Maxwell-MacMillan Int., 1992.

Farkas, I., Rendik, Z., Toyoda, K. and Van Boxtel, A.J.B., Comparison of different models used in fixed bed drying processes, this workshop (CAPPT'95, Ostende, B)

Ioslovich, I., Seginer, I., Gutman, P.-O., Borshchevsky, M., Sub-optimal CO_2 enrichment of greenhouses, J. agric. Engng. Res. (1995), 60, 117-136.

Kessler, H.G., Molkereitechnologie, Verlag A. Kessler, Freising (D), 1988.

Kwakernaak, H. and Sivan, R., Linear Optimal Control Systems, Wiley-Interscience, 1972.

Lewis, F.L., Optimal Control, Wiley-Interscience, 1986.

Lievense, L.C., The inactivation of *Lactobacillus plantarum* during drying, Ph. D. thesis, Wageningen Agricultural University, 1991.

Luyben, W.L., Process Modelling, Simulation and Control for Chemical Engineers, McGraw-Hill, 1990.

Maciejowski, J.M., Multivariable Feedback Design, Addison-Wesley, 1989.

Mayne, D.Q. and Polak, E. Optimization based design and control, Preprints IFAC World Congress, 1993, 3, 129-138.

Michalska, H. and Mayne, D.Q., Robust Receding Horizon Control of Constrained Nonlinear Systems, IEEE Transactions on Automatic Control, 1993, 38 (11), 1623-1633.

O'Callaghan, J.R., Menzies, D.J. and Bailey, P.H., Digital simulation of agricultural drier performance, J. agric. Engng. Res. (1971) 16(3), 223-244.

Stephanopoulos, G., Chemical Process Control, Prentice-Hall Int., 1984.

Tap, R.F., van Willigenburg, L.G., van Straten, G. Experimental results of receding horizon optimal control of greenhouse climate. Submitted to Acta Horticulturae, 1995.

Toyoda, K., Study on intermittent drying of rough rice in a recirculation dryer, Sixth Int. Drying Symposium IDS'88, Versailles, 1988, OP.171-178.

Van Straten, G. and Challa, H., Greenhouse Climate Control Systems, In: J.C. Bakker, G.P.A. Bot, H. Challa and N.J. van de Braak, Greenhouse Climate Control, an integrated approach, Wageningen Pers, 1995, pp. 249-265.

MATHEMATICAL MODELLING ON STORAGE OF FRUITS AND VEGETABLES IN MODIFIED ATMOSPHERE PACKAGING (MAP) AND CONTROLLED ATMOSPHERE STORAGE (CA).

Marcel HERREGODS

V.B.T. - I.W.O.N.L.
Tiensevest 136
B-3000 Leuven (Belgium)
Tel. : 0032/16/20.12.71
Fax : 0032/16/20.30.35

ABSTRACT

In modified atmosphere packaging (MAP) our aim is to find by mathematical models a steady state where the CO_2 production, O_2 consumption and H_2O transpiration rates of the commodity are proportional to the permeability rates of the respective gases through the packaging at a given temperature to reach maximum storage life.
In this way we abtain optimum concentrations of O_2, CO_2 and relative humidity, avoiding the limits put by fermentation, too fast ripening, senescence or appearance of disorders and decay.
Commodity characteristics (maturity, dry matter, size), temperature, O_2 and CO_2 are known to infuence respiratory activity in a package. We have to recognize an important interrelation between these different parameters and the changings in time.
Nature, thickness, surface, holes or micropores as well as gradients of CO_2, O_2 and H_2O at a certain temperature and relative humidity determine permeability for gases.

In controlled atmosphere storage (CA). Next the fruit properties, also refrigerating characteristics (cooling capacity, delta t, air circulation and defrosting systems), and the used CO_2 regulating system, (CO_2 absorbers or N_2 flushing apparatus) are important parameters to realize the optimum temperature, O_2, CO_2, ethylene and relative humidity to maintain freshness of the produce.

Mathematical modelling on CA and MAP could help to frind not only the longest storage time, to reduce losses but also to find an optimal equilibrium between costs and benefits.
New sensors are yet available or are to be developed.

INTRODUCTION

Proper preservation methods are necessary in order to reduce the very high percentage of post harvest losses of fruit and vegetables, to deliver the quality required to a particular market, to store the products for a longer period, to reach higher-priced markets further afield and to achieve a sufficiently low temperature during transport, trade and retail.
New environmentally friendly storage techniques are available, as well as pre harvest applications in the field as post harvest systems during storage.

- new genetically manipulated long-life varieties
- new data on the influence of climate, soil, fertilization, drougth or heat stress and pathogenic chemical treatments
- new criteria for determining optimum harvesting time, and non-destructive methods for determining quality and maturity
- more efficient refigerating techniques
- the developments of computer controlled equipment to accurately maintain storage conditions (temperature, R.H, CO_2, O_2, ethylene)
- humid forced air precooling techniques
- an optimum combination of the storage parameters maturity, temperature, relative humidity, CO_2, O_2 and ethylene, closely related to the desired storage time and the post harvest quality required, also for an increasing number of specially crops and for lightly processed fruit and vegetables
- minimum impacts on bruising-artificial apple
- new films for modified atmosphere packaging with micropores or with layers absorbing water or ethylene
- new technics to optimize conditions during transport, trade and retailing
- C.A.-storage as quarantine treatment

RESULTS

OPTIMUM STORAGE CIRCUMSTANCES

The optimum storage circumstances (temperature, R.H., O_2, CO_2, ethylene) each commodity vary with in orde to avoid quality deterioration and the appearance of disorders and decay. (tabel 1).

An atmosphere optimal for the retential of firmness may not be optimal for retention of colour or flavour.

Chilling and freezing

Most fruits and vegetables have the longest storage period near freezing point. Some temperate fruits, most tropical fruits and fruit-vegetables show chilling injury at a too low temperature and must be stored different degrees above 0°C. The lowest temperature, causing chilling injury, is dependent from the maturity stade (more chilling with unriper fruits) and can be reduced with heat or ethylene treatment, lower oxygen or higher CO2 concentration.

Optimum relative humidity

The higher relative humidity inside the package reduces water loss and shriveling, but can stimulate decay if not counteracted by a higher CO_2 content for some commodities.

The differences in weight loss between commodities are very important (table 2).

By the same relative humidity, weight loss is higher at higher temperatures and higher air circulation rates.
For some vegetables as mushrooms, the optimum relative humidity and weight loss are very critical. A too high humidity conducts to browning of the skin after condensation, a too low humidity to a brown skin after drying out.

A little lower relative humidity (90% to 92% R.H.) improves storage life of tomatoes, sweet peppers and cumcumbers; a very high relative humidity (96% to 98% R.H.) on the contrary improve storage life of carrots and cabbages.

Optimum O_2, CO_2 and ethylene concentration

For most fruit and vegetables the optimum O_2 concentration is 1% to 5% O_2, the optimum CO_2 concentration 0% to 25% CO_2 and the optimum ethylene concentration < 0.01 ppm.
The internal O_2 and CO_2 concentration inside the product can deviate from the conditions outside in the package.

The internal O_2 and CO_2 concentration is related to the air volume in the tissue, the skin resistance, the thickness of the peel, the number of open lenticells, the wax deposits and the internal respiration rate and must be taken into consideration in the determination of the critical external O_2 and CO_2 levels.

Fermentation (anaerobic respiration)

The limits of low O_2 and high CO_2 concentration are determined by the anaerobic compensation point where respiration changes to alcoholic fermentation and is depending from an interrelated combination between maturity, temperature, O_2 and CO_2, typical for each commodity.
Riper fruits are more sensitive to low O_2 and high CO_2. At higher O_2 concentrations fruits resist more to high CO_2.
It has been reported that RQ (respiratory quotient) can be used to determine the lower oxygen limits.

Decay

Suppression of growth of moulds by a 2% O_2 atmosphere is modest

An elevation of CO_2 above 5% suppresses growth of moulds cinerea. At 40% CO_2 the growth is completely inhibited.
A high CO_2 concentration with 20% reduces decay for strawberries and currant berries.

A too higher CO_2 concentration conducts to more decay for tomatoes and cucumbers (>5% CO_2).

2. OPTIMIZING STORE VOLUME, REFRIGERETING TECHNIQUES, AND C.A. CIRCUMSTANCES

Optimizing store house volume

We try to fill or empty a store sufficiently quickly. For long-term storage of apples we need to fill an apple store in 5-6 days and empty in no longer than 8-10 days.
The dimensions of the room are depending on the dimensions of the pallets. We have to consider a free space of 30 cm between the walls and the pallets, 10 cm between the rows of pallets in the circulation direction and 80 cm to 100 cm from the ceiling.
In smaller rooms there is a profitable optimum depth from 9 to 12 cm.
For sufficiently rapid filling and emptying of the rooms there was a tendency to move to smaller rooms (100 to 160 tons per room of 400 to 700 m^3).

With the possibility of renewing the controlled atmospheric composition very quickly by N_2 flushing there is again a tendency to choose bigger rooms.

Sufficient insulation determines weight loss and consumption of electricity. As thickness for expanded polystyren for walls 12 - 15 cm with a density of 18-20 kg/m^3, for expanded polyurethane for walls 8 - 10 cm with a density of 35 - 40 kg/m^3 is needed.
Floor insulation - 5 cm - is a must for long-term storage time.
For C.A.-storage a sufficient gas-tightness means that the water pressure after a suppression of 10 mm takes more than 10 minutes to come to 7 mm.

Refrigerating techniques

The cooling capacity depends on different parameters such as load per day, temperature and kind of product, specific heat, heat production and external temperature and must be optimized. In our circumstance we try to reduce the temperature of one fifth of the apple capacity of a room with 15°C day. The cooling capacity of the compressor is in our circumstances for apples and pears in non-decentralised system aproximalety 50 - 55 Watt/m^3/h at -5°C and +35°C and for a centralized unit 35 - 40 Watt/m^3/h at -5°C and + 35°C.
For the cooler a capacity of 45 - 50 Watt/m^3/h at -5°C and 5°C delta t has been used in our circumstances.
There is tendency to go to coolers with a minimum free space between wall and cooler (< 1,80 m).
For capacity of the fans 50 m^3/m^3/hour, discontinuous with an adjustable after cooling time is recommended. Defrosting systems - heat resistances or hot gas - need a defrosting stop systems.
The range of the thermostat may not be too narrow (± 0,8°C).
Problems due to ozone layer depletion and green-house effects stimulate the use of new cooling media such as R135A, R404A and others with their specific problems of oil, capacity, pressure, COP and glide.

A profitable storage : equilibrium costs and benefits

A small increase of storage costs can result in higher benefits through less losses from disorders and decay, better post storage quality after storage and a longer shelf life. There are still possibilities of reducing costs. For a complex of 5 rooms of 100 ton the cost of storing apples for 6 months show the following differences in cost/kg (30 BEF = 1 US $) :

- cost difference in size of the room : 160 ton/room 0,64 BEF/kg, 77 ton, 0,93 BEF/kg.
- insulation thickness cost difference between 8 cm PUR and 12 cm PUR 0,06 BEF/kg
- cost difference in hangar characteristics 0,38 BEF/kg, 0,60 BEF/kg
- proper sizing and flexibility in range of compressors, coolers, condensors, discontinue air circulation, 0,30 BEF - 0,50 BEF/kg cost reduction
- air circulation discontinue stopped : 0,25 BEF - 0,30 BEF/kg energy saving
- electricity peak capacity regulation 0,10 BEF - 0,25 BEF/kg energy saving.

New storage methods

A proper maintenance of the optimum range of temperature and relative humidity is the basis of good storage conditions. Not a substitute, but supplementary improvements are :

- modified atmosphere storage (MA) or modified atmosphere packaging (MAP) conditions
- controlled atmosphere storage (CA) with CO_2 scrubbing or N_2 flushing
- computerized ultra low oxygen storage (ULO)
- quick pull down of O_2 with N_2 flushing
- scrubbing or N_2 flushing in order to reduce ethylene, farnesene and other unwanted volatile components
- intelligent, dynamic control systems using biological, physical and mathematical models

Humid forced air precooling

There are high differences in product price between different markets. To reach more distant markets fast cooling after harvest especially for strawberries, small fruits and vegetables are necessary requirements.
In addition to conventional cold rooms other precooling systems are available as vacuumcooling, hydrovacuumcooling, hydrocooling, top-icing, package liquid-icing and forced air cooling. Forced air-cooling with as cold source ice-banks, is mostly used, giving many benefits.

C.A. storage conditions

Last years the interrelated combination maturity, temperature, CO_2, O_2, was optimised. More mature fruits has to be stored at a higher temperature, with a lower CO_2 and an increased O_2 concentration than less mature ones.
The production of CO_2 is about 15 to 30 kg CO_2/100 ton/day with apples and pears. We can apply active charcoal CO_2 absorbers or dry lime (20-30 kg calcumhydroxyde per tons) to absorb CO_2. Stabilizing the CO_2 concentration at the optimum level by flushing with 4 to 7 liter N_2/minute per room volume of 100 m^3 is possible.

Heat treatment, very low O_2 or very high CO_2, just above the level of irreversible damage, has promising results, but is not easy to apply without phytotoxic risks. There are promising results as quarantine treatments.
An increase of CO_2 concentration for storage of currant berries and blueberries (0,5°C, 20% CO_2, 21% O_2) gives benefits.

3. OPTIMIZING MODIFIED ATMOSPHERE PACKAGING (M.A.P.) CIRCUMSTANCES

The combination between the permeability of the film (material, thickness, holes), the absorbed O_2 and produced CO_2 (commodity, maturity, temperature), the relationship quantity and volume product-package size and the duration of storage determine the composition of the internal atmosphere in M.A.P.

Influence product characteristics on respiration

The respiration rate varies with the kind of commoditity, maturity stade, dry matter, size, the relationship quantity product to peckage valume.

- Between the kind of commodities a difference in respiration rate from 1 to 10 is possible (table 4).
- During maturity evolution for climateric fruits an increase of the rate of respiration from five to ten times is possible. Non-climacteric fruits show no clear-cut respiratory climacteric and evolve very little during ripening. We can abserve again an increase of the respiration rate for non-climacteric fruits treated with ethylene.
- Smaller, early harvested closed shape cichory endives with a 15% higher dry matter show a 20% higher respiration rate than bigger, late harvested more open ones.

Influence of storage circumstances (temperature, O_2, C.O.$_2$ storage duration) on respiration rate.

Between 5°C and 20°C there is an increase of the respiration rate from 3 to 4 times (table 5). Increasing O_2 concentration from 5 to 21% O_2 results in little changes in respiration. Below 5% O_2 the respiration rates drops quickly.

An increase of CO_2 concentration from 1 to 20% CO_2 decreases repiration rate, but depending from commodity there are limits in this range from 1 to 20% CO_2, where CO_2 produktion again increase as a result of fermentation.

The rapid establishment of altered atmosphere within a few minutes could produce commodity responses that would differ from those observed after the relative slower extablishment of altered atmosphere within days.

Data on gas and water vapour permeability of selected packaging fim material

O_2, CO_2 and H_2O permeability of different film materials varies with the plastic material, the thickness of the film, the presence of holes, the temperature, the difference in humidity and gases outside and inside the film. (tabel 4).

The relative oxygen and carbon dioxide permeability of the current films is also not the best for extablishing optimum atmospheres.

There is an increase of the transmission for O_2, CO_2 and H_2O with increasing temperature. Realistic conditions with relative humidities near 100% inside the packaging and about 50% outside differ largely from the normal conditions in gas permeation as commercially available.

New films for modified atmosphere packaging

New packaging films with micropores improve storage life enhacing water permeability and avoiding condensation. This is also the case with new plastic films with a fungistatic layer (sorbic acid), impregnation with ethylene oxydance (silicium oxyde) or water absorbances. Packets containing chemicals to alter gas composition (R.H., CO_2, O_2, ethylene), insert within the film package, are intriguing but can add costs to an already expensive package (MgO to absorb CO_2, iron powder to achieve O_2 deplation, baking powder to produce CO_2).

Combination product respiration and M.A.P. film gas transmission

If we know the optimum modified atmosphere circumstances (RH, O_2, CO_2) we can combine facts obtained on products and packaging in an appropriate date base and develop a computer program capable of simulating the combined product / packaging beraviour in dependence of the actual temperature (table 6).

This balance is discrupted when changes in temperature effect diffusivity and respiration differently. Temperature control during marketing is ofter more precise enough to maintain the required balance. In addition to temperature control the respiratory activity of climacteric fruit can dramatically change during storage us they riper.

4. INTEGRATORS AND SENSORS

To measure temperature O_2, CO_2 during storage and transport the actual sensors are sufficient accurate and reliable. New, non destructive techniques are promising to follow quality maturity and senescence. Following the O_2 concentration inside a fruit should be useful.
Inclusion of low O_2 or high CO_2 integrators in film packages may be a prudent safety promotion.

Table 1 : O P T I M A L   S T O R A G E C I R C U M S T A N C E S

	Regular Storage	*Modified Atmos phere*	*Controlled.- Atmosphere*	*Ultra Low Oxygen*
Apples				
Golden, Gloster, Mutsu	0,3°C 17,5 - 16,5 O_2	0,8°C 3,5 - 4,5% CO_2 2 - 3% O2	0,8°C 2 - 3% CO_2 2% O2	0,8°C 2% CO_2
Elstar, Idared	1°C 2 - 3% CO_2 18 - 19 O_2	1°C 1 - 2% CO_2 2 - 3% O2	1°C 2% CO_2 2% O2	1°C
Jonagold	0,8°C	0,8°C 3,5 - 4,5% CO_2 17,5 - 16,5 O2	0,8°C 2 - 3% CO_2 1 - 2% O2	0,8°C 2% - 2,5% CO_2 1% O2
Cox's Orange, Boskoop, Suntan, Karmijn, Melrose	3 tot 3,5°C		3 - 3,5°C < 1% CO_2 2 - 3% O_2	3 - 3,5°C < 0,8% CO_2 2% O_2
Pears				
Confér.	-0,5 tot -1°C			-0,5°C tot -1°C < 0,7% CO_2 2 - 2,5% O_2
Doyenné du Comice	0 tot -0,5°C		2 - 2,5% O_2	0 - -0,5°C < 0,7% CO_2

Table 2 Weight loss (%/day) for different commodities

	20°C 60% R.H.	5°C 80% R.H.
Radishes	34	17
mushrooms	31	10
Brussel sprouts	13	3
Oinons	0,5	0,5

Costs (5 rooms each 100 ton) (BEF/kg) (10 kg apples/bruto m3) (30 BEF = 1 US $) (table 3)

Investment costs		**Year costs BEF/kg 6 month storage**	
standard walls + doors	8 BEF/kg		
insulated floor	2 BEF/kg		
refrigerating systems	6 BEF/kg		
active charcool scrubber	2 BEF/kg	investments :	2.89 BEF/kg
regulation and control systems	2 BEF/kg	energy :	0.70 BEF/kg
hangar	6 BEF/kg	maintenance :	0.10 BEF/kg
bins	4.5 BEF/kg	labour :	0.10 BEF/kg
fork-lifts	1 BEF/kg		
Total	31.5 BEF/kg	Total :	3.79 BEF/kg

Table 4 - Typical rates of respiration (mg CO_2/kg tissue/hour) of some vegetable crops in air (after Robinson et al., 1975).

	5°C	20°C
Asparagus	44	127
Brussel sprouts	30	90
Cabbage (winter white)	7	20
Calabrese (Broccoli)	58	240
Cauliflower	34	126
Leeks	28	110
Lettuce (soft)	11	37
Mangetout peas	36	139
Onion (bulb)	5	8
Peppers (sweet)	11	35
Potatoes (maincrop)	3	6
Sweetcorn	55	210
Tomatoes	9	30

J.D. Geeson, Modified atmosphere packaging of fruits and vegebables. Acta Horticulturae, I.S.H.S. 1989 - 163 - 150.

Table 5 (for 100 µm film thickness, basis: 23 °C)

Material	CO_2 / O_2 permeability ratio	O_2-transmission (cm^3/m^2 d bar)	CO_2-transmission	H_2O-transmission (g/m^2 d at 85 % → 0 % r.h)
PE	3 ... 4	2000 ... 3000	6000 ... 12000	1 ... 5
PVC	5.7 ... > 6	2000 ... 2300	12000 ... 13000	30
EVA/PE-Copolymer (Yuka Wrap)	6.4	2300 ... 3700	15000	2
Copolyether-ester (Sympatex)	22	490	10700	107
Poly-(ε) caprolactone	3.4	2070	ca. 7000	69
Cellulose acetate	5.3	450 ... 1000	2800 ... 5600	65
Cellulose	not known	0.5	below det. limit	133 ... 150
Polypropylene, microperforated (P+)	1	900 200000	900 ... 200000	0.7 ... 5 (calculated)

H.C. Langowski, Progress Report, 1994, E.C.-AIR project, AIR2 - CT - 1326, MASTER.

VISUALISATION AND QUANTIFICATION OF THE AIR FLOW PATTERN IN VENTILATED SPACES BY USING IMAGE ANALYSIS

M. Van Pee, D. Berckmans, K. Van De Weyer, M. De Moor

Laboratory for Agricultural Buildings Research, Katholieke Universiteit Leuven,
Kardinaal Mercierlaan 92, B-3001 Heverlee, Belgium

Abstract: The process of air mass transport has an important influence on the resulting micro-environment within a storage place or cooling room. To achieve an efficient transfer of energy and mass around the products, the air flow pattern should be controlled. To study the air flow pattern in a more quantitative way, a practical measurement technique should be available. In this paper a method is presented to visualise and to quantify the airflow pattern in a ventilated room. The use of a low cost camera is combined with an algorithm on personal computer to calculate automatically the co-ordinates of an air jet and to visualise the (dynamic behaviour of the) air flow pattern.

Key words: Air flow pattern modelling, image analysis, visualisation of air flow, conservation, storage, climate control.

1. PROBLEM: A QUALITATIVE VALIDATION OF MODELS

Man, animal, plant and product are responding to their physical micro-environment. The physical micro-environment (temperature, humidity, gas-concentration, velocity, dust, etc.) is strongly depending on the fluid flow rate and the fluid flow pattern. Many agricultural, environmental and industrial processes are dealing with imperfectly mixed fluids. In several agricultural processes the ventilation rate and the fluid flow pattern are of crucial importance (e.g.: in natural or mechanical ventilated livestock buildings; in storage places for grain, potatoes, fruits; in drying processes etc.). The ventilation rate and the air flow pattern play a very important role in the basic mechanism of energy and mass transfer within a fluid. Because of the importance of the moving fluid problem, expensive and sophisticated techniques (super computers, numerical techniques, etc.) and a lot of manpower are used to study these phenomena. Advanced techniques are used to model the air flow pattern in ventilated spaces (Murakami, 1992). To validate the theoretical models there always is a need for experimental data.

Sophisticated and expensive techniques (scale models, tracer techniques) are applied to generate data. When using an appropriate illumination a moving tracer can be visualised and photographic or video techniques can be applied. Although the use of expensive methods, the data obtained by such techniques are used in a qualitative way. The validation of models by comparing simulation results with visualised data, is a qualitative approach (Murakami and Kato, 1987; Tsutsumi, 1990).

The technique of velocity measurement through image processing has been developed since 1979 and described in literature (Kaga and Yoshikawa, 1989). Most of the available techniques however are rather expensive and focused on fluid flow velocity and most of the reported developments where done in water as a fluid.

2. OBJECTIVE: QUANTIFICATION OF VISUALISED AIR FLOW PATTERNS

A first interest in studying the fluid flow problems is to quantify and to localise the fluid flow pattern within the three-dimensional space. The objective of this work was to do a low cost image analysis to visualise and to quantify the air flow pattern. More specifically it is achieved to develop a useful algorithm for visualisation of an air jet within a ventilated space and for automatic registration of the two-dimensional co-ordinates of the centreline of a visualised air jet within a space.

3. METHOD: THE LABORATORY TEST INSTALLATION WITH IMAGE ANALYSIS TECHNIQUES

The general principle is to visualise the air by using a tracer and to monitor the visualised air flow pattern by video registration. The images are digitised and analysed by an image analysis algorithm. To realise the objective an experimental installation was used, data were collected, an image analysis algorithm was developed and tested on the data obtained from laboratory experiments.

3.1 The experimental room

The achieved algorithm should be applicable for different types of air flow patterns such as falling air, horizontal air and the case of high instability in the air flow pattern. The type of resulting air flow pattern will be determined by the characteristics of the used test room such as (Croom-Gale and Roberts, 1975; Randall, 1975): geometry, dimensions and position of the air inlet, temperature of the surrounding walls, temperature difference between the incoming air and the air in the test room, inlet velocity of the incoming air, turbulence of the air. Using these principles the test room was constructed. The purpose of this room was to generate stable air flow patterns and to measure the transient behaviour from one steady state air flow pattern to another. As been mentioned before, different parameters must be considered to guarantee stable air flow patterns. Considering the conclusions in literature (Barber and Ogilvie, 1981; Barber and Ogilvie, 1984; Katz, 1974; Leonard and Mc. Quitty, 1985; Mullejans, 1966; Randall and Battams, 1979; Randall, 1975; Rodahl, 1985; Timmons, 1980) the following room geometry was taken: length: 3m; width: 1.5m; height: 2m. As mentioned before, the dimensions of the supply opening are dictated by the desired maximum air velocity. The influence of the Coanda-effect and the effect of buoyancy are function of the inlet velocity. In literature (Randall and Battams, 1979) a velocity of 5 m/s is proposed to achieve a stable, horizontal air flow pattern. The value of 3 m/s is suggested as a lower limit (Granier et al., 1989). With a given maximum air flow rate in the test installation of 320 m^3/h, it was decided to take 0.02 m^2 as size for the supply opening. Based on recommendations in literature (Leonard and Mc. Quitty, 1985; Mullejans, 1966; Randall and Battams, 1979; Regenscheit, 1974) the horizontal dimension of the rectangular supply opening was set to 1 m. The vertical dimension of the air inlet section was 0.02 m. Based on the experimental test chamber of Leonard and McQuitty (1985) the air supply opening is positioned 1.55 m above the floor level. To minimise the disturbing influence of the walls on the air flow pattern as caused by the surface temperature of the walls, a second building envelope (also in Plexiglas) was built around the primary test room. The temperature in the space between the two test chambers was controlled to reduce the conduction of heat through the walls of the internal test chamber. A general overview of the test installation is given in Figure 1. The test room has been described in a more detailed way in literature (Berckmans et al., 1992).

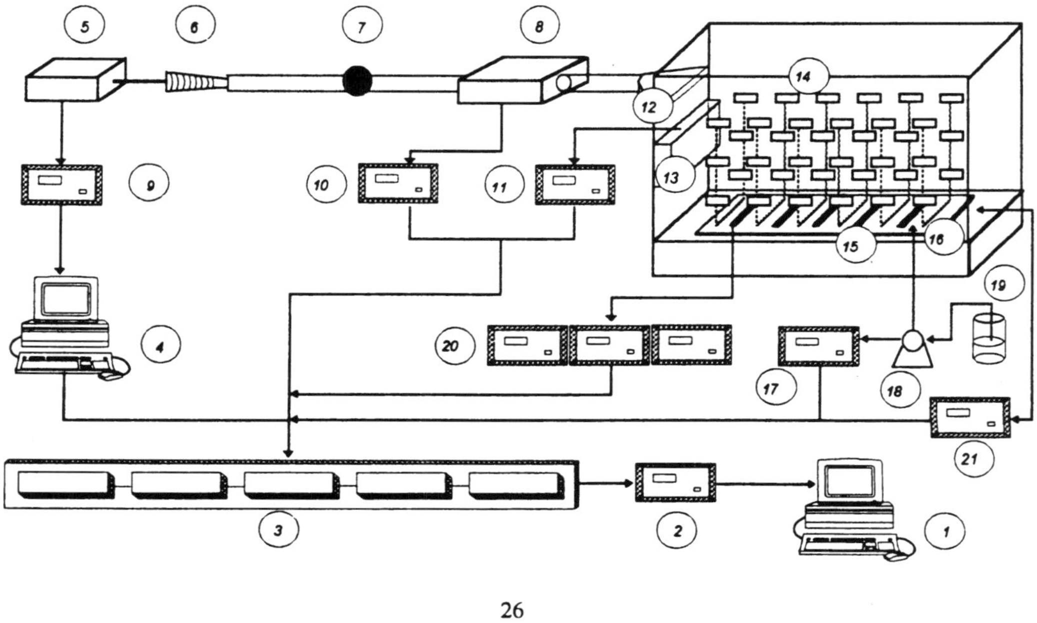

1. Minicomputer (monitor, floppy disc, to store and visualise the measured data). **2.** Parallel-interface for digital and analogue signals. **3.** Scan- and measurement unit. **4.** Minicomputer (to control and measure the produced air flow rate). **5.** Stepmotor to control the position of the cone, used as diaphragm. **6.** Cone, used as diaphragm, to produce the desired air flow rate. **7.** Centrifugal fan, to generate a ventilating rate. **8.** Cooling installation to control the inlet temperature. **9.** Differential pressure transducer to measure pressure difference between the test chamber and the envelope. **10.** Control- and measurement unit of the cooling installation. **11.** Control- and measurement unit of the heating element. **12.** Air inlet (slot inlet). **13.** Heating element. **14.** Three dimensional grid of temperature and humidity sensors. **15.** Aluminium semi conductor heat sinks to provide internal heat production. **16.** Shallow water reservoir with a streamer containing hot water to generate the internal moisture production. **17.** Unit to control and measure the amount of water supplied to the shallow water reservoir. **18.** Water pump. **19.** Water supply reservoir. **20.** Power supplies for internal heat production. **21.** Pressure difference measurement used to control the outlet fan.

Figure 1: Different parts of the test room

3.2 Smoke as a tracer

There are different visualisation methods (Wen, 1989): wall tracing, tufts, tracer methods and optical methods. For the visualisation of moving air the tracer method is the most popular one. Different products have been used as a tracer: aluminium or metaldehyde particles (Rolloos, 1978; Murakami, 1992), a smoke (Mullejans, 1966), (electro)chemical or photochemical reactions or helium bubbles. As a tracer many authors use a smoke which can be produced in different ways (Mullejans, 1966; Tsutsumi, 1990). One possibility is to produce smoke by heating organic or mineral fluids (Mueller, 1989). An example of this method is the use of saturated vapour (Merkzirch, 1987). The advantage of this technique is that the air can be recirculated and the saturated vapour can be removed in an easy way in contrast with other methods to produce smoke. This method however is not most appropriate when studying the thermodynamic state of the air during experiments. Another way to produce smoke, is the chemical, electrolytic or photochemical reaction in the fluid. An example of this method is the use of $TiCl_4$. This product reacts with moisture in the air and generates a white smoke of TiO_2 (Merkzirch, 1987; Leonard and Mc. Quitty, 1985). A third method is the so called *smoke-wire-method*. In this method an electrical wire resistance is rubbed with mineral oil. By heating the wire a two dimensional smoke plane is generated. In Table 1 an overview is given of different smoke generators.

3.3 Illumination of the tracer smoke

According to Mueller (1989) the illumination of an object to be filmed is one of the crucial aspects regarding the visualisation and photography of visualised fluid flow patterns. When image analysis is used to study fluid flow patterns much attention has to be paid to the *quality* of the original shots, because it determines the different techniques that can be applied (Smith, 1989). In literature (Mueller, 1989) three possible positions of the camera and the light source are described. When front illumination is used, the major problem is the reflection of the light on the test room; when illumination from the background is used, it must be avoided that the camera receives direct incidence of light, but only the light diffused by the smoke pattern. Illumination from the top of the test room always can be used.

For visualising three-dimensional volumes the selection of an appropriate light source is important. Different possibilities are mentioned in literature

Table 1: Overview of different commercial available smoke generators and there properties.

	Generator 1	Generator 2	Generator 3	Generator 4
smoke-production	evaporation	chemical	evaporation	evaporation
product	alcohol	based on water	?	?
range	0 to 2.8 m^3/h pure smoke	0 to 22 m^3/h smoke with 1.5 m visibility	1000 m^3 in 5 min.	3780 m^3 in 4 min.
control flow rate	available	available	control of smoke pulse duration	control of smoke pulse duration
operation period	15 min. max. capacity	50 min. max. capacity	pulses of a few minutes	pulses of a few minutes
toxic	no	no	no	no
particle dimensions	controllable	no	no	no

(Mueller, 1989): stroboscopic lamps, continuous lamps or laser lights. When stroboscopic lamps are used, the flash time has to be synchronised with shutter time of the camera (Mueller, 1989). When only a plane in the test room has to be illuminated a light source in combination with lenses is used. In that context laser light is often used because of its precision and variety of power levels.

To develop the algorithm for air flow pattern visualisation two halogen lamps (500 Watt each) were used to illuminate the smoke pattern, positioned near the wall with the air outlet (see Figure 1). The camera used for the experiment was a simple Hi-8-video camera .

4. IMAGE ANALYSIS TECHNIQUE

4.1 Digitalisation

The digitalisation of the images is performed using a PC-controlled digitisation card (*Frame Grabber*). In series of 16 images each, the images are digitised with a time interval of one second. When the frame grabber has been used on the video images, a 256 x 256 data matrix is obtained. Each element of this matrix is a grey value with a value varying from 0 to 255. Each element is corresponding to a rectangular part of the original video image. Eventually these grey values can be related with a colour scale to obtain an artificial colour image. The colour is then a measure for the intensity of the light as reflected by the smoke particles.

4.2 Segmentation

Segmentation is the process of isolating the visualised air flow pattern from the rest of the digitised image. Two methods can be used: either the borders of the smoke pattern are determined or the surface is calculated. In this application the second method was chosen because of its simple implementation. When applying a threshold value, it is possible to distinguish two types of pixels: the one with a value bigger than the threshold (the pixel is a part of the object) and the one with a value smaller than the threshold (the pixel is not a part of the object). Thus besides a grey value, each pixel has a *logical* value: *object* or *non-object*. By doing so it becomes possible to have a clear view of the visualised air flow pattern (Figure 2). To avoid *discontinuities* and *isolated areas* in the resulting image two processes are used: the *dilatation* and *erosion* technique. By applying an *erosion* algorithm each object-pixel, from which the 4 neighbour pixels are non-object, is no longer considered as being an object-pixel. By using such an erosion technique on the digital image, some individual pixels disappear from the object (Figure 2). The *dilatation* algorithm is doing the opposite action. Individual gaps in the object are filled up (Figure 2). The result of successively installing a threshold value, erosion and dilatation gives a full closed area. The resulting image can be applied to calculate the centreline of the isolated air flow pattern as explained further on.

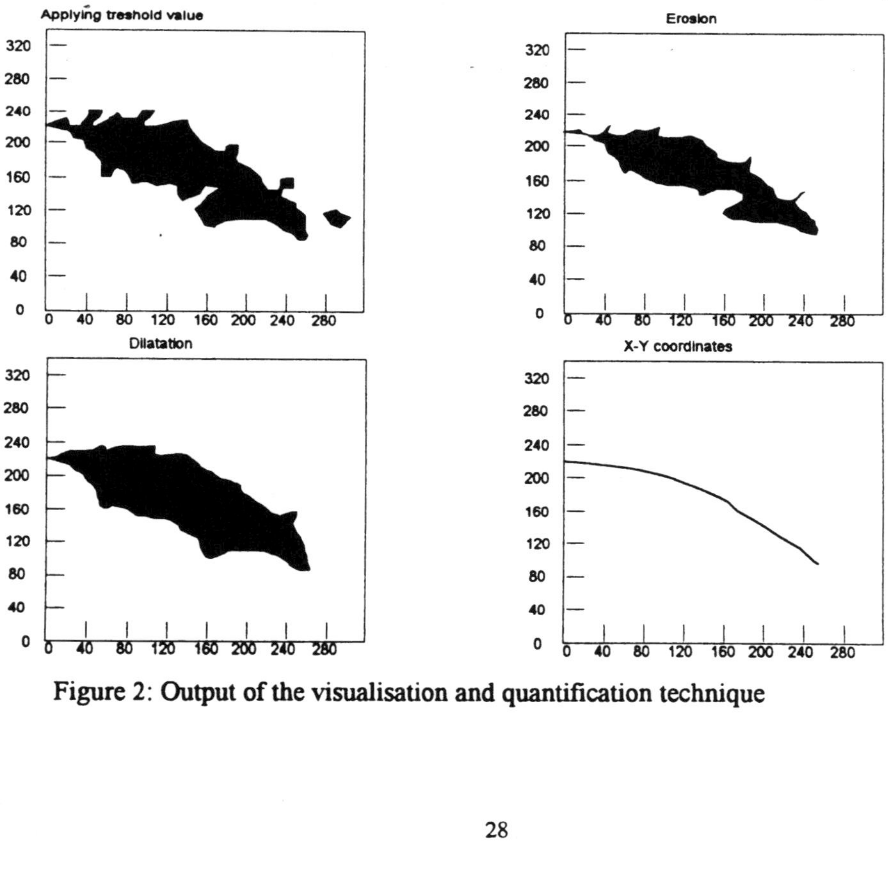

Figure 2: Output of the visualisation and quantification technique

4.3 Results and discussion: determination of the centreline of the visualised air flow pattern

In literature a method has been provided to solve the problem of determination of the centreline of an object, the *Medial Axis Transform (MAT)* (Castleman, 1979). This technique is a method to store the image information of a visualised object without the need to use the whole object in further processing. In literature two ways have been mentioned to apply the object description:

- A first method is based on the assumption that a point, belonging to the object, is located on the centreline (medial axis) of the object only then when that point is the centre of a circle that is limited by the boundary line of the object in two non-tangent points. When applying this criterion, each point of the centreline corresponds to a value equal to the radius of the above mentioned circle. This value is the minimum distance between the considered point and the boundary line of the object.
- The second technique is the so called *boundary peeling* (Castleman, 1979). Successively a layer with thickness of one pixel is removed from the object, starting with the boundary layer of the object. When in a point, the removed layer is tangent to itself, the point is considered to be a possible object of the centreline. An illustration of the boundary peeling algorithm is shown in Figure 3. Each point of the centreline has a value equal to the number of layers that were removed at that moment as shown in Figure 3. As can be seen on Figure 3 it is possible that more than one centreline can be determined (for example in Figure 3 the pixels with value equal to 3). To solve this problem a *smoothing filter* (Van De Weyer, 1991) can be applied on the peeled image. When applying this technique, local maxima are smoothed because of the low values surrounding the local maximum in the image. By doing so, a unique centreline is calculated.

After applying one of these techniques, the original image can be reconstructed using the centreline and the values of each point of this centreline. In this application the *boundary peeling* technique was used because of its analogy with the above mentioned *erosion* technique. In figure 4 an overview is given of the whole technique of visualisation of the air flow pattern is given. It is possible to calculate the centreline of the air flow pattern in the test room with an accuracy of ± 0.8 cm.

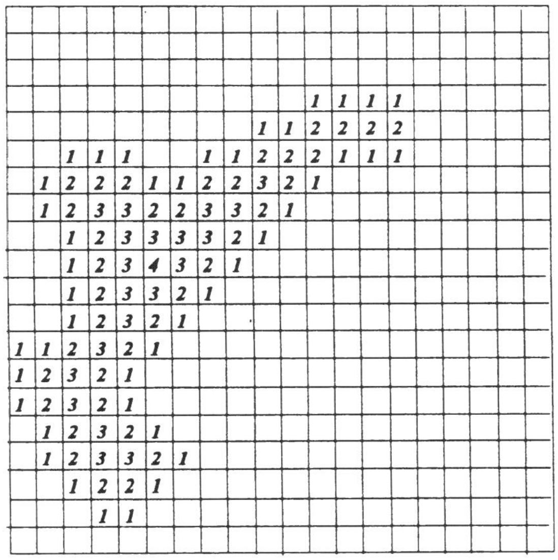

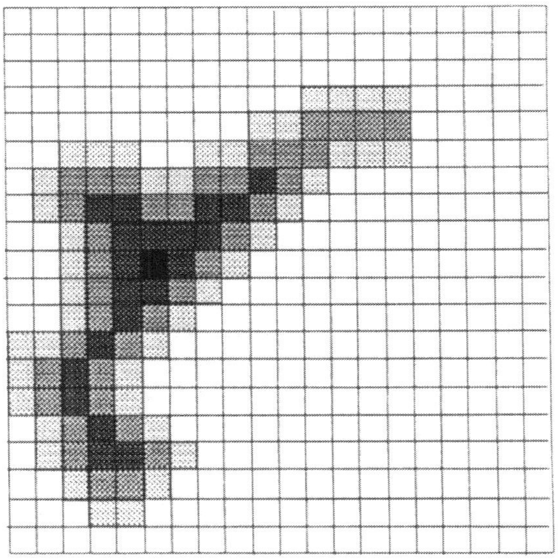

Figure 3: The result of applying the boundary peeling

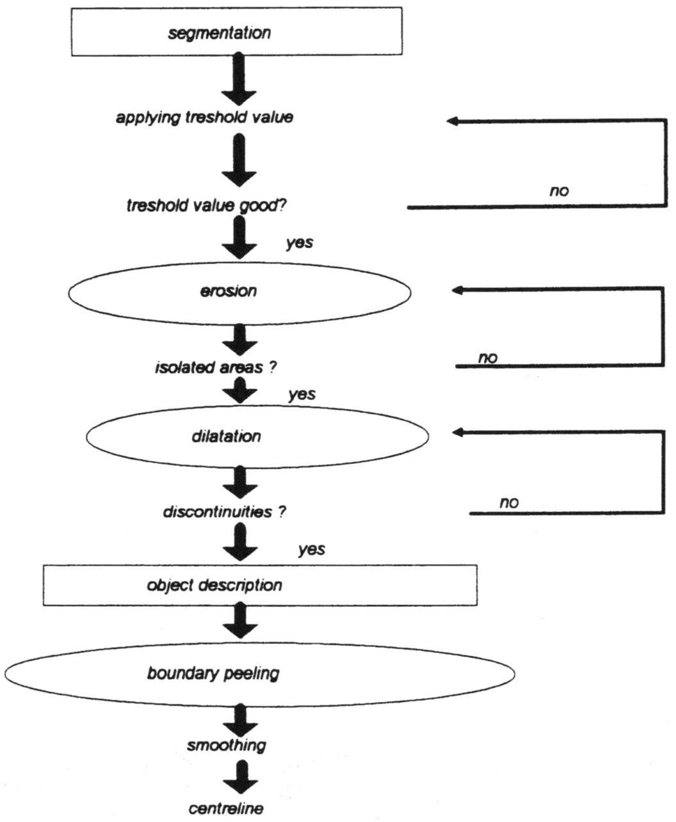

Figure 4: Overview of the technique for visualisation of the air flow pattern by using image analysis

5. CONCLUSIONS AND POSSIBLE APPLICATIONS OF THE TECHNIQUE - EXPERIMENT

As indicated by Saunders and Albright (1989) the accuracy of quantitative results depends on the validity of the assumed linear relationship between reflected light intensity and tracer particle density. After this work this still remains an open question. The objective however of this paper was to find out how accurate a low cost system could quantify the centreline of an air jet in case we assume the higher mentioned linearity.

1. The described method allows to determine automatically the two- dimensional centreline of a visualised air jet in a ventilated room in relation to the position of the objects in the room (e.g. fruit boxes).
2. On a total of 13 experiments, the algorithm was successful in an automatic way in 70 % of the experiments. With manual interaction in the *smoothing* actions, another 20 % of the experiments could be solved.
3. The theoretical possible accuracy to determine the co-ordinates of the centreline of an air jet is depending on the resolution used during digitisation. Normally this can be 1/256 of the dimensions of the room at condition that the appropriate video equipment is used so that the video images cover the whole front surface of the test room under study. In our test room with a height of 2.0 m the centreline was defined with an accuracy of ± 0.8 cm.
4. This method permits to quantify other characteristics of the visualised air jet such as: the two-dimensional shape of the air jet, three-dimensional composition of an air jet when scanning the illuminated surface, detection of well-mixed zones, etc.
5. It should be possible to extend this technique to the three-dimensional co-ordinates of the centreline of an air jet. The principle can be applied for two cameras as well and for different illuminated planes.
6. This technique can be used for a variety of different scales of the studied air flow patterns.

More work is needed to establish a more precise relationship between tracer particle concentration and reflected light intensity (Saunders and Albright 1989).

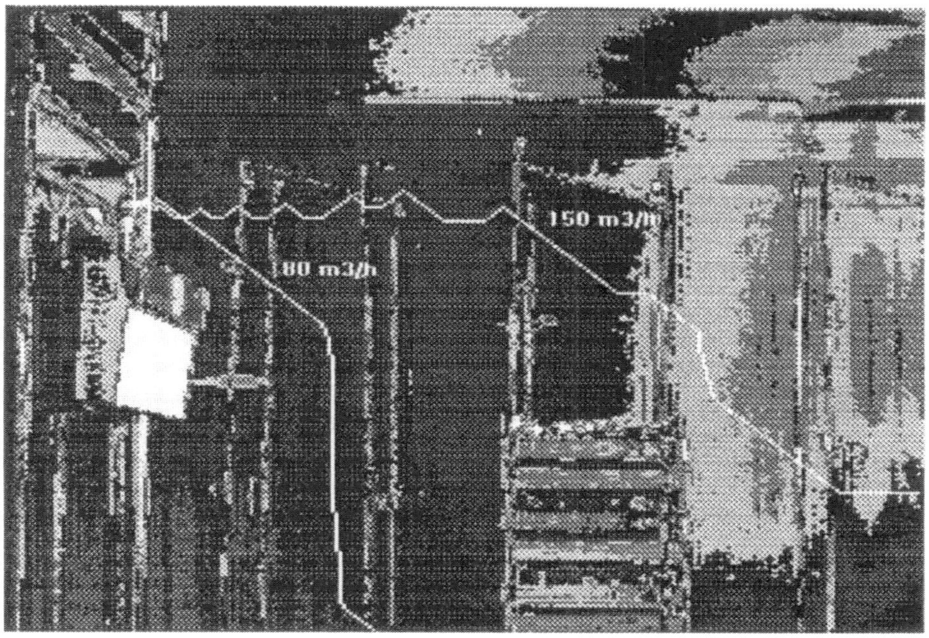

Figure 5. Processed image with centrelines of the air flow surrounding fruit boxes at different flow rates. (Mark the inlet opening in the left wall, the fruit boxes down right and the vertical bars which carry the sensors.)

Apart from the many necessary experiments for the development of the technique, for the purpose of this paper a demonstrative experiment was carried out to show an application of the technique. Apple boxes were placed in the test chamber and different flow rates were applied. The temperature of the incoming air was 9°C and the temperature in the chamber was 16°C. After filming the smoke patterns, the images were processed and the centrelines were calculated (Figure 5).

Although this test installation was not designed for fruit conservation (9°C was the lowest possible temperature), the experiment shows that the technique for calculating the centreline by image analysis can be applied in cooling cells to check whether they are well functioning. It is clear that in a malfunctioning cooling cell temperature gradients can occur which are too high to guarantee good conservation.

Furthermore, it must be possible to control the air flow in such a way that fruit and vegetable boxes are completely surrounded by air of the same temperature. Such climate control algorithms based on mathematical identification modelling are also being developed on the same test installation (Berckmans et al, 1992). These algorithms can then, in turn, be checked for well functioning by using the same image processing technique.

6. REFERENCES

Barber, E.M. and J.R. Ogilvie (1981). *Incomplete mixing in ventilated airspaces. Part I. Theoretical considerations.* Can. Agric. Engng., 24, 1, 25-29.

Barber, E.M. and J.R. Ogilvie (1984). *Incomplete mixing in ventilated airspaces. Part II. Scale model study.* Can. Agric. Engng., 26, 2, 189-196.

Berckmans, D., M. De Moor and B. De Moor (1992). *Test installation to develop a new model concept to model and control the energy and mass transfer in a three dimensional imperfectly mixed space.* Contribution to Roomvent '92: Air distribution in rooms, 2-4 September, Aalborg, Denmark.

Castleman, K.R. (1979). *Digital Image Processing.* Prentice-Hall, Inc., Englewood Cliffs, N. J. 07632, pp. 429.

Croom-Gale, D.J. and B.M. Roberts (1975). *Air-Conditioning and ventilation of buildings.* Pergamon Press, pp. 446, 365-395.

Granier, R., C. Chosson, E. Retif and P. Rousseau (1989). *Appréciation des circuits d'air et des gradients thermiques dans une porcherie pour différents systemes de ventilation.* Journées Rech. Porcine en France, 21, pp. 245-252.

Katz (1974). Luftstromung in Raumen. In: *Lehrbuch der Klimatechnik*, Bd.1 Grundlagen, pp. 141-

200.

Kaga, A. and A. Yoshikawa (1989). Velocity distribution measurement through digital image processing of visualised flow images. In: *Building systems: Room air and air contaminant distribution*. By L.L. Christianson, ASHRAE publications, ISBN 0-910119-64-6, pp. 91 - 97.

Leonard, J.J. and J.B. Mc. Quitty (1985). *Criteria for the control of cold ventilation jets*. ASAE-paper 85- 4014, 19 pp.

Merkzirch, W. (1987). *Flow visualisation*. Universitat Essen, Federal Republic of Germany. Academic Press, Inc. pp. 286.

Mueller, T.J. (1989). Part 2: Fundamentals: Gases; Smokes. In: *Handbook for flow visualisation* by Wen-Yei Yang. Hemisphere publishing coorporation pp. 672, 45-62.

Mullejans, H. (1966). *Uber die Ahnlichkeit der nicht-isothermen Stromung und den Warmeubergang in Raumen mit Strahlluftung*. Westdeutscher Verlag, Koln, pp. 59.

Murakami, S. (1992). *New scales for ventilation efficiency and their application based on numerical simulation of room air flow*. Proc. of the Int. Symposium on Room Air Convection and Ventilation Effectiveness, pp. 22-38.

Murakami, S. and S. Kato (1987). Current status of numerical and experimental methods for analysing flow field and diffusion field in a room. In: *Building systems: Room Air and Air Contaminant Distribution.*, by L.L. Christianson, Amercian Society of Heating, Refrigerating and Air-Conditioning Engineers, Inc. pp. 263.

Randall, J.M. (1975). *The prediction of air flow patterns in livestock buildings*. J. Agric. Engng. Res., 20, 199-215.

Randall, J.M. and V.A. Battams (1979). *Stability criteria for air flow patterns in livestock buildings*. Reprinted from J. Agric. Engng. Res. 21, 361-374.

Regenscheit, B. (1974). *Die Archimedes-Zahl*. Gesundheits-Ingenieur Heft 6, 91, 172-176.

Rodahl, E. (1985). *The point of separation of cold jets flowing along the ceiling*. Proc. Clima 2000 conference on heating, ventilation and air-conditioning, vol. 2, pp. 251-254.

Rolloos, M. (1978). *Schaalmodelonderzoek. Over de toepasbaarheid van schaalmodellen van grote ruimten voor het vinden van de meest wenselijke luchtverdeling*. TH-TVVL Lezingendag 14 maart 1978 pp. 61-94.

Saunders, D.D. and L.D. Albright (1989). A quantitative air mixing visualisation technique for two-dimensional flow using aerosol tracers and digital imaging analysis. In: *Building systems: Room air and air contaminant distribution*. By L.L. Christianson, ASHRAE publications, ISBN 0-910119-64-6, pp. 84 - 88.

Smith, C.R. (1989). Part 3: Image processing and Computer Assisted Methods: Computer Aided Flow Visualisation. In: *Handbook for flow visualisation* by Wen-Yei Yang. Hemisphere publishing coorporation pp. 672, 375-394.

Timmons, M.B. (1980). *The use of models to predict fluid motions in mechanically ventilated structures*. ASAE-paper.80-4018.

Tsutsumi, J. (1990). *Numerical simulation of thermal convection in a room with natural ventilation caused by buoyancy*. Proceedings of the international CIB W67 symposium on energy, moisture and climate in buildings, 3-6 September 1990, Rotterdam, NL, pp. 6.

Van De Weyer, K. (1991). *Kwantificatie van gevisualiseerde luchtstromingen met behulp van beeldverwerking*. Thesis of agricultural engineer, K.U.Leuven, pp. 96.

Wen, Yei-Yang (1989). Fundamentals: Introduction. In: *Handbook for flow visualisation* by Wen-Yei Yang. Hemisphere publishing coorporation pp. 672, 3-6.

OPTIMIZATION FOR FRUIT QUALITY DURING A STORAGE PROCESS

T. Morimoto, W. Purwanto, J. De Baerdemaeker* and Y. Hashimoto

Dept. of Biomechanical Systems, Ehime University, Tarumi, Matsuyama 790, Japan

*Dept. of Agric. Engineering, Katholieke Universiteit, Leuven, Belgium

Abstract: Plant control systems are characterized by complexity and uncertainty. This paper studies the application of an intelligent control technique, including neural networks and genetic algorithms, to the optimization of fruit storage process. Two optimization problems were considered. One is to maintain the water loss of fruits at its desired response by relative humidity. The other is to determine the setpoints of relative humidity which minimize the objective function F(h) consisted of the water loss and the degree of disease of fruits (h: relative humidity). In the system, F(h) as affected by relative humidity is first identified using neural networks and then the optimal setpoints of relative humidity are searched for through simulation of the identified model using genetic algorithms. This intelligent control technique was quite useful for the optimization of such complex systems as fruit storage process.

Keywords: Optimization, fruit storage process, water loss, relative humidity, neural network, genetic algorithm

1. INTRODUCTION

In recent years, advanced agricultural production systems such as a plant factory have been significantly developed. These systems consume much electrical energy for artificial light and air conditioning. From the view of cost-performance, the optimization (or optimal control) technique for a mass production and higher quality of plants are necessary in these systems (Challa and van Straten, 1991; Tantau, 1993; Hashimoto, 1991 and 1993).

Plant production process is mainly composed of cultivating process and storage process. The storage process is more simple and controllable than the cultivating process. Many types of fruits are matured and colored while maintaining freshness and their final commercial prices are determined during the storage process (De Baerdemaeker *et al.*, 1987 and 1994). Hence, the optimization of storage process would be very important for fruit production.

For the optimization of such a complex system as storage process, intelligent control techniques are more suitable than conventional mathematical approaches. Neural networks have the capability to identify complex systems with their own high learning ability (Chen *et al.*, 1990; Hunt *et al.*, 1992). Nowadays, the identification of such arbitrary behaviors as plant responses (Hirafuji, 1991; Morimoto *et al.*, 1991) and fruit responses (De Baerdemaeker and Hashimoto, 1994) are vigorously carried out by using neural networks. Genetic algorithms are one of the combinatorial optimization techniques by which an optimal value can be globally obtained. It simulates the evolutionary process based on a crossover and mutation in genetics (Goldberg, 1989). Applications of the genetic algorithm to the optimal control of an agricultural production system have been observed in recent years (Morimoto *et al.*, 1993 and 1995)

The present work is an attempt to apply a new intelligent control technique based on neural networks and genetic algorithms to the optimization of the cold storage process of apples. The manipulating variable is relative humidity and the control variables are the quality of fruits given by objective functions.

2. OPTIMIZATION PROBLEM

Fruit used for the experiment is apples which allow a long term storage (Kitagawa, 1989). The most severe disorder during the storage process seems to be the wilting of fruit's surface. Even after the harvest, fruits continues to transpire from surface through stomata and epidermis. In the present study, we considered two optimization problems.

Optimization problem I; If the water loss caused by evapotranspiration of fruits can be well controlled, the adjustment of the shipping day will become easy. The control aim here is to maintain the water loss at its desired response. This can be achieved by minimizing the error between the desired values and the current values of water loss. The objective function $F_1(h)$ (h: relative humidity) is given as follows:

$$F_1(h) = \Sigma (W_h(k) - W_{h,ref}(k))^2 \quad \text{.............. (1)}$$

$$W_{h,ref}(k) = c_1 * k \ (c_1\text{: constant value}) \quad \text{.........(2)}$$

where $W_h(k)$ (≥ 0) represents the response of the water loss of fruits as affected by relative humidity h and $W_{h,ref}(k)$ is the desired response of the water loss which is given as a linear function (2) (k=1, ..., N: Sampling time). Here, the values of $W_h(k)$ are estimated from the neural network model. The storage process was divided into six stages. The optimization problem I is to determine the 6-step setpoints of relative humidity (h_1, h_2, h_3, h_4, h_5 and h_6) which minimize the objective function $F_1(h)$.

Optimization problem II; To prevent the evapotranspiration, it is necessary to keep the relative humidity at high levels. However, a continuously high humidity induces diseases related to mold fungi. These disorders severely affect their commercial prices. The objective function $F_2(h)$ is as follows.

$$F_2(h) = \sum_k (\alpha \cdot W_h(k) + \beta \cdot D_h(k)) \quad \text{..........(3)}$$

where $D_h(k)$ (≥ 0) represents the degree of disease as affected by relative humidity, and two coefficients, α and β, are respectively weights for both evaluators. Here, the degree of disease means the occurrence rate of disease among many fruits after storage. For the reduction of $W_h(k)$, a higher humidity is desirable. For the reduction of $D_h(k)$, a lower humidity is desirable. These relations are contrary to each other. Here, the values of $D_h(k)$ are estimated from the nonlinear model given by equation (4). The storage process in this case was divided into four stages. The optimization problem II is to determine the 4-step setpoints of relative humidity (h_1, h_2, h_3 and h_4) which minimize $F_1(h)$. The temperatures in the optimization I and II are kept at constant (2℃).

3. METHODOLOGY FOR OPTIMIZATION

3.1 Strategy of a skilled grower

A skilled grower can well deal with such complex systems as fruit storage and cultivating systems with his own intuition and experience. In order to realize the storage optimization, it is better to imitate a skilled grower's technique. Considered a skilled grower's techniques in the storage planning, he tries to make the best strategy for optimization based on his own past experiences. Such experiences seem to be analogous to the formation of mental models newly created in his brain through his learning activity. We imitate these manners to realize the optimization of fruit storage using neural networks and genetic algorithms.

3.2 Design of control system for optimization

The optimization of physiological processes of fruits is accomplished by controlling the environmental factors in a storage house. Figure 1 shows a block diagram of the control system for the optimization of fruit storage process. In the system, fruit responses (water loss of fruits) as affected by environmental factors (relative humidity) are first identified using neural networks, and then the optimal setpoints of relative humidity which minimize two objective functions are searched for through simulation of the identified model using genetic algorithms.

The first action (modeling) using neural networks seems to be the same as the action that a skilled grower has experienced. Furthermore, the way for obtaining an optimal value in the second step corresponds to the action that a skilled grower newly

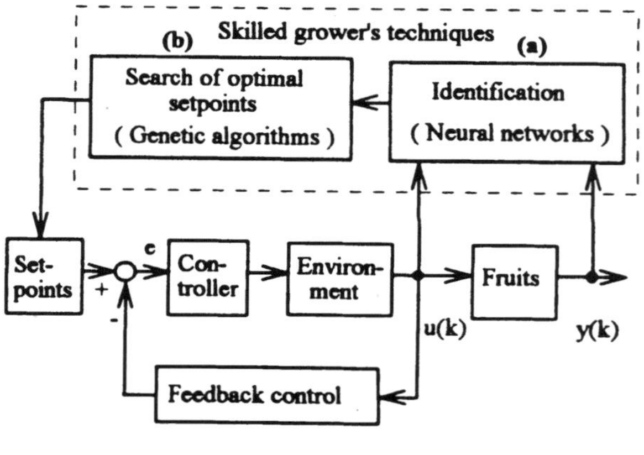

Fig. 1 Block diagram of a control system for optimization of fruit storage system

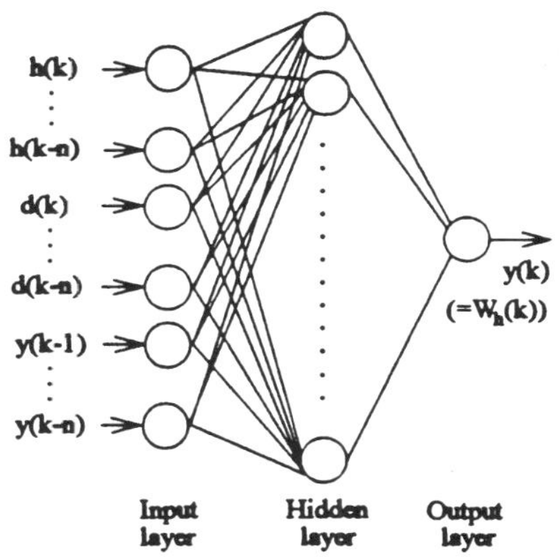

Fig. 2. Three-layer neural network architecture for process model identification.

selects a better (or best) strategy from among his own experiences and imagination. The identification and the search of an optimal strategy are thought to be essential for human mental activity. These procedures should be periodically carried out in order to adapt the time-varying characteristics of the physiological dynamics of fruits.

3.3 Neural network for model identification

In this control technique, a model with high accuracy is severely required because an optimal value is selected from model simulation. The model of $W_h(k)$ to relative humidity is built based on an identification. For the effective identification of daily changes in $W_h(k)$, we adopted the day as an input and supposed a multi input (relative humidity and days after storage) single output ($W_h(k)$) system. As for this treatment, it will be discussed in the section 4.1. So, the $W_h(k)$ as affected by both relative humidity and days after storage is identified by the neural network (part (a) in the Fig.1). Figure 2 shows the neural network architecture for process model identification. The (n+1)th time series of two input variables, relative humidity {h(k), ..., h(k-n)} and days after storage {d(k), ..., d(k-n)}, and nth past time series of output variable, $\{W_h(k-1), ..., W_h(k-n)\}$, are applied to the input layer and the current output $W_h(k)$ is applied to the output layer as a training signal (k=1, 2, ..., N: sampling time). Hence, neuron numbers in the input and output layer are given by (3n+2) and 1, respectively. The learning algorithm of the neural network is error back propagation (Rumelhart *et al.*, 1986).

3.4 Genetic algorithm for optimization

Coding of decision variable; In order to use the genetic algorithm, we have to define an individual which means a decision variable to be obtained. The decision variables in the optimization problem I and II are respectively the 6-step setpoints and 4-step setpoints of relative humidity. Hence, the individuals in the problem I and II are given as follows:

(Optimization problem I):
Individual = $(h_1, h_2, h_3, h_4, h_5, h_6)$
=(0001001, 1100110, 0101010, 1100110, 1110001, 0110010)
(Optimization problem II):
Individual = (h_1, h_2, h_3, h_4)
=(0001001, 1100110,0101010, 1100110)

It is noted that all parameters of humidity are coded as 7-bit binary strings. The set of individuals is called a "population" and the binary digit 0 or 1 is called a "gene".

Procedure of genetic algorithm; The flow of genetic algorithm is as follows. At first, the initial population P(0) consisting of N_i sorts of individuals are generated at random. Second, genetic operators (crossover and mutation) are applied to these individuals. In this case, for a directed evolution, the selection are applied two times, after the crossover and after the mutation. Through the crossover, N_c sorts of individuals are newly created according to the crossover rate P_c and furthermore N_m sorts of individuals are newly generated according to the mutation rate P_m. Third, the fitness of all individuals are calculated using the identified neural network model. Here, the fitness is the values of objective functions $F_1(h)$ and $F_2(h)$. Finally, several individuals with lower fitness are retained for the next generation by a selection. A new population P(t+1) can be obtained through these operations. An optimal value, which is an individual with the lowest fitness, is obtained by repeating these procedures.

Methods of crossover and mutation; For the crossover, two individuals (0000101 and 0011111) are first mated at random. Next, these binary strings are cut at the 4-bit position along these strings and then two new individuals (0001111 and 0010101) are generated by swapping all binary characters from the 1-bit to 4-bit position. For the mutation, on the other hand, one individual (e.g., 0000011) is first selected at random, and then a new individual (0100001) is created by inverting two characters (genes) from 0 to 1 or 1 to 0.

4. IDENTIFICATION OF FRUIT RESPONSE

4.1 Response of water loss of fruits, $W_h(k)$, and its identification

During the storage process, the water content of fruits gradually decreases with evapotranspiration. Figure 3 shows observed daily changes in the water loss (%) of apple fruits under different humidity

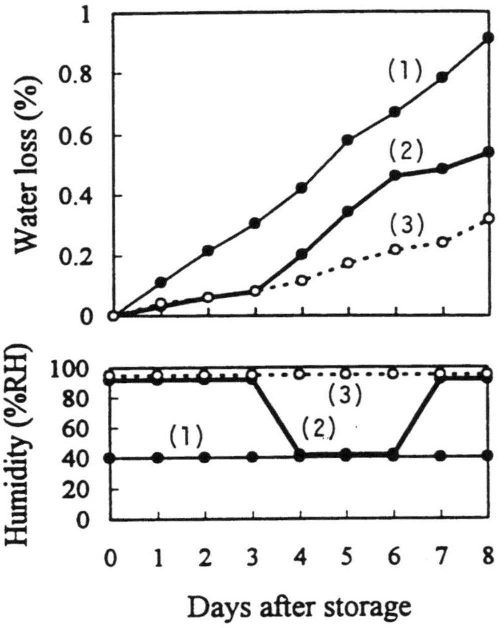

Fig. 3. Observed daily changes in water loss of apples $W_h(k)$ under different humidity ((1): 40%RH, (2): 95→40→95%RH, (3): 95%RH).

conditions. This storage type is a cold storage where the air temperature is kept at 2℃. Curves (1), (2) and (3) are responses in cases when the relative humidity are respectively controlled to 40%RH, 95 →40→95%RH and 95%RH. The water loss were calculated from the weight loss. Even after the harvest, the evapotranspiration of fruits is active. The fruit weight decreases with the evapotranspiration. As shown in the figure, the water loss increased almost linearly with time. It is also found that the water loss varied according to the change of relative humidity. From this figure, we found that the water loss of fruits can be successfully controlled by relative humidity. Since the rapid water loss induces severe damages in quality, it is important to suppress the water loss by controlling the relative humidity.

Next, these responses were identified using the neural network. At first, we examined the better way to identify the daily changes of water loss. Figure 4 shows the comparisons of the estimated error under different treatments of day for input variable. The system parameter number n is 1st. The error (A) shows the case that the input did't include day (without day). The input neuron number was (2n+1)th. In the case of (B), the input included (n+1)th time series of day (with time series of day). The input neuron number is (3n+2)th. In the case of (C), it contained only one day (with only one day). The input neuron number is (2n+2)th. From the figure, the treatment like a (C) showed the smallest error. This reason seems to be because the data of day, which increases linearly, was well suited for the daily change in the water loss. Figure 5 shows the

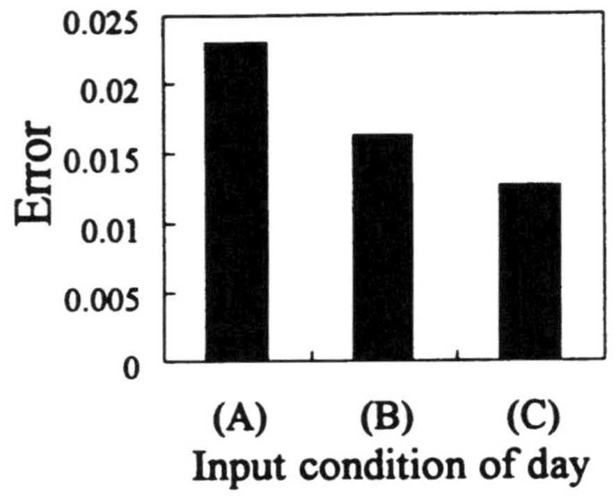

Fig. 4. Comparisons of estimated error of water loss for treatments of day ((A): without day, (B): with time series of day, (C): with only one day).

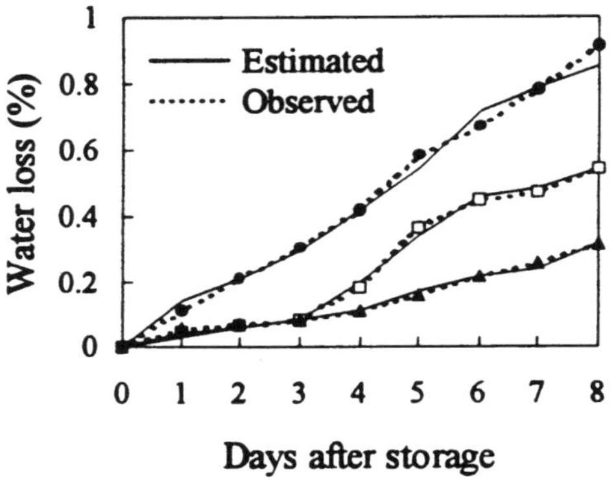

Fig. 5. Comparisons between estimated and observed responses.

identification results obtained through the treatment (C) in Fig.4. The system parameter number n and the hidden neuron number are 1 and 5, respectively. These values were determined to be the best from the viewpoints of computing time saving and error minimization. The iteration number for learning is 5000 times. It is clear that all estimated responses are closely related to the observed responses. Thus, we could obtain a successful model for calculating the $W_h(k)$ as affected by any combination of the 6-step or the 4-step setpoints of relative humidity.

4.2 Relation between relative humidity and the degree of disease, $D_h(k)$

Relative humidity in the storage house becomes high due to the continuous evapotranspiration of many fruits. Excessively high humidity near 100%RH accelerates the activities of mold fungi and pathogenic bacteria. The storage method in the paper is a cold storage which is an effective way for preventing the propagation of mold fungi. However, some mold fungi can live and propagate in spite of low temperature (Kitagawa, 1989). In order to prevent the disease, therefore, an adequate drop in relative humidity seems to be effective. The relation between relative humidity and the degree of disease

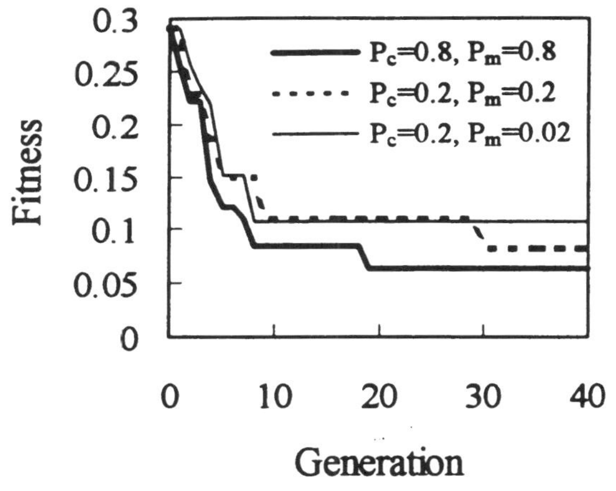

Fig. 6. Search evolution curves of an optimal value under different crossover and mutation rates.

$D_h(k)$ was expressed by sigmoid function as follows:

$$D_h(k) = 1 / (1+ e^{(-h(k)+92)1.5}) \quad (4)$$

The values of $D_h(k)$ in the objective function (2) is calculated from this nonlinear model.

5. OPTIMIZATION RESULTS

5.1 Search evolution characteristics

Next, we'll search for an optimal value through simulation of the identified model using genetic algorithms. Figure 6 shows a search evolution curves of an optimal value. The horizontal axis is generation, and vertical axis is the fitness of the best individual in each generation. In all cases, the fitness dramatically dropped with the generation. However, the degree of drop is larger for the higher crossover and mutation rates than for the lower crossover and mutation rates. For example, the fitness reached a minimum value at the 18th generation under the high crossover and mutation rates (e.g. P_c=0.8 and P_m=0.8). Under the lower rates (P_c=0.2 and P_m=0.02), however, the fitness fell into a local optima.

These results suggest that both crossover and mutation rates should take higher values rather than lower values in order to efficiently search for an optimal value. This consideration is confirmed from the recent studies of molecular biology. According to the Darwinian survival process, it is assumed that natural evolution proceeds gradually through the accumulation of the point mutation rarely occurred at random. The mutation rate is significantly small. With such a small mutation rate, however, it is impossible to evolve from unicellular organism to human being only during 3.5 billion years (Gould,

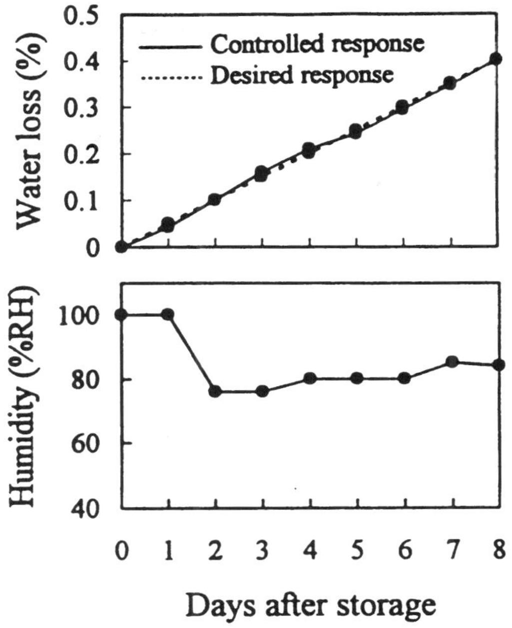

Fig. 7. Optimal control performance of the water loss of fruits in the optimization problem I.

1994). For faster evolution, the high diversity of population is essential. Some genes, which is called "transposons", in a DNA have the ability to move actively (Rennie, 1993). Also, retroviruses, such as cancer virus and AIDS virus, have the ability to change the DNA structure of a host cell by invading the cell nucleus of it (Toh *et al.*, 1985). These assumptions lead to the considerations of directed mutation and high mutation rate.

5.2 Optimal control performance

Figure 7 shows the control result in the optimization problem I obtained from simulation. The control aim is to follow the desired response of the water loss. Upper figure shows the control performance and lower figure denotes the 6-step setpoints of relative humidity. It is found that the water loss is well controlled to the desired values. This control result is applicable to the adjustment of shipping day of fruits.

Figure 8 shows the control results in the optimization problem II under different weights, α and β, in the objective function. The lower figures show the optimal 4-step setpoints of relative humidity and the upper figures represent the control performances of the water loss of fruits, $W_h(k)$. The optimal 4-step setpoints are estimated to be 93→93 →100→100%RH for α=1.0 and 82→87→82→ 82%RH for α=0.4, respectively. When the α takes a lower value, the optimal setpoints were recommended to be lower level. This is because the lower values of α weakened the power of the water loss in the objective function. There are no contradiction in this result.

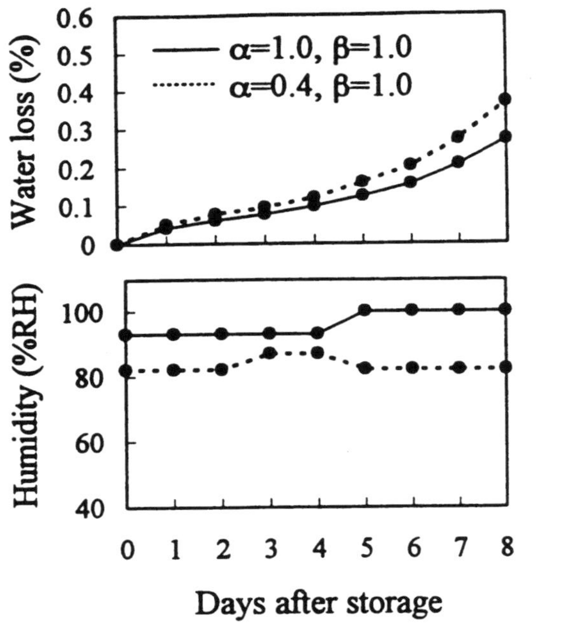

Fig. 8. Optimal control performance of the water loss of fruits in the optimization problem II.

6. CONCLUSION

We proposed a new intelligent control technique based on neural networks and genetic algorithms for the optimization of fruit storage process. In the system, the water loss of fruits as affected by relative humidity was first identified using neural networks, and then the optimal setpoints of the relative humidity were searched for through simulation of the identified model using genetic algorithms. These procedures are quite similar to a skilled grower's planning. A three-layer neural network was effective for the identification of multi-input and single output system. Furthermore, the genetic algorithm allowed the optimal value to be quickly and successfully sought from the simulation under high crossover and mutation rates. Thus, it seems that this intelligent control technique suitable for the optimization of such complex systems as fruit storage systems.

REFERENCES

Challa, H. and G. van Straten (1991). Reflections about optimal climate control in greenhouse cultivation. *Proc. 1st IFAC/ISHS Workshop on Mathematical and Control Applications in Agric. and Hort.*, pp.13-18, Pergamon Press, Oxford.

Chen, S., S.A. Billings and P.M. Grant (1990) Non-linear system identification using neural network. *International Journal of Control,* 51(6), 1191-1214.

De Baerdemaeker, J., A: Wouters (1987). Mechanical properties of apples I. In: Jowitt et al. (eds.) Physical properties of foods 2, Elsevier applied science, London, pp.403-416.

De Baerdemaeker, J and Y. Hashimoto (1994). Speaking fruit approach to the intelligent control of the storage system. *Proc. of 12th CIGR World Congress,* Vol.1, pp.190-197.

Goldberg, D. (1989). Genetic algorithms in search, optimization and machine learning, Addison-Wesley.

Gould, S.J. (1994). The evolution of life on the earth. *Scientific American,* October, pp.63-69.

Hashimoto, Y. (1991) Computer integrated plant growth factory for agriculture and horticulture. *Proc. 1st IFAC/ISHS Workshop on Math. and Control Applications in Agric. and Hort.,* pp.105-110, Pergamon Press, Oxford.

Hashimoto, Y. (1993). Computer integrated system for the cultivating process in agriculture and horticulture. In: The Computerized Greenhouse (Hashimoto, Y. eds.). 175-196, Academic Press.

Hirafuji, M. (1991) A plant growth model by neural networks and L-system. *Proc. 9th IFAC Symp. Identification and System Parameter Estimation, Vol.1,* pp.605-609.

Hunt K.J., D. Sbarbaro, R. Zbikowski and P.J. Gawthrop (1992) Neural networks for control systems - Survey. *Automatica, 28(6)*, 1083-1112.

Kitagawa, H. (1989). Storage of fruit. *Refrigeration,* 64(741), 745-759.

Joyce, G. F. (1992). Directed molecular evolution, *Scientific American*, December, pp.48-55.

Morimoto, T., I. Cho and Y. Hashimoto (1991) Identification of hydroponics in an advanced control system of the greenhouse. *Proc. 9th IFAC Symp. on Identification and System Parameter Estimation, Vol.1,* pp.610-615.

Morimoto, T., T. Takeuchi and Y. Hashimoto (1993). Growth optimization of plant by means of the hybrid system of genetic algorithm and neural network. *Proc. Inter. Joint Conf. on Neural Networks*, Vol.3, pp.2979-2982.

Morimoto, T., J. De Baerdemaeker and Y. Hashimoto (1995). Optimization of storage system of fruits using neural networks and genetic algorithms. *Proc. 4th IEEE Inter. Conference on Fuzzy Systems*, Vol.1, pp.289-294.

Rennie, J. (1993). DNA's new twists. *Scientific American,* March, pp.89-96.

Rumelhart, D.E., G.E. Hinton and R.J. Williams (1986). Learning representation by back-propagation error. *Nature,* 323(9), 533-536.

Tantau, H.-J.(1993). Optimal control for plant productions in greenhouses. In "The Computerized Greenhouse (Hashimoto, Y. *et al.*, eds.), Academic Press, San Diago, pp.139-152.

Toh, H., M. Ono, K. Saigo and T. Miyata (1985). Retroviral protease-like sequence in the yearst transposon Ty*1*. *Nature,* 315(20), pp.691.

ENERGETICAL OPTIMALISATION OF THE GRAIN DRYING PROCESSES BY THE CONTROL OF AIRFLOW RATE

Neményi, M. - Czaba, I. - Kovács, A. - Jancsók, P.

PANNON Agricultural University, Faculty of Agricultural Sciences
Department of Agricultural and Environmental Engineering
H-9200 Mosonmagyaróvár

Abstract: The control of the grain driers is very important, both to save the quality of the agricultural products and to reduce the energy consumption of drying. The drying process of the different kernel products has to be divided into phases. The influence of the air flow rate on the drying characteristics of the maize kernels was investigated. The subject of this research work was to keep the relative humidity of the output drying air constant and hereby to increase the power efficiency of grain driers. The mathematical model described in this study is helpful in designing the energy and quality saving grain driers.

Keywords: Air-flow rate control, Drying, Grain.

1. INTRODUCTION

Since in Hungary, maize is harvested at higher moisture content of kernels (25-35 %, wet basis) than the required limit of storage, yearly 4-5 million tons of shelled corn should be dried artificially. Furthermore, corn is the most important feeding crop, so care must be taken in to account during drying process of corn kernels, for both purposes energy saving and quality preservation. Generally two main matters are considered during formulating the objectives of the investigated aspects:

1. To develop new theoretical techniques of drying;
2. To accomplish theoretical and practical investigations needed for improving both power efficiency and drying quality of the running driers.

Since in Hungary, there are nearly 1000 driers with high workability (2000-2500 kg water/h), the high capacity of these drying units arises the importance of the problem under consideration.

Many previous studies handled with the artificial drying process of cereals (corn was involved). These articles separately investigated the kinetic of the dehydration of kernels in thin layer or deep bed. One shortcoming of these articles was that the results were obtained under statical conditions. Hence the parameters of the input drying air (temperature, relative humidity, flow rate etc.) were constant during drying process, where the pile remained in statical bed. In contrast, it is well known that during drying the heat physical characteristics of shelled corn change continuously. Departing from this fact the relative humidity of outdoor air at constant air flow rate gradually decreases, and at the same time the temperature of the grains increases, which causes energy efficiency reduction and quality degradation. On the other hand, usually in the practice the corn kernels are mixed continuously to achieve a uniform dried medium.

To keep the relative humidity (RH) of the outdoor air constant by means of improving the energy efficiency, the drying conditions must be changed as a function of decreasing of the drying intensity (drying speed). There are three different possibilities:

1. to increase the bed depth,
2. to decrease the temperature of drying air, and
3. to reduce the mass flow of drying air.

Such questions as the optimal air temperature and air-flow rate for the energy and quality saving (and economic, resp.) drying of grain have been only partially discussed.

The main aim of this study was to answer the question, how the drying air at constant exhaust relative humidity influences the drying process of shelled corn. Since the outdoor air flow is intensively polluted during drying, it is difficult to measure the relative humidity continuously, the required measured values for the control are to be taken on at expedient time intervals. Convenient computer programs are required for permanent regulation.

2. BACKGROUND

The most general model which describes the drying of thin layer of corn kernels is the well known Newton's exponential model:

$$MR = \frac{X_g - X_e}{X_i - X_e} = \exp(-k\tau) \qquad (1)$$

where:

- X_g: average moisture content of the grain (decimal, dry basis),
- X_i: initial moisture content of the grain (dec., d.b.),
- X_e: equilibrium moisture content of the grain (dec., d.b.),
- k: drying constant (h^{-1}), and
- τ: time (h).

Equation (1) has been further adapted utilizing the experimental experiences of other researchers (i.e. Henderson and Pabis, 1961) as:

$$MR = a\exp(-k\tau) \qquad (2)$$

where:

a: constant, dependent on particle shape

$(a = \frac{6}{\pi^2}$ for a sphere).

The factor k can be expressed as a function of the diffusion coefficient in the next form:

$$k = \frac{\pi^2 D}{r^2}$$

where:

- r: radius of kernel (m),
- D: moisture diffusivity (m^2h^{-1}).

The moisture diffusivity depends on the local moisture content and temperature. On the other hand it depends on the chemical, morphological characteristics of different corn hybrids.

For maize hybrids with different characteristics Neményi (1988, 1991) gave diffusion equations as a function of temperature and moisture content. Further equations were given by him for the diffusion coefficient (at a given moisture content and temperature) as a function of the kernel fat and fiber content.

Westerman, White and Ross (1973) depending upon Newton's exponential model (1) have determined k as a function of the temperature and relative humidity of the drying air. An Arrhenius type relationship:

$$k = c_1 \exp(-c_2 / R) \qquad (3)$$

where R is temperature (in the form given by the authors R is in degrees Rankine) and c_1 and c_2 are experimental constants, were used to establish the k values as a function of air relative humidity (RH) and temperature. They found the next equation:

$$k = \exp\left(13{,}328 - 0{,}0115(RH) - 8255{,}9(R^{-1})\right) \qquad (4)$$

where RH is percent relative humidity, and R is temperature in degrees Rankine.

(The form of equation (4) is in SI Units as:

$$k = \exp(13{,}328 - 0{,}0115(RH) - \\ - 8255{,}9(1{,}8T + 0{,}021)^{-1})$$

where T is the air temperature in K.)

They considered, that the equilibrium moisture content is different under static and under dynamic circumstances.

$$X_{DE} = 14{,}052 + 0{,}106(RH) - (7{,}8 \cdot 10^{-4})(RH)(T) \qquad (5)$$

where X_{DE} is percent dynamic equilibrium moisture content (d.b.) and T is the air temperature in degrees F. (According to the SI Units the function (5) must be described in the following form:

$$X_{DE} = 14{,}052 + 0{,}106(RH) - \\ - (7{,}8 \cdot 10^{-4})(RH)(1{,}8T - 459{,}67)$$

where T is the air temperature in K.)

Another form of equation (1) was used by different researchers and specified for drying of thin layer, the equation is:

$$MR = \exp(-k\tau^{n}) \qquad (6)$$

White, Ross and Westerman (1973) gave for the above described equation the values of the empirical drying parameters n and k as a functions of drying air temperature and dew point temperature as:

$$n = 0,464 + 0,000778T + 0,00169T_{dp} \qquad (7)$$

and

$$k = \exp\left(-2,794 + 0,00113T + 0,00397T_{dp}\right) \qquad (8)$$

where T_{dp} is the dew point temperature of the drying air in degrees F. (In SI Units are the forms of the last two equations:

$$n = -1,7185132 + 0,0014T + 0,007146T_{dp}$$

$$k = \exp\left(-9,81316 + 0,02034T + 0,007146T_{dp}\right)$$

where T and T_{dp} in K)

Misra és Booker (1980) conducted an equation for drying of shelled yellow corn. They found a significant correlation between n and relative humidity (RH=90%....100%).

Morey and Li (1984) gave coefficients k and n for two grown years experiment. They investigated how drying equations for a thin layer could be utilized for each layer of deep bed piles. No significant differences were reported between the measured and predicted values. From the representing figures, one may visually notice the considerable divergence between the measured and predicted values. These differences might be attributed to the unconsidered changing in relative moisture content of the drying air.

3. THE MODEL OF DRYING KINETICS

The basic idea of the model was the assumption, that the drying parameters of mixed pile (deep bed) are to be calculated by the medium values of air (between the input and output values of the air parameters, Fig.1.) as at the thin layer drying.

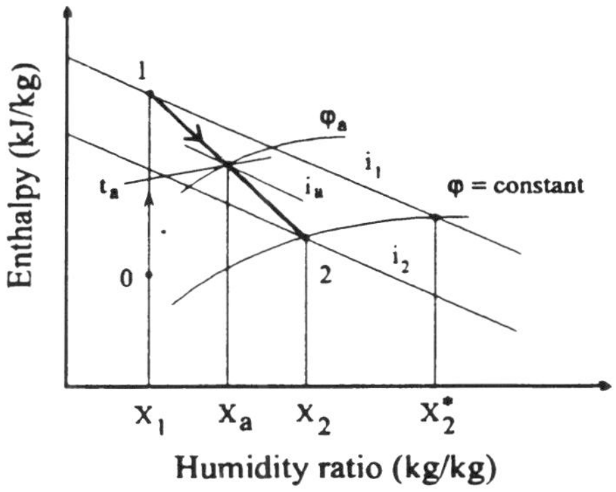

Fig. 1. Calculation scheme of the drying air parameters.

In addition it was further assumed that the enthalpy loss only issues from the heating of kernels. Because the enthalpy loss changes during drying, the parameters of the drying air also change and so the drying process does not run under constant circumstances.
The change of air flow rate is:

$$\frac{dA}{d\tau} = \frac{1}{\Delta X_a}\frac{dX_g}{d\tau} \qquad (9)$$

where A is the air flow rate (kg/kg dry grain) and

$$\Delta X_a = X_2 - X_1 \text{ (kg/kg)}.$$

The increase of kernel temperature was calculated by the equation:

$$\frac{dt_g}{dt} = -\frac{1}{\Theta}\left[\frac{c_{tg}}{c_{tg} + X_g c_w}\right](t_g - t_a) \qquad (10)$$

where Θ is the time constant of heat transfer in h (Baughman, Hamdy and Barre, 1973), the value of Θ was 1 h in these study.

The required drying curves were determined by Newton's function (1).
The parameters k and X_{DE} were computed by the equations (4) and (5), respectively. The enthalpy loss of one kg drying air resulted from the heating of corn kernels in a given time interval is expressed as:

$$\frac{\Delta i_{loss}}{\Delta\tau} = \frac{X_2 - X_1}{\Delta X_g}\frac{\Delta t_g}{\Delta\tau}(c_{tg} + X_g c_w) \qquad (11)$$

Symbols used in equation (11) are defined below: t_g: average kernel temperature (K), c_{tg}: specific heat of dry grain (1,548 kJ/kg K), c_w: specific heat of the water (4,186 kJ/kg K).
The enthalpy of the exhaust drying air is given by the difference:

$$i_{loss} - i_1 = i_2 \quad kJ/kg$$

Since the increasing speed of the grain temperature continuously changes, the parameters of the drying air change during the drying process as well.

The main drying air parameters were computed by the following equations (Wilhelm, (1976) and Szentmarjay et al. (in Print)):

Saturation vapor pressure:

$$P_{vs} = \exp(23,4 - \frac{3917,6}{231 + t}) \quad (Pa)$$

Humidity ratio at saturation:

$$X_s = 0,62198 \frac{P_{vs}}{P_{atm} - P_{vs}} \quad (kg/kg)$$

Humidity ratio:

$$X = \frac{(2501 - 2,411t_w)X_s - 1,006(t - t_w)}{2501 - 1,775t - 4,186t_w} \quad (kg/kg)$$

Vapor pressure at X:

$$P_{vx} = \frac{P_{atm} X}{0,62118 + X} \quad (Pa)$$

Relative humidity

$$RH = \frac{P_{vx}}{P_{vs}} \quad (dec., -)$$

Enthalpy:

$$E = 1,006t + X(2501 + 1,775t) \quad (kJ/kg)$$

where t is the temperature of the air (°C), t_W is the wet temperature of the air (°C), P_{atm} is the atmospheric pressure (10132,5 Pa).

4. EXPERIMENT AND PROCEDURE

Drying apparatus (Fig.2):

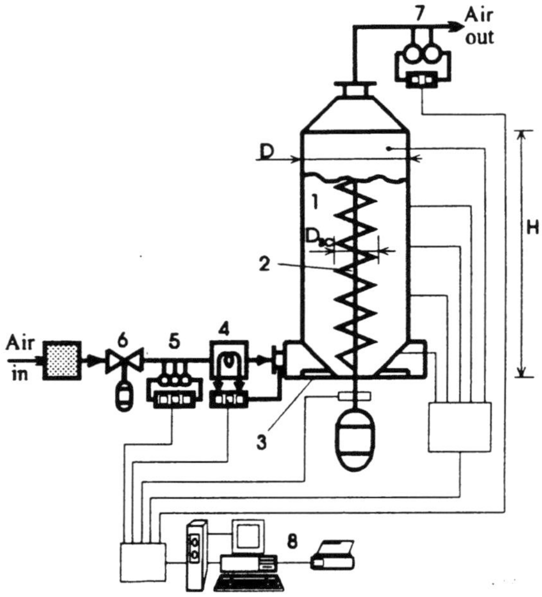

Fig. 2. Scheme of the experimental apparatus; (1. column; 2. conveyor screw; 3. air inlet; 4. air-heater; 5 TETO-452 temperature, air flow and humidity meters; 6. valve; 7. humidity meter; 8. PC/AT)

The mechanically mixed bed drying apparatus constructed in the Research Institute of Chemical Engineering of the Hungarian Academy of Sciences (Veszprém) was described in detail by Szentmarjay et al. (1992), Szentmarjay (1994) and Szentmarjay et al. (in Print).

The drying air enters tangentially the bottom of the apparatus through slots. The kernels are mixed by a housless screw of vertical axis. Temperature, relative humidity of the input and output air were measured by a multi-function measuring equipment (TESTO-452) supplied with an PC adapter for reading the data measured for the computer. These data are transfered to an RS-232 of the PC. The programmable amplifier and channel multiplexing daugther board card transfer the signals to a PCL-812 PG multi function data acquisition card, and from there to an PC/AT computer, which controls the drying process and the dryer.

On the monitor and record its possible to watch the momentary values of the measured and controlled parameters. The control is issued by the three terms PID controllers. Their time constants can be varied.

The output air relative humidity values were kept with an indirect control constant.

Main dimensions of the experimental equipment:
apparatus: D=0,138 m, H=0,65 m,
screw: D_{SC}=0,04 m,
diameter of the axis: d_{SC}=0,016 m, pitch: s=0,028 m.

Experimental parameters:
The parameters of the ambient air are:
Temperature: 23 °C, Humidity ratio: 0,0027 kg/kg,
Temperature of the indoor air: 80 °C, 100 °C, 140 °C. RH_{out}=60 %.
Rotation speed of the screw: 215 rpm,
Mass of the corn pile: 2,3 kg, 5 kg,
Initial moisture content of the corn kernel: 22,3 %, 25,9 % (d.b.),
Time interval of the recording data: 2 min

5. RESULTS AND DISCUSSIONS

When the experiments were running the exerted efforts were not able to measure the moisture content of the grains continuously. Instead, the required measures were taken on only at given times. Fig (3). presents the comparison of the measured and calculated drying curves.

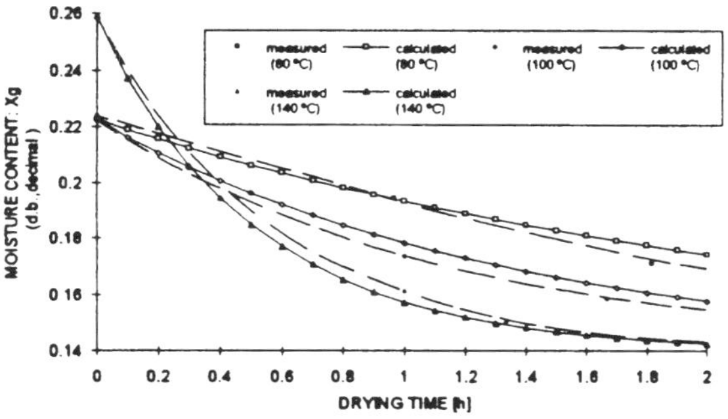

Fig. 3. The measured and calculated drying curves

A comparison of the simulated and measured curves of the input-exhaust absolute humidity (humidity ratio) differences of the drying air with time represented by Fig (4), shows a good agreement which proves the accuracy of the applied model.
Differences appeare in Fig (5) between the calculated and measured air flow rate as a function of drying time. These divergences may be attributed either to the inaccurat measuring of the relative humidity of the exhaust air, or to the indirect controlling of air flow during the experiment.

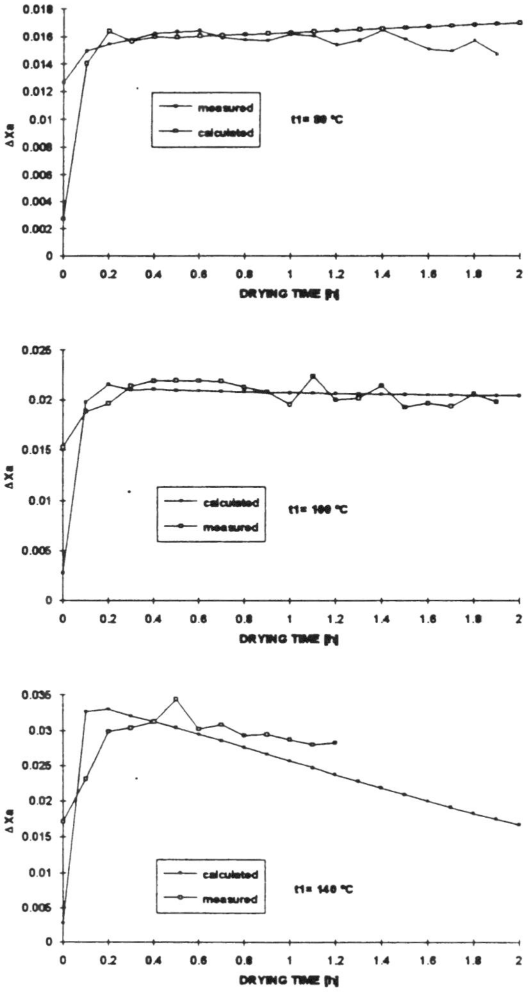

Fig. 4. The measured and calculated differences of the absolute humidity between indoor and outdoor drying air.

Fig 5 The air flow rate as a function of the drying time.

Utilizing given data the specific heat consumption (q) is to be calculated as:

$$q = \sum_{1}^{n} \left[\frac{i_1 - i_o}{X_2 - X_1} \right]_n \quad \text{kJ/kg water}$$

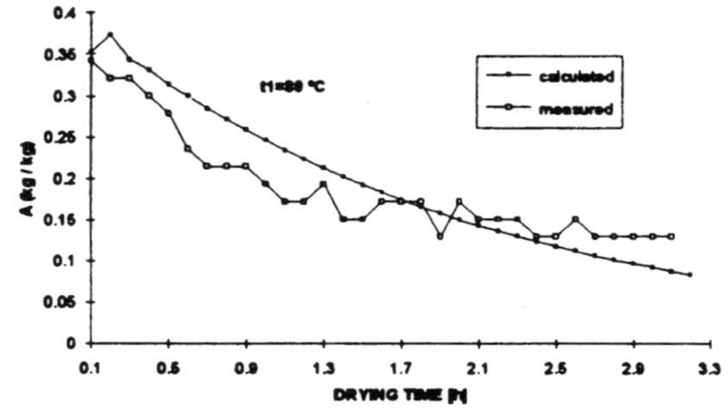

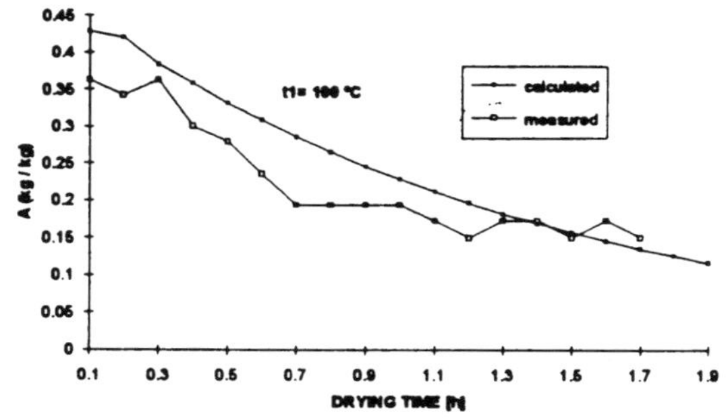

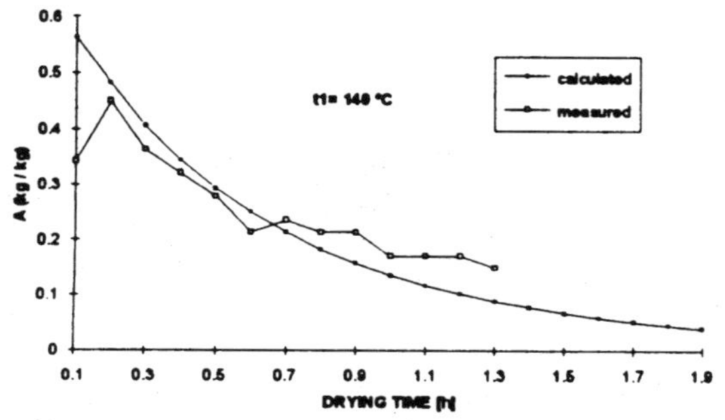

Table 1 The change of the specific heat consumption as a function of the drying air temperature:

Temperature C°	Specific heat consumption (kJ/kg water)	
	measured	calculated
80	3941	4220
100	3647	4084
140	3065	3463

Inverse relationship can be concluded from Table (1) between the two parameters (drying air temperature and specific heat consumption). While the temperature of drying air increases the specific heat consumption, of both the measured and computed values decreases. Furthermore, the maximum temperature of the drying air is restricted by the maximum temperature value of grain which causes quality damages.

6. CONCLUSIONS

The experiments proved the following:

1./ The drying characteristics of the mixed grain piles can be described by utilizing of thin layer equations.

2./ By controlling of the air flow rate in connection with the drying of corn kernels, the power efficiency of drying process may be increased to approach a maximum value.

3./ By the constructed model, the drying process including the energy and quality saving could be successfully controlled.

7. ACKNOWLEDGMENT

The authors are grateful to Dr. Tibor Szentmarjay (head of Department of Gas- and Solid Heattechnics of the R.I. of Ch. E., Veszprém) for technical aid and to PhD Student Abdul Mounem Mouazen for his contribution in translating this paper into English.

REFERENCES

Baughman, G.R., Hamdy, M.Y. and Barre, H.J. (1973): Experimental study and simulation of concurrent - flow dryers. *TRANSACTIONS of the ASAE,* **Volume 16**, pp. 890-894.

Henderson, S.M. and Pabis, S. (1961): Grain drying Theory. *Journal of Agricultural Engineering Research.* **Volume 6**, No.3:pp. 169-174.

Misra, M.K. and Brooker, D.B. (1980): Thin-Layer drying and rewetting equations for shelled yellow corn. *TRANSACTIONS of the ASAE,* **Volume 23**, pp. 1254-1260.

Morey, V.R. and Li, H. (1984): Thin-Layer equation effects on deep-bed drying prediction. *TRANSACTIONS of the ASAE*, pp. 1924-1928.

Neményi, M. (1988): *Characteristics of corn hybrids dryable energy and quality saving.* Academic Press, Budapest (in Hungarian).

Neményi, M. (1991). Drying characteristics of several corn hybrids of different FAO numbers. *Acta Ovariensis*, Mosonmagyaróvár, pp. 37-49.

Szentmarjay, T. (1994): Drying of materials of high moisture content in mechanically spouted bed drier. *Hungarian Journal of Industrial Chemistry.* **Volume 22**, pp. 57-66.

Szentmarjay, T., Szalay, A. and Pallai, I. (1992): Hydrodynamical measurements in mechanically spouted bed. *Hungarian Journal of Industrial Chemistry*. **Volume 22**, pp. 219-224.

Szentmarjay, T., Szalay, A., Bencze, I. and Vass, J.: Control of drying process in mechanically spouted bed Dryer. *Drying Technology* (in Print).

Westerman, P.W., White, G. M. and Ross, I.J. (1973): Relative humidity effect on the high-temperature drying of shelled corn. *TRANSACTIONS of the ASAE*, **Volume 16**, pp. 1136-1139.

White, G.M., Ross, I.J. and Westerman, P.W. (1973): Drying rate and quality of white shelled corn as influenced by dew point temperature. *TRANSACTIONS of the ASAE*. **Volume 16**, pp. 118-120.

Wilhelm, L.R. (1976): Numerical calculation of psychometric properties in SI Units. *TRANSACTIONS of the ASAE*, **Volume 19**, 2:pp. 318-321.

OUTDOOR TEMPERATURE MODELLING FOR GRAIN COOLING CONTROL

Sz. Szöke, V. Wertz, E. Persoons

Unité de Génie rural
Place Croix du Sud 2 Bte 2, B-1348 Louvain-la-Neuve, Belgium
Unité d'Automatique, de Dynamique et d'Analyse des Systèmes
Bâtiment Euler, Av. G. Lemaître 4, B-1348 Louvain-la-Neuve, Belgium

Abstract: A good knowledge of the outdoor air temperature is necessary in models computing grain temperature changes during a cooling process conducted in a bin. In countries like Belgium, the outdoor temperature evolution may be described by a classical sinusoidal curve. As temperature increases faster in the afternoon than it decreases during the night, non-orthogonal axes must be considered. On a particular situation, using the measured temperature from the last two days and considering the minimal and maximal temperatures given by the weather forecast, it was possible to compute accurately temperature for the next 24 hours. This result was used to model grain temperature changes during a cooling process.

Keywords: Post harvest modelling, Grain Cooling, Air temperature prediction.

1. INTRODUCTION

In practice, grain managers need to cool and/or dry their product. Grain is often harvested at a temperature above 25 °C and fans must be used as soon as possible to decrease its temperature. This cooling process takes place during the night if the outdoor temperature is low enough and the air humidity not too high. Many models have been developed to compute heat and water exchanges occurring in this case.
They all need the air characteristics as input. The better these characteristics are known, the better the cooling prevision will be. A reasonable accuracy is needed because condensation may occur in some cases and therefore important damages may be observed in the product.

If the grain moisture content is too high, a drying may be necessary. In this particular case, air is warmed and the original properties of the air are not so important.

After drying, it is necessary to cool the grain again before it is stored definitely in bins. The final stage of the process will need again a good knowledge of the air characteristics for at least 24 hours.

2. MATERIEL AND METHODS

When a decision to cool a grain bin must be taken, the following information is supposed available.

The air temperature has been measured hourly for at least two days.

Weather forecasts give the minimal temperature of the coming night and the maximal temperature of the following day. The exact times when these two temperatures occur are NOT known.

The objective of this paper is to predict, with the elements given above, the temperature wave for the next day.

Analysing weather data of Belgium shows that a daily variation of temperature may be considered nearly all year long (Fig. 1).

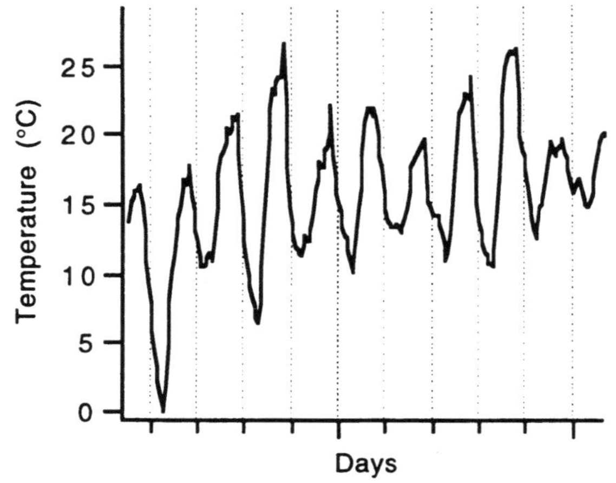

Fig. 1 : Daily variation of temperature

The following basic equation has been chosen:

$$T = K_0 + K_1 \sin(K_2 \ \ i + K_3) \qquad (1)$$

where:
K_0 is the mean temperature,
K_1 is half of the daily amplitude,
K_2 is the angular speed, $2\pi / \lambda$, where λ represents a 86400 seconds period,
K_3 corresponds to a time offset,
i is the temperature measurement increment.

This equation, fitted on the last two days temperatures, gives the results shown in Fig. 2.

It is clear that the temperature increases faster on the afternoon than it decreases during the night.
Maximal values show a left offset and minimal values show a right one. The proposed equation is modified to take this remark into account.

Each point of the curve may be moved by a value di, related to the temperature by the equation:

$$di = K_4 \ \ T = K_4 \ \ \sin(K_2 \ \ i). \qquad (2)$$

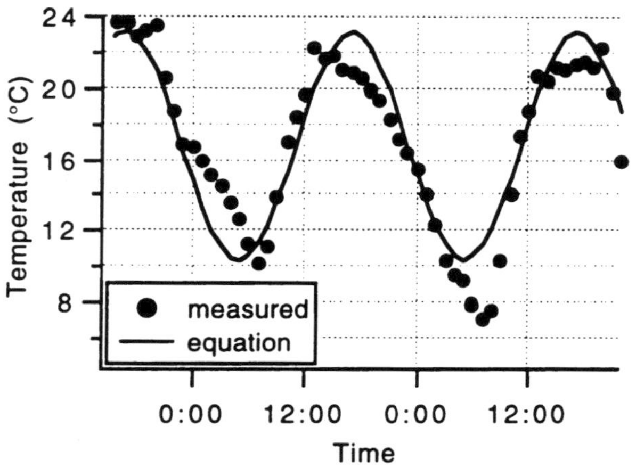

Fig. 2: First equation fitted on 2 days

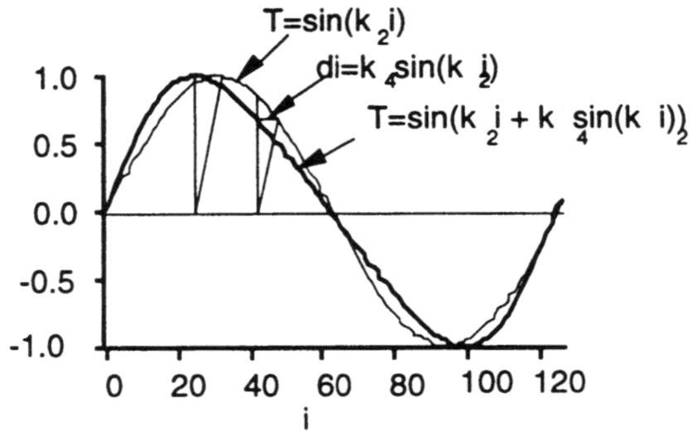

Fig. 3: Modified sinus wave

The new relation is:

$$T = K_0 + K_1 \ \ \sin(K_2 \ \ i + K_3 + |K_4| \ \sin(K_2 \ \ i)). \qquad (3)$$

This relation is shown in Fig. 3.

The K_2 parameter is used a second time in the sinus brackets because the original wave and the extrema must stay in phase. Moreover, as the maximum point must move forward and the minimum point backward, an absolute value is used to ensure that the K_4 parameter stays positive during the fit.

This new relation fitted to the last two days gives results shown in Fig. 4.

A last improvement has been tried. The mean temperature may change according to a second order polynomial form. The final equations is:

$$T = K_0 + K_1 \ \ \sin(K_2 \ \ i + K_3 + |K_4| \sin(K_2 \ \ i)) + K_5 \ \ i + K_6 \ \ i^2 \qquad (4)$$

It is clear that this equation is particularly adapted to the temperature evolution under the climate in Belgium. Results must be taken with care under other climate conditions.

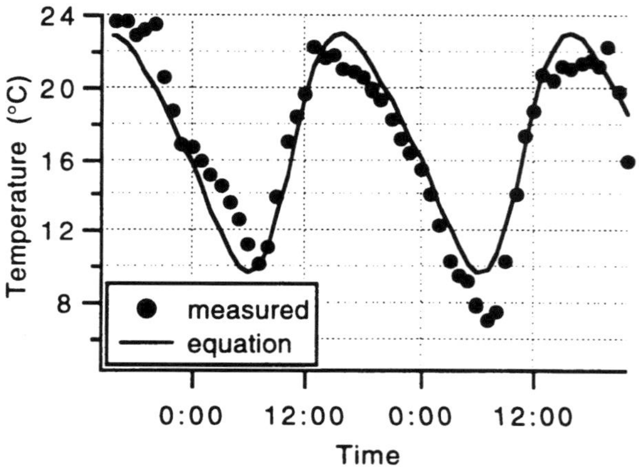

Fig. 4: Second equation fitted on 2 days

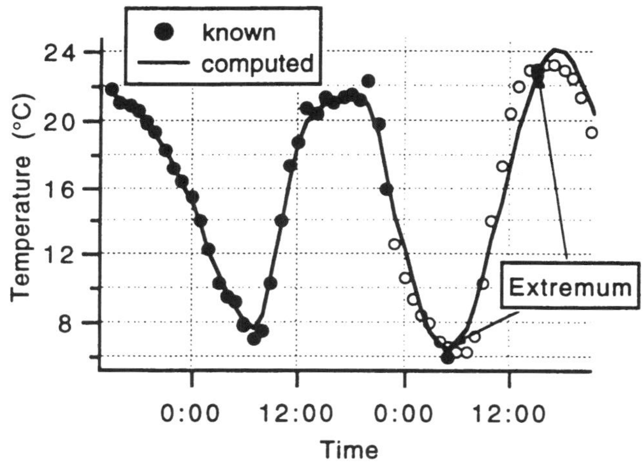

Fig. 5 : Final result

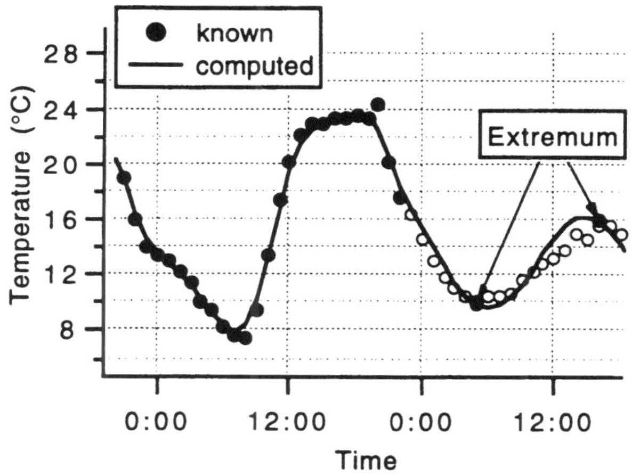

Fig. 6: Important changes in the temperature wave form.

Supposing that equation 4 represents the temperature wave well, a prediction may be computed as follows:
- the K_2 parameter is based on a 24 hours period, corrected by the temperature measurement interval,
- the mean parameter K0 takes the value of the actual temperature,
- a value of 0.2 is assigned to parameter K4,
- parameters K_1, K_3, K_5 and K_6 are fitted using the temperature of the last two days.

As the time offset K_3 is computed, it is possible to define when exactly the minimal and maximal temperatures of respectively the next night and the next day, will take place.

A second fit is realised using these two new points and the temperature for the next 24 hours is therefore easily computed.

A typical result is shown in Fig. 5.

In practice, this kind of computation is always running on the evening because cooling is always conducted during the night. The procedure explained in this paper is running perfectly well if started after the moment the maximal temperature of the day has been reached.

3. RESULTS AND DISCUSSION

This model has been tested on data taken in 1989. Even if an important change occurs in the general aspect of the temperature wave form, the results are accurate, as can be seen in Fig. 6.

It is clear that temperature at time τ is not only related to temperature at time τ-1. Therefore it is almost impossible for this kind of model to be accurate in absolutely any situation. Finally, the accuracy of the results is significantly decreased if the minimal and maximal temperatures given by the weather forecast are erroneous.

4. APPLICATION ON COOLING MODELS

A model developed in the laboratory was used to compute grain temperature measured in a bin during cooling.
Air is blown in the bin from the bottom to the top. A temperature front will move upwards at a speed given by equations 5 and 6.

$$s_{top} = K_2 \ \Delta T^{K_5} \ \varphi_g^{K_6} \ s_a \quad (5)$$

$$s_{bottom} = K_3 \ \Delta T^{K_5} \ \varphi_g^{K_6} \ s_a \quad (6)$$

where:
- s_{top} is the speed of the top of the temperature front (m/s),
- s_{bottom} is the speed of the bottom of the temperature front(m/s),
- K_i are parameters,
- ΔT is the difference between the initial grain temperature and the air temperature.
- φ_g is the grain moisture content.
- s_a is the air velocity (m/s),

Top and bottom speed of the front must be computed separately because its thickness increases with time.
The bin is divided in small layers about 1 cm thick and time is divided into periods of approximately 10 minutes. For each time step the speed of the front is computed. The front position is given by equations 7 and 8.

$$Z_{top} = \Delta\tau \sum_i s_{top_i} \quad (7)$$

$$Z_{bottom} = \Delta\tau \sum_i s_{bottom_i} \quad (8)$$

The grain temperature is defined by the following algorithm:
- If the grain is above the front, nothing has occurred and grain temperature is the initial temperature.
- If it is below the front, an equilibrium is reached depending on grain moisture content and air characteristics. The final temperature T_f is given by equation 9.

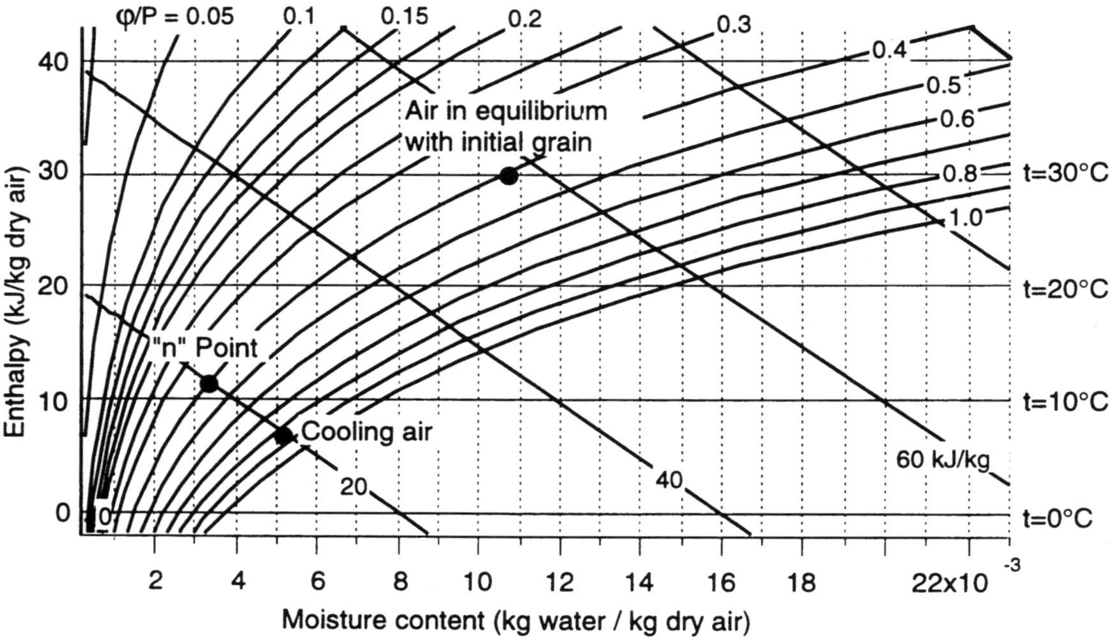

Fig. 7: "n" point on the Mollier diagram for humid air

$$T_f = Coeff\ T_n + (1 - Coeff)\ T_a \qquad (9)$$

where T_a is the air temperature, T_n a temperature shown in Fig. 7, *Coeff* a coefficient calculated by equation 10 and displayed in Fig. 8.

$$Coeff = \frac{\varphi_g^{K_0}}{\left(K_1^{K_0} + \varphi_g^{K_0}\right)} \qquad (10)$$

The "n" point is defined by the intersection of the following curves: the constant relative humidity of air in equilibrium with initial grain and the enthalpy of cooling air.

For a very dry grain, *Coeff* is equal to zero. This means that dry grain will reach the temperature of blown air: $T_f = T_a$. In case of a humid grain, *Coeff* takes a value of one and the final temperature is T_n. It is evident that a dry product will have the air temperature at the end of the process.

An example of application on a 6 m high bin cooled during 16 hours is shown in Fig. 9.

As it is now possible to predict the air temperature and then the heat exchanges occurring in the grain bin, an optimisation procedure has been written to help the manager to switch on and off the fans.
The value being minimised must be chosen between the followings:
- the maximal temperature of the grain,
- the mean temperature of the grain,
- the maximal temperature found in the lowest half of the bin, in case of a cooling process spread on two nights,
- the mean grain temperature in the lowest half of the bin.

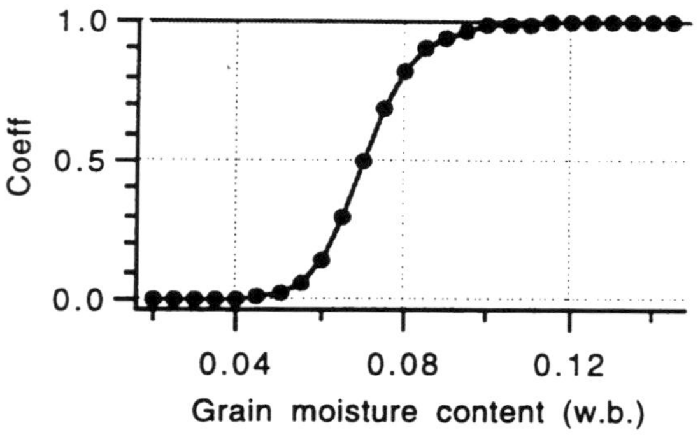

Fig. 8: Coeff value vs. grain m.c.

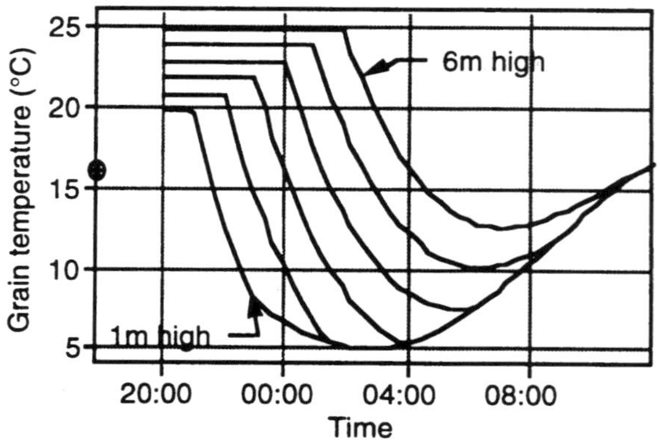

Fig. 9: Modelled grain temperature

The software can then easily define the best times to switch on the fans in the evening and switch them off in the morning.

5. CONCLUSIONS

A model has been developed for the cooling and drying processes that take place in a grain bin after harvest. (Szöke, 1994). Based on that model, a computer simulation program has been developed which can help to optimise the operation of the drying and cooling installation (reference to software). In this model, an accurate prediction of the air temperature for the next 24 hours is necessary. This part of the model has been presented in more detail in this paper.

REFERENCES

Szöke Sz. (1994). Modélisation de la ventilation et du séchage des céréales en couche épaisse. Thèse de Doctorat, 206 pg. Université catholique de Louvain. Unpublished.

COMPARISON OF DIFFERENT MODELS USED IN FIXED BED DRYING PROCESSES

I. Farkas[1] - Z. Rendik[1] - K. Toyoda[2] - A.J.B. van Boxtel[3]

[1]Department of Physics and Process Control
University of Agricultural Sciences
H-2103 Gödöllő Páter K. u. 1, Hungary
[2]Department of Environmental Information & Bio-production Engineering
Kobe University, Kobe, Japan
[3]Department of Agricultural Engineering
Wageningen Agricultural University
6703 HD Wageningen, The Netherlands

Abstract: Different black-box and white-box models were used to compare results of fixed bed drying processes. The main purpose of this report is to demonstrate the differences of these methods and to show which one to use under different circumstances. Well defined and known models were used as a basis. Calculations have been performed with the help of i/o models (ARX, ARMAX, etc.), deep bed modeling technique with neglecting some dynamic changes and a block oriented way. It was found that block oriented approaches have advantages comparing to other methods. Solving differential equations is impossible with traditional methods, but it is very simple with the help of block oriented technique. In other case, if the purpose is only to control the dryer, the black-box modeling is much faster, but it may be solved by using physically based modeling with neglecting some dynamic changes.

Keywords: modeling, physical models, optimization, model reduction, drying

1. INTRODUCTION

Modeling is an old tool in the researchers' hand. Appearing of the computers, it started his quick development and today it is a separated field in the research. It has several methods, but basically it has two approaches black-box and white-box modeling. Both modeling has advantages and disadvantages. Every researchers have met with at least one of them. Some of them propose a black- box modeling i.e. i/o modeling, the others propose a kind of white box method, the physically based modeling. There is big difference between them, but there are several methods to apply their advantages together. In the recent paper the different methods are compared and their results are shown in the field of drying, especially in fixed bed drying.

During drying process of agricultural crops there are opposite requirements should be fulfilled at the same time, such as to reduce the energy consumed and to get a high quality of end-product. For that reason it is needed to describe the whole drying process and also the control strategy applied during the course of the process has an important issue, which leads up to solve a multipurpose operation problem. So, it should be kept a reasonably good balance between the requirements.

The physically based modeling of a drying problem can be described with complicated differential equations, i.e. heat and mass balance equations. In order to solve easily of these equations normally, they should include also empirical and partly empirical parts. The definitions of such equations is time consuming and requires a lot of measures, therefore they are very expensive and make a lot of possibilities of error sources, but very useful for studying the entire process. In the field of designing, there is enough time to compute the operational and optimal parameters of a randomly structured drier, so in this case the physically based modeling is a good chance.

The black-box modeling technique i.e. i/o modeling can be identified on the basis of measurements. In that case, if it used for operating of a drier, it is not necessary to know all the physical parameters during the drying. It is possible to determine one or more significant parameters, which one is suitable for the operation, it may control the whole drying process. Disadvantage of such method that it is not suitable to examine inside parameters, they can serve only the output parameters. This method is very useful for processing stage, because it is very fast, but they are not available in design phase as it was mentioned above.

To solve an actual problem, the defining of requirements is needed in the first step. This point is very important in choosing the best method in the modeling, because it determines all the work and its predicted results. If a black box modeling is chosen, its result will be applicable in the field where the high speed is basic requirements and not necessary all the parameters. In case of physically based modeling very detailed and time consuming results are expected. In this case there are several methods like simplifying the complicated differential equations, neglecting the changing of some parameters, etc. to make faster the calculation, but in many cases even these simplifications are not enough.

2. PHYSICALLY BASED MODELING

The physically based modeling technique is a kind of white-box method which gathers a-priori information and examines the smallest details in relation to physics through the definition of the parameters applied in the equations.

The main advantage of the physically based modeling is that with the help of it, there is a chance to receive a full knowledge of the details of the process. This way the inside parameters can be easily defined in case of any input and it is possible to carry out parameter optimisation and identification for the process. At the same time this method cannot be used directly for purposes of process control because to generate the output signals from the input ones takes too much time due to complicated relations and equations. That is the reason why this method is more advisable for planning and analysing purposes, but it should be simplified and reduced along with decreasing the running time to be appropriate for a given purpose, perhaps for process control, as well.

2.1. Thick layer modeling

Generally agricultural materials is dried with a thick layer after harvest. There are several well-known drying models which show the changes in moisture content and temperature in the drying bed. These models have steady state error, because drying characteristic of a thick layer depends on state parameters (drying air temperature, air humidity, grain temperature and moisture content, etc.) and it is too complex to investigate directly. That is why a thick layer is generally devided into thin layers and parameters are calculated from heat and mass balances for the drying process on the basis of a thin layer drying characteristic. However, such a thin layer drying characteristic is applied to calculate parameters, it causes the complication of controller or an increase in calculation time. A thick layer modeling is required to be not only accurate but simple. The requirements mentioned above are analysed and carried out in the modeling of alfalfa drying. Just as the different components in several other agricultural products, stems and leaves of alfalfa have different drying characteristics. During drying the water concentration of the leaves falls faster than that of the stems. Moreover, it has been observed that during break periods of drying the water concentration of the stems still decreases while that of the leaves increases. It means that during the break the water concentration of the components equalises. From the equalisation of process during the breaks it may be expected that intermittent drying is an attractive way. Because the intermittent drying utilizies the potential to proceed the drying of the stems during the breaks while the differences in water concentrations are limited, resulting in a more homogeneous moisture content in the product. To estimate the true optimal operation a physical model of stems and leaves drying has to be combined with optimization methods for dynamic processes and using an objective function related to the drying requirements.

A sensitivity study was carried out in order to determine the optimal size of thickness of an alfalfa layer can be applied during the simulation. A half metre thick fix-bed of alfalfa was considered as an initial parameter. The bed was devided into 1, 3, 9 and 27 layers to keep the comparability in the middle layer. The "1 layer" represents the total thick bed. On the basis of the given example it could be observed that 9 layers seemed to be sufficient, i.e. Δz=0.05 m can be used in drying calculations.

Since the calculation time was too long afterwards, three different stages of model reduction were considered in analysing the possibility to reduce the calculation time along with keeping the calculation accuracy in moisture distribution. The study of reduction steps showed that neglecting the dynamics of temperature of material components gives a reasonably good result in all sense. The calculation time is about ten times less and the accuracy stayed within 0.5 % compared to the original model.

Making further simplifications in the thick bed model it yields a way to use it in optimization calculations. The results of such an optimization of a 0.05 metre thick layer are given in Fig. 1. In the first 10 hours water transfer from stems and leaves is relatively high and requires gradually decreasing air flow rates above the minimum value (Fig. 1a). After this period the minimum air flow rate satisfies to minimize the costs for energy consumption by the fan and costs for heating.

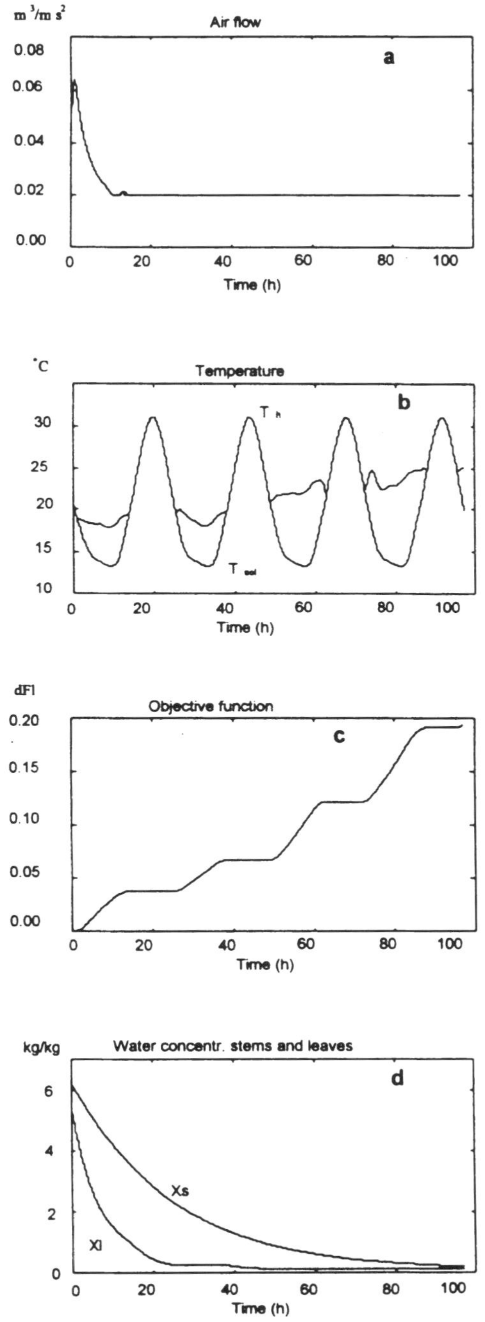

Fig. 1. Optimization results

Mass and heat balances are good starting-points to build models of production systems. Basic knowledge on transport phenomena, transformation kinetics, general data and physical constants given in handbooks can be used. Models of subsystems can easily be combined to simulate the systems behaviour and the total outcome. In order to describe the moisture distribution of alfalfa during the drying process a two-component (leaf and stem) physically based model is essential. During the breaks in drying rewetting in stem components could appear which can be calculated with the knowledge of sorption and desorption isotherms of both components.

The physically based models serve a much more complex information on the status of the thick drying bed. However, their solution causes calculation troubles sometimes. The model reduction is a way to get an easily applicable model which can be used even for control purposes along with keeping the physically based concept.

2.2. Thin layer modeling

Physically based modeling requires the creating of heat and mass balances. As it was mentioned earlier, in case of thick layer equations, the solution causes calculation troubles sometimes and it is too complex to investigate directly. That is why the simplest method is to create the thin layer differential equations where there are no similar problems, and it is possible to build up the drying bed from these thin layers. Applying the method the output parameters of a thin layer is assumed to be the input parameters of next thin layer and in this way the state values and output parameters of each thin layer can be calculated (see in Fig. 2).

As a result of appearance of high capacity computers several methods came into light to help solving differential equations with increasing also the calculation time.

One of the methods for solving the differential equations is the block oriented approach. This method is not completely new, but it is not so widely used, on the contrary of the fact that randomly structured technological processes can easily be examined with it. There are many other advantages of block oriented simulation systems like their clear hierarchical structure.

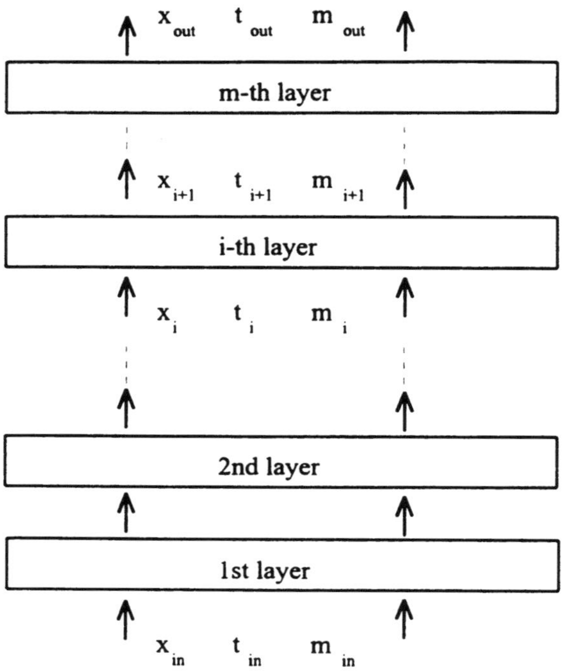

Fig. 2. Arrangement of layers in a fixed-bed model

After creating the differential equations for heat and mass balances, the structure of the drying bed is needed to create from the thin layers. With the help of the block oriented system it is possible to connect these layers one after the other as it is determined by the length of the drying bed and thickness of the layer (see in Fig. 2).

The structure of the thin layer block can be determined on the basis of differential equations describing the heat and mass transfer between the drying air and material. Taking into consideration that four unknown variables (air temperature, air humidity, material temperature, material moisture content) change during the drying, four balances, i.e. four differential equations must be compose:

$$\frac{\partial x}{\partial z} = f(m, \frac{\partial X}{\partial \tau}) \qquad (1)$$

$$\frac{\partial t}{\partial z} = f(x, t, m, T) \qquad (2)$$

$$\frac{\partial X}{\partial \tau} = f(x, m, T) \qquad (3)$$

$$\frac{\partial T}{\partial \tau} = f(\frac{\partial x}{\partial z}, t, m, X, T). \qquad (4)$$

The solution of the above differential equation set in a block oriented way is shown in Fig. 3.

The block oriented approach modeling of a grain dryer like above was developed and detailed. With

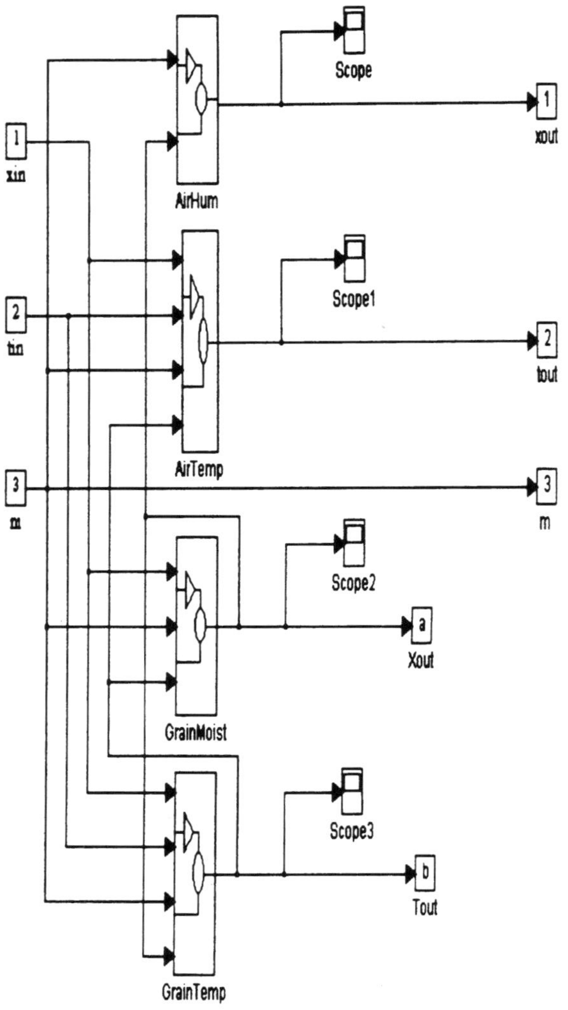

Fig. 3. Block oriented model of a thin layer of crop

the help of this method it is possible to build up a randomly access dryer structure and to carry out several tests. The main advantage of this structure is that there is no need long and expensive tests, but in case of a well-formed model fast and exact results may be reached. Taking into consideration that in the design phase of dryer it can not be determined the operational parameters with the help of traditional methods. This approach gives new perspectivity to determine the optimal operational parameters, too.

The calculation results show the moisture distribution along the depth of dryer (see in Fig. 4) and a comparison in average value between here calculated data and those one provided by Bakker Arkema, et al. (1970) as it is shown in Fig. 5. In the latter case the difference is less than 3 percent.

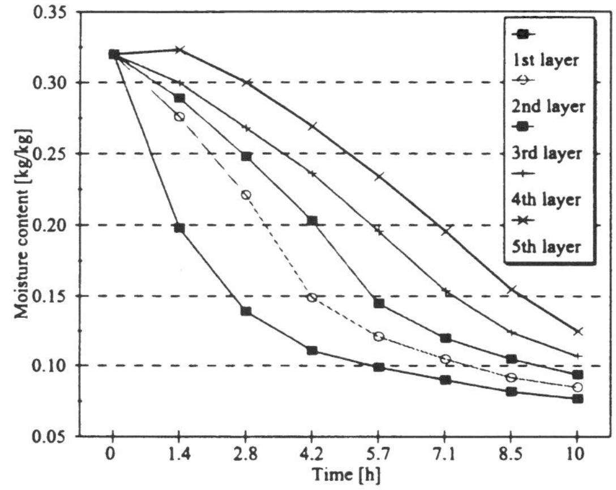

Fig. 4. Moisture distribution in a fixed bed dryer

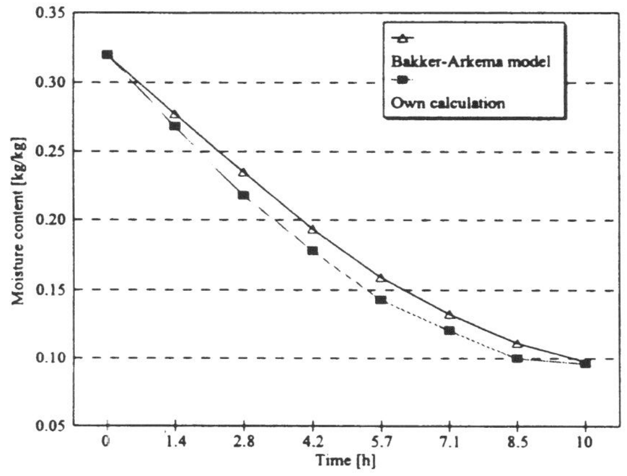

Fig. 5. Comparison of simulated results

3. BLACK BOX MODELING

During the last decades the role of decision support systems for production control increased. For these systems it is essential to have disposal of a model of the production process. The large family of models for production control can be characterized in different ways. One way is to make distinction between a class of white-box models (physical or first-principles models) as it was mentioned earlier. The other way is the class of black-box models (transfer models, neural nets and fuzzy models). The applicability and suitability of these models depends on the purpose and the situation where the models will be used. For existing systems (existing crop production and processing technologies) and the design of traditional regulators black-box models are suitable. For different applications the model parameters can be estimated so that the model fits well to existing data and, in the future, the model can be used for prediction of the considered system.

The black-box method does not require such a knowledge, but the definition of the relations of the input and output is quite an expensive process and needs accurate and time consuming measurements.

There are several methods to measure the material moisture content, since this is the significant parameter of the material leaving the drying bed. For example, DC conductance, AC capacitance, microwave resonant cavity, etc. have been already applied in practical instruments.

There is an other indirect method for estimating the material moisture in the drying bed. For that purpose the outcoming air temperature is proposed by Nybrant (1986), because air temperature can be easily measure and without any time lag. As it was proved, this is valid for crossflow dryers, as well. At the same time, control of the material moisture content, based on a change in the exhaust air temperature, has an inaccuracy as because of the variation in function relationship between them at the drier outlet. Accordingly, a moisture meter should be adopted to provide the on-line measurement at the inlet and outlet of a grain column.

To predict relation between grain moisture and exhaust air temperature there are several methods. The mean value and its standard deviation of the measured data can be used to predict the parameters of different i/o models, i.e. ARX, ARMAX, Output Error and Box-Jenklins as it was suggested by Toyoda et al. (1994).

In such a way a linear system model based on a parametric approach can be introduced, where the dynamic from the input u(t) to the output y(t) can be described in the following generalized model structure:

$$A(q)\, y(\tau) = q^{-n} \frac{B(q)}{F(q)} u(\tau) + \frac{C(q)}{D(q)} e(\tau). \qquad (5)$$

After the estimation of the parameters in the model mentioned above it was found that the ARMAX model was capable for estimating the mean grain moisture from the air temperature measured at the bottom of the drying column. The distribution of grain moisture caused by thick layer in a drying column was evaluated by the standard deviation of the measurement per minute and it can also be estimated from the air temperature by the ARMAX model.

4. CONCLUSION

Different modeling techniques can be used successfully for describing the behaviour of dryers.

The proper selection of the model should based on the purpose what it is applied for.

As a case study, the alfalfa drying systems and their multipurpose optimisation was discussed. A physically based thick bed model was applied. A comparison was carried out with the different reduction steps. It was found that there is no significant difference in the results between the reduction on the contrary of the neglecting of the dynamics of some variables. The control algorithm contains the simplified reduced order model of the drying derived from a more sophisticated physically based one. A block oriented approach was applied to solve the model where a thin layer block was defined on the basis of the relevant discretized drying equations.

A thin layer model was used for grain drying, where it was found that physically based modeling is a good possibility to examine drying process where there are not available test dryers e.g. in the design phase, etc. and the detailed knowledge of drying process is necessary. Disadvantage of this modeling, that it is time consuming, but it is possible to reduce calculation time with reduction of differential equation, neglecting dynamics of the changing parameters, as it was shown also in case of alfalfa drying.

Finally a black box modeling was reported. It was found that it is very usable for controlling purpose, it does not require detailed knowledge of the process, but its use is limited.

REFERENCES

Bakker-Arkema, F W., L.E. Lerew, S.F. De Boer and M.G. Roth (1974). Grain dryer simulation, Research Report, Michigan State University, Michigan, USA.

Farkas, I. and Z. Rendik (1994). Block oriented modeling of drying processes. (To be publihed).

Nybrant, T.G. (1986). Modeling and control of grain dryers. Report UPTEC 8625R, Institute of Technology, Uppsala University, Uppsala, Sweden.

Toyoda, K., I. Farkas and H. Kojima (1994). Estimation of moisture content of grain during drying in a recirculation dryer. (To be published).

van Boxtel, A.J.B., I. Farkas, L. Lukasse and Z. Rendik (1993). Physically based modeling and optimal operation for product drying during post-harvest processing. Mathematical and Control Applications in Agriculture and Horticulture, Second IFAC-ISHS International Workshop, Silsoe, UK

APPENDIX

e	white noise
m	air flow rate, kg/hr m^2
n	time delay
q	shift operator
t	air temperature, °C
u	input signal
x	humidity ratio, kg/kg
y	output signal
z	bed depth, m
T	product temperature, °C
X	moisture content, decimal
τ	time, hours

Subscript

i	time-step coordinate
in	inlet
out	outlet

MODELLING THE FINAL DRYING PROCESS FOR HAY IN A DEEP LAYER

J. STENCL and P.HOMOLA

Mendel University of Agriculture and Forestry, Brno
Zemedelska 1, 613 00 Brno, Czech Republic

Abstract: The paper deals with automation of hay drying in a deep layer. Major factors controlling drying processes in a barn are temperature of a stored material, its moisture content (estimated from hay impedance/bioimpedance) and the relative humidity and temperature of the drying air. Measurements were made of these four factors during fan ventilation of the hay. Algorithms were then developed based on these measurements. In addition, the paper gives information on the low temperature drying process of hay samples under laboratory conditions.

Keywords: hay; drying; bioimpedance; temperature; humidity; fan

1.0 INTRODUCTION

The importance of hay for cattle feeding has been assessed differently in the past. As a result, the quantities produced in individual years differed markedly. Hay production depends on geographical and climatic conditions, however, hay remains an irreplaceable component of cattle feed, particularly for dairy cows and calves. Efforts to increase its production are motivated by attempts to increase the ratio of dry feed in winter with the aim to reduce the amount of acids in silage per unit of dry matter. The importance of vitamin value of good-quality hay is not negligible, either. Hay production is, therefore, expect to continue or increase again in the future.

The analysis of main problems concerning hay production shows, that from the viewpoint of quality and energy demands, the low drying capacity of near ambient air blown in by fans to wilted hay layer on grates in hay barns is the decisive adverse factor. Moreover, this problem cannot be overcome by increasing the amount of the air blown in. In case of grass, the optimum moisture content at harvest is considered to be 30 to 35%, in alfalfa 35 to 40%. Using all the latest information, this level can be achieved by pre-drying the hay in the field.

A tonne of hay of 16% moisture is produced by removing approximately 4350 kg water from about 5350 kg green fodder. Of this amount of water, about 18% is removed on grates in a barn, the remaining in the field after mowing. With the average water-holding capacity of near ambient air of 0.7 g water per m^3, 1 mil. m^3 of air must be blown in the hay barn for each tonne of hay. As a rule, hay barns have 6 to 2 fans usually with an output of 30000 $m^3.h^{-1}$ each. With an average consumption of about mil. m^3 of air to produce 1 tone of hay. The output of hay barns per hour is then 0.18 to 0.36 $t.h^{-1}$.

Of the methods which can exhibit beneficial effects on the air saturation complement and thus also its drying capacity, the method most readily available today is moderate heating. Other possible ways of making final drying of fibrous material in barns more effective include the application of tested algorithms to control the process of blowing-in the air in real time (Stencl, Skyrik, 1991) and in chemical treatment (Rotz, Thomas, 1988; Stencl 1992) of the material stored. In all these cases basic requirements of high quality must be met.

1.1 Influence of temperature on hay drying rate in the area of low temperature drying

At the atmospheric temperature of 10 °C, the maximum water content in 1 m^3 of air is about 10 g while at 20 °C it is already about 17 g. Under natural conditions with gradual heating of the air during the day, this phenomenon cannot be fully utilized because entering the hay barn. Only rarely does drying capacity the air exceed 2 g. m^{-3}, viz only a few hours daily. From this point of view, it is interesting to note the increase in the rate of drying d(MCwb)/dt of fibrous fodder in relation to the increasing temperature of the drying environment. During laboratory measurements it was observed that a temperature increase from 20 to 37 °C results in an increase of up to 100% in the rate of drying.

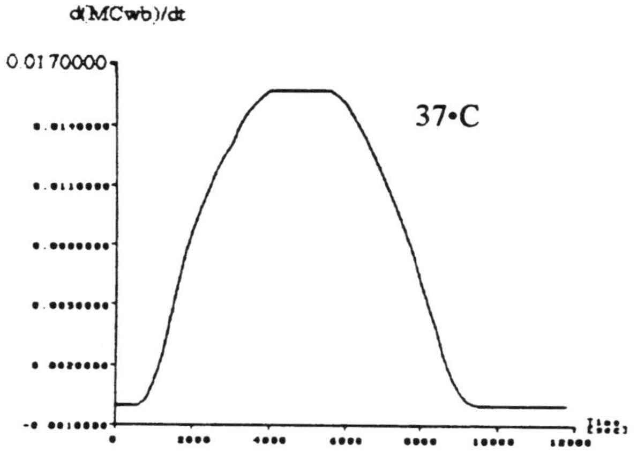

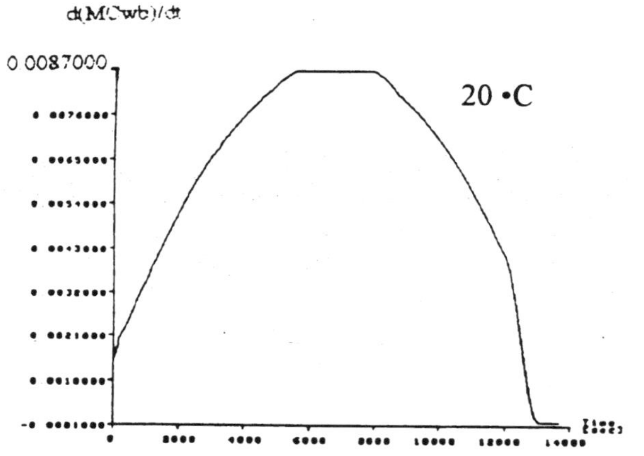

Fig. 1 Rate of drying of grass at 20 °C and 37 °C

These and other facts point to the advantages of moderate increases in the drying air temperature. Depending on local conditions, waste heat from industrial plants, geothermal and in some places solar energy, waste wood or straw etc. can all be used for this purpose.

2.0 REAL TIME CONTROL OF HAY BARN VENTILATION

One of the prerequisites for successful studies of the problems is good knowledge of biochemical processes as they relate to physical conditions existing at the time of final drying. There are still extensive untapped possibilities not only as regards the preservation of biological value of the feed dried in hay barns but also in energy savings. Today, it is unthinkable to control the technological equipment for final drying on the basis of subjective opinions of the operator only. Final drying itself should be viewed as a dynamic process, where the physical and biochemical values are changing constantly. A significant factor in increasing the quality and effectiveness of final drying is its automation, which is also one of the main prerequisites of energy savings. The introduction of automated control systems supported by electronic elements with a high degree of integration allows the creation of complex algorithms which are relatively easy for the user and comprehensible to the operator. These questions required solution of the following research tasks:

- sorption behaviour of the material being dried in relation to the control process;
- suitable temperature and particularly moisture sensors;
- proposal and optimization of the control algorithm in relation to energy demands, degree of technical complexity and preservation of a high biological value of the dried material.

Control algorithms for final drying of meadow hay were tested experimentally in one compartment of a hay barn (Stencl, 1991). The following values were continuously measured and recorded (see the Fig.2):

- near ambient air temperature;
- relative humidity of near ambient air;
- temperature of the material;
- moisture content of the material (as a function of bioimpedance);
- intervals of fan operation.

Graphic relations clearly show a gradual increase to 37 °C in the temperature of freshly stored hay caused by spontaneous heating. This temperature was reached in about 90 minutes. At that time, the fan was switched on automatically (the maximum temperature of the material was preset at 37 °C). When the fan was switched on, the hay gradually cooled and the dry matter content increased markedly because the water, released during the short-term controlled spontaneous heating, was removed. This was due to removal of the free moisture content at a temperature higher than that of the ambient air. This process was repeated once more after 5 hours. Now, worth mentioning is the increase in the dry matter of hay expressed as bioimpedance (during the course of two ventilation cycles at night bioimpedance increased by 25 kOhm). During

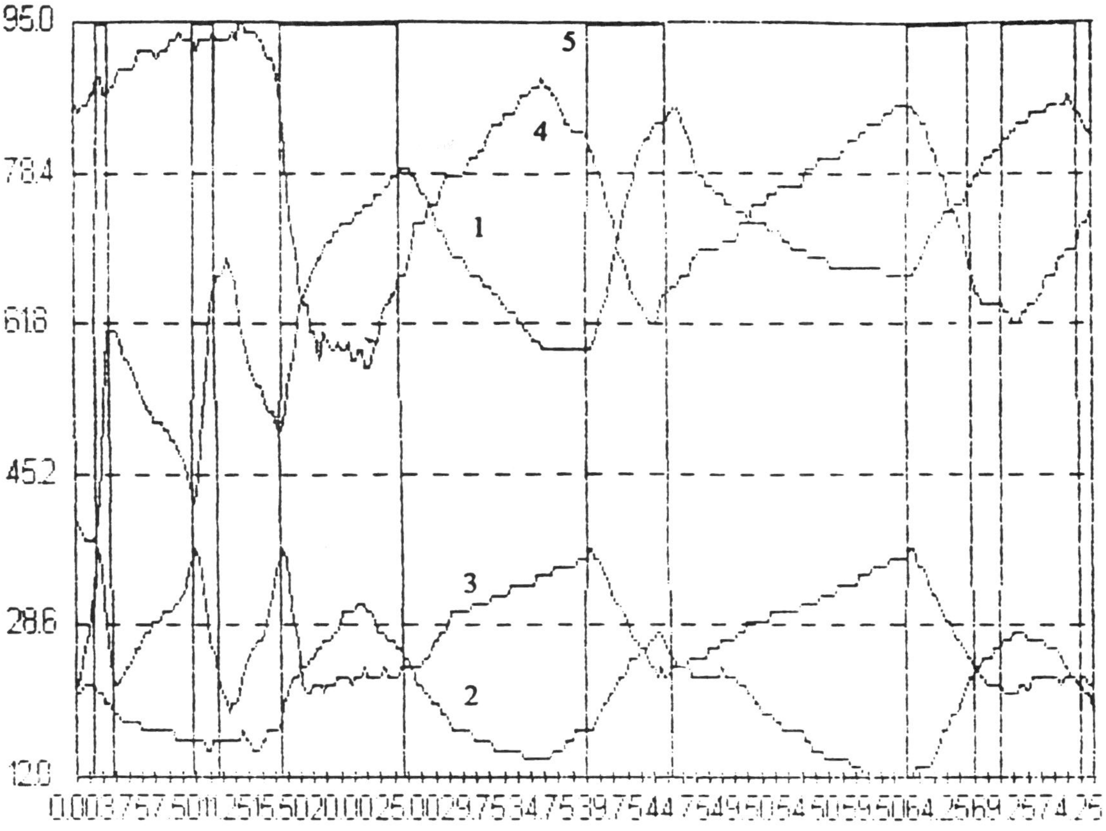

Fig.2 Diagram of drying processes of hay in a barn.

1 - Bioimpedance of hay
2 - Temperature of atmospheric air
3 - Temperature of hay
4 - Relative air humidity
5 - Fan ON/OFF

another interval of fan operation the high temperature of hay, owing to spontaneous heating, rapidly decreased and in the end fell below the value of the temperature of the air blown into the deep layer. At that time intensive evaporation took place and the temperature of the material partly dropped. With the increase in the dry matter of hay under given relative humidity, the difference between the temperature of hay and the temperature of the air gradually decreased. This was confirmed during another interval of fan operation. The diagrams of the quantities suggest the relationship between the temperature of the stored material and its bioimpedance. It can be said that they create a mirror image.

2.1 Control system of hay deep layer ventilation in a barn

The control system for ventilation (intervals of fan operation) of hay deep layer in barn has been developed on the basis of numerous practical measurements and laboratory tests. Main parameters followed were temperature of near ambient air and stored material , relative air humidity and moisture content of material (as a function of bioimpedance). Main menu (Fig.3) covers all these parametres inclusively printing the values measured as a graph (Fig.2) or a table (Fig.6) and also price calculations (Fig.5). It is possible to specify the basic conditions for fan operation (Fig.4).

It is recommended to preset the maximum temperature of the hay at 37 °C, because of drying efficiency and a risk of ignition.

3.0 CONCLUSION

The process of hay drying in barn has been controlled predominantly by the temperature of the material loaded or sometimes by the relative humidity of the drying air. When the preset limit values of these quantities are reached, a signal is produced to switch on/ff the fan. This procedure is suitable predominantly with respect to fire prevention though occational unpredectable local rises in temperature cannot be excluded. With respect to the hay drying process and hay quality (nutritive value) it would be best to control ventilation by changes in the moisture content of the material layer.

GACR 510/95/1214

Number of Measurement		*1*	*Time 0,00*	*< Go Top >*
Bioimpedance	*40,00*	*Rel. Air humidity*	*85*	*< Go Bottom >*
Hay Temp.	*22*	*Near Ambient Air Temp.*	*21*	*< Next >*
Ventilation F	*Real Time*	*16,00*		*< Previous >*
< Update >	< Price >	< Table >		Print table
< Setup >	< Graf 1 >	< Graf 2 >		Exit

Fig.3 Main menu for hay drying process control

Settings

1. Condition

Fan ON
Near ambient air temperature 5°C higher than hay temperature
AND relative air humidity lower than 70%

Fan OFF
Near ambient air temperature equal hay temperature
AND hay temp. lower than 30°C

2. Condition

Fan ON	*=>*	*Critical hay temperature:*	*37°C*	*< Cancel >*
Fan OFF	*=>*	*Hay temperature:*	*23°C*	*Exit*

Fig.4 Parameter setting for fan switch on/off

Total Costs

Fan power consumption *([kWh])* *7 Price per 1kWh* *1,80*

Total costs *0*

< Re-count > *Exit*

Fig.5 Cost calculation

Row	Time	Real_time	Bioimp	Hay_temp	Near_amb_t	Rh	On	Price
5	1.00	17,00	38.00	28	22	86	F	0.00
6	1.25		38.00	30	22	87	F	0.00
7	1.50		38.00	33	22	87	F	0.00
8	1.75		40.00	37	21	89	T	3.15
9	2.00	18,00	46.00	35	21	89	T	3.15
10	2.25		53.00	31	20	87	T	3.15
11	2.50		61.00	27	20	87	T	3.15
12	2.75		61.00	23	19	88	F	0.00
13	3.00	19,00	61.00	22	19	88	F	0.00
14	3.25		61.00	22	19	89	F	0.00
15	3.50		60.00	23	18	90	F	0.00
16	3.75		58.00	24	18	90	F	0.00

Fig.6 Table of parameters of the final drying process in a hay barn

REFERENCES

Arinze, E.A, Sokhansanj, S., Schoenau, G.J. (1993). Simulation of natural and solar-heated air hay drying system. Computers and Electronics in Agriculture, 8/4, p. 325-345

Lamond, W.J., Smith, E.A, Bowden, P.J. (1985). The effect of conditioning on hay drying. Dep. note SIN/464, p. 34. Penicuik, Scot. Inst. Agr Engineering.

Lamond, W.J., Glasbey, C.A., Spencer, H.B. (1986). The effect of conditioner throughput rate on the drying rate of grass. Dep. note SIN/464, p. 11. Penicuik, Scot. Inst. Agr. Engineering.

Patil, R.T., Sokhansanj, S., Arinze, E.A. (1993). Methods of expediting drying rates of chopped alfalfa. Transactions of the ASAE, 36/6 , p. 1799-1803.

Rotz, C.A., Thomas, J.V. (1988). Comparison of chemicals to increase alfalfa drying rate. Applied Engineering in Agriculture, 3/88, p. 47.

Stencl, J, Skyrik, J. (1991). Automatic control of final drying process of hay in barns. Proceedings of the 9th IFAC/IFORS Symposium , p. 601-604, Akaprint Budapest, Hungary.

Stencl, J. (1992). The influence of Ammonia on the drying rate of hay during final drying. Proceedings of IFAC workshop on Expert Systems in Agriculture, p. 226-230, International Academic Publishers, Huangshan, China

DYNAMIC SIMULATION OF PLANT OPERATION IN THE WINE INDUSTRY

Pierre Grenier (1), Philippe Vilette (2), Christophe Durand (1), Jean-Michel Roger (1), Francis Sevila (1), Yvan Racault (3)

(1) Cemagref, BP 5095, 34033, Montpellier cedex 1
(2) IMECA, BP 94, 34800, Clermont-L'Hérault
(3) Cemagref, BP 3, 33611 Gazinet cedex

Abstract: A model of simulation of wineries operation has been proposed. Each elementary set of knowledge used in the model has been defined as an entity gathering constant data, variable parameters, and expert calculation functions. The model has been structured hierarchically. It has been implemented in object oriented programming. This expert simulation software produces dynamic balances of resources. This paper presents an application of the software to estimation of water consumption and organic matter rejections. At the winery Caves du Sieur d'Arques in Limoux, both predictions proved to be accurate with a maximum error of 20 % for the 10-day period of the peak of activity of grape-picking.

Keywords: Dynamic modelling, Object-oriented programming, Food processing, Water pollution

1. INTRODUCTION

During the period of grapepicking, Enologists have to manage resources as varied as human power, grapes, process equipments, and refrigeration power. A simple idea for optimising the use of these resources is the dynamic simulation of the winery operation (Grenier et al, 1992).

In the winery, refrigeration is one of the few resources being taken care. But even so, the management rules remain quite rough. An inputs calendar is translated into refrigeration needs by making the hypothesis of an average fermentation activity with an imposed temperature setpoint. Cautiously the refrigeration power is overdesigned. Graphics permit to determine the refrigeration requirements. Recent softwares contain more precise calculations. But they do not simulate process lines.

In order to simulate the operation of a winery, flexibility is required. It is necessary to integrate the winemaker's expertise for going into more detail. As the Enologist directs winemaking, implementing his knowledge and reasonning can make the simulation more realistic. In addition, from one winery to another, the enological conditions may change and one must be able to modify the simulator by a simple change in the winery configuration or in the implemented expertise. The techniques which permit to maximize the flexibility of programming and to facilitate expertise implementation are relevant to Artificial Intelligence (AI).

Hitzmann et al (1991) insist on the interest of implementing expertise by means of expert systems methology mixed to other algorithmic approaches that might seem more natural. According to Konstantinov and Yoshida (1992), the results so far achieved in the field of knowledge based control of fermentations lead to the conclusion that modern expert systems should typically integrate various methodologies including rule-based approaches, fuzzy sets, mathematical models, and neural nets. AI is consequently interesting for our simulation scope.

In the Food Industry, optimization of the plant as a whole is a new concern while chemical engineering has more experience about it. For instance, the software SpeedUp permits to instanciate complex plant situations (Lucas et al, 1988; Daubas et al, 1991). But the cost of decision aiding systems developped for manufacture or chemical plants is too high for most food plants, especially wineries. An original dynamic simulation methodology has been implemented by building a knowledge based simulator (Nivière et al, 1994). This generator has been structured around a hierarchical model as defined by Oussalah (1988) and Mozetic (1990). A special attention has been dedicated to winemaking, and especially to equipments use and energy expenditure.

Since 1994, wineries processing more than 20000 hl/year need an authorization for carrying on their activity, and this authorization depends on strict environmental conditions. In addition, such wineries will have to be equipped with a depollution unit before the end of the year 2000, either individually or within an industrial community. Present associations with urban depollution units will be prohibited. Wineries processing less than 20000 hl/year will just have to declare their activity. But they will have to pay increased environmental taxes (a seven-fold increase from 1990 to 1994) unless they treat their wastes.

These constraints push winemakers to invest in depollution. In order to reduce investment costs, it is necessary to look for reducing rejections on one hand, and saving water on the other hand. This corresponds to the general notion of clean and sober processes. Each equipment should be controlled. Knowledge about what they reject is necessary, as well as about how to reduce these rejections. The same remark can be made with water consumption and saving. The simulation of these parameters and relevant knowledge should be very usefull to the winemaker for decision aiding. The more detailled the simulation, the more the winemaker will be interested in using the software instead of rough estimations, provided the ergonomics be good enough.

The dynamic simulator presented by Nivière et al (1994), here called by its Imeca name P3^{TM}, can be a good tool for aiding the winemaker to find appropriate solutions of lowering the depollution cost. It has been modified and tested at the winery Caves du Sieur d'Arques at Limoux (France).

Wineries are taken in this research as an example representative of a large sector of the food industry, with a high seasonality. The whole food industry should benefit from this research especially for environmental aspects, as it represents in France 24 % of the industrial rejections of suspended solids, 45 % of the organic matters, and 36 % of the matters containing nitrogen (Ministère Délégué à l'Environnement et à la Prévention des Risques Technologiques et Naturels Majeurs, 1991).

2. BIBLIOGRAPHY

At present time, no software has been presented in the litterature for aiding decision about waste reduction prior to depollution, excepted the suggestion made by Nivière et al (1994). But many engineering solutions have been proposed for achieving such reduction.

Racault (1992) suggested separating rain from used waters, saving water, and fractionning water between categories of waste waters more or less charged with pollution. Adams (1986) proposed to retain non soluble solids with grids and filters, to rinse barrels with wine prior to water, to recycle refrigeration water, and to save water through pressure rinsing, ultrasounds, automatic washing of presses. Mayer et al (1992) has compared various solutions for reduction of water consumption. By an automatic washing of drains in presses, Rochard could reduce the water consumption from 600 liters to 80 liters (1994). The same author showed that muds and lees represent up to 50 % of a winery pollution and that they can be recuperated.

Filtration adjuvants are an important source of pollution (Rochard, 1992). Clean filters dry the adjuvants after use by means of a gas and they are recuperated by centrifugation. But the top solution for pollution reduction is the use of tangential filters which do not need adjuvants. Chemical tartar removal can be done by recovery of both soda and tartrate. Mains recover water of bottles washing. This water is neutralized and stored for settling clarification (Mayer, 1993). Bottles can be sterilized by ultraviolet exposure without any chemical (Geiser, 1994).

Rochard (1994) has emphasized the state of floors in water consumption. Concrete floors are difficult to clean, while epoxy are easy and carrelages even better considering their resistance to shocks and chemicals.

Reduction of wastes rejection and water consumption can be really significant. In the dairy industry,

Massette (1989) has observed that with the 50 most important dairy factories of the region Adour-Garonne in France, pollution was reduced by 13,4 tons of organic matter per day between 1977 and 1987 for a total rejection of 38 tons per day in 1987, while the volume of milk processed passed from 6,3 to 10,3 millions of liters in the same period. Weise and Trantolo (1994) proposed the following solutions for the dairy industry: fully emptying of tanks and pipes before washing, fractionning of first washing waters, automatic taps at end of each hose, recycling solid matters, linkage and dripping protection, by-products recovery.

Alosi (1990) has conducted a similar approach at the Angoulême plant of Sanofi-Bio-Industrie for production of gelatine. A first step was to recycle refrigeration water, separate rain from washing waters, recycling washing water. A second step was to eliminate suspended solids by grids, filtration of muds and drying, settling, filtration and chemicals for grease removal and neutralization. A third step was the building of a biological depollution unit. The overall methodology has permitted to reduce the Oxygen Chemical Demand (OCD) from 5900 to 270 kg per day.

In the distilled beverages industry, pretreatment of distillery rejections has been achieved by ceramics membrane before aerobic treatment, with a cut down of 50 % in OCD and nearly 100 % of suspended solids. Some distilleries could valorize the rejections by means of concentration until 70 % solids content, followed by precipitation and potassium removal (Leroy, 1993).

In the brewery industry, collecting separately residual beer permits to cut down by 50 % the rejections. A good choice of labels and glue associated to separative techniques gave interesting results with a very significant drop of pollution (Schumann, 1986). In Obernai, the Kronenbourg plant separates clean refrigeration waters going to the river from charged used waters to be treated (Leroy, 1993).

Reverse osmosis is used in Dieppe where the Coffee plant of Soprad Nestlé concentrates effluents from marc juice. Concentrated effluents are incinerated (Leroy, 1993).

3. RATIONALE

Present european regulation encourages wineries to equip for reducing the volume of used waters and the pollution charge. The technologies actually used by the winemakers, as well as novel technologies or methods, should be characterized for insertion in the model of the simulator.

In a first step, the winemaking equipments and processes implemented in the wineries should be characterized. In the case of white winemaking, the pressing area is the most polluting zone of the winery during grape-picking and contains operations difficult to characterize as far as washing is concerned. It should therefore be emphasized. A wahings model for this area of the winery should be calibrated and tested.

In a second step, it will be necessary to study the variability of wineries or food plants and to suggest a methodology for adapting easily the washings model from in-situ measurements and a general data bank.

The work reported in this paper presents preliminary results of the first step.

4. MATERIAL AND METHODS

The winery Caves du Sieur d'Arques has been chosen as an experimental site for the application of the software P3^{TM} to water and effluents study. This winery processes in average 90000 hl per year and makes essentially sparkling wines with Crémant as the best selection. All operations are important for final quality, but pressing has a special importance. Juice selection starts with pressing: the best juices make the cuvée, then the first taille and the second taille. For each 17 tons of grapes press, 85 hl of cuvée juices are obtained, 27 hl of first taille and 9 hl of second taille.

The difficult point of this work is the fitting of calibration curves for water and rejections prediction. Racault (1992) has suggested a methodology permitting to estimate the daily pollution charge of a winery. The basic assumption is that the winery is made of a succession of elementary operations which specific pollution charge can be measured.

The main operations are: reception and handling of grapes, pressing, temporary storage, settling preclarifying, settling clarifying, centrifugation, fermentation, ... For each operation, the rejected volume of water is estimated and the average concentration in OCD of the sample is measured. The pollution charge per unit operation is thus estimated, yielding specific charges: g of MES (suspended solids) and g of OCD. This method is not very precise because of size effect of the tanks and the variability of musts and washings. But it gives orders of magnitude.

Water consumption is easy to estimate through installation of counters. Concerning used waters, global measurements at the exit of the winery were made during grape-picking and the results were compared to the pollution calculated on a daily basis. Global rejections were measured by automatic uptakes of samples and flowrates measurements in canals exits.

In a first campaign (1993), two 48 hour periods of observations were achieved with instrumentation sets at the three main exits of polluted waters in the winery.

In a second campaign (1994), the global consumption of water was measured at two levels: the whole winery with general counters, and the pressing area with specific counters. The pollution charge has been measured only at the pressing area, in connection with the fractionning of used waters: inspection hole inside of the area collecting part of the used waters, inspection hole outside the area collecting all waters minus the most charged separated by a fractionning equipment, and a tank collecting the most charged waters. The tank of charged waters was studied through average daily uptake after mixing by an immerged pump. In the two mentionned inspection holes, automatic on-line sample uptake and flowrate measurement was set up.

5. RESULTS

5.1. Estimation of data for the knowledge base of the simulator

The unit operations considered are presented at Table 1. According to the methodology described, ratios of consumed water to processed volumes of must were determined as well as rejected organic matter per hl of processed must. The units are respectively liters of water per hectoliter of must and gramms of OCD per hectoliter of must (Racault and Védrenne, 1994).

Table 1. Estimations of water consumption and organic matter rejection per hl of processed must for some unit operations at the Caves du Sieur d'Arques in Limoux in 1993.

Unit operation	water (l/hl)	OCD (g/hl)
automatic press washing	3	80
cuvée tank	8	60
taille tank	12	40

5.2. Modelling of water consumption and organic matters rejection

The configuration of the winery in the software P3™ describes equipments and technologies.

The washing model of the pressing area has been established with the following assumptions: automatic washings took place at end of each cycle as soon as they were empty and they consumed 450 liters for 170 hl of grapes (131 hl of must) containing after use 80 g of OCD per hl of must. It was assumed that temporary storage tanks of extracted musts were cleant with 8 l of water per hl of must for the cuvées and 12 l for the tailles, with respective charges of 60 g and 40 g of OCD per hl of must (Table 1).

Washings of the equipments for last juices extraction (rebêches, sold to fruit juice industry), belt conveyors, floors, and transportation tools, as well as the manual rinsing of presses and all circuits, were considered in a first approximation as being made on the morning from 8 to 13:00. To each hour of this period, the quantity Q_L corresponded to the overall washing volume in the pressing area divided by 6 hours, minus automatic washing volume of presses and minus manual inside washing of temporary storage tanks. To this quantity Q_L was associated $Q_{OCD\,L}$ which was the pollution charge transported by these washing waters.

Quantities M_1 and M_2 were defined as the respective amounts of washing water and OCD calculated by P3™ on the basis of the data of Table 1 for automatic washing of presses and manual inside washing of temporary storage tanks. For the period of time delimited by 9/19/94 at 12:00 to 9/21 at 12:00, M_1 = 33733 l and M_2 = 425060 g.

Using these two days for calibrating the model, the following equations are set:

$$Q_{water} = x_1 \,.\, J + y_1 \,.\, (J\text{-}1) - M_1$$

$$Q_{OCD} = x_2 \,.\, J + y_2 \,.\, (J\text{-}1) - M_2$$

where x_1 and y_1 are in liters of water per hectoliter of must, and x_2 and y_2 in grams of OCD per hectoliter of must.

$$x_1 = 8{,}884 \text{ and } y_1 = 19{,}621$$
$$x_2 = 17{,}23 \text{ and } y_2 = 300{,}\ 17$$

with J and (J-1) the input grapes of the two dates corresponding to each 24 hour period from noon to noon, Q_{water} and Q_{OCD} the estimations of Table 2.

Table 2. Consumed water and rejected organic matter at the pressing area for 24 hour periods from noon to noon ending on September 20, 21, 22 and 23 at the Winery Cave des Sieurs d'Arques in 1994.

Date	Water (m^3)	OCD (Kg)
Sept 20	129	1632
Sept 21	105	975
Sept 22	124	1557
Sept 23	101	967

5.3. Prediction

The consumed water and and pollution discharge for the pressing area in 1994 have been predicted for the whole period of grapepicking by steps of 1 hour. The graphic can be read with a cursor. For a better lisibility, the dynamic balances have been transposed to Excel™. Figure 1 compares the prediction of consumed water in the pressing area to extrapolated measurements. The general trend is well respected. However, morning washings have been overestimated to the expense of washing made during the rest of the day, with significant differences. These differences are quite systematic, which encourages to think that the model could be improved. Moreover, when cumulating the predicted water consumptions from 9/13 at 11:00 am to 9/23 at 18:00, the relative error is only of 4.5 %.

Getting into more detail with the period from 9/19 to 9/23, Table 3 compares predictions of consumed water and OCD rejections to field measurements. It appears that for this four-day period, the predicted cumulated consumed waters were 21 % lower than measured consumptions (Figure 2) and predicted rejected OCD were 10 % above measured rejections (Figure 3). Taking into account that the measurements of OCD are corrected values which are upper limits, the error might be close to 20 %.

Table 3. Comparison between measured and predicted amounts of consumed water and rejected organic matter at the pressing area for 24 hour periods from noon to noon ending on September 20, 21, 22 and 23 at the Winery Cave des Sieurs d'Arques in 1994.

Date	Predicted water (l)	Measured water	Predicted OCD (g)	Measured OCD
9/20	113638	129240	1735347	1632190
9/21	74912	105160	1178051	975120
9/22	107302	124500	1638259	1556690
9/23	68739	101000	1091023	967280
Total	364591	459900	5642680	5131280
Error	20,7 %		10 %	

6. CONCLUSIONS

In spite of all variability factors of washings and organic matters rejections per unit operation and per hl of processed must, it is reasonnable to think that correct balances of water consumption and organic rejections are feasible through the methodology developed. Results are quite encouraging, as the punctual errors are rather systematic and as global differences are around 20 % and lower when considering long periods. Improvement of the washings model is possible, and this should lead to a good accuracy for a given situation defined by a winery and its washings processes.

In a near future, the use of a software for dynamic simulation of resources in the winery should help the enologist to use its equipments in a cleaner and more sober way. He will be able to visualize the impact of different processes and equipments on water consumption and pollution, to know the exact level of pollution for calculation of environmental taxes, to reason about how to reduce this pollution before treatment, and to size depollution units.

ACKNOWLEDGEMENTS

We thank Mrs Gayda and Leclercq, Heads of the technical service at the winery Caves du Sieur d'Arques in Limoux, for their welcome during grapepicking, and the Regional Centers for Industrial Transfers of Technology (CRITT) Trial and Verseau in Languedoc-Roussillon, France.

REFERENCES

Adams K., 1986. Techniques favorables à l'environnement, utilisables en cave vinicole. Der Deutsche Weinbau, 15, 725-732.

Alosi A., 1990. Sanofi-BioIndustrie, vingt ans de dépollution à l'usine d'Angoulême. Magazine Adour-Garonne, 46.

Daubas B., Pingaud H., Koehret B. Développement d'un simulateur de procédés discontinus, semi-continus et continus: ProSim. Récents Progrés en Génie des Procédés, Ed GFGP (France), 5, 307-311 (1991).

Geiser A., 1994. Une nouvelle technique de stérilisation des laveuses de bouteille dans les entreprises des boissons. Documents Rehman Process Industrie.

Grenier P., Sablayrolles JM, Chabas J, Barre P., 1992. Réfrigération des moûts en fermentation: progrés récents. Revue générale du froid, 5, 31-35.

Hitzmann B., Lübbert A., Shügerl K., 1991. An expert system approach for the control of a bioprocess. I: Knowledge representation and processing. Biotechnology and Bioengineering, 39, 33-43.

Konstantinov K. B. and Yoshida T, 1992. Mini Review. Knowledge-based control of fermentation processes. Biotechnology and Bioengineering, 39, 479-486.

Leroy C., 1993. Deux solutions pour les effluents: séparer et valoriser. R.I.A., 505, 78-80.

Lucas P., Isambert A., Depeyre D., Cuille P., Rossiny P., 1988. Dynamic simulation and control strategy of a continuous separation system with unsteady inputs. Chemdata Congress, June 13-15, Goteborg (Suède).

Massette M., 1989. Industrie laitière: dix ans de dépollution. Adour-Garonne, 41.

Mayer A., 1992. Indices de calcul et simulation des flux de matière en brasserie. Brauwelt, 46, 2427-2432.

Mayer A., 1993. Nettoyage écologique et ingénieux des bouteilles. Brauindustrie, 4.

Mozetic Igor, 1990. Hierarchical model-based diagnosis. International Journal of Man-Machines Studies, 35, 329-362.

Ministère Délégué à l'Environnement et à la Prévention des Risques Technologiques et Naturels Majeurs, 1991. Pollution des eaux industrielles en France. Rapport annuel.

Nivière V., Grenier P., Roger J.M., Sévila F., Oussalah M., 1994. Intelligent simulation of plant operation in the wine industry. Journal of Food Control, 5, 2, 91-95.

Oussalah M., 1988. Modèles hiérarchisés multi-vues pour le support de raisonnement dans les domaines techniques. Thèse de l'Université d'Aix Marseille.

Racault Y., 1992. Les effluents des caves vinicoles, évaluation de la pollution, caractéristiques des rejets. Informations techniques du Cemagref, 92, 4, 1-8.

Racault Y., Védrenne J., 1994. Etude de la charge polluante des Caves du Sieur d'Arques à Limoux. Rapport Cemagref.

Rochard J., 1992. Réduction de la charge polluante et du volume des rejets dans les caves vinicoles. Revue Française d'Oenologie, 134.

Rochard J., 1994. Oenologie propre, réduction de la charge polluante et des volumes de rejets. Actes du Congrés International sur le traitement des effluents vinicoles, Ed. Cemagref, 261-266.

Schumann G., 1986. Minimisation des rejets de papier d'étiquette et de bière résiduelle. Brauwelt, 49, 2440-2443.

Weise D.L., Trantolo D.J., 1994. Réduction de la pollution dand l'industrie laitière. Procédés industriels pour le contrôle de la pollution et la minimisation des rejets, 705-713.

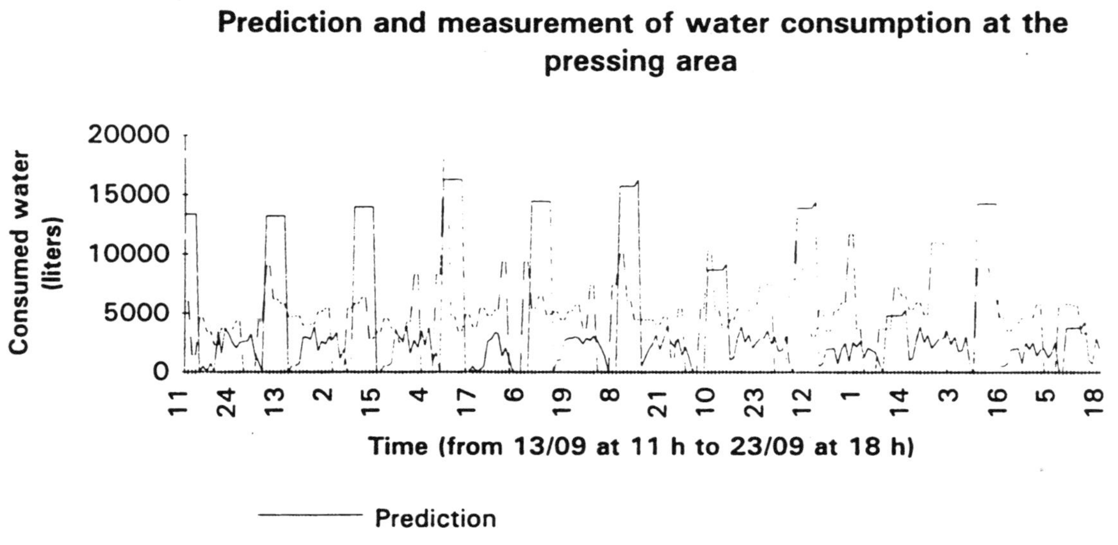

Figure 1. Prediction and measurement of water consumption at the pressing area of the Caves du Sieur d'Arques winery in 1994 at each hour from Sept 13 at 11:00 to Sept 23 at 18:00.

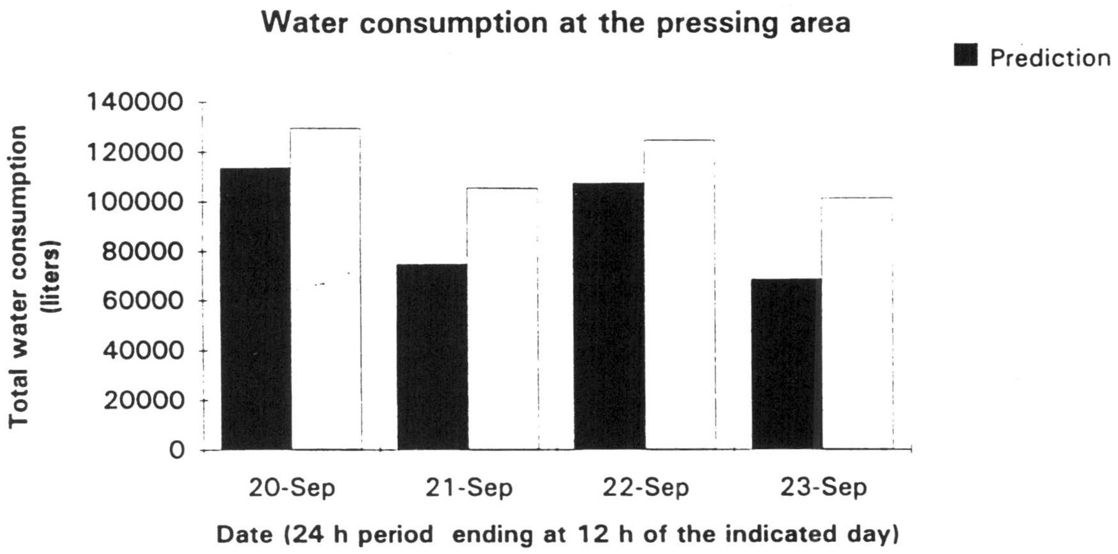

Figure 2. Prediction and measurement of water consumption at the pressing area of the Caves du Sieur d'Arques winery in 1994 on a daily basis from Sept 19 at 12:00 to Sept 23 at 12:00.

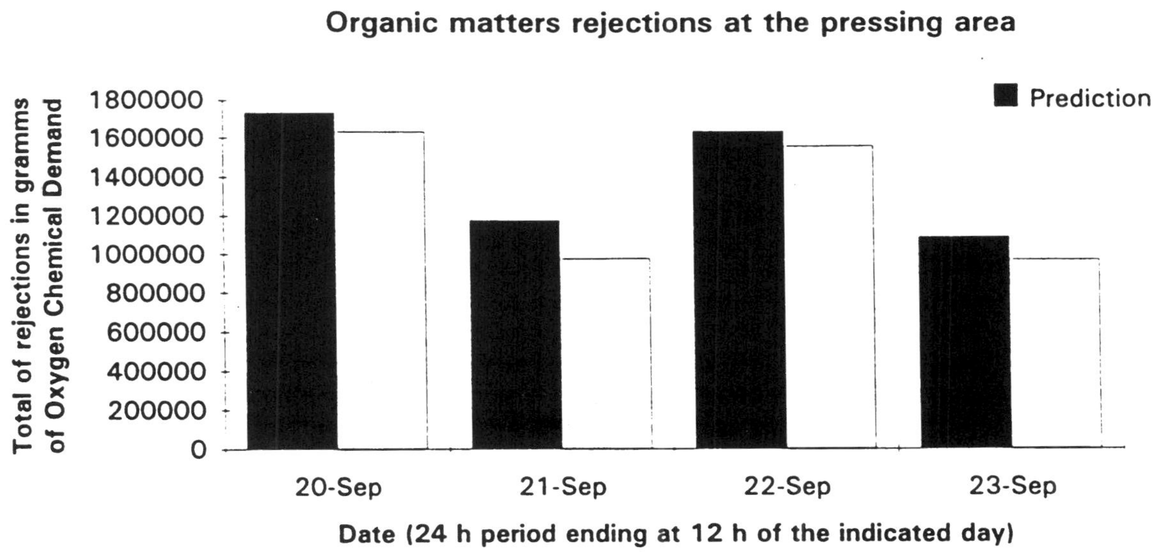

Figure 3. Prediction and measurement of organic matter rejection at the pressing area of the Caves du Sieur d'Arques winery in 1994 on a daily basis from Sept 19 at 12:00 to Sept 23 at 12:00.

PREDICTION OF CONDENSATION ON TOMATOES

Y.Q. Wei, B.J. Bailey, A.K. Thompson*, B.C. Stenning*

Silsoe Research Institute, Bedford, MK45 4HS (England)

**Silsoe College, Cranfield University, Bedford, MK45 4DT (England)*

Abstract: Some diseases of greenhouse tomatoes are encouraged by condensation, since moisture on the surface of fruits provides a suitable environment for the growth of micro-organisms. The condensation period can be predicted using an energy balance model which has been devised to simulate the temperature of fruits and predict when condensation might occur. This model was tested experimentally using a flexible resistance wetness sensor, based on the change in electrical resistance of a grid element when moistened. A series of experiments was undertaken to measure the temperature of fruits and air humidity, and to determine the condensation on tomato fruits using the wetness sensors. The experimental results showed that the model represented accurately the occurrence of condensation on tomato fruits.

Keywords: tomato fruits, condensation, disease, energy balance, model

1. INTRODUCTION

Surface wetness of greenhouse plants results mainly from condensation, which occurs when the surface temperature is lower than the dew point temperature of the air. The presence of condensation and the persistence of free water on plant surfaces provides conditions in which fungal spores can germinate and infect the host (DeWeille, 1965; Sutton et al., 1978; Lhomme and Jimenez, 1992). Arauz and Sutton (1990) found that 94% of *Botrytis obtusa* spores germinated in water after 4 hours at 24 °C. Evans et al. (1992) showed that disease severity increased with increasing wetness duration.

Modelling dew occurrence requires the accurate computation of the temperatures of different parts of the plant. Predictions of plant temperatures, either a whole canopy temperature or a single leaf temperature, have previously been carried out (Pedro and Gillespie, 1982a; Yang et al., 1989; Bailey et al., 1993; Scherm and van Bruggen, 1993; Papadakis et al., 1994). The simulation of crop temperature depends on a knowledge of the energy transfer processes such as sensible heat and latent heat transfer exchange between the plant and its surroundings. This depends on the plant organ (leaf, stem or fruit), and is affected by the micrometeorological conditions. In fact, condensation lasts much longer on fruits than on leaves of the same plant because of the difference between their physical and thermal properties. Dew accumulation varies significantly depending on the location within the canopy because of fluctuations of air velocity, solar radiation and air temperature in greenhouses.

The simulation of leaf or canopy wetness duration using an energy budget approach has been studied (Monteith, 1957; Monteith and Butler, 1979; Pedro and Gillespie, 1982a). Collins and Taylor (1961) developed a method to predict the onset of dew on the surface of a thin, large leaf, using atmospheric variables such as relative humidity, temperature and radiation; their prediction of the onset of condensation had a mean error of 0.8 h. Mintah (1977) developed a numerical energy budget model to estimate surface wetness duration at a weather station. The results showed that duration was underestimated by 1.5 h on average. Crowe et al (1978) used a multiple regression approach, based on relative humidity, air speed and minimum temperature, to forecast dew duration with an accuracy of ± 3 h. Pedro and Gillespie (1982a) proposed a simplified surface energy balance technique, combined with heat transfer theory for flat plates, to estimate dew duration using microclimatic data. They found that there was generally a 1 h difference between computed values and observations for shaded leaves and sometimes a maximum difference of 2 h occurred. A mean absolute difference of 0.97 h was found between observed and predicted dew periods when Scherm and van Bruggen (1993) used Pedro and Gillespie's model (1982b) in coastal fogbelt, coastal valley and interior valley locations. Little work has been done on fruit wetness with tomatoes during growing seasons in greenhouses.

The objective of this study was to develop a fruit energy balance model which can be used in predicting the occurrence of condensation on greenhouse plants. This provides a scientific understanding of the transient heat and mass transfer to the fruits on greenhouse tomato plants and of how the formation of condensation is influenced by the physical properties of the plant organ, the surrounding climate and its rate of change. Verification was undertaken by comparing the simulation with experimental measurements.

2. MATERIAL AND METHODS

2.1 *Model structure*

When the dewpoint of the surrounding air rises above the temperature of greenhouse plants in the early morning they become covered with dew, either in a film or as droplets. This occurs especially in well shaded portions of the crop canopy where low air speed and high humidity are normally found. The process of condensation is the reverse of the evaporation process. The model describes the thermal energy exchanges at the fruit surface by radiation, convection, condensation, evaporation and conduction. Neglecting the minor contributions involving the production of metabolic energy and energy storage, as suggested by Gates et al (1965), the energy balance at a tomato fruit was expressed by the relationship:

$$E_c + E_v + E_r + E_d = 0 \qquad (1)$$

where, E_c is the mixed convection heat transfer flux (W m^{-2}); E_v is the latent heat flux (W m^{-2}); E_r is the net transfer of heat by radiation (W m^{-2}); and E_d is the conduction heat flux (W m^{-2}) between the fruit surface and the centre. The boundary conditions were the dry bulb and dewpoint temperature of the air in the plant canopy.

Convective heat transfer is influenced by microclimate factors such as air speed, the temperature difference between the plant and air, the angle between forced airflow and natural airflow (Tang et al., 1991), and physical properties of the crop. According to the discussion of air movement by Fernandez and Bailey (1991 and 1994) and the recommendation of Stanghellini (1987), mixed convection heat transfer was assumed in this model.

$$E_c = \lambda_a \frac{m\,(R_e^*)^q}{d_{eff}}(T_a - T) \qquad (2)$$

where, λ_a is thermal conductivity of air (W.m^{-1}.K); m is a coefficient in the forced convection equation (dimensionless); q is the exponent in the forced convection equation (dimensionless) (Tang et al., 1991); d_{eff} is the effective diameter (m) available for developing natural convection (Tang et al., 1991);T_a is the air temperature (°C); T is the fruit surface temperature (°C); R_e^* is the effective Reynolds number for mixed convection (scalar addition), which combines a forced convection Reynolds number (R_e) and an equivalent Reynolds number (Re_{eq}) for natural convection:

$$R_e^* = R_e + R_{e_{eq}} \qquad (3)$$

The exchange of long-wave radiation depends on the fruit geometry, location within the canopy and on the difference in temperature between the radiating surfaces. Net radiation also depends on the incoming solar radiation. A geometric factor η was used to represent the fraction of the fruit surface exposed to solar radiation. This had a value of 0.5 for fruit fully exposed to solar radiation and 0 for fruit completely shaded by the canopy. The fruit exchanged thermal radiation with nearby leaves. The net radiation flux can be written:

$$E_r = \varepsilon_1 \sigma\, X_{12}\, \frac{T_c^4 - T^4}{T_c - T}(T_c - T) + \eta \xi I_m \qquad (4)$$

where ε_1 is the emissivity of fruits (dimensionless); σ is the Stefan-Boltmann constant (W.m^{-2}.K^{-4}); X_{12} is the fraction of radiation emitted by surface 1 which reaches surface 2 (dimensionless); ξ is the reflection coefficient of fruits for solar radiation (dimensionless); η is a geometric factor of fruits (dimensionless); T_c is canopy temperature (K) (which assumed a single canopy temperature with the same value as air temperature before sunrise dominated the exchange); I_m is net incoming long-wave solar radiation ($W.m^{-2}$).

The latent heat flux is the rate at which energy is transferred by the evaporation and condensation of water. Equation (5) provides the basis for simulating condensation and evaporation of dew droplets on fruits (Monteith and Unsworth 1990):

$$E_L = \frac{\rho_a C_P \, [e_a - e_s(T)]}{r_d \gamma} \qquad (5)$$

where ρ_a is the density of air ($kg.m^{-3}$); C_p is the specific heat of air ($J.kg^{-2}.K^{-1}$); r_d is the boundary layer resistance to vapour transfer between the wet surface and the air ($s.m^{-1}$); γ is psychometric constant ($Pa.K^{-1}$); e_a and $e_s(T)$ are vapour pressure of air and saturated vapour pressure at T respectively.

2.2 *Measurements*

Accurate measurements of wetness duration were made using a wetness sensor, which was constructed from a thin and flexible copper-coated polyamide film (Wei et al., 1995), based on the electrical conductivity of a polymer. The flexible sensor made good thermal contact with the surfaces of different shapes and sizes of tomato fruits. Moisture was indicated when dew droplets bridged the gap between the electrodes and so reduced the sensor resistance.

The wetness sensor was attached, using rubber bands, to the surfaces of tomato fruits in a greenhouse. The fruits were still attached to the plants during the experiments which lasted about 10 days depending on the maturity stage of the fruit chosen. The surface temperature of the fruit was measured with a fine wire copper-constantan thermocouple (0.11 mm diameter) (**Figure 1**). Microclimatic data such as air temperature (T-type thermocouple), dewpoint of air (dew point sensor) and solar radiation flux (solarimeter) were recorded by a data logging system having a resolution of 0.6 μV.

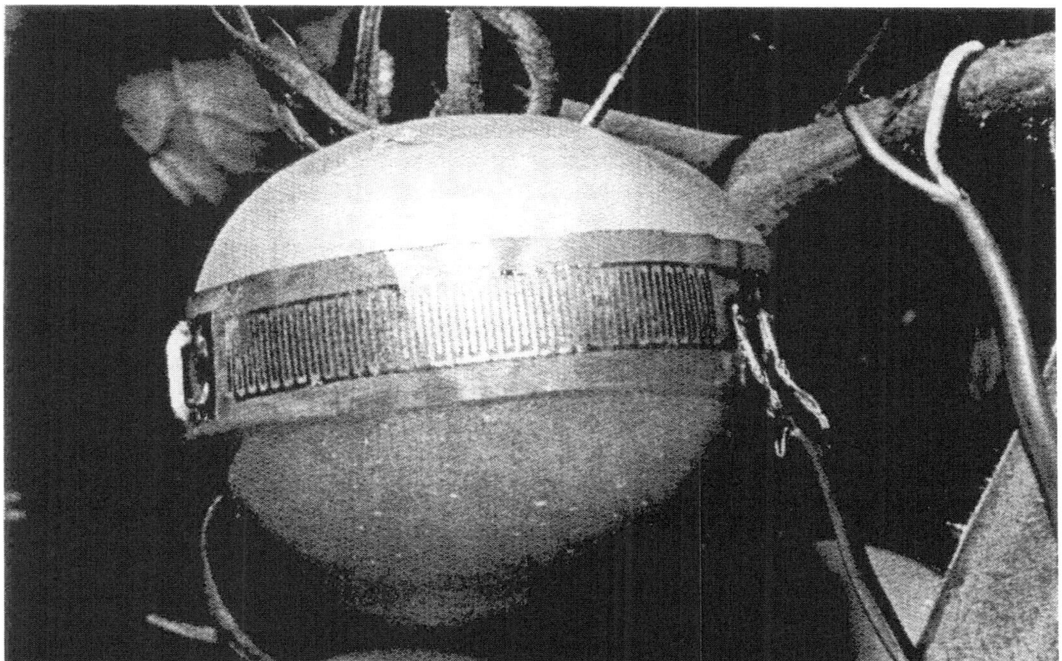

Fig.1 The deployment of the wetness sensors and thermocouple to detect the onset of condensation and the temperature history of tomato fruit surfaces and centres. A hypodermic thermocouple sensor used to measure the fruit centre temperature can also be seen (not discussed in this paper).

3. RESULTS AND DISCUSSION

3.1 *Model response to a step-increase in air temperature*

An initial temperature for fruits was taken as 16 °C (tomato blueprint at night minimum temperature (MAFF, 1981)). The model was evaluated by making a stepwise increase of air temperature from 20 °C to 25 °C at a constant relative humidity of 88%, where the air dewpoint temperature was 16.0 °C and 22.8 °C respectively.

The model predicted that condensation occurred on the fruit surface when the surface temperature was below the dew point temperature of air during the warming up stage (**Figure 2**). Dew accumulated on the fruit surface until the surface temperature was above the dewpoint when evaporation started. During the evaporation stage, the exposed surfaces of fruit maintained the same temperature as the air wet-bulb temperature of 23.5 °C until they had become dry. The fruit surface temperature then started to increase again and approached the air temperature of 25 °C.

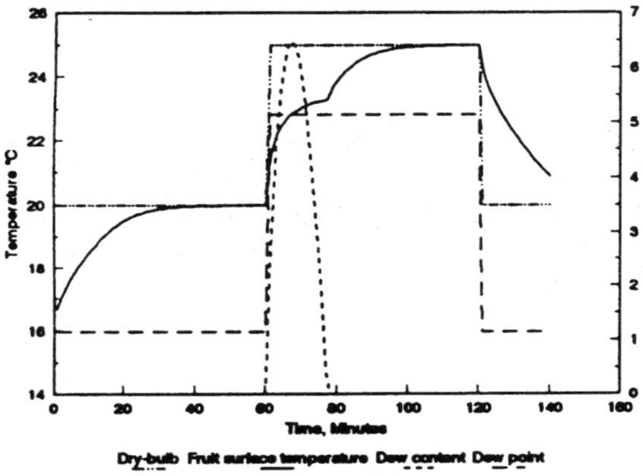

Fig. 2. Tomato fruit surface temperature and formation of condensation modelled under a stepwise change of air temperature between 20 to 25 °C.

3.2 *Comparison of model prediction with values measured in the greenhouse*

The diurnal temperature changes of fruit were simulated using microclimate data from experiments in a greenhouse sited at Silsoe Research Institute (SRI) in England. The period investigated included sunny, cloudy and rainy days and included day and night periods. **Figure 3** shows the predicted temperature history of the fruit surface and the comparison with measurement on two fruits. Good agreement was revealed through both the day-time and night-time in this comparison. As the dew point temperature of air was always below the fruit surface temperature during the measurements made 9th on April, the model showed that no condensation took place, which agreed with the records from the wetness sensors. The r.m.s. difference between the simulated and measured surface temperatures of five fruits was 0.4 °C with a standard deviation of 0.4 °C (**Table 1**). The average maximum temperature difference during 24 h was only 0.9 °C. This shows that the model predicted the fruit temperatures well and that it correctly represented the fruit energy transfer processes.

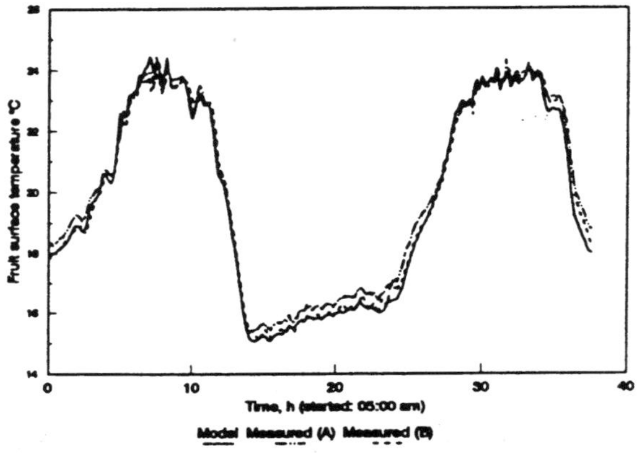

Fig.3. Temperature of tomato fruit surfaces as predicted by the model and measured on two fruits (A and B). Measurement taken on 9-10th April 1994.

Table 1 Differences between simulated and measured fruit surface temperature over 24 h

$T_{model} - T_{measurement}$		Max. Diff. °C	r.m.s. diff. °C	Standard deviation
9th April, 1994	A	0.8	0.4	0.3
	B	1.0	0.3	0.3
29th April, 1994	C	0.6	0.4	0.4
	D	0.8	0.4	0.4
19th June, 1994	E	1.3	0.5	0.4
Total	**Mean**	0.90	0.4	0.4

3.3 *Prediction of wetness period*

The microclimatic data, obtained from measurements in the greenhouse on 25th June 1994, were used as input to the model to test prediction of the wetness period. In the practical greenhouse measurements the data were recorded every 5 minutes and the measurement accuracy of the dew point was ± 0.6 °C. **Figure 4** shows that the fruit surface was predicted as becoming wet at 04:00 am, five minutes before the resistance of the wetness sensor began to fall. This delay may be attributable entirely to the five minutes measuring interval or it may reflect the time required for the condensed droplets on the sensor to bridge the gap between the grid elements and so change the resistance. The model does not take account for any effect of the dew droplet size.

A short period of disappearance of condensation predicted by the model was due to ± 0.6 °C of measurement errors of air dew point and air temperature used in the model. The peak of the condensation measured by the sensor occurred about 10 minutes later than that predicted by the model. In the evaporation stage, the model predicted that the dew droplets began to evaporate from the fruit surface when the air dew point temperature fell below the surface temperature. The drying rate detected by the wetness sensor ended at a similar time to the model prediction (**Figure 4**).

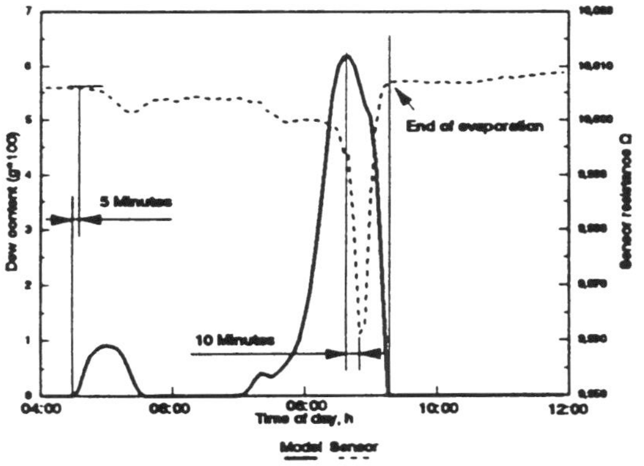

Fig. 4. Comparison of measured and predicted wetness periods on the surface of tomato fruits on 25th June 1994.

4. CONCLUSION

A model based on heat transfer theory was developed to describe the energy balance of tomato fruits. The model simulated the temperature of fruits and predicted when condensation could occur on fruit in greenhouses during the tomato growing season using greenhouse micro-meteorological information. The computed values of fruit surface temperature agreed with measured values to within 0.3 °C to 0.5 °C with a standard deviation of 0.4 °C . The difference in time between the simulated and the measured onset of condensation was only 10 minutes, which was smaller than results published by other authors. This confirmed that the model represented the energy transfer processes to individual fruits inside a canopy and their surrounding environment well. This will help in predicting the occurrence of condensation on tomato fruits accurately and in determining the microclimate conditions when condensation occurs.

ACKNOWLEDGEMENT

This research was undertaken for, and supported by, the Horticultural Development Council. The help and advice of Dr A Reynolds and Dr T Norris in the modelling are gratefully acknowledged. The photograph was kindly supplied by A. Hilton of Silsoe College.

REFERENCES

Arauz, L.F.; Sutton, T.B. (1990). Effect of interrupted wetness periods on spore germination and apple infection by Botryosphaeria obtusa. *Phytopathology*. **80** (11), 1218-1220.

Bailey, B.J., Montero, J.I., Biel, C., Wilkinson, D.J., Anton, A. and Jolliet, O., (1993). Transpiration of *Ficus benjamina*: comparison of measurements with predictions of the Penman-Montheith model and a simplified version. *Agricultural and Forest Meteorology*. **65**, 229-243.

Collins, B.G. and Taylor, R.J., (1961). Conditions governing the onset of dew on large leaves. *Aust. J. Appl. Sci.*. **12**, 23-29.

Crowe, M.J., Coakley, S.M. and Emge, R.G., (1978). Forecasting dew duration at Pendleton, Oregon, using simple weather observations. *J. Appl. Meteorol.*, **17**, 1482-1487.

DeWeille, G.A. (1965). The epidemiology of plant disease as considered within the scope of agrometeorology. *Agricultural and Meteorology*, **2**, 1-15.

Evans, K.J., Nyquist, W.E. and Latin, R.X. (1992). A model based on temperature and leaf wetness duration for establishment of Alternaria leaf blight of muskmelon. *Phytopathology*, **82**(8), 890-895.

Fernandez, J.E. and Bailey,B.J. (1991). The use of fans in greenhouses. International symposium of plant environment control for quality production under mild climate. *Tunisia*, 9-13 December. 8pp.

Fernandez, J.E. and Bailey,B.J. (1994). The influence of fans on environmental conditions in greenhouses. *J. Agric. Engng. Res.* **58**, 1-10.

Gates, D.M., Tibbals, E.C. and Kreith, F. (1965). Radiation and convection for ponderosa pine. *AM. J. Bot.*, **52**(1), 66-71.

Lhomme, J.P. and Jimenez, F.O. (1992). Estimating dew duration on banana and plantain leaves from standard meteorological observations. *Agric. and Forest Meteorol.*, **62**, 263-274.

MAFF, (1981). Tomato production. Part 5: Crop environment. *MAFF(Publications)*. 14pp.

Mintah,C.N. (1977). A numerical model to estimate leaf wetness duration. *M.Sc. thesis, University of Guelph, Ont.*, 101 pp.

Monteith, J.L. (1957). Dew. *Q.J.R. Meteorol. Soc.*, **83**, 322-341

Monteith, J.L. and Butler, D.R. (1979). Dew and thermal lag: a model for cocoa pods. *Q.J.R. Meteorol. Soc.*, **105**, 207-215.

Papadakis, G., Frangoudakis, A. and Kyritisis, S. (1994). Experimental investigation and modelling of heat and mass transfer between a tomato crop and the greenhouse environment. Journal of Agricultural Engineering Research **57**, 217-227.

Pedro, M.J. and Gillespie, T.J. (1982a). Estimating dew duration. I. Utilizing micrometeorological data. *Agric. and Forest Meteorol.*, **25**, 283-296.

Pedro, M.J. and Gillespie, T.J. (1982b). Estimating dew duration. II. Utilizing standard weather station data. *Agric. and Forest Meteorol.*, **25**, 297-310.

Scherm, H. and van Bruggen, A.H.C. (1993). Sensitivity of simulated dew duration to meteorological variations and different climatic regions of California. *Agric. and Forest Meteorol.*, **66**, 229-245.

Stanghellini, C. (1987). Transpiration of greenhouse crops. *Instituut Voor Mechanisatie, Arbeid en Gebouwen Wageningen*. 150pp.

Sutton,J.C., Swanton,C.J. and Gillespie, T.J. (1978). Relation of weather variables and host factors to incidence of airborne spores of *Botrytis squaniosa*. *Can. J. Plant Sci.*, **56**, 2460-2469.

Tang, L., Johnson, A.T. and McCuen R.H. (1991). Empirical study of mixed convection about a sphere. Journal of Agricultural Engineering Research, **50**, 197-208.

Wei, Y.Q., Bailey, B.J. and Stenning, B.C. (1995). A wetness sensor for detecting condensation on tomato plants in greenhouses. Accepted by *Journal of Agricultural Engineering Research*, Unpublished.

Yang, X. Short, T.H., Fox, R.D. and Bauerle, W.L. (1989). The microclimate and transpiration of a greenhouse cucumber crop. *Transaction of ASAE*, **32**(6), 2143-2150.

AUTOMATION OF TECHNOLOGICAL PROCESSES OF GRAIN DRYING AND TREATMENT IN THE CZECH REPUBLIC

Rudolf Pawlica, Jan Bouček

Research Institute of Agricultural Engineering [1]
K šancím 50, 163 07 Praha 6, Czech republic

Abstract: A major proportion of the grain crops of the Czech republic is harvested in moist conditions. This fact led to the installation of hot-air dryers with continuous operation in post-harvest treatment lines well equipped with grain handling equipment. The contribution evaluates widely disseminated system of applied automation in medium capacity post-harvest lines based on the indication of the heating-up of grain. It shows the results of the experiments applying automated systems for control of moisture content of grain based on microwave and capacitance moisture meters.

Keywords: Process automation, Quality control, Measuring elements, Agriculture.

1. INTRODUCTION

The geographic location of the Czech republic in the centre of continental Europe, on the watershed of European rivers, and the climate of the country imply that the major part of grain production is harvested in wet condition. The second premise related to our topic is the economic and political development that, in former years, consisted in a forced areal concentration of farm production and in the government subsidies on behalf of its intensification. This led to the construction of a large number of modern state owned grain elevators and later also to the development of farm elevators and treatment plants. Originally the principal point was this of cutting quantitative, later also qualitative storage losses. The equipment was solely of domestic origin, it was on a very good level as far its conception is concerned, whereas the quality of machinery components was not always satisfying.

[1] This paper has been prepared by support of the Czech Grant Agency in Prague.

2. GRAIN PRESERVATION METHODS

Cereal grains, especially wheat and barley are harvested at 13 to 19 % MC (MC = moisture content), in some years even over 20 %. About 2/3 of production have higher MC than is allowed by the storage standard and has to be preserved. Besides of marginal methods, like artificial cooling and chemical preservation, the principal approaches are methods of active ventilation and heated air drying.

2.1 Active ventilation

Active ventilation is applied as conditioning measure (forced intensive aeration of grain in silos) in the large storage and further as moisture decreasing measure in smaller storage spaces on farms. Conditioning ventilation uses axial fans dimensioned for approx. 10 mcub/h per 1 ton for batches up to 16 % MC, drying ventilation uses radial medium pressure fans dimensioned to approx. 20 mcub/h per 1 ton for batches up to 18 % MC. Systems of drying ventilation are always equipped with automatic control of motor switching. Ventilation will be activated depending of the indication of:

- ambient air temperature
- ambient air relative humidity
- stored grain temperature.

The control system is designed to intervene in the case of suitable combination for slow drying by impairing the equilibrium moisture content in the intergrain space and in the case of nests with higher than safe grain temperature with the vital objective to cool the grain down without respect to the risk of partial moistening the batch.

2.2 Heated air drying

Heated air drying as the most perfect conservation type (Pticyn, 1964), especially for grains with MC above 18 %, is vastly applied and has a long tradition in the Czech republic. After dissatisfactory experience with the obsolete type of former Soviet ZSPŽ-8 dryer (there were 1500 units in operation here), only after 1975 appearance of modern Czech dryers LSO, SSZ and TS, the farmers confidence in drying was restored and the advantages of the high grain quality resulting from this preservation method have contributed to its wide application. The respective types have been stationary continuous flow dryers with 20 and 40 t/h capacity with 4 % of water removal. The air is heated only indirectly, the dryers are equipped with various automation levels, the last modifications being equipped with energy saving systems. There are about 600 of such units in operation in the Czech republic.

3. AUTOMATION OF POST-HARVEST TREATMENT AND STORAGE PLANTS

Disproportions between the moist grain reception from trucks in batches and the continuous dryer operation have led to the construction of post-harvest lines and buffer storage to equalize the large material flows during the harvest. They facilitate the starting of dryer operation and ensure the buffer capacity before grain shipments. Automation ensures the harmonization of different material flows and storage dimensions and purposes.

3.1 Stores and plants of commercial companies

In major cases, these are typified stores of a unit capacity of 21.000 and 50.000 tons. They are constructed of reinforced concrete. The capacity of one cell is 1.000 t. The capacity of the reception line equals 75 t/h or 2 x 75 t/h. The drying line has the capacity of 40 t/h. There are approx. 200 of such stores in the Czech republic.

The commercial company stores are equipped with necessary elements of material flow automation, beginning with locking interlinked transport lines, locking depending of full cell level, indicating operation states (run - rest - failures) of machinery etc. The store is operated from one center - control room, where the operation is indicated graphically and by lights and the operator starts and supervises each activity. The older stores are equipped with relay control centres, the newer ones with electronic control centres. All stores are equipped with cell thermometers based on thermistor probes and with the evaluating central station.

3.2 Stores and plants on farms

The equipment of farms is more heterogeneous - some stores in the dryer areas have only ventilated stores, some stores in the more humid areas have only drying plants and in the transition areas there are stores and dryers combined in one plant, usually a silo battery. Arid areas have floor stores. The storage volume usually ranges from 3.000 to 10.000 t, the reception capacity from 40 t/h to 60 t/h, the typical dryer capacity being 20 t/h.

The material flow of most of the combined lines has three circuits:

- grain reception to store
- drying
- intensive cleaning and shipment.

The major automation element in floor stores is the automatic ventilation control, whereas in the vertical elevators there are the following: locking of transport ways, simple signalling of operation states and equipment control from the control room. The dryer circuit is usually designed as independent of the reception circuit and the material flow of variant intensity is controlled directly from the standardized dryer control system.

4. PROCESS AUTOMATION IN CZECH DRYERS

The dryer is the key equipment in a post-harvest plant and its throughput is decisive for managing the material flows during harvest. It is especially important for farms that are dealing with different kinds of crops and where the rapid initial treatment is most important to keep the quality of the harvested product. A typical representative of our modern dryers is the type TS-5 developed for specific local conditions. From the point view of automation, following features of its design are significant:

- original partitions and deflectors
- dryer filling and discharge
- pre-selection of the material flow
- semi-automatic process control

Original partitions and deflectors built-in in the shaft channels (Czechoslovak patent), that ensure perfect uniformity of grain flow. Besides of improving dryer capacity and also energy economy, they have also a meaning for automatic control of drying process.

Dryer filling and discharge. The standardized dryer control system controls the intervals of opening the feeding duct from the selected bunker following the grain level variances as signalled by the operation level indicators. The operation state is moreover ensured by the emergency level indicators (overfilled - no grain) with quick locking and audio signal. This circuit ensures also the balancing of levels in serial connection of the drying shafts. The dryer discharge, as the principal active control element of the process, is interlocked with the reliable function of the multi-element discharge way.

Pre-selection of the material flow. The operation mode of the TS-5 dryer can be selected by four flaps with actuating mechanisms by a key selected combination of following factors:

- parallel/recirculation - starting or stopping operation, or batch drying
- parallel/continuous - both shafts operated simultaneously
- serial/recirculation - starting or stopping operation, or batch drying in series
- serial/continuous - successive drying in two shafts (for high drying rates with maize or when drying seeds)

Thus, transporter connection, level monitoring system, grain refilling, and audio signalling will be changed.

Semi-automatic process control. When drying, the decisive process variable is the value of warming up the grain (i.e. highest temperature of the kernel [caryopsis] that is achieved during drying). The preceding work of the authors (Pawlica, 1982 and 1985) quantified the link between the kernel warm up and the outgoing air temperature. Impact of kernel warm up on its mechanical and other properties was investigated by Lichtfield and Okos (1988).

After mechanical averaging of an exactly defined shaft section, the outgoing air temperature can be measured exactly and without delay. This has been utilized for the TS-5 dryer control system, that is shown in the schematic drawing Fig. 1.

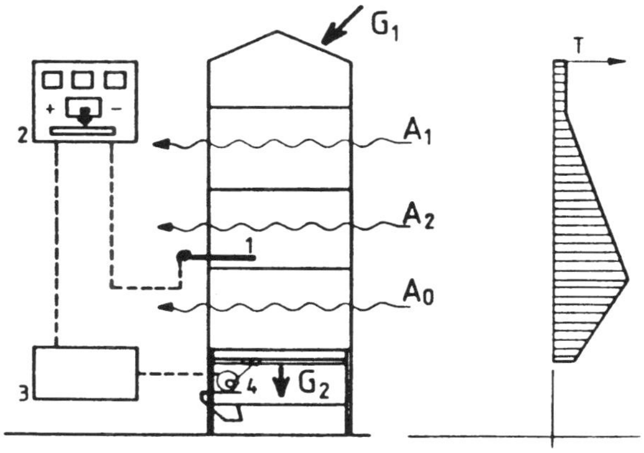

Fig. 1 Schematic view of a continual flow grain dryer including its control system. On the right, procedure of gradual grain heating and subsequent grain cooling when passing through the dryer. (The scheme corresponds to the dryer types SSZ-2 and TS-5 in the Czech republic).
G_1 - grain input
G_2 - grain output
A_1, A_2 - heated air
A_0 - cool air
1 - temperature probe
2 - electronic regulator
3 - power distributor
4 - grain discharge drive
T - grain temperature in the dryer shaft

Following the instruction manual, the operator will set the desired value of kernel warm up on the TRS regulator. The regulator will compare it to the actual value and issue a pulse for regulation. In the case of a negative deviation it will issue a command to decrease, in the case of a positive deviation to increase the speed of the grain throughput. The TRS regulator is a three-positional device PID type and the design of the dryer discharge unit has been adjusted to this system. The discharge unit has a grate design with three-stage operation: "0", "1/2", and "1". The desired throughput is modulation set by the regulator, viz. between (0 - 1/2) at a small throughput, (1/2) at stabilization to about 20 t/h, and (1/2 - 1) modulation when higher throughputs are needed.

This control system can also be used with advantage to maintain stable output MC values independently of the input MC variances: input of grain with higher MC means also the drop of the output temperature and vice versa. The temperature probe is very sensitive and records quickly even slight changes which result in increasing/or decreasing the throughput by the regulator. A regular control of output MC by a moisture meter in intervals of about 4 to 6 hours is recommended. If the operator anticipates a change of the grain input MC, he will accelerate the control process by an approximative change of the value desired.

Practical verification has shown that this system maintains reliably the pre-set admissible kernel warming up and optimizes the dryer capacity. When operating below of the biological warm-up limit, this system is able to damp efficiently the input MC variances to 1/4 - 1/3 in the output MC; it is apparent from one of the recordings shown in Fig. 2. Considering that the system is very simple, reliable and low in costs, it is justified to apply it as serial feature in the TS-5 dryers.

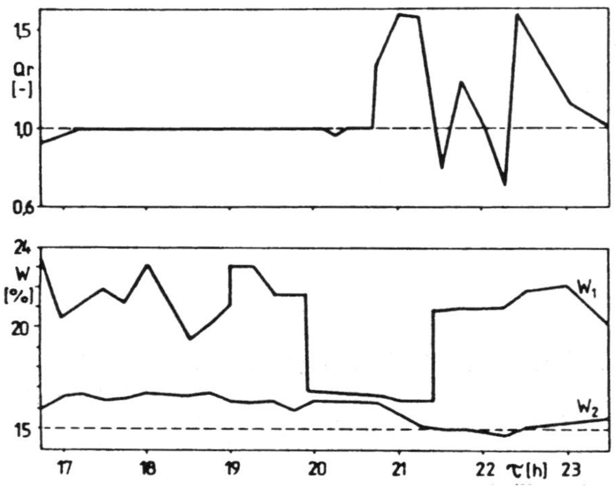

Fig. 2 Course of operation variables in the SSZ-2 dryer under conditions of large variances of grain input MC. (Barley, drying air temperature max. 72°C, warm up of grain 48°C).
Q_r - relative grain throughput of the dryer (-) (instantaneous and nominal throughput ratio)
W_1 - grain MC at the entry of dryer (%)
W_2 - grain MC at the dryer outlet (%)
τ - real time (hours)

5. AUTOMATIC CONTROL OF THE OUTPUT MOISTURE CONTENT

The most exacting task when automating grain dryers is to maintain a stable product output MC. It is, at the same time, the most important issue, because of its close relation to economics: overdrying means losing money for fuels, underdrying means a risk to the quality of stored grain. As it will be shown later, the problem is not so much involved in the control algorithm as rather in the reliability of MC probe data. The reason is that the grain MC is not directly a physical variable but a contractual value and the requirements on the identity with the reference method are great, notwithstanding many other influences.

Testing methods and devices described Pawlica and Bouček (1991).

5.1 Microwave moisture meter MFG-G

After having resolved the semi-automatic process control in dependence of the output temperature for the TS dryers and after its introduction to commercial production, the possibility of a improved control of output MC was sought for. The primarily problem was a reliable continuous moisture meter, that was at that time (1987) inaccessible. There was found a microwave moisture meter, adjusted for discontinuous measuring of continually transported construction materials, a produce of the Bauakademie Berlin. The measuring set was recalibrated for cereal grains and field tested. The measuring cycle included probe filling, stabilization, measurement of the radiation damping, probe discharge and pause. In one probe, alternatively input and output MC were measured with 2 minutes intervals.

The measuring results showed a good consistency when compared to gravimetrical measurements, but its functional reliability and highly demanding calibration presented problems. These and several more technical problems did not allow to use this moisture meter in practical operation as a component of the then developed system of computer controlled regulation of DIAMO dryers. This technically complex system was not suitable for the difficult conditions of the dust loaded dryer ambiance.

5.2 COMPU-DRY 4715/PCP control system

This Danish system was purchased, installed in dryers and tested. It is a microprocessor control device equipped with two probes for continuous MC measuring. It records the grain input MC chronologically and layer after layer and with the aid of output MC data performs the evaluation with the output of the active intervention for the regulation of discharge unit. The grain MC measurement is based on dielectric constant measurement in a high frequency electromagnetic field. The probe calibration according to the reference methods showed some defaults, viz. casual variances of the probe data, increasing of the differences between the measured and actual values with increasing grain MC and shift of approx. 2 % MC between the readings of the upper and lower probes.

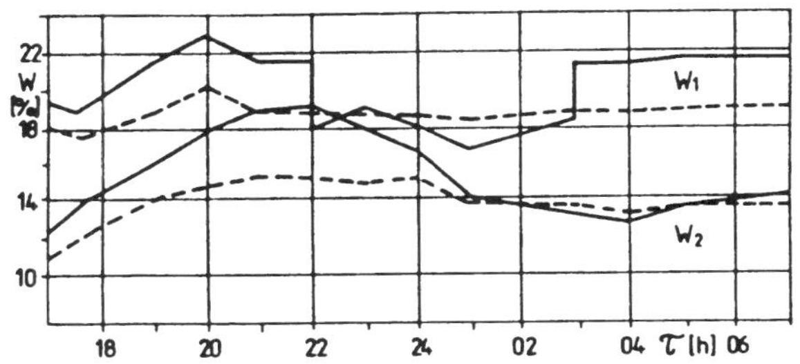

Fig. 3. Course of maize grain MC during the experimental testing of the COMPU-DRY control device at the continuous flow dryer TS-5.

\- - - - - - COMPU-DRY

________ gravimetric method

W_1 - grain MC at the entry of dryer (%)

W_2 - grain MC at the dryer outlet (%)

τ - real time (hours)

The recording of one field measurement of maize drying - see Fig. 3 - shows the situation, when after 5 hours operation with maize (22 % MC), grain already dried down to 18,5 % MC was reintroduced in the dryer, and after other 5 hours again the harvested material with 22 % MC was fed to the dryer. If the transport delay is taken in account, it is apparent that during the first two hours the control device collected data only and did not regulate the operation, the middle part of the test was impacted by the change of the batch and the regulating function was still not apparent, and in the last third the control could not regulate because the input MC probe indicated values entirely different from the actual state.

Verifications were undertaken also with other grains and in different locations. It can be said, that the more the grain MC on the input or output differs from the MC value during calibration, the larger is the error of probe data. The COMPU-DRY is able to stabilize the output MC value in a certain way, but easily the MC would be stabilized on a different value than it is desired - evidently this is due to the fact that the device does not measure correctly except for the two points calibrated. It is capable to function acceptably when the drying process is stabilized, with minimum variances of input MC, or with very slow changes, whereas it reacts inadequately to more abrupt changes.

5.3 DIAMO - KVO-50 control system

The continuous KVO-50 moisture meter is installed in the vertical material flow in the dryer. A part only of the grain flow is directed into the measuring chamber and the chamber itself is maintained steadily full by mechanical flaps. The grain moisture content is indicated as change of the capacitor capacity and is digitally evaluated.

Many comparative tests in field operation were performed. Records from one measurement may show differences against the DICKEY-JOHN moisture meter, nevertheless the absolute value difference is not important for regulation purposes, as it is possible to correct it by shifting calibration points. What is important, is the reaction to the MC changes and this ability was proven for the KVO-50. Nevertheless nearly all tests showed that after a longer operation run (after some 3 to 5 hours), the indicated values suffered from "deviations" - see example in Fig. 4.

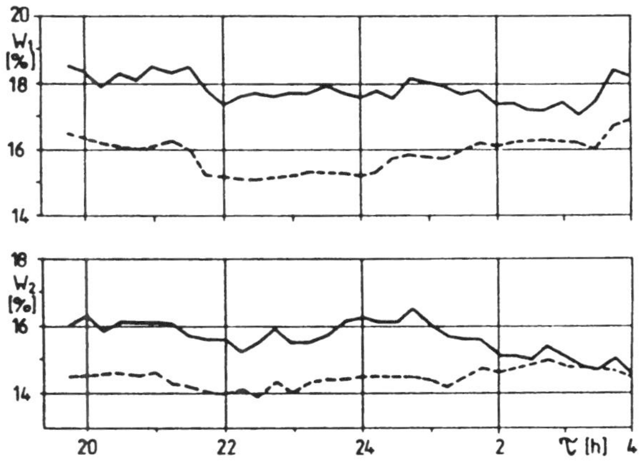

Fig. 4. Course of the experimental testing of the continual grain moisture meter KVO-50, installed on the entry and on the outlet of the dryer. Grain samples were taken simultaneously and grain MC was measured comparatively.

\- - - - - - KVO-50

________ DICKEY JOHN

W_1 - grain MC at the entry of dryer (%)

W_2 - grain MC at the dryer outlet (%)

τ - real time (hours)

It can be stated, that KVO-50 measures generally in a satisfactory way materials that pour easily and have not a very high moisture content, and that it reacts well to changes. Materials that pour with difficulties, have a high MC, or light materials are measured dissatisfactorily, not reliably. For reasons, not found, KVO-50 requires to perform a probe calibration, according to the user manual, each 4 hours.

DIAMO-L is a universal computerized control system for regulation applications that has been applied to control the entire operation of the TS dryer. To keep even the possibility to operate the dryer with the original relay system, it was applied a transfer electrical distributor that enables to switch the relay control to the electronic system. As there is possible to gain certain control constants during dryer operation only, the system was "tuned" during nearly three campaigns. This work was hampered by extraordinary dry periods during the recent years, when the dryer was run only occasionally. A partial experience might be mentioned, viz. that some control circuits are easier to be managed by analog control devices (e.g. continuous burner control). Further - an algorithm is scarcely capable to embrace every, even the uncommon operation states and then inertial processes (like continuous drying) can be stopped sometimes without necessity. Further - in many cases, the operator has a bad access for a manual intervention, and also the sophisticated computer technique and insufficient experience can lead to complications. Last but not least, it is also necessary to compare computerization to partial, semi-automatic control from the economic point of view.

5.4 Conclusions

To summarize this stage concisely: it has to be taken in account, that an on farm dryer will be operated (for grains) some 20 days a year only. Nevertheless, during these 20 days it must not occur, that troubles with the control system would decrease the dryer capacity or even cause a shut off or outage. The knowledge resulting from the preceding work shows, that the full computerization of the drying process control will require a certain simplification and also improvement, especially to approach better the phenomenon of the operating man. It also requires substantially higher operation reliability, so as to offer a decisive weapon for the short period of fight for saving the crops when harvest time is rainy.

6. ANTICIPATED TRENDS OF FUTURE DEVELOPMENT

Czech agriculture makes its way through the transformation period and is preparing to adapt to the situation anticipated after joining the European structures. Practically it will mean the differentiation of development of farm businesses whatever be their property form. From the point of view of automation, it will mean a retreat from the mechanically understood unique development to more liberal differentiation of technical equipment. Extensively oriented farms will be rather interested in simple equipment functions and low prices, whereas a smaller proportion of farms oriented toward intensification will feature the demand for higher equipment quality and possible substitution of the unreliable human factor. And here a space opens for entries of automation systems of new generation, which might link the high-tech automation of highly developed countries to the specific needs and conditions of the farm production in the Czech republic.

REFERENCES

Pticyn, S. D. (1964) *Zernosushilki.* Mashinostrojenie, Moskva.

Pawlica, R. (1982) Uneven speed of grain descent through the shaft of grain drier. *Zeměd. Techn.*, Vol. 28, pp. 321-332.

Pawlica, R. (1985) Indication of a decrease in grain quality and the method of the control of drying process in a grain drier. *Zeměd. Techn.*, Vol. 31, pp. 719-729.

Lichtfield, J. B. and Okos, M. R. (1988) Prediction of corn kernel stress and breakage introduced by drying, tempering and cooling. *Trans. ASAE*, Vol. 31, pp. 585-594.

Pawlica, R. and Bouček, J. (1991) Continuous measurement of moisture content and automation of drying process in grain driers. *Zeměd. Techn.*, Vol. 37, pp. 283-292.

STUDY FOR AN EXPERT SYSTEM ON PRESERVATION OF THE QUALITY OF POST-HARVEST GRAIN : PLANNING OF STORAGE TECHNICAL ROUTE

A. Ndiaye

INRA - LIDS, Domaine de la Grande Ferrade, B.P. 81, 33883 Villenave d'Ornon Cedex, France
Tel. : (33) 56 84 32 94, Fax : (33) 56 84 32 89, E-mail : amadou.ndiaye@bordeaux.inra.fr

Abstract : The planning space is represented on 3 different abstraction levels : *grain*, *operation*, and *silo*. The planning is hierarchical and non linear and involves several techniques of artificial intelligence : causal diagnosis of grain quality at *grain level* ; selection of the storage operations needed to preserve grain quality at *operation level* ; ordering the selected operations, allocating and reasoning on resources, at *silo level.*
The main originality of this study lies in the introduction of a preventive approach to stored grain management which is guided by the preservation of the grain quality.

Keywords : Expert system, Artificial intelligence, Hierarchical planning, Causal diagnosis, Qualitative reasoning, Reasoning on resources, Grain quality.

1. INTRODUCTION

Developing expert systems to help maintain the quality of stored grain is one of the major concerns of post-harvest research. Storage proves to be the weak link in foodstuff quality chain. Modern farming technics have allowed an increase in the world production of cereals ; and, consequently, an increase in storage capacities and duration. It so happens that any lengthening of storage duration means an increase in the risk of grain degradation. Grain is a living organism, and the normal outcome of its vital activity is germination and oxidation phenomena. A stock of grain is the natural environment of insects, mites, and micro organisms : insects and mites consume and spoil the grain (Fleurat-Lessard *et al.*, 1994a), while micro-organisms deteriorate and contaminate it (Cahagnier *et al.*, 1993).

Degradation in grain quality is a normal and irreversible process. Degradation speed depends on the environment factors : temperature, moisture, impurity rate and CO_2 content. The question we are asking is the following : **how to store grain in order to preserve its quality ?** Of course one should start with a definition of the idea of grain quality.

Grain quality is a complex entity in which four different specific qualities take place (intrinsic & physico-chemical, sanitary & security, technological and nutritional).These specific qualities are in function with the environment factors and the degradation agents which are measurable data. The grain state of quality can be appreciated through grain quality tests. These tests are put together depending upon which qualities they can evaluate (Ndiaye and Fleurat-Lessard 1994).

The interest of using expert systems (ES) to take up this problem can be explained by 1) the lack of any grain quality preservation model ; 2) the strongly experimental and granulary nature of the available knowledge. For example, the population growth rates are known only for certain degradation agents, at certain temperatures and moisture and for certain grains (Beckett *et al.*, 1994). During the last years, several pest control ES have been developed to help make decisions within the context of fight against degradation agents present in stored grain (GPA ; Wilkin, *et al.*1991), (SGA ; Flinn and Hagstrum, 1990), (PestMan ; Longstaff, 1993) (PEST ; Pasqual and Mansfield, 1988). These ES have confirmed the interest of ES to help preserve grain on one hand, and, on the other hand, have shown the limits of pest control as far as grain quality maintenance is concerned.

Our aim is to build a preventive system of stored grain quality maintenance guided by the measured values of the environment factors. The studies on environment factors show that they give an early information on the risks of quality degradation (Fleurat-Lessard *et al.*, 1994b ; Wrigley *et al.* 1994 ; Juliano, 1994). Our approach differs from existing systems in that we give priority to the prevention of grain quality degradation, and therefore to predictive signs of degradation. The fight against degradation agents remains a mean to decontaminate grain.

Our strategy of preventive preservation of stored grain quality is based on a good appreciation of instantaneous grain quality in order to be able to determine the optimal storage technical route allowing us to monitor the quality of grain. It involves the following 4 stages : 1) to establish the state of grain quality before storage, 2) to determine an optimal storage preventive technical route, 3) to monitor the evolution of grain quality during storage, 4) to react to any event representing a foreseen or effective quality degradation and readjust the storage technical route accordingly.

This paper presents the different planning stages of the stored grain preventive maintenance : causal diagnosis of grain quality, selection of the operations needed for grain storage, and the ordering of these operations in storage preventive technical routes. These two stages constitute the phase of storage technical route planning.

2. PLANNING OF STORAGE TECHNICAL ROUTE

Within the planning space of the storage technical route, we distinguish 12 different entities (Table 1) which are linked by causality relations of a *influence* or *give_information_on* type (Steyer *et al.*, 1993).

Table 1. Entities present in the space of the storage technical route planning

Entity	Variable	Abbreviation	Type
Available Resources	Consumables	Rc	Observation
	Equipments	Re	
Degradation agents	Presence of insects	PIns	Measurement
	Presence of mites	PMit	
	Proliferation of micro-organisms	ΠμO	
Degradation Risks	Insects	Ins	Calculation
	Germination	G	
	Micro-organisms	μO	
	Mites	Mit	
Environment factors	Impurity rate	ImpR	Measurement
	Moisture Content	MC	
	Temperature	θ	
	% CO2	CO2	
Grain aspect	Color	Color	Observation
	Smell	Smell	
Grain identity	Harvest year	HY	Observation
	Production area	PA	
	Variety	Va	
Grain Quality	Grain Quality	Q	Calculation
Quality test	Bread making quality	BQ	Measurement
	Enzymatic activity	EA	
	Grain hardness	GH	
	Germinative power	GP	
	Heavy metals	HM	
	Lipid quality	LQ	
	Mycotoxins	MT	
	Nutritional value	NV	
	Pesticide residues	PR	
	Protein content	PC	
	Protein quality	PQ	
	Radioactivity	Rad	
	Specific weight	SW	
	Starch quality	SQ	
	Weight of 1000 grains	W1000	

Specific Qualities	Intrinsic & physico-chemical Quality	Qipc	Calculation
	Nutritional Quality	Qn	
	Sanitary & security Quality	Qss	
	Technological Quality	Qt	
Storage operation	Possible storage Operation	O	Calculation
Storage Technical Route	Storage Technical Route	STR	Calculation
Utilisations of Grain	grain possible destination	U	Calculation
	maximal length of storage	Δt	

These entities are gathered in 3 different knowledge levels (Fig. 1) :

- The *grain level* is the level of grain quality diagnosis. The quality of grain is appreciated from the environment factors, presence or proliferation of degradation agents, quality tests, grain identity and aspect.
- The *operation level* is the level of selection of storage operations. The operations are selected on the basis of risks of grain degradation, and possible utilisation of the grain.
- The *silo level* is the level of the setting up of the Storage Technical Routes (STR). The STR are made of storage operations which can be executed in the silo.

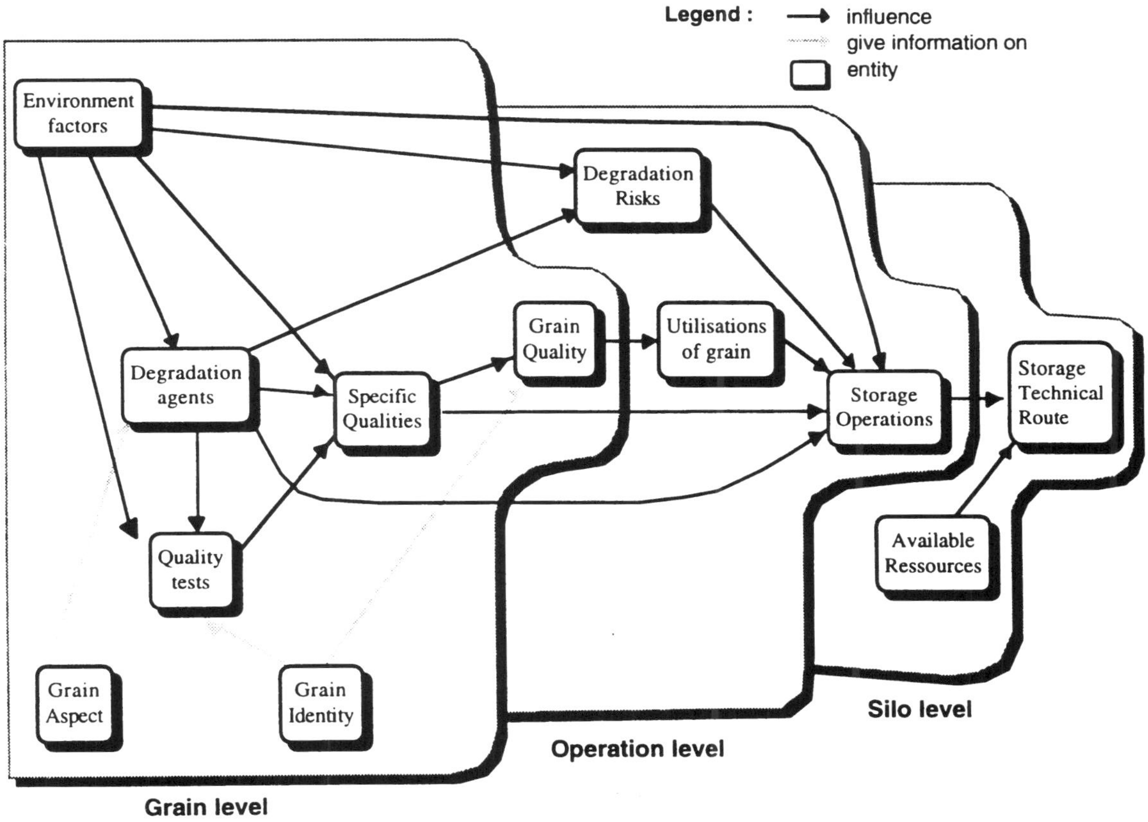

Fig. 1. The relations between the entities present in the space of the storage technical route

The planning is hierarchical and non linear : the grain quality is appreciated at grain level, the operations of quality preservation are selected at operation level, and the ordering of operations in storage technical routes is made at silo level. Each of these knowledge levels is an abstraction level of the cognitive process of determination of a storage technical route. (Fig. 2)

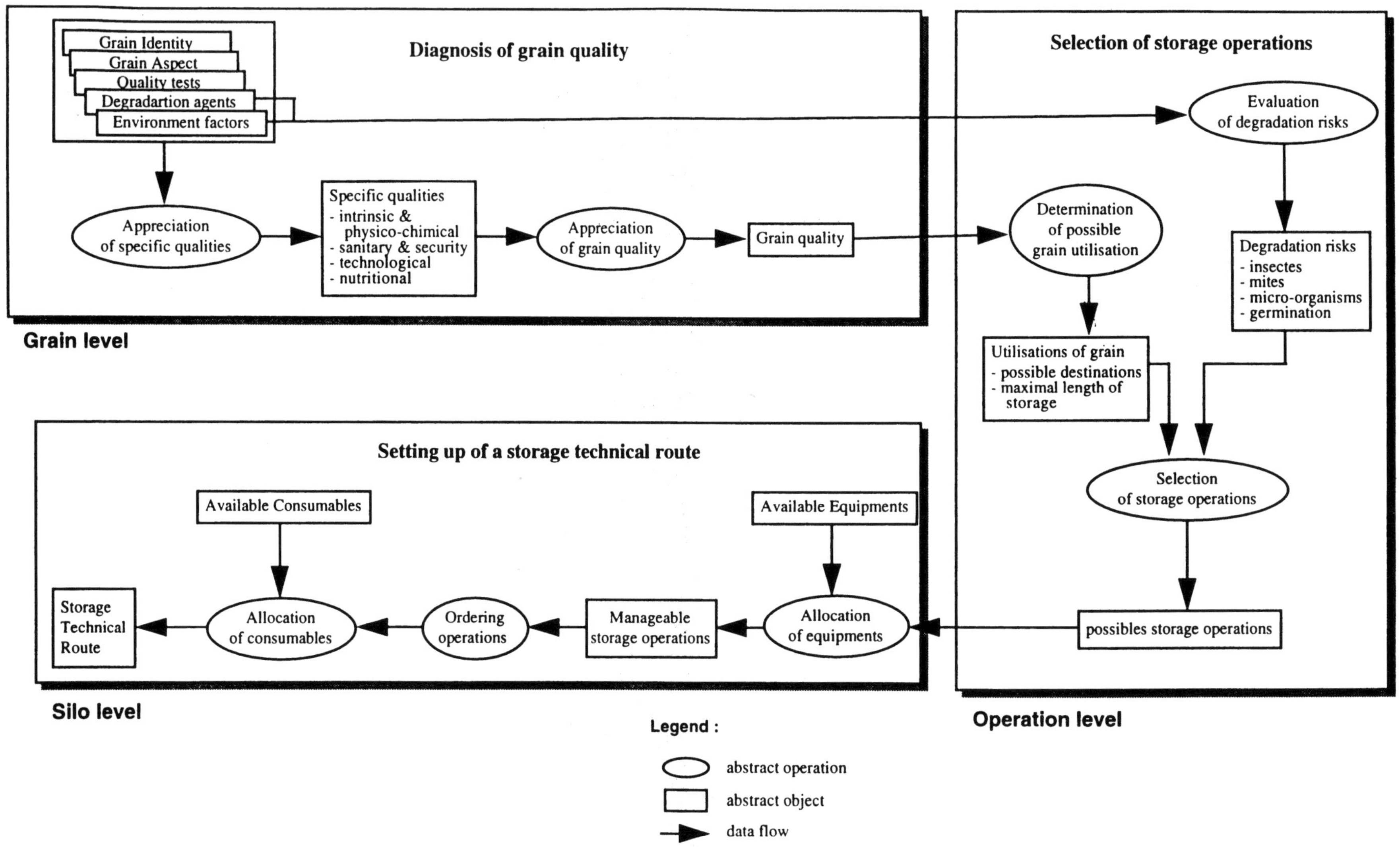

Fig. 2. The process of the planning of storage technical route

2.1. diagnosis of grain quality

In literature, grain quality is defined nowadays in function with the grain foreseen utilisation : *"the word 'quality' means suitability for the specific process or utilisation for which it's destined"* (Wrigley *et al.*, 1994). Such a definition infers an inversion in causality between *quality* and *future utilisation* of grain. Each utilisation implies a given quality of grain. This required quality must be available at the time of utilisation to make it possible. Aiming at preservation of grain quality during storage on one part, and at assigning it to an economically optimal utilisation on another part, it is necessary to respect the causality direction : from quality to grain utilisation. Therefore it is imperative to determine the quality in function with the characteristics and not in function with the foreseen utilisation of the grain which is, after all, only a bet on future.

It is still difficult nowadays to give a formal definition of grain quality, so we will limit ourselves to the following operating definition :

grain quality comes out of its specific qualities : intrinsic & physico-chemical, sanitary & security, nutritional and technological.

We determine the quality of grain through a process of causal diagnosis, the data of which are : the environment factors, the degradation agents, the quality tests and the identity of grain. On the basis of these measurements or observations which are not necessarily all known, we appreciate the specific qualities of grain ; and on the basis of these specific qualities, we appreciate the quality of grain.

At grain level the data and the appreciations are reduced to quantitative values. the formalism for qualitative reasoning developed by Guerrin (1991) in the domain of ecology gives us a sufficient framework for grain quality diagnosis. Our quantity space (QS) needs seven symbols (vvl, vl, l, m, h, vh, vvh) (Table 2)

Table 2. Quantity space symbols and some interpretations

vvl	vl	l	m	h	vh	vvh
very very bad	very bad	bad	average	good	very good	excellent
very very low	very low	low		high	very high	very very high
	very damp	damp		dry	very dry	

This formalism devised to interpret heterogeneous data, allows us to :

- Express the grain heterogeneous characteristics into homogeneous qualitative valuation.

  Example : a moisture content (MC) between 13 and 14% is considered *average*, and a specific weight (SW) over 76kg/hl is considered *high* ; and therefore MC and SW will have as respective values *m* and *l*.

- Convey the experience of the human experts through qualitative calculation rules and tables.

Example : valuation table of the intrinsic & physico-chemical quality (Qipc) of wheat from grain specific weight (SW) and moisture content (MC) and impurity rate (ImpR) :

	X	**MC** vl	l	m	h	vh
SW	l	vvl	l	vl	vl	vl
	m	vvl	m	m	m	l
	h	vvl	vh	h	h	m

	Qipc	**X** vvl	vl	l	m	h	vh
ImpR	l	vvl	vl	l	l	l	l
	m	vvl	vl	l	m	h	vh
	h	vvl	vl	l	h	h	vvh

- Represent the causal dependencies between variables representative of the grain quality domain under the form of an oriented graph called causal graph (Fig. 3) : the nodes represent the variables, the arcs in full line represent the *influence* relations (cause /effect) and the arcs in streaked line represent the *give_information_on* relations.

Examples :

1- baking wheat with a moisture content (MC) above 15% will be of an *average* or *bad* intrinsic & physico-chemical quality (Qipc) except if it benefits from a *high* specific weight (above 76 kg/hl). Moisture content is a cause for bad intrinsic & physico-chemical quality, which is expressed by the influence relation MC → Qipc

2- wheat of Soissons variety (Va) harvested anywhere in France (PA) in 1994 (HY) belongs to the category *superior baking wheat* averaging a W of 222 and a P/L of 0,67 (for bread making quality), a protein content (PC) of 11,1%, a *Med-hard* P.S.I. grain hardness (GH) and a Hagberg of 344 seconds (for enzymatic activity). The information on cereal quality is given each year after a "Cereal Quality" survey made by ONIC and ITCF* in every French production area and on all varieties with a significant tonnage. The results of this survey are available as early as October for July/August harvests. A prediction of cereal quality covers the period from harvest time to the publication of the results of the survey. This example illustrates the *give_information_on* relation linking the variables Va, PA and HY (grain identity) to each of the following variables : Q, W, P/L, PC, GH, Hagberg.

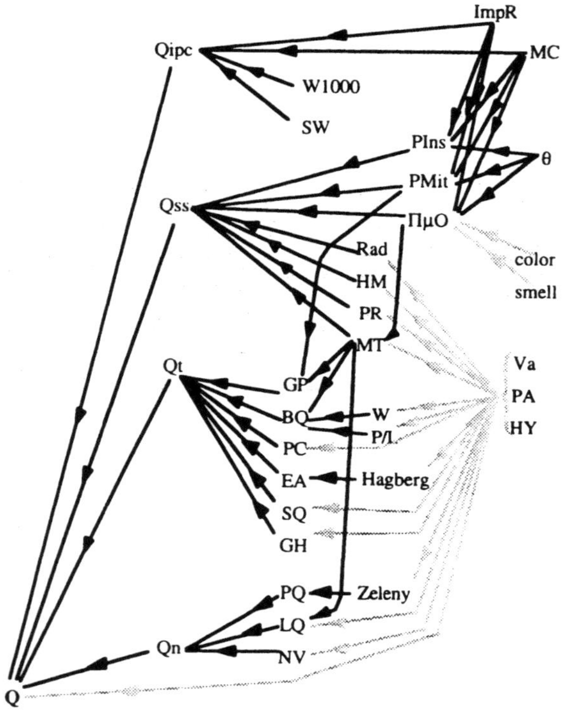

Fig. 3. Causal graph of variables present at grain level

At grain level we can distinguish five types of data :

- the data which are immediately available for a first appreciation of grain quality : θ, MC, ImpR, SW, PIns, color, smell, Va, PA, HY. These data, which are systematically measured or observed in all storing organisms, are sufficient for a first appreciation of quality. The grain identity (Va, PA, HY) is used to obtain the average values of the quality tests and the knowledge on the security which is specific to the variety, the harvest year, and the grain production area. The average values of the quality tests are used to appreciate the technological and nutritional qualities (Qt) and (Qn) ; and the knowledge on grain security is exploited to have good indications on the safety aspects of the sanitary & security quality (Qss). The production area exposure to radioactivity at a given period is an example of the knowledge on grain security. The *color* and *smell* data are sensorial indicators of micro-organism proliferation.

- the data indicating that additional tests should be carried out : PMit, ΠμO, MT, PR, HM. Presence of mites (PMit), in particular *Acarus siro*, implies checking the germinative power (GP). Proliferation of micro-organisms (ΠμO) implies an ergosterol measuring out to precise the levels of degradation and contamination by grain mycotoxins. The presence of mycotoxins (MT) implies an investigation on types and species, and eventually checking the bread making quality (BQ), the germinative power (GP) and the lipid quality (LQ). The presence of pesticide residues (PR) implies a measuring out of the active residues in order to improve the proportion of any pesticide application and to determine a minimal grain storage duration in relation with the residues activity period. The presence of heavy metals (HM) implies a measuring out of Pb and Cd.

- the data for fast confirmation of the quality obtained during the first appreciation : PC, GP, GH, PQ. These tests which can be done rather quickly (within a 24 hour delay) will complement the available data for a better valuation of the technological and nutritional qualities (Qt) and (Qn) and therefore of the grain quality.

- the data indicating a redhibitory defect at grain quality level : Rad, MT. Radioactive grain is not edible. Cancer-producing mycotoxins (MT) such as *aflatoxin B1* have a tolerance threshold in the region of ppb ; which in fact means that their presence in grain cannot be tolerated.

- data on additional information to honor a specific schedule of conditions : W1000, ΠμO, MT, PR, HM, Rad, BQ, EA, SQ, LQ, NV. For example the nutritional value (NV) will be measured only on grain intended for controlled food such as baby food or diets for sick people.

Our strategy to diagnose the grain quality is progressive and data driven. It can be split up into five stages :

1) To make a first causal diagnosis of grain quality from the data that are immediately available.

* ONIC : Office National Interprofessionnel des Céréales
ITCF : Institut Technique des Céréales et Fourrages

2) To identify the necessary additional tests from the data indicating additional tests to be carried out.
3) To check that there is no redhibitory defect from the data indicating redhibitory defects.
4) To confirm grain quality from the data of fast confirmation of quality.
5) If necessary, to confirm grain quality from the additional information data.

2.2. *Selection of storage operations*

We can distinguish nine different storage operations :

- *Airtight storage* : storage of grain under modified or confined atmosphere.
- *Apply pesticides* : application of any pesticide by : nebulisation, spraying or fumigation.
- *Clean* : any grain cleaning operation due to high impurity rate.
- *Cool* : cooling down ventilation or refrigeration of grain.
- *Dry* : drying of grain with heat supply.
- *Mix* : mixing grains of equivalent quality in order to improve their respective specific qualities Qipc or Qss
- *Move* : any mechanical turn over operation of grain.
- *Select bin* : selection of the most adapted bin for preservation of grain quality. In the selection of the bin comes the fact that it can be filled with grains from different origins, but of same quality.
- *Stabilize* : stabilization of grain with use of organic acid.

The role of each of these operations is to bring an answer to a specific need for improvement or preservation of grain quality and for fight or prevention against degradation agents (Table 3)

Examples :

1- to cool the grain favours good preservation of its quality and prevents degradations due to insects and mites. The rate of isomultiplication of insects and mites decreases in proportion with temperature, and the sprouting and oxidation processes slow down. The combination of these phenomenon allows an increase in the period of grain preservation (Lasseran and Fleurat-Lessard, 1991, 1994).

2- to mix a very damp wheat with a very dry wheat of equivalent specific weight improves the intrinsic & physico-chemical quality (Qipc) of grain and reduces the risk of degradation due to micro-organisms (ΠμO). The resulting grain will have a larger specific weight and *damp, average* or *dry* moisture content according to the proportions of the mixture .

Table 3. storage operations

Storage operations	Incidence on quality	Effects on degradation agents	Selection criteria
Airtight storage	preservation	prevention	Ins, Mit, μO, U, Δt
Apply pesticides	preservation	fight, prevention	Ins, Mit, PIns, PMit, U, Δt
Clean	preservation	fight, prevention	Qipc, ImpR
Cool	preservation	prevention	Qipc, Ins, Mit, U, Δt
Dry	improvement	prevention, fight	μO, Mit, MC, ΠμO
Mix	improvement	prevention	Qipc, Qss, MC, SW, HM, MT
Move	improvement	prevention, fight	μO, Mit, MC, PMit, ΠμO
Select bin	preservation	prevention	U, Δt
Stabilize	preservation	prevention	μO, U, Δt

We select the storage operations mainly in function with specific qualities (Qipc, Qss), degradation risks (Ins, Mit, μO) and the foreseen utilisation of grain (U, Δt). These main criteria are backed up by the secondary criteria which are the environment factors (ImpR, MC, SW) and the degradation agents (PIns, PMit, ΠμO). Our selection mechanism is causal, it is guided by the cause/effect relation linking the main criteria to the secondary criteria. It can be expressed by a rule of this type :

<u>If</u> grain specific quality (Qipc) is *very bad* <u>and</u> the MC is *very damp* <u>and</u> the SW is <u>average</u> or <u>high</u>, <u>then</u> you have to *Mix* grain with grain of equivalent quality but with *dry* <u>or</u> *very dry* MC <u>and</u> *high* SW.

Such a mixture will bring Qipc to *good* and MC to *average* or *damp* and SW to *high.*

We have just seen an example where the choice was made by selecting a storage operation. In the selection process we can also be led to exclude the selection of a storage operation. For example, if the corn is intended for human consumption, stabilization with an organic acid is excluded (it would leave a disagreeable taste).

2.3. Setting up of a storage technical route

A storage technical route is a succession of storage operations that should be undertaken in order to preserve grain quality. A storage operation is manageable only if it has been allocated the resources (equipment and consumables) necessary to its accomplishment. An equipment is a fitting such as : dryer, cooling aeration system, airtight storage system, etc.... A consumable is a pesticide or stabilizing product. The resources that are taken in account are the resources available at the storage silo.

Our strategy to set-up a technical route is as follows:

1) to allocate to each storage operation the equipment necessary to carry it out;
2) to order the manageable storage operations thus obtained in a plan ;
3) to allocate to each operation of this plan the products that are best adapted to its execution within the plan.

2.3.1. Allocation of equipment

The problem involved in the allocation of an equipment to a storage operation is the problem of checking whether the equipment exists, is available and is specifically adapted to the operation that should be carried out. For example, if you want to dry corn, you will first check if there is a dryer in the silo, then if it is available, and, last, if it is adapted to corn drying. Such an approach is easy to formalize in terms of preconditions to allocation rules (Wilkins, 1988).

2.3.2. Ordering of manageable storage operations

A manageable storage operation is caracterized by its process, its effects on grain, and the energy it consumes.

Examples :

1- To clean grain, you have to move it to the cleaning machine and to move it to the storage bin or to be weighed.

2- The *Dry* operation uses a dryer and its effects on grain are a decrease in grain moisture and an increase of its temperature.

3- The *Cool* operation uses a bin equipped with a cooling aeration or refrigeration system, and its effect on grain is a noticeable decrease of its temperature.

The example 1 shows that there exists a *teleological relation* (Sacerdoti, 1979) between *Move* and *Clean*. This teleological relation is due to the process of *Clean*. It also links *Move* to *Apply pesticide*, *Mix* and *Stabilize*. It implies that each of these operations is usually a double operation, except in some very precise cases. For example, Apply pesticide by fumigation does not need *Move*.

With examples 2 and 3 we can see that if you apply to grain a *Cool* operation followed by a *Dry* operation, the effects of the second operation would annul the effects of the first one. Therefore *Dry* should always precede *Cool*. The relation linking these two operations is a *relation of order*, determined by their effects on grain.

If we consider the operations *Dry* and *Cool*, one can precede the other and inversely. No relation of order links these two operations. They can be considered as equivalent as regard the order, and therefore linked by a *relation of equivalence*.

We set up the order of the storage operations by using the 3 relations, teleological, of order and of equivalence ; and by taking into account the fact that some equipments use a great amount of electricity. There are usually hours in the day during which the electricity costs less (off-peak periods). Operations such as *Dry* which use up a lot of electricity can be preferably carried out during the off-peak periods. This constraint due to optimisation of storage cost influences the order of the operations.

2.3.3. Allocation of the consumables

Consumable resources are products and therefore can be shared and still considered as available. The problem involved in the allocation of consumables is selecting the product which is the most adapted to the storage operation process, to the equipment to be used, to the quality of grain and eventually to the concerned degradation agents.

3. CONCLUSION

The development of expert systems for stored grain management is a very young field of research (around 10 years). These last five years have been marked by the coming of pest control systems guided by the fight against insects and mites already present in the stored grain. These systems proved themselves to be little efficient for grain quality preservation.

Preserving the quality of stored grain is a new approach to stored grain maintenance implying the planning of the storage technical route. The data present in the planning space are heterogeneous. The planning process can be broken down into three stages : diagnosis of quality, selection of the storage operations and ordering of the operations. We approach them respectively through a causal qualitative reasoning, inference rules, and a reasoning on resources.

This first approach, which is sufficient to build a specific system, has to be developed in order to achieve a planning model of the storage technical route. One of our main difficulties to-date lies in the existence of significant areas in the field of stored grain maintenance for which knowledge is lacking.

ACKNOWLEDGEMENTS

The experts on grain quality preservation who took part in this study are : Francis FLEURAT-LESSARD and Bernard CAHAGNIER.

Part of this project has been financially supported by the CONSEIL REGIONAL D'AQUITAINE.

REFERENCES

BECKETT S.J., LONGSTAFF B.C., EVANS D.E., (1994). A comparison of the demography of four major stored grain coleopteran pest species and its implications for pest management. *Proceedings 6th Int. Working Conf. Stored Product Protection, Canberra, April 94,* **1**, 4991-497. CAB International, Wallingford.

CAHAGNIER B., LESAGE B., RICHARD-MOLARD D., (1993). Mould growth and conidiation in cereal grains as affected by water activity and temperature. *Letters in Applied Microbiology,* **17**, 7-13.

FLEURAT-LESSARD F., ANDRIEU A.J., WILKIN D.R., (1994a). New trends in stored-grain infestation detection inside storage bins for permanent infestation risk monitoring. *Proceedings 6th Int. Working Conf. Stored Product Protection, Canberra, April 94,* **1**, 397-402. CAB International, Wallingford.

FLEURAT-LESSARD F., JUST D., BARRIEU P., LE TORC'H J.M., RAYMOND P., SAGLIO P., (1994b). Effect of modified atmosphere storage on wheat seed germination vigour and on physiological criteria of the ageing process. *Proceedings 6th Int. Working Conf. Stored Product Protection, Canberra, April 94,* **2**, 695-700. CAB International, Wallingford.

FLINN P.W., HAGSTRUM D.W., (1990). Stored Grain Advisor : a knowledge-based system for management of insect pests of stored grain. *AI Applications,* **4**, 44-52.

GUERRIN F., (1991). Qualitative reasoning about an ecological process : interpretation in Hydroecology. *Ecological Modelling,* **59**, 165-201.

JULIANO B.O., (1994). Concern for quality maintenance during storage of cereals and cereal products. *Proceedings 6th Int. Working Conf. Stored Product Protection, Canberra, April 94,* **2**, 663-665. CAB International, Wallingford.

LASSERAN J.C., FLEURAT-LESSARD F., (1991). Aeration of grain with ambient or artificially cooled air : a technique to control weevils in temperate climates. *Proceedings 5th Int. Working Conf. Stored Product Protection, Bordeaux, Sept. 90,* **2**, Fleurat-Lessard & Ducom Ed., 1221-1231.

LASSERAN J.C., FLEURAT-LESSARD F., (1994). Quality enhancement of stored grain by improved design and management of aeration. *Proceedings 6th Int. Working Conf. Stored Product Protection, Canberra, April 94,* **1**, 296-299. CAB International, Wallingford.

LONGSTAFF B.C., (1993). Expert System for Managing Grain Pests in Store. *Postharvest News and Informations,* **4**, 117N.

NDIAYE A., FLEURAT-LESSARD F., (1994). Research on an expert system for appropriate management of the quality of stored grain for food and feed processing. *Proceedings 94 Int. symp. and exhib. on quality of foodstuffs from cereal grains and oil seeds. Beijing, Nov. 94,* Xie Guifang & Ma Zhongdeng. Ed.,537-540.

PASQUAL G.M., MANSFIELD J., (1988). Development of a Prototype Expert System for Identification and Control of Insect Pests. *Computers and Electronics in Agriculture,* **2**, 263-276.

SACERDOTI E.D. (1979). Problem solving tactics. *Proceedings Int. Joint Conf. on Artificial Intelligence, Tokyo, 79,* 1077-1085.

STEYER J.P., QUEINNEC I., POURCIEL J.B., GOMA G., (1993). On the interest of a qualitative approach in biotechnological process modeling and control. *Procedings Artificial Intelligence for Agriculture and Food, Nîmes, Oct. 93,* 221-230.

WILKIN D.R., MUMFORD J.D., NORTON G., (1991). The role of expert systems in current and future grain protection. *Proceedings 5th Int. Working Conf. Stored Product Protection, Bordeaux, Sept. 90,* **3**, Fleurat-Lessard & Ducom Ed., 2039-3046.

WILKINS D.E. (1988). Practical planning : extending the classical AI planning paradigm. Morgan Kaufmann.

WRIGLEY C.W., GRAS P;W., BASON M.L., (1994). Maintenance of grain quality during storage - prediction of the conditions and period of 'safe' storage. *Proceedings 6th Int. Working Conf. Stored Product Protection, Canberra, April 94,* **2**, 666-670. CAB International, Wallingford.

VARIABILITY OF THE STRENGTH PROPERTIES OF WINTER RAPE STEMS AS AN EFFECT OF CONTROLLED STRATEGY IN PLANT GROWTH CONDITIONS

Grażyna Skubisz

Institute of Agrophysics, Polish Academy of Sciences, 20-236 Lublin, Poland

Abstract: The study determines the strength properties of winter rape stems of Jupiter variety, originating from plots of different sowing density values (20, 40, 60, 80 plants per m^2) on which a somewhat increased dosage of nitrogen fertilization - N=120 kg/ha - was applied. The strength properties were interpreted by means of mechanical parameters determined in the static test, in the process of bending (rigidity, maximum bending stress), and in the dynamic test, in the process of shearing (dynamic shearing energy). The study showed that, as a result of controlled strategy of rape cultivation, an enormous differentiation of rape stems was obtained, both from the viewpoint of their morphology and their strength properties.

Keywords: rape stems, lodging, rigidity, bending stress, shearing energy.

1. INTRODUCTION

Winter rape is one of the basic industrial plants of our climatic zone. It suffers from lodging, and this process, apart from lowering the crop yields, makes harvest difficult. The resistance of plants to lodging depends clearly on the mechanical properties of the stems of individual plants. For years now this fact has been the cause of close attention on the part of researchers, this attention being reflected in numerous studies on the mechanical properties of plant stems (Skubisz, 1982; Skubisz, *et al.*, 1989; Skubisz, 1991; Skubisz, 1993). Moreover, studies on the mechanical properties of stems revealed the hereditary character of features (Doliński, *et al.*, 1989; Jeżowski, *et al.*, 1988), and therefore provide valuable information for breeders of new varieties.

This study uses statistical analysis to determine stem rigidity and the maximum bending stress, and dynamic tests to determine the stem shearing energy. These parameters were used to characterize the strength properties of rape stems obtained as a result of controlled strategy in the conditions of plant growth.

1.1 Material and method.

The study was conducted on winter rape stems of the Jupiter variety, originating from a sowing density experiment. The stems were collected at the end of the blooming phase and in the phase of total silique filling.

The experiment was set up using varied agrotechnical measures, including an increased dose of nitrogen fertilizer of 120 kg/ha, conducive to plant lodging, and differentiated sowing density, of 20, 40, 60 and 80 plants per meter square. In order to obtain plots with the required number of plants per m^2, the plants were counted many times in spring, and excess plants were removed. This procedure was necessary, as during the three years of the experiment the same number of seeds sown did not constitute a replicable factor with relation to the number of plants per m^2, which means that a different number of plants survived the winter each year. To obtain a high accuracy and replicability of results, the plots were prepared in three replications. For studies, 25 plants were taken from each plot at a time. To provide a reference point for the results obtained, it was necessary to determine the mechanical properties of the stems of winter rape plants grown under standard conditions, plants which did not lodge (N dose was 60 kg/ha and the lodging index was 10).

Measurements were taken at a characteristic point on the stem, right at the first fork of the plant. The mechanical parameters were determined in static and dynamic tests. Statistical determinations concerned the maximum bending stress (σ_{max}), in the process of stem bending, and the rigidity (EI), by means of an Instron strength tester. Sample subjected to bending was freely supported at both ends and a force (load) was applied at the centre point between the two supports (Skubisz, 1991/1992; Skubisz, 1993). The results were recorded by means of a computer. Dynamic tests were used to determine the shearing energy (E_d) by means of a Dynstat apparatus. The area of stem cross section (S) was determined using a ΔT-areometer. Plant lodging was assessed by means of a 10-degree scale, where 10 means no lodging, and 1 means most advanced lodging (stem close to the ground).

1.2 Results.

In the study, as a result of the controlled strategy of winter rape cultivation, the author obtained differentiation of stems both with respect to their morphology and their strength characteristics (Figs. 1-4). All of the parameters under analysis showed considerable variability in the course of plant growth and development.

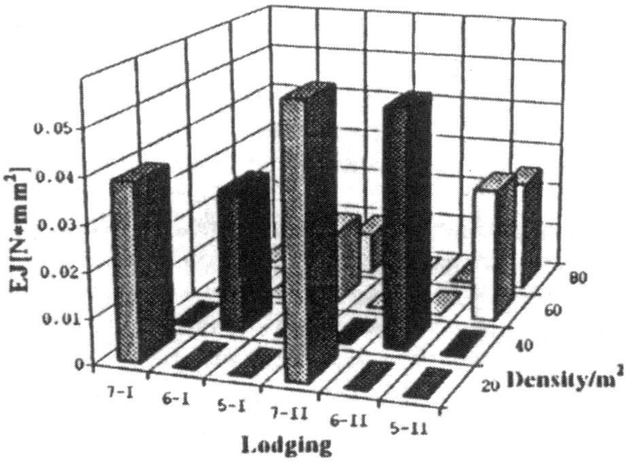

Fig. 1. Variability of the values of rigidity (EI) of winter rape of the Jupiter variety with relation to sowing density (number of plants per m^2) (I - end of blooming, II - total silique filling).

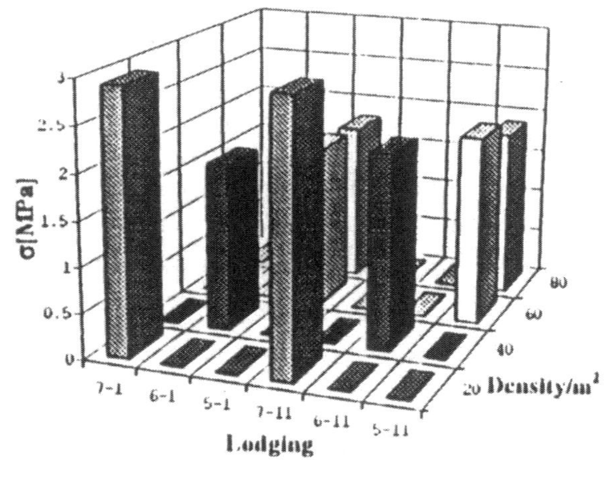

Fig. 2. Variability of the values of maximum bending stress (σ_{max}) of winter rape of the Jupiter variety with relation to the sowing density (number of plants per m^2) (I - end of blooming, II - total silique filling).

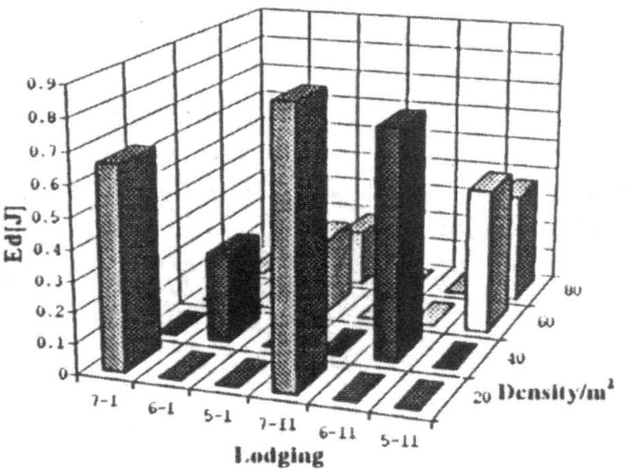

Fig. 3. Variability of the values of dynamic shearing energy (E_d) of winter rape of the Jupiter variety with relation to the sowing density (number of plants per m^2) (I - end of blooming, II - total silique filling).

There is no doubt that the application of differentiated agrotechnical measures resulted in the obtaining of rape stems with considerable variability of mechanical properties, caused by the widely varied area of stem cross section. The need for the application of differentiated agrotechnical measures was suggested by earlier studies, in which a wide spread of results concerning the values of the mechanical parameters was observed, caused by the differentiation of the dimensions of stems under study, as expressed by the stem cross section (Skubisz, *et al.*, 1989; Skubisz, 1993). In this study, the stem cross

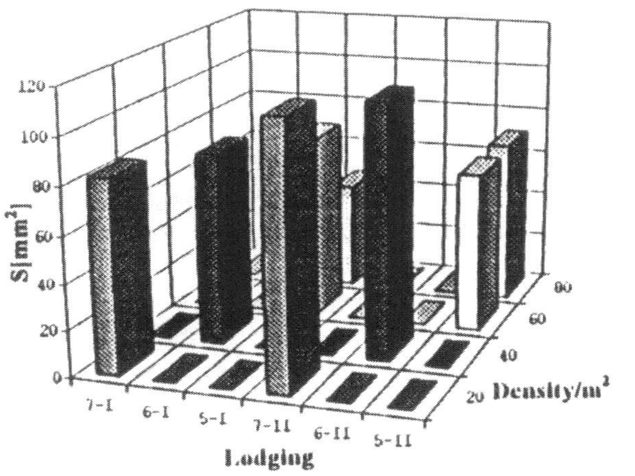

Fig. 4. Variability of the values of stem cross section area (S) of winter rape of the Jupiter variety with relation to the sowing density (number of plants per m^2) (I - end of blooming, II - total silique filling).

section areas varied from 46.1 to 82.7 mm^2 in the blooming phase, and from 69.1 to 112.1 mm^2 in the total silique filling phase, which corresponds to the values obtained in earlier studies.

The study showed that the maximum bending stress, rigidity, as well as the dynamic shearing energy, decreased in values with increasing number of plants per m^2, especially the mean values of maximum bending stress were high for the stems of plants originating from plots with 20 plants/m^2 as compared to the values of bending stress for the stems of plants growing more densely. This regularity was observed both in the phase of blooming and in the phase of total silique filling. Moreover, it was found that plants very resistant to lodging were characterized by very high values of the maximum bending stress (1.86 to 2.92 MPa in blooming, 2.16 to 2.98 MPa in total silique filling) as compared to plants susceptible to lodging (1.79-1.81 MPa in blooming, 1.88- 2.11 MPa in total silique filling), high values of rigidity EI (resistant plants: 0.0317-0.0386 Nmm^2 in blooming, 0.0524-0.0582 Nmm^2 in total silique filling; susceptible plants: 0.0094-0.0165 Nmm^2 in blooming. 0.0248-0.0294 Nmm^2 in total silique filling), and by high values of dynamic shearing energy E_d (resistant plants: 0.29-0.66 J in blooming. 0.75-0.89 J in total silique filling; susceptible plants: 0.18-0.26 Nmm^2 in blooming, 0.36-0.48 Nmm^2 in total silique filling). Comparing the results obtained with the results concerning the strength properties of stems of plants grown under standard conditions it was determined that the resistance of plants to lodging is strictly related with the strength characteristics of the plant stems. It was found that the mean values for plants grown under standard conditions were high and close to the mean values obtained for stems of plants growing fairly loosely, i.e. about 40 plants/m^2, when the nitrogen dosage conducive to lodging was 120 kg/ha. The mean values of the mechanical parameters under analysis for stems of winter rape grown under standard conditions were as follows: maximum bending strength (σ_{max}) - 1.84 MPa in blooming and 2.14 in total silique filling, rigidity (EI) - 0.0196 Nmm^2 in blooming and 0.0364 Nmm^2 in total silique filling, and dynamic shearing energy (E_d) - 0.28 J in blooming and 0,48 J in total silique filling.

The assessment, carried out within the scope of this study, of the degree of lodging showed that the most resistant to lodging were those plants which were characterized by high mean values of the mechanical parameters studied (index 7) and originated from plots with a low number of plants per m^2 (20 plants/m^2). The most strongly lodging (index 5) were those plants whose stems had both a low rigidity and low values of maximum bending stress and dynamic shearing energy, and grew on plots with high sowing density (60, 80 plants/m^2).

The study showed that as a result of controlled cultivation of winter rape plants a great differentiation of plant stems was obtained, both with respect to their geometrical dimensions and their strength properties. At the same time the study showed that the resistance of plants to lodging is closely related to the strength characteristics of rape stems. As a final result of the study, an enormously variable material was obtained - rape straw which can be used by industry. Thus, the results obtained, apart from characterizing the susceptibility of plants to lodging which lowers the crop yields, provide also information necessary in the processing of stems as an industrial material.

1.3 Conclusions.

The study showed that the application of controlled conditions of plant cultivation permits the control of the obtained range of variability of the strength properties of rape stems.

The study confirmed the relationship, suggested by earlier studies, between the variability of the mechanical properties of stems and their morphological structure as expressed by their cross section area.

A significant correlation was found between the mechanical properties of stems and the resistance of rape plants to lodging.

The study showed that those plants which were resistant to lodging were characterized by very high values of maximum bending stress, rigidity, and dynamic shearing energy, and originated from plots with low numbers of plants per m^2.

It was found that the mean values of the mechanical parameters decreased with increasing sowing density (number of plants/m^2), while their lodging degree deteriorated, which means that the resistance of the plants to lodging clearly diminished.

REFERENCES

Doliński R., Tarkowski Cz., Bichta J. (1989). Studies on the heredity of selected physical properties of winter wheat stalk. Zesz. Probl. Post. Nauk Roln., 383, 133-142.

Jeżowski S., Surma M., Adamski T. (1988). Diallel analysis of characters determining lodging resistance of barley (Hordeum vulgare L.) III. A genetic analysis of lodging grade and physical properties of the stem. Genet. Pol., 29, 275-280.

Skubisz G. (1982). Elasticity of cereal stalk (in Polish). Problemy Agrofizyki, 38.

Skubisz G., Tys J., Blahovec J. (1989). Mechanical properties of the stems of winter rape. International Agrophysics, 5(3-4), 205-220.

Skubisz G. (1991). The variability of mechanical properties of winter rape stems during plant vegetation period. 8th International Rapeseed Congress Saskatoon, Canada, Proceedings, vol. 6 of 6, 1795-1800.

Skubisz G. (1991/1992). Determination of the rigidity of winter rape stems (in Polish), Annales Universitatis Mariae Curie-Skłodowska, Lublin-Polonia, v. XLVI-XLVII, 38, section AAA, 1991/1992, 391-398.

Skubisz G. (1993). A method for the determination of rape stalk bending stress. Zesz. Probl. Post. Nauk Roln., 399, 213-217.

Skubisz G. (1993). Determination of the mechanical properties of winter rape stalks. Zesz. Probl. Post. Nauk Roln., 399, 219-225.

EXPERT SYSTEM TECHNOLOGY APPLICATION IN SEED PROCESSING

Peteris Rivzha and Irina Arkhipova

Institute of Informatics, Latvian University of Agriculture, 2 Liela street, Jelgava, Lv 3001, Latvia

Abstract: The work was aimed at realizing the problem of seed processing. The work offers to use the expert system technology which allows to solve problems connected with practical knowledge of experts in the region of investigation effectively. The method of knowledge structuring is worked out in the work. As the result applying the worked-out method the prototypes of the expert system (ES) are created:
- ES "Sieves" : research prototype for the sieve choice for the seed sorting machines,
- ES "Diagnostics" : displaying prototype for the technological determination of action disrepair of seed sorting machines produced by the company "Petkus".

Keywords: Decision trees, domain analysis, expert systems, knowledge acquisition, rule-based systems, rules.

1. INTRODUCTION

The processing of seeds after harvesting is one of the most labour-consuming seed production processes. It includes different technologically connected operations in which sorting and cleaning of seeds are the most important. Cleaning and sorting the seeds special attention is turned to the action quality of seed sorting machines which depends on that how the machine is ready to act and input action regime. To set action regime of seed sorting machines the operating staff uses the sieve choice tables. But these tables are formed for the standard conditions of seed humidity, measures and littering and therefore they allow to determine the necessary sieve cells' form and measures only approximately. Any deflection of seed showings from the standard demands modifications in the sieve cells' form and measures, otherwise the effect of segregating lowers. The user determines the optimal activity regime for the seed sorting machines basing on his experience and carrying out an experimental sorting (the so called method of experiments and errors). Obviously, because of the lack of experience of the operating staff, setting process can be long and this may cause baseless dissipation of the material.

2. METHOD OF STRUCTURING OF KNOWLEDGE

The research purpose - to work out ES prototypes which allow to choose sieves for the seed sorting machines produced by the company "Petkus" in all the technological line stages of the seed processing. Because of technological disrepair's of seed sorting machines the user often cannot achieve the optimal action regime of the machines irrespective of the chosen sieves. Then it is necessary to analyze the disrepair and to elicit the ways of their elimination.

Working out ES prototypes the knowledge acquisition about the domain is one of the main tasks. In the work the process of knowledge acquisition is regarded to be a process consisting of three stages: elicitation, analysis and processing. The elicitation is work directly connected with the knowledge acquisition from the expert. In the analysis' stage the process structure of problem solution is determined. The further interpretation of elicited data is in the processing stage. The method of knowledge structuring consisting of 5 stages in the processing stage has been worked out (Arkhipova and Rivzha, 1994).

2.1 The formation stage of the AND-OR tree of variants.

The expert must single out the objects that are essential for the given problem and give them conventional names. In turn, the totality of possible meanings must be defined for each object. Basing on these data the AND-OR tree of variants is formed. At the first level of the hierarchic AND-OR tree lie the peaks-objects, but at the second level - peaks-meanings. Among the second level peaks can be both peaks-meanings and peaks-objects in which the initial object is divided. Peaks can be AND-peaks that unite different objects in the domain or OR-peaks. In this case they contain alternative objects or their meanings characterizing individual features of each object.

2.2 The formation stage of the context tree.

At this stage the expert knowledge about domain structure and relations among its objects is necessary. The context tree can be interpreted as a succession of sub problems in which the initial problem is divided by the help of the method understood for the user. The peak of the context tree is the target of problem, meanings of which are to be determined in the decision process.

2.3 The formation stage of the decision tree.

The decision tree is formed based on the AND-OR tree of variants and the context tree. The following steps are performed in the formation of the decision tree:

- seeking for decisions in the AND-OR tree of variants among all the possible AND-OR trees stating all the admissible AND-trees. Admission is determined basing on expert knowledge about the peak coherence in a separate AND-tree,
- giving the expert - stated meaning to each admissible AND-tree which can be the decision of a problem. Each admissible AND-tree is formed in accordance with structure of the context tree. The meaning of the concrete problem decision peak - is added to each AND-tree,
- uniting all the subtrees with identical problem decision peaks-meanings, the decision tree is formed. The designed decision tree forms hierarchy of objects' peaks-meanings and determines the structure of the domain, in which the root peak is a problem target.

2.4 The formation stage of rules.

The formation of rules starts from the leaves of the decision tree. In the conditioned part of the rule the meaning of one peak-progeny is encoded in case their coherence is in OR-peak, or meanings of all peaks-progenies in case their coherence is in AND-peak. In the final part of the rule the meaning of the peak-progeny is encoded. The formulated rules form the knowledge basis.

2.5 The formation stage of the conclusion tree.

The conclusion tree reflects the application sequence of the elements of knowledge (rules). Peaks of the trees are the rules which are coherent. Coherency determines the sequence of the rule conclusion. The root peak is the rule which determines the target of the problem but its peaks-progenies - rules determine meanings of the decision (target) of the problem. In turn progenies of these rules are rules determining the meanings of the conditioned part - progeny, etc., until all the rules for formation of the conclusion tree at the last stage in the structuring of the field of objects are exhausted .

3. THE PLANNING OF ES "SIEVES" RESEARCH PROTOTYPE AND ES "DIAGNOSTICS" DISPLAYING PROTOTYPE

The worked - out method of knowledge structuring was used to plan the ES research prototype for sieve choice for seed sorting machines produced by the company "Petkus". The AND-OR tree of variants, the context tree (Fig. 1.), decision tree, conclusion tree were formed.The consultation performance starts with the user's demand on purpose of sorting. The user puts the necessary data into the machine. Further the system depending on the user's answer asks about:

-the seed type: small or big,

-the seed and admixture form: round (rv) or oblong (lv),

-ability to pour: easy or badly pouring,

-the seed measures: length(Kl), width(Kb), thickness(Kd),

-the nominal productivity of the machine (Qk%),

-the amount of the segregated mixture (Rk%),

and comes to conclusions about machine type, structure of sieve mechanism, the sizes and forms of all sieve cells. The result was formed as the ES prototype "Sieves" knowledge basis which contains 200 rules.

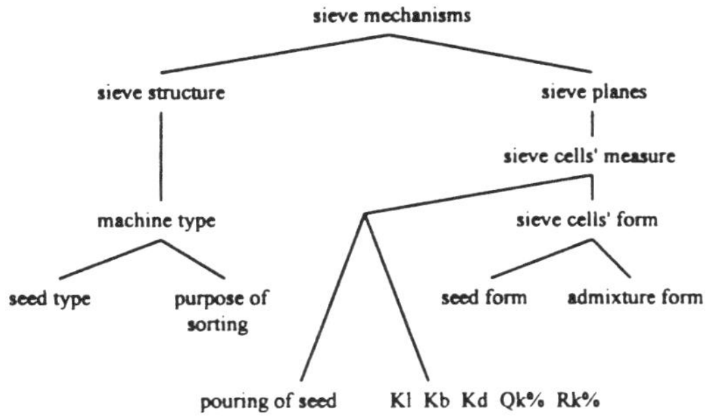

Fig. 1. The context tree of the sieve choice problem.

The ES for technological determination of seed sorting machines "Diagnostics" is logically connected with the ES "Sieves". If in the dialogue with the ES "Sieves" inexact data are input and the sieve scheme is formed from the given data, the sorted seed quality will be bad. The seed quality may be also bad because of the seed sorting machine imperfection. The ES "Diagnostics" must be used to find out the reason lowering the seed sorting quality.

All the possible sorting machine imperfections, reasons for their occurring and the ways of their eliminating were summarized to form the ES "Diagnostics". The reasons could be divided into several groups:

1) the constructive disrepair of the machine,

2) incorrect service of the machine,

3) unqualitative material for processing.

The consultation performance starts with the choice of the possible action disrepair of seed sorting machines. When the user determines the disrepair reason the system will give a method eliminating it.

4. CONCLUSION

The ES "Sieves" can be used for the user's consultations as to the choice of the structure of sieve mechanism and to help the user to cope with the methods of the choice of sieve mechanism. The ES "Diagnostics" helps to discover disrepair in seed cleaners and sorters and the ways of removing it. The ES "Diagnostics" to get acquainted with the possible kinds of disrepair in seed cleaners and sorters caused by constructive and assembly imperfection, wrong technical service and low seed quality.

REFERENCES

Arkhipova, I., and P. Rivzha (1994). Method of structuring of knowledge taking as an example the problem of seed material processing. In: *XII C.I.G.R. World Congress and AgENG'94 Conference on Agricultural Engineering,* Vol.2,pp.845-846. Milano.

Intelligent Monitoring for Quality of Fresh Citrus Fruit from Packing House to Supermarket

L. Quinn, S. Yaxley and J.A.G. Knight

De Montfort University
The Gateway
Leicester LE1 9BH
United Kingdom

The quality of fresh fruit is of vital importance, not only to the retailers and customers, but also to distributors, packing houses and producers. A system to help that the produce reaches the market place in the best possible condition has been developed, aiming specifically at citrus fruit. The developed systems incorporates and integrates sensors, databases and expert knowledge to help with every aspect of the citrus fruit life cycle.

Keywords: Quality control, Sensors, Data acquisition, Database systems, Monitored control systems

Introduction

Each year, vast quantities of citrus fruit are grown and harvested in the southern states of the European community and transported to states in the north. The importance of this produce to the economies of these southern states is great. Since the trade depends on the quantity and quality of the fruit, it is important that the maximum quantity of fruit at the highest possible quality arrives at its destination, the retailer. To improve the quality and quantity of fruit arriving at the point of sale, a research project commenced to achieve this through the use of information technology and novel sensors.

Citrus fruit has a number of stages in its life from

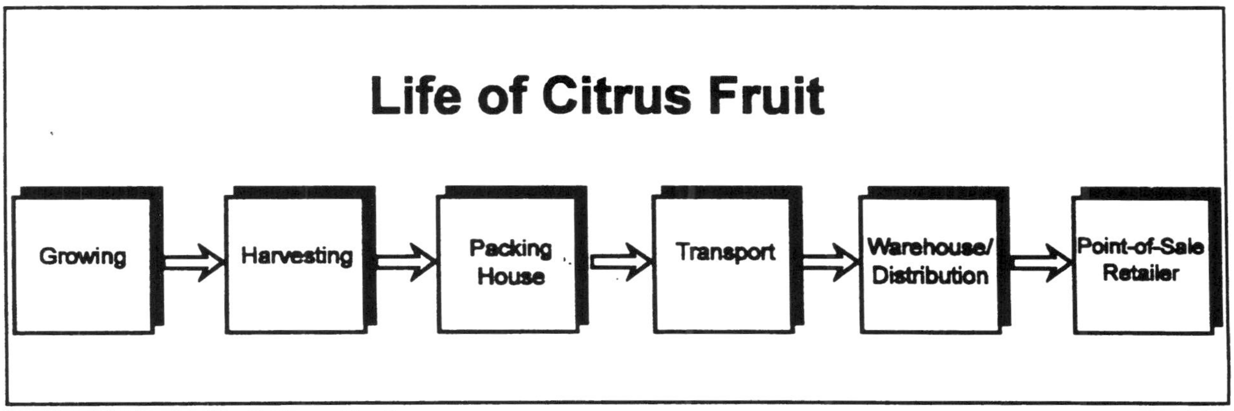

Figure 1 Life Cycle of Citrus Fruit

harvesting to the point of sale as shown in figure 1. The work undertaken was to implement life tracking from harvesting to point of sale similar to that in an manufacturing environment such as textiles, cars etc but with the further requirement of the need to monitor and determinate the rate of decay of the produce.

Aims of FRUIT ESPRIT 5379

The Fresh Fruit Produce Life Tracking System project or FRUIT was funded by the European Communities ESPRIT programme to investigate methods to monitor the quality of the citrus from harvesting to the point of sale allowing appropriate action to be taken where necessary.

A study was undertaken of the current practices within the citrus fruit industry concentrating on

- Production and consumption
- Growing, handling and treatment
- Transportation modes and practices
- Sales and consumption.

This revealed that the practices used in the citrus processing and transportation were far from state-of-the-art. Little monitoring of the processing and transportation is carried out at any stage and information that is collected is rarely passed through the supply link with no feedback to earlier stages. Frequently, the only information to be conveyed is origin, variety, name and grade. The main system users in the citrus life cycle were found to be

- Producers
- Packing Houses
- Transporters
- Importers/Wholesalers/retailers.

The research was undertaken in three main areas; developing the life tracking system, sensor development and development of the monitoring and analysis equipment.

Quality Requirements of Citrus Fruit

The quality requirements of citrus fruit is specified by governments throughout the world, in Europe, these regulations are set by the European Commission. The regulations cover such requirements as

1. External quality highlighting size, colour and exemption from defects.
2. Internal quality trying to obtain the grade of maturation and its palatability.
3. Presentation quality referring to packaging, decoration, marking and labelling.

In addition to the regulations set down by government, many retailers, particularly the large chain stores set further requirements specifying the fruit to be of a higher quality to that required by law. For example, in EC regulations for produce to be graded as class I, then no more than 10% of the fruit by count can fail the requirements, but a number of large retailers reduce this tolerance, some to 8% by count.

Life Tracking System

The Life Tracking System was developed for the four main areas of the life cycle of the fruit; Producers, Packing Houses, Transporters and the Importers/Wholesalers/Retailers. At each of these users, specific data to the fruit belongs. The majority of the data held at the individual sites was found to be only of importance or interest to that actual user, with previous or subsequent users in the life cycle only requiring a small portion of the data. Occasionally it was found that subsequent users in the life cycle would require further information on the produce.

The system developed was developed on a distributed database scheme, each site or user holding the entire information that added to the produce at the site with other users/sites having limited access to data further to that already received.

At the growers, information on aspects of the growing of the fruit was incorporated into the database. This information included climatical, physiological, irrigation. geographical, fertilisation and planting data. Data on the producer, harvest method and criteria and harvest conditions were also included.

Transportation was found to be an important aspect of the life cycle and had a great effect on the quality of the fruit through damage. Information on the transportation of the fruit from the harvest to packing houses was also included due to its importance holding travel/loading/unloading time, type of transport and container information.

The database at the packing house was designed to hold information on which operations took place and when, environmental conditions and amount of wax, cleaning foam used.

Finally the transportation stage to the retailer held data on loading/unloading times, destination, journey times, route schedule and environmental conditions that were present during the stage. The environmental conditions were collected using sensors that were present on each vehicle and down loaded to the database on arrival at the destination (FRUIT Deliverable 2.1.3).

Sensor Development

A detailed study of defective fruit found that particular gases were emitted from the orange when a particular defect was present.

The sensor research was divided into two categories, chemical and biochemical.

The biochemical sensors operate when the volatile emissions react with enzyme. These types of sensors are highly specific but can be slow to react. To produce an reaction, the sensors need to kept at high temperature which is the opposite to that required in the transportation of fruit.

A more robust sensor than biochemical are chemical sensors. The sensor require an electrochemical reaction on a specially prepared surface. The negative side to these sensors are that they are less specific than biochemical sensors.

Work on the sensors is continuing and further sensors are being developed for use in the meat industry to detect defects in the curing of ham (ESPRIT Project EP 8095).

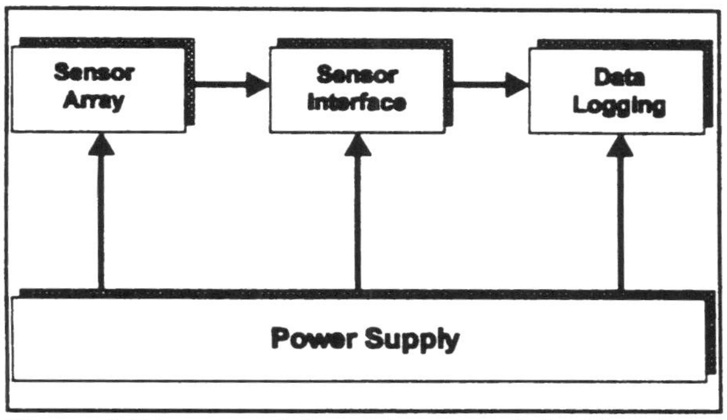

Figure 2 Diagram of the main parts of the 'Fruit Box'

Monitoring and Analysis Equipment

The monitoring and analysis equipment was developed for two different stages; the transportation stage between packing house and retailer and for hand held use at various stages in the life cycle.

The transportation data-logger was developed to log information such temperature and humidity when the fruit is being transported as these parameters are important factor related to the quality of the fruit. The sampling rate of the environment could be varied as the journey progressed.

A 'fruit box' system was developed to monitor the gases released and collect information on temperature and humidity. The box consists of three main parts as shown in figure 2 together with communications to a Pc via serial 232 link.

The device used Taguchi TGS 800 commercially available sensors. Two channels were used for the senors, one for limonene and the other for alcohol/ethanol emissions. Temperature and humidity and were included on two further channels.

Fruit Sniffer

Figure 3 Picture of 'Fruit Sniffer'

The fruit sniffer is a portable hand-held device capable of analysing the gases released from the fruit and determining whether the fruit is good, bad or borderline. A red light represents bad, green light, good fruit and yellow light represents borderline. The device used sensors to detect limonene and alcohol\ethanol and housed in a hand-held car vacuum cleaner as seen in figure 3.

The sniffer sucks air over the sensors and at the same time eliminating any gases that could be in the housing from previous tests. The output of the sensors is proportional to the amount of that particular gas found. The sniffer determined good, bad and borderline produce by setting thresholds for the output of the sensors.

The sniffer can decide whether or not there is a problem in a batch of fruit but cannot give a detailed analysis of the problem. In practical situation this would not be sufficient as regulations regarding citrus fruit and standards set down by the retailers dictate the number or percentage that can be defective. Therefore, the batch would have to manually analysed.

Further work would have to be undertaken to enhance the sniffer to operate effectively in a real situation. One major element that must be developed is how the analysis is carried out. Currently result is determined by the voltage output from the sensor, eg a voltage greater than 1 volt represents bad fruit. It is obvious, for example, that a large batch of fruit with one percentage of defective fruit may have a larger concentration of the gas being detected than say a small batch of fruit with fifteen per cent of defective fruit.

The sniffer has created a great deal of interest from the retailers, importers and distributors of citrus fruit since it would allow fast checks on all fruit arriving at their premises. Work will continue on the sniffer to produce a device which is more capable of working in real environment.

Overall Operation

The overall operation of the system was envisaged that each of the elements described earlier would be capable of communicating between each other giving maximum information and decision support to the users. It was capable of being implemented in stages and each element of the total system could operate independently from the other elements, thus giving greater versatility and marketing potential.

Using past data, an expert system was developed to predict shelf life of the fruit at certain points in the fruit's life cycle. As was already briefly discussed, particular gases are emitted when certain changes are taking place in the fruit, eg decay, mould, etc and these gases are detected by the sensors and give a reading corresponding to the type of gas and its concentration. This information is used by the expert system and the remaining shelf life can be determined. The implication of such a system can clearly be seen. If a consignment of fruit during transportation was deemed to have only 1 weeks shelf life left and transportation was to take another 3 days, then it would be sensible to reroute the consignment to a destination closer thus maximum the shelf life at the end user. The advantage of such a system to relevant to not only fruit but to all perishable products.

Conclusion

The FRUIT project was about implementing information technology in non-traditional sectors for use of IT. The project highlighted the need and demonstrated the advantages of using IT in the processing and transportation of fresh produce. Computer technology is used in packing houses for every day business tasks such as finance but not in processing of the produce.

Retailers and distributors have seen the need for the implementation of computer technology into their operations. Bar coding systems are common place in such operations. It is clear as the end user becomes more computer orientated, the supplier will also have to become computer orientated as seen in traditional manufacturing.

When sensors are developed that can detect emissions from a number of defects in numerous food stuffs, then the production of food will change a great deal. The automation of numerous manual quality checking tasks will become common place.

The use of hand held quality analysing devices is an exciting avenue to pursue and will revolutionise both the way produce is checked by retailers and distributors but also shoppers themselves. Today, quality and value for money are increasingly important issues for the shopper. It is envisaged that soon devices will be available, that are carried by the shopper around supermarkets allowing the shopper to check the quality of the produce for themselves. Certainly, the fruit sniffer went some way of demonstrating this and bringing this day closer.

The importance of IT in the production and handling of fresh produce will increase in the future. The collection of data as described earlier will set down standards for produce which would of previously been impossible due to the nature of some parameters.

References

ESPRIT Project E 8095, Ham Automated Monitoring System (HAMS), Funded by European Commission under the ESPRIT program.

ESPRIT Project 5379, Deliverable 2.1.3, System Specification Access and Security Requirements, November 1991.

Knight, J., Savage, B.M, Emmanoulopoulos, G, and

Puig Gomez, J, Fresh Fruit Product Life Tracking, Proceedings of IFAC workshop on Expert Systems in Agriculture, 134-139, 1992.

Long, J., Yaxley, S. and Karia, N., Detection Devices for Citrus Fruit Quality States, Proceedings of the t International Congress on Condition Monitoring and Diagnostic Engineering Management, 140-145, 1993.

Hoddell, S.E.J, Long, J.L, and Savage, B.M., A Quality Monitoring System for Fruit, Proceedings of the t International Congress on Condition Monitoring and Diagnostic Engineering Management, 134-139, 1993.

MECHANICAL RESISTANCE OF STORAGE APPLE*

Bohdan Dobrzański, Rafał Rybczyński, Adam Kuczyński

Institute of Agrophysics, Polish Academy of Sciences
ul. Doświadczalna 4, 20-236 Lublin, Poland

Abstract: Apple's resistance to bruising and the potential for good storability are related to its firmness. Although it is well known that bruising results from excessive force on the fruit surface, it is still not clear which factors determine the differences in susceptibility of fruit to given force. Four apples varieties: Gloster, Idared, Jonagold and Red Boskop held in cold storage at temperature 0-2°C, 6-8°C and 14-16°C were subjected every 5 weeks to mechanical tests. The firmness tests have been performed using a puncture test of apple, a compression test of cylindrical specimen of flesh and a tension test of skin.

Keywords: mechanical properties, damage, force, elasticity, apple, tissues, storage, temperature.

1. INTRODUCTION

Apple quality at purchase and consumption is dependent on the degree of ripeness and the absence of mechanical damage and decay. The mechanical resistance of fruit generally decreased with storage time and maturity. Mechanization of horticulture production has subjected fruits to situations which often cause mechanical damage, mainly bruising.

Fruit bruises are caused by three types of loading: compression, impact, and vibration (Brusewitz and Bartsch, 1989). Bruise damage usually results when an apple impacts either a hard surface or another fruit, and is a major cause of quality loss for fresh market and storage apples (Siyami, et al., 1988; Schoorl and Holt, 1978; Holt and Schoorl, 1984). Although it is well known that bruising results from excessive force on the fruit surface, it is still not clear which factors determine the differences in susceptibility of fruit to given force (Topping and Luton, 1986).

The apple,s resistance to bruising and the potential for good storability are related to its firmness. The conventional firmness tests have been done primarily using a puncture test with a Magness-Taylor probe and the compressive stress was used as an indicator of the fruit firmness. Firmness measurements in fruit have been used for many years as a guide to the quality of the product. Firmness meters attempt to record a value which represents how easily the product can be deformed under a pressure applied to a limited area of its surface (Studman and Boyd, 1994). Over the past 60 years a range of penetrometer devices have been developed in various parts of the world, and they range from laboratory systems, to simple hand held devices, which can be used in the field (Studman and Boyd, 1994). The devices have been applied to a wide range of fruits and are often the main test specified to establish the acceptability of the product for a particular market or storage condition.

The penetration test is a destructive test, however, it has a number of practical benefits. The success of the penetrometer, despite its problems, indicates that there remains a need for a test which is fast, and inexpensive, and yet which gives an improved assessment of product properties.

Non-destructive versions of the penetrometer test have been described, but as yet none have found widespread approvall (Duprat et al., 1994).

The attempts have been made by several researchers to provide mechanical characteristics of the apple,s skin (Clevenger and Hamann, 1968; Rybczyński et al., 1994; Rybczyński and Dobrzański, 1994a; Szot et al., 1989) and the modulus of elasticity has been determined from the force-deformation curves, which

* This work was supported by the State Committee for Scientific Research, Poland under the Grant No. PBZ-51-02.

result from subjecting the material samples to quasi-static loading tests.

(Chappell and Hamann, 1968; Szot, et al., 1989; Dobrzański and Rybczyński, 1994; Rybczyński and Dobrzański, 1994a,b) studied the mechanical resistance of apple's flesh.

The problem is always to correlate the reading from the device with an appropriate quality parameter such as bruising resistance or texture determined by taste panels. In general the correlations for most devices have been quite poor. However, better results have been obtained if more than one parameter is measured, and correlations determined after principle component analysis (Studman, 1994).

This study was part of a more extensive project concerned with physical characteristics of fruits evaluating their quality.

The objectives were to determine the elastic behavior of storage apple and to quantify the effect of temperature, duration of storage and variety on the study material mechanical resistance variability.

2. MATERIAL AND METHOD

Apples of the following varieties: Gloster, Idared, Jonagold and Red Boskoop for this study were hand-picked from the Lublin Agriculture University orchard. The same size fruits (7-8 cm diameter) for experiments, at harvest ripeness, were sorted and held in cardboards in the refrigerated storage at temperature 0-2°C, 6-8°C and 14-16°C. The apples were removed from storage every 5 weeks, and stabilized at room temperature before experiments.

The maximum force, the deformation and work deformation caused damage of the apple tissue during various static tests were noticed. The values for modulus of elasticity (an indicator of apple mechanical resistance) have been determined from the force-deformation curves which resulted from subjecting the material samples to quasi-static, loading tests.

The flesh samples, cut out perpendicularly to the stem calyx axis of the fruit, had a cylindrical shape with both diameter and length of 13mm. Four samples were cut out from each apple. Thirty samples were used as one combination.

The flesh cylinders were loaded with the Instron 6022 apparatus at 5 mm/min speed of crosshead. The samples were compressed between two parallel plates to a point of a rapid decrease of the reaction force. The energy, force and deformation at the point of damage of the apple flesh were noticed. Forces-caused deformations: 0.86 and 1,73mm were used to calculate the modulus of elasticity using following formula:

$$E = \frac{(F_2 - F_1)}{A\,(e_2 - e_1)} \qquad (1)$$

where:

E - modulus of elasticity [MPa],

F_1 - force causing deformation 0.86mm,

F_2 - force causing deformation 1,73mm,

A - cross-sectional area [mm^2],

e_1 - relative strain for the deformation 0.86mm,

e_2 - relative strain for the deformation 1,73mm,

Five skin belts (2mm width and 0.2-0.3mm thick) were cut out from each apple for the tension test. The skin belts were placed in a special holder fixed to cross-head. Tension tests were performed using Instron apparatus at the same rate (5mm/minute) of cross-head move.

The maximum force, the deformation and work deformation were also noticed, however, in this case forces-caused deformation: 0.4 and 1mm were used to calculate the modulus of elasticity. The values of force and deformation for the calculation were chosen from the straight-line part of the curve.

The penetration test with a 6mm diameter cylindrical plunger was also performed at the same speed rate of cross-head move. All values such as force, deformation and work deformation connected to the fruit firmness were recorded at the point when the penetrometer squeezed in flesh after the skin damage. Only values corresponding to the modulus of elasticity were determined at a set point before skin damage from the elastic range of deformation.

All routine calculations (average value, standard deviation and coefficient of variance) were made.

3. RESULTS

3.1. Tension test

Mean values of tensile strength (F) at the break of apple skin for Jonagold variety are shown in Figure 1.

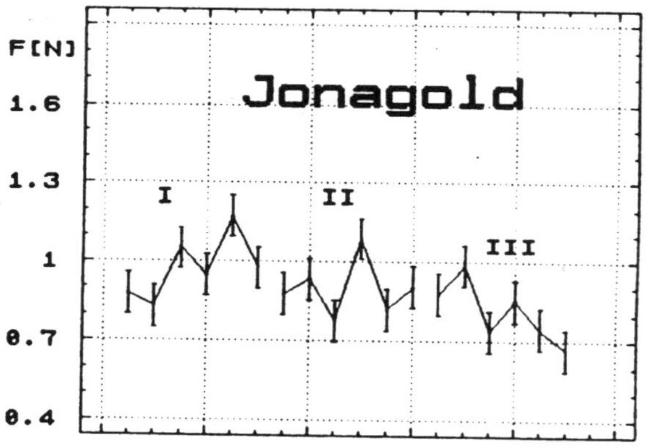

Fig. 1. Tensile strength (F) at break of the apple skin for Jonagold variety storage 30 weeks (6 time intervals) at temperature: 14÷16°C-I, 6÷8°C-II and 0÷2°C-III.

In all cases the time intervals and the temperature of storage made not significant difference to tensile strength (F), deformation (D) and work deformation (Wd) at break at the 5-percent level. However, higher (growing) deformation tendency after the storage is presented in Figure 2. The modulus of elasticity was the best indicator of skin strength results at tension

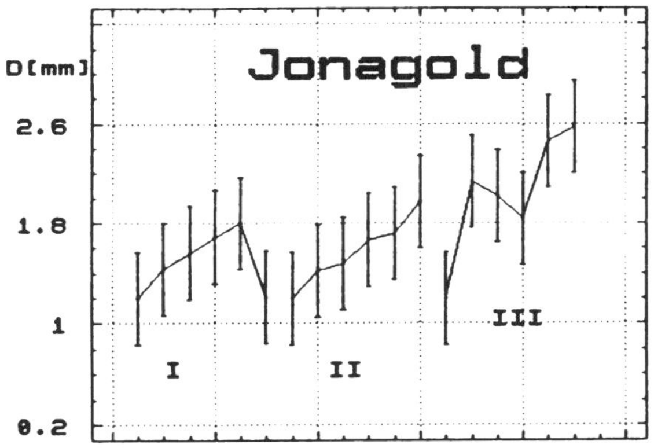

Fig. 2. Tensile deformation (D) at break of the apple skin for Jonagold variety storage 30 weeks (6 time intervals) at temperature: 14+16°C-I, 6+8°C-II and 0+2°C-III.

test. Figure 3 shows the highest values of modulus of elasticity (17.55 MPa) after 25 weeks of storage at temperature 14-16°C. In both cases during storage at

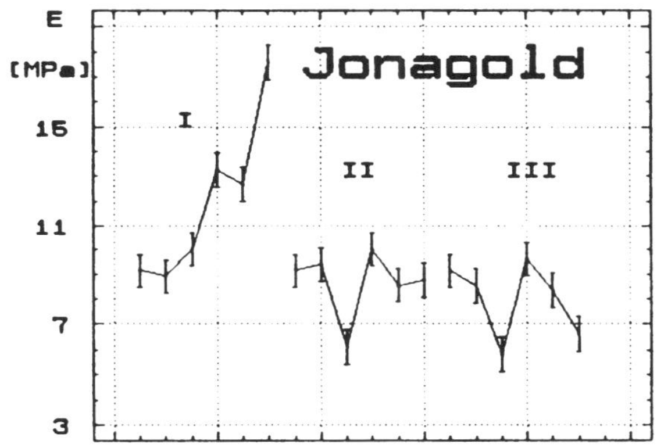

Fig. 3. Modulus of elasticity (E) of apple skin for Jonagold variety storage 30 weeks (6 time intervals) at temperature: 14+16°C-I, 6+8°C-II and 0+2°C-III.

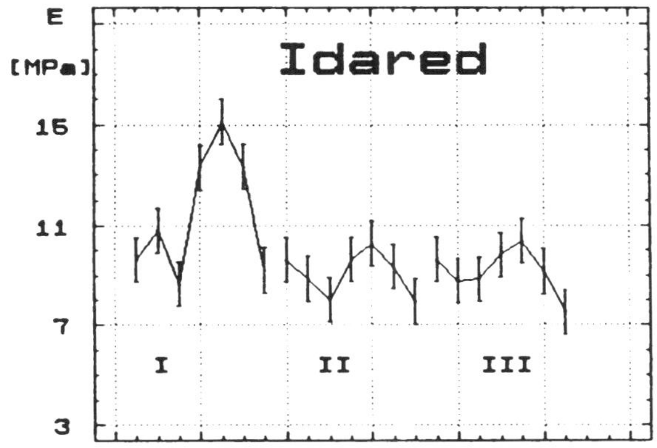

Fig. 4. Modulus of elasticity (E) of apple skin for Idared variety storage 30 weeks (6 time intervals) at temperature: 14+16°C-I, 6+8°C-II and 0+2°C-III.

temperature 0-2°C and 6-8°C modulus of elasticity of Jonagold apple skin reached values of the range (5.80-9.61MPa). A similar tendency for Idared variety (Figure 4) was observed. The modulus of elasticity of apple skin reached values of the range (7.54-10.38MPa) at storage temperature 0-2°C and 6-8°C. The highest values (15.11MPa) were obtained for a more dry skin of apple storage at temperature 14-16°C. The values determined for modulus of elasticity of apple skin of Gloster variety showed that skin keep its firmness only at cool storage at temperature 0-2°C.

3.2. Compression test

The mean values of compression force (F) at flesh damage for Jonagold variety are shown in Figure 5.

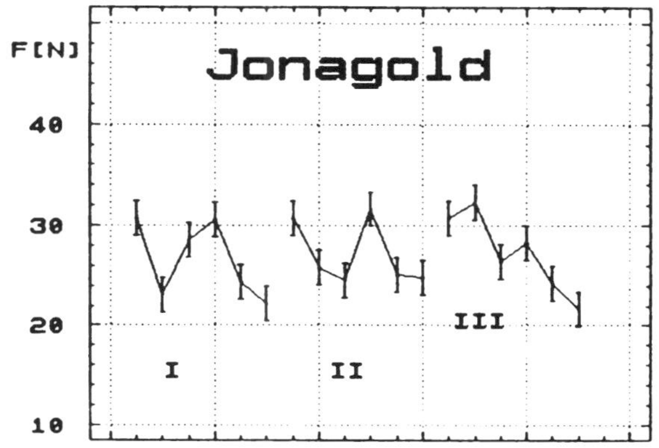

Fig. 5. Compression strength (F) of apple flesh for Jonagold variety storage 30 weeks (6 time intervals) at temperature: 14+16°C-I, 6+8°C-II and 0+2°C-III.

In all cases the time intervals and the temperature of storage had not significant difference on compression strength (F). The range (21.68-32.26N) covered all values reached in the compression test. A slight fall of modulus of elasticity during storage was observed.

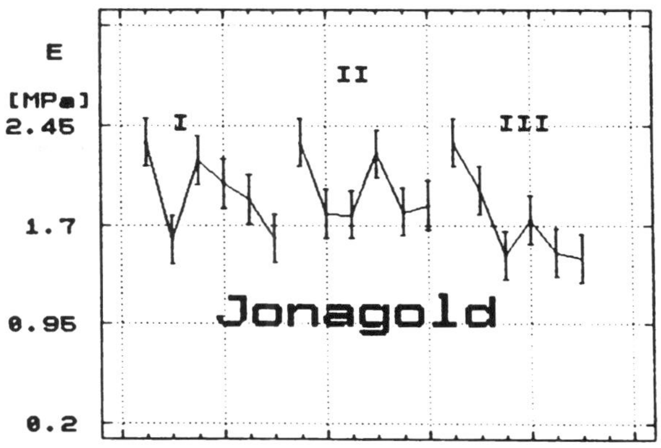

Fig. 6. Modulus of elasticity (E) of apple flesh for Jonagold variety storage 30 weeks (6 time intervals) at temperature: 14+16°C-I, 6+8°C-II and 0+2°C-III.

The average values of modulus of elasticity (E) of the flesh for Jonagold variety determined during compression test are presented on Figure 6. There were not significant (p=5%) differences between the tempera-

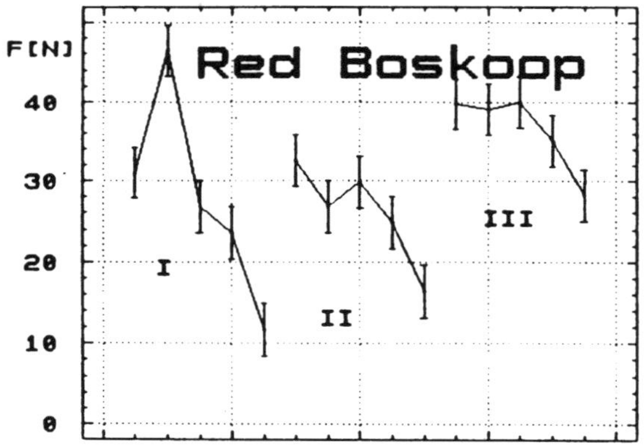

Fig. 7. Compression strength (F) of apple flesh for Red Boskoop variety storage 30 weeks (6 time intervals) at temperature: 14+16°C-I, 6+8°C-II and 0+2°C-III.

ture of storage and the highest value of the modulus of elasticity (2.32MPa), which were obtained after harvest and the lowest (1.45MPa) after 30 weeks of storage.

The values obtained in mechanical tests for other studied varieties showed more distinctly the effect of storage conditions on fruit firmness. The force registered for Red Boskoop variety (Figure 7) fall from 46.51N (10 weeks storage) to 11.70N (30 weeks storage) at 14-16°C and from 39.68MPa (5 weeks storage) to 28.21N (30 weeks storage) at 0-2°C. Similar differences were observed for the modulus of elasticity of cylindrical flesh of Red Boskoop variety. The values fell down from 2.69MPa to 0.50MPa at 14-16°C and from 2.93MPa to 2.19MPa at 0-2°C (Figure 8).

The force obtained for Idared variety fell down from 38.11N after harvest to 28.75N, 23.96N and 22.60N after 30 weeks storage at 0-2°C, 6-8°C and 14-16°C respectively (Figure 9).

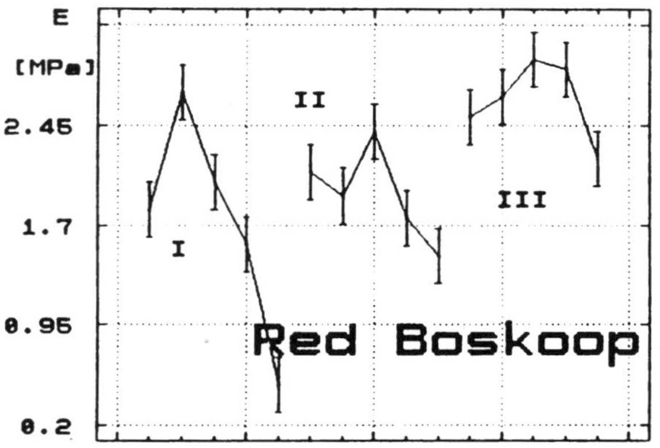

Fig. 8. Modulus of elasticity (E) of apple flesh for Red Boskoop variety storage 30 weeks (6 time intervals) at temperature: 14+16°C-I, 6+8°C-II and 0+2°C-III.

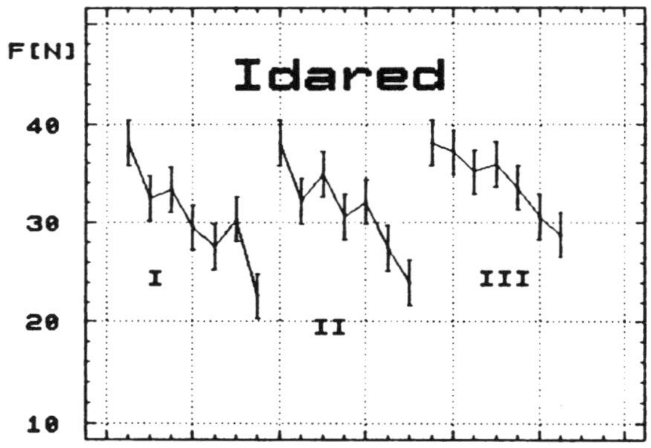

Fig. 9. Compression strength (F) of apple flesh for Idared variety storage 35 weeks (7 time intervals) at temperature: 14+16°C-I, 6+8°C-II and 0+2°C-III.

The force observed for Gloster variety fell down from 46.25N after harvest to 31.03N, 19.03N and 16.59N after 30 weeks storage at 0-2°C, 6-8°C and 14-16°C respectively (Figure 10).

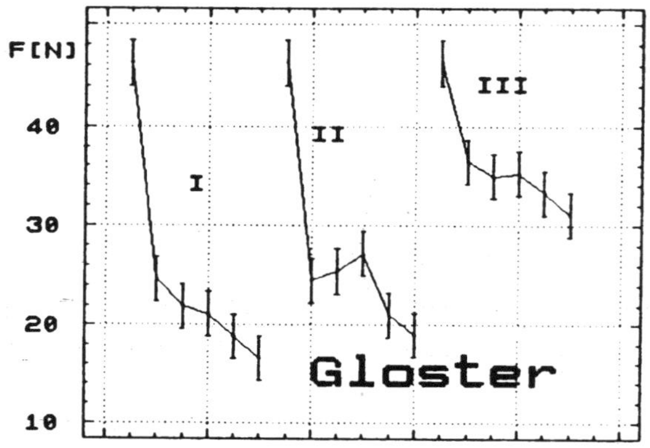

Fig. 10. Compression strength (F) of apple flesh for Gloster variety storage 30 weeks (6 time intervals) at temperature: 14+16°C-I, 6+8°C-II and 0+2°C-III.

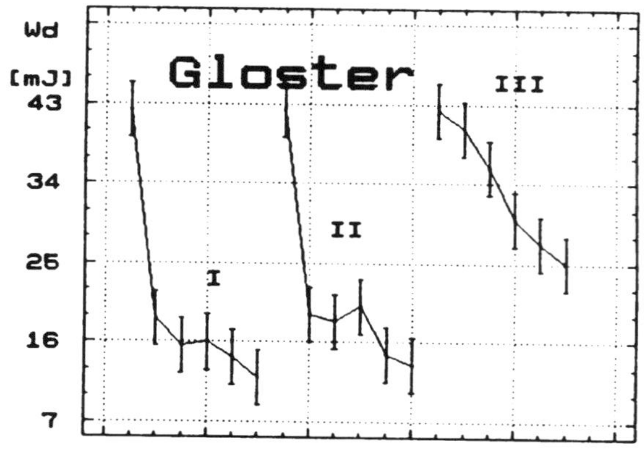

Fig. 11. Work deformation (Wd) of apple flesh for Gloster variety storage 30 weeks (6 time intervals) at temperature: 14+16°C-I, 6+8°C-II and 0+2°C-III.

The firmness after storage was low and the values of work deformation (Figure 11) reached after 30 weeks 13.40mJ at 6-8°C and 11.95mJ at 14-16°C. A higher value of work deformation 25.02mJ was obtained for apple flesh of Gloster variety storage at 0-2°C.

3.3. Penetration test

The average values of force (F) at damage received in the penetration test of apple for Jonagold variety are presented on Figure 12. After harvest, the average force was equal at 24.68N. After 5 weeks, the squeez-

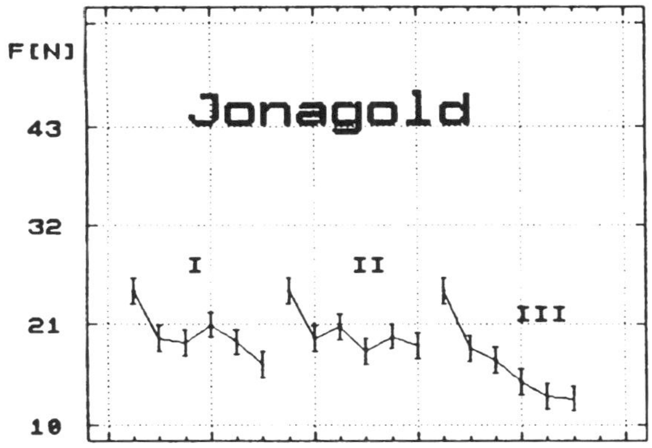

Fig. 12. Compression strength (F) at penetration test of apple for Jonagold variety storage 30 weeks (6 time intervals) at temperature: 14+16°C-I, 6+8°C-II and 0+2°C-III.

ing force fell down up to 18.72N after 30 weeks of storage at 6-8°C and up to 16.67N at 14-16°C. In both cases the forces were at similar level at all storage periods. However, not clearly understood lowest value of 12.88N was obtained for apple storage after 30 weeks at 0-2°C.

Figure 13 shows the deformation of apple of Jonagold variety storage at different conditions. The duration of storage had statistically significant effect (p=5%) after each tested term on the deformation of apple

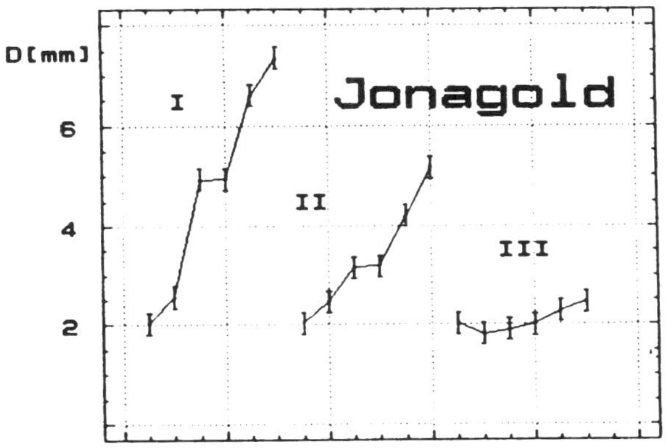

Fig. 13. Deformation (D) at penetration test of apple for Jonagold variety storage 30 weeks (6 time intervals) at temperature: 14+16°C-I, 6+8°C-II and 0+2°C-III.

storage at 6-8°C and 14-16°C. The lowest deformation 2.02mm was obtained after harvest. After 30 weeks the deformation was two times higher (5.17mm) for storage apple at 6-8°C and more than three times higher at 14-16°C (7.35mm).

However, values related to the elasticity, determined with penetration test for Jonagold variety (Figure 14),

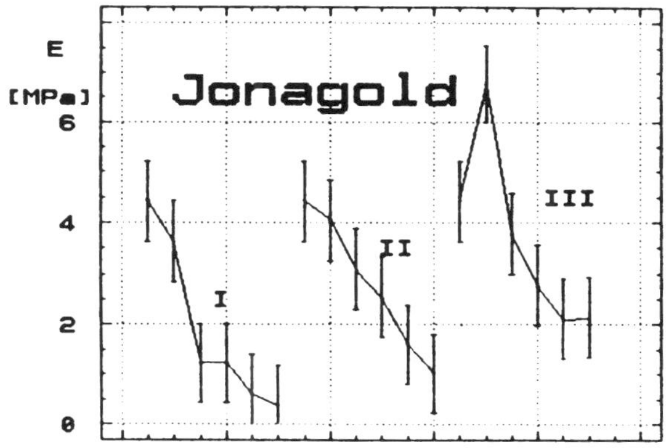

Fig. 14. Elasticity (E) at penetration test of apple for Jonagold variety storage 30 weeks (6 time intervals) at temperature: 14+16°C-I, 6+8°C-II and 0+2°C-III.

showed that firmness had fallen down at storage at all studied temperatures. The elasticity of harvested apples was at 4.42MPa. After 30 weeks the elasticity was 2.10MPa, 1.02MPa and 0.37MPa at 0-2°C, 6-8°C and 14-16°C respectively for storage apples of Jonagold variety.

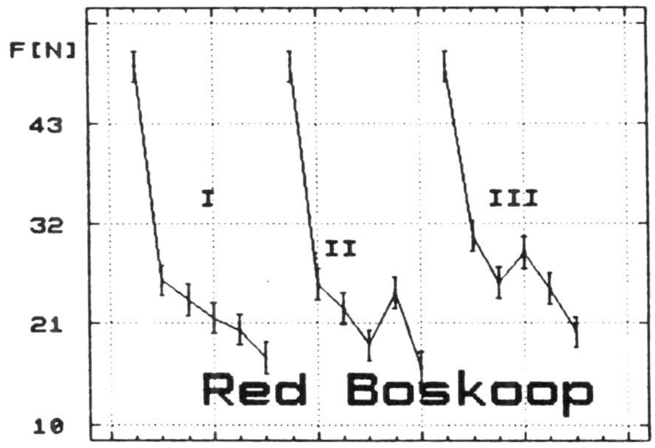

Fig. 15. Compression strength (F) at penetration test of apple for Red Boskoop variety storage 30 weeks (6 time intervals) at temperature: 14+16°C-I, 6+8°C-II and 0+2°C-III.

Red Boskoop variety (Figure 15) was stronger than Jonagold (24.68N) and after harvest force observed at the penetration test reached 49.23N. After 5 weeks storage at all temperatures, the force rapidly fell down up to 17.23N, 16.23 and 20.06N at 14-16°C, 6-8°C and 0-2°C respectively (after 30 weeks storage).

Different characteristics of storage apple of Idared variety are presented on Figure 16. The force

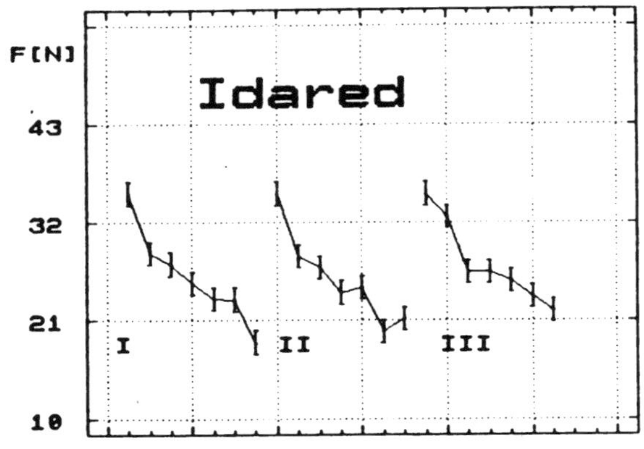

Fig. 16. Compression strength (F) at penetration test of apple for Idared variety storage 35 weeks (7 time intervals) at temperature: 14+16°C-I, 6+8°C-II and 0+2°C-III.

continually fell down during storage at all temperatures from 35.24N to: 18.01N (14-16°C), 21.23N (6-8°C) and 22.13N (0-2°C). A similar tendency of the deformation of Idared variety presented Jonagold variety as well. The duration of storage had statistically significant effect (p=5%) after each tested term on the deformation of apple (Idared variety) storage at 6-8°C and 14-16°C. The lowest deformation 2.34mm was obtained after harvest. During 30 weeks of storage at 0-2°C the deformation was at range from 2.34 to 2.82mm.

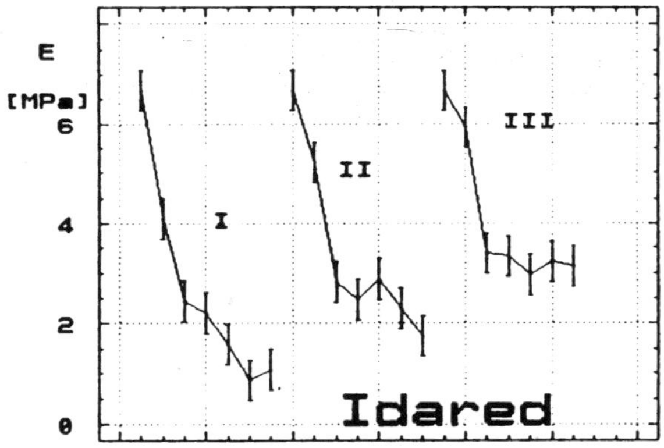

Fig. 17. Elasticity (E) at penetration test of apple for Idared variety storage 35 weeks (7 time intervals) at temperature: 14+16°C-I, 6+8°C-II and 0+2°C-III.

The values related to elasticity determined with the penetration test for Idared variety (Figure 17) showed that firmness had fallen down during storage at all studied temperatures. The elasticity (6.68MPa) of Idared variety after harvest was higher than for Jonagold (4.42MPa). After 30 weeks the elasticity was 3.14MPa, 1.74MPa and 1.07MPa at 0-2°C, 6-8°C and 14-16°C respectively for storage apples of Idared variety.
A similar to Idared variety results of the penetration force were observed for Gloster variety. After harvest 36.33N of force was noticed. After 30 weeks of storage force covered values in the range 12.7-18.4N. The deformation of Gloster variety was on a similar level to other studied varieties at all storage temperatures.
The duration of storage had highest statistically significant effect (p=5%) on the modulus of elasticity of

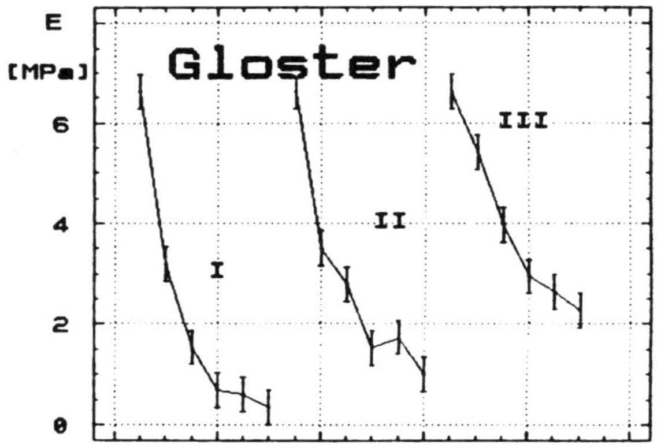

Fig. 18. Elasticity (E) at penetration test of apple for Gloster variety storage 30 weeks (6 time intervals) at temperature: 14+16°C-I, 6+8°C-II and 0+2°C-III.

Gloster variety. Figure 18 shows the differentiation of these values obtained for six storage periods, ranging from 6.62MPa (maximum after harvest) to 0.61MPa, 1.01MPa and 2.26MPa at 14-16°C, 6-8°C and 0-2°C respectively (minimum after 30 weeks of storage).

4. DISCUSSION

Considering the effect of storage and maturity on bruise susceptibility of apples, information in the literature appears to be contradictory. Diener et al. (1979) reported at harvest, more mature apples were less easily bruised, whereas Hyde and Ingle (1968); Tsukamoto (1981); Klein (1987) found just the opposite. They also noted a decrease in damage after longer storage periods following harvest. However, Schoorl and Holt (1978) found increased bruise susceptibility with increasing storage time.
The mechanical tests performed on apple and apple specimens of flesh and skin showed different behavior of apple firmness during storage at all studied temperatures. The highest differences were obtained in each case for storage apple at 14-16°C.
For all values of: force, deformation and elasticity obtained at tension test for skin of Gloster variety, after harvest and after 30 weeks of storage at 14-16°C, a slight fall of skin firmness was observed.
The force observed at compression test for apple of Gloster variety after harvest (46.25N) was close to being three times higher than after 30 weeks storage at 14-16°C (16.59N). The deformation was on the same level (1.98mm after harvest and 1.99mm after

30 weeks of storage). A modulus of elasticity of cylindrical specimen apple flesh was only two times higher 2.74MPa (after harvest) than 1.34MPa (after 30 weeks of storage at 14-16°C).

Comparing values of force, deformation and elasticity obtained at penetration test for Gloster variety, in each case the statistical differences were observed. However, for apple storage at 14-16°C after 30 weeks, the values of force fell three times, down from 36.33N to 12.84N. The values of deformation increased also three times from 2.21mm to 6.61mm but the elasticity fell dawn more than ten times from 6.62MPa to 0.61MPa for the same storage period.

All results obtained using penetration, compression and tension tests suggest that the penetration test was the best to compare time and temperature of storage for all studied varieties. The values related to a modulus of elasticity more distinctly show changes of apple firmness during storage.

5. REFERENCES

Brusewitz, G.H. and J.A. Bartsch, (1989). Impact parameters related to post harvest bruising of apples. *Trans. of the ASAE*, **32**, 2, 953-957.

Chappell, T.W. and D.D. Hamann (1968). Poisson's ratio and Yung's modulus for apple flesh under compressive loading. *Trans. of the ASAE*, **11**, 5, 608-610, 612.

Clevenger, Jr. J.T. and D.D. Hamann (1968). The behavior of apple skin under tensile loading. *Trans. of the ASAE*, **11**, 1, 34-37.

Diener, R.G., K.C. Elliott, P.E. Nesselroad, M. Ingle, R.E. Adams and Bliggard S.H. (1979). Bruise energy of peaches and apples. *Trans. of the ASAE*, **22**, 2, 287-290.

Dobrzański, B. and R. Rybczyński (1994). Bending test for study of mechanical properties of apple tissue. *Journal of Food Physics*, **2**, 13-15.

Duprat, F., E. Pietri, M.G. Grotte and C.J. Studman (1994). A multi-purpose firmness tester for fruits and vegetables. *Journal of Computers and electronics in Agriculture*.

Holt, J.E. and D. Schoorl (1984). Mechanical properties and texture of storage apples. *Journal of Texture Studies*, **15**, 377-394.

Hyde J.F. and M. Ingle (1968). Size of apple bruises as affected by cultivar, maturity and time in storage. *Proceedings of the American Society for Horticultural Science*, **92**, 733-738.

Klein J.D. (1987). Relationship of harvest date, storage conditions, and fruit characteristics to bruise susceptibility of apple. *Jouranal of the american Society for Horticultural Science*, **112**, 1, 113-118.

Rybczyński, R. and B. Dobrzański (1994a). Mechanical resistance of apple in different place of fruit. *International of Agrophysics*, **8**, 3, 455-459.

Rybczyński, R. and B. Dobrzański (1994b). Modulus of elasticity of apple flesh. *Journal of Food Physics*, **2**, 77-79.

Rybczyński, R., B. Dobrzański and I. Szot (1994). Mechanical resistance of apple skin in different fruit area. *AgEng Milano*, Report N. 94-G-079.

Siyami, S., G.K. Brown, G.J. Burgess, J.B. Gerrish, B.R. Tennes, C.L. Burton, and R.H. Zapp (1988). Apple impact bruise prediction models. *Trans. of the ASAE*, **314**, 1038-1046.

Schoorl, D. and J. Holt (1978). The effects of storage time and temperature on the bruising of Jonathan, Delicious and Granny Smith apples. *Journal of Texture Studies*, **8**, 409-416.

Studman, C.J. and L. Boyd (1994). Measurement of firmness in fruit and vegetables. *AgEng Milano*, Report N. 94-G-066.

Studman, C.J. (1994). Quality in fresh fruit - meaning, measurement and maintenance. *AgEng Milano*, Report N. 94-G-067.

Szot, B., R. Rybczyński, B. Dobrzański and A. Stępniewski (1989). Some mechanical properties of skin and under-skin layer of apple. *4th ICPPAM*, Rostock, **2**, 834-838.

Topping, A. J. and M.T. Luton (1986). Cultivate differences in the bruising of English apples. *J. Hort. Science*, **61**, 9-13.

Tsukamoto M.(1981). Studies on the mechanical injury of fruit: II. Susceptibility to impact and compression in apple fruit as related to storage periods and fruit portions. *J. Jpn. Soc. Hort. Sci.*, **49**, 571-575.

LASER LIGHT BASED MACHINE VISION SYSTEM FOR NONDESTRUCTIVE RIPENESS SENSING OF GOLDEN APPLES

F. Duprat, H. Chen, M. Grotte, D. Loonis, E. Pietri

INRA-LAMPE, Domaine Saint Paul, Site Agroparc, 84914 Avignon Cedex 9, France

Abstract: A laser light image on the apple surface was captured and digitized by a machine vision system to a 256 pixels x 256 pixels x 256 gray levels digital image. It was then smoothed by using a linear low pass filter, converted to a binary image at a threshold gray level of 15, and calculated for the size. The image size increases with the ripening of the apple. Measurements on 120 Golden apples showed a nonlinear negative relation between the image size and the firmness with a correlation coefficient (r) of 0.84. During shelf display, the image size increases linearly with time. This system can be used to evaluate the apple ripeness.

Keywords: apple, image, laser, machine vision, ripeness.

1. INTRODUCTION

Since the appearance of the third generation computers in 1960s, machine vision systems have been developed and applied to various fields such as astronomy, medicine, biology, geography, cartography, meteorology, geology, archaeology. During the last one or two decades, there has been an increasing demand on automatic processing machines in the fresh fruit and vegetable industry both to release the labor workers from their tedious work and to have a consistent high processing quality. Considerable research on machine vision systems has therefore been carried out for inspection of agricultural products during processing. The major application of machine vision systems has been to grade apples (Graf, 1981; Rehkrugler and Throop, 1986, 1989; Davenel et al., 1988), peaches (Miller and Delwiche, 1989, 1991), potatoes (Pathare et al., 1993; McClure and Morrow, 1987; Marchant et al., 1988; Tao et al., 1990), tomatoes (Sarkar and Wolfe, 1985a, b), raisins (Okamura et al., 1991), Cucumbers (Nakahara et al., 1979), asparagus (Rigney et al., 1992) and bell peppers (Wolfe and Swaminathan, 1987) by their size, shape and surface features such as color, scar, bruise, wrinkle quality. Machine vision systems have also been used to detect the orientation of peaches in a standard pitting machine (Delwiche and Drews, 1992), the orientation of bell peppers in packing lines (Wolfe and Swaminathan, 1987), and the location of tomatoes (Whittaker et al., 1987), peaches and apples (Sites and Delwiche, 1988) on trees in harvest. Some machine vision systems have been developed to detect the stem of blueberries in quality determination (Wolfe and Sandler, 1985) and the stem/root joint of carrots in cutting stem (Batchelor and Seary, 1989). Furthermore, Throop et al. (1989, 1994) developed a machine vision system to detect apple watercore. Storlie et al. (1989) used the machine vision system to monitor the growth of whole plants by estimating leaf area and wet and dry weights. At present, some machine vision based sorting lines are commercially available.

2. OBJECTIVES

So far, only the surface color has been used as a ripeness indicator in the machine vision systems to evaluate the fruit quality. The surface color is actually a representative of the fruit surface feature. In this work, a low power laser beam is used in a machine vision system to illuminate a small part of a fruit. Laser radiation has the properties of being monochromatic, coherent and directional as well as having a well defined beam diameter, which make it distinct from all conventional light sources. The object of this work is to develop a machine vision system which may detect more the texture under the skin for ripeness estimation.

3. IMAGE ACQUISITION SYSTEM

Fig.1 shows the image acquisition system. The whole system is placed in a dark room. An apple rests in an apple holder in which the distance of the apple from the camera can be adjusted. A LDM145/670/1 compact laser diode module fixed over the apple projects a laser beam at an incident angle of about 30° onto the top of the apple. A WV-CD20/C camera equipped with a Wild M3C stereomicroscope captures the laser image on apple surface. The video signal from the camera is then digitized by a Synaps A/D image board into a 256 row pixels x 256 column pixels image with each pixel having eight bits of gray level, i.e., a gray level between 0 (black) and 255 (white). The digitized image from the Synaps board is both transferred to a 386 personal computer where C programming language is used for image processing and saving, and meanwhile displayed via a VT-110s magnetoscope on a 14-in. black and white monitor. The digitized image can also be saved by the magnetoscope on a video cassette for replaying later.

LDM/145/670/1 laser diode module is a low power (1 mW output) red (670 nm wavelength) laser light source with the beam diameter of 5 mm at aperture. Such an output power may avoid the image to be saturated when the diaphragm level of the stereomicroscope is properly set. It is not desirable to have such a big beam size. Choosing properly the distance of the laser source from the apple surface may get a small size of the beam due to the divergence property of the beam. But this may be accompanied with the increase of the light intensity in the beam area, causing the image to be saturated. In our research, the laser source was fixed about 30 cm from the apple. To have a smaller beam size, a cap with a central hole of 2 mm in diameter was mounted on the head of the laser source to let the laser beam pass through the hole. The actual beam size in our measurement was 2 mm in diameter.

The stereomicroscope has a five-step magnification changer. Each step of the magnification corresponds to one field-of-view of the camera. Due to the body reflectance of the laser light inside the apple, the laser image on the apple surface is larger than 2 mm diameter, changing with apple ripeness. To ensure the overall view of the laser image to be captured, a magnification of 6.4 was selected, at which the camera field-of-view is about 5.95 mm x 5.95 mm. It corresponds to 23.4 μm x 23.4 μm per pixel for the digitized image. In this magnification, the laser image size of an unripe apple accounts to 15 to 20 percent of the total field-of-view, while that of a ripe apple accounts to 60 to 75 percent of the field-of-view. The stereomicroscope has five levels of diaphragm to chose. Level 3 was selected in our measurement. Fig.2 shows the gray level histogram of two images with curve 1 corresponding to an unripe apple and curve 2 to a ripe apple. It can be seen that pixel numbers decrease sharply with the gray level. The unripe apple has the highest gray level about 100 and the ripe apple has the highest gray level of about 160. None of the images is saturated. It verifies that the laser source and the diaphragm level have been properly selected.

The working distance or focal length between the laser image on the apple top and the lens of the stereomicroscope is a very important factor in image capture. In the initial adjustment, a small piece of paper printed with small black grids was sticked on the apple top and illuminated by normal light. The working distance was changed by turning the bilateral focusing knob of the stereomicroscope till a most clear image was obtained. The position of the laser source was then adjusted to have the image displayed on the center of the monitor. During whole measurement, the relative position of the laser source and the stereomicroscope was fixed. The working distance for apples of different size was kept constant by adjusting the height of the apple holder. As the laser beam is projected at about 30° from vertical direction onto the apple surface and reflected to the lens, a little change in the working distance may cause the image to shift on the monitor. Therefore, the working distance can be visually estimated and adjusted by the position of the image on the monitor. That method is easier and faster than measuring directly the distance between the apple top and the lens.

4. IMAGE PROCESSING ALGORITHMS

The laser light reflectance image on the apple surface can be described by the light intensity at the coordinates x, y in the measurement area. Fig.3 shows the three-dimensional view of the image of an unripe apple (a) and a ripe apple (b). The images have 256 pixels along the X (row) and Y (column) axis respectively, which means that each image contains 256x256 pixels. Each pixel has a gray level between 0 (black) and 255 (white). Within the laser beam projection area (upper part of the images), both images exhibit many irregular sharp peaks. Outside this region, gray level drops sharply outwards. The laser light image outside the laser beam projection area is supposed to be formed by the body reflectance or light scattering inside the apple beneath the skin. When a laser light beam falls onto the apple surface, due to the high transmission properties of the laser source, the laser light transmits through the skin, and encounters randomly oriented internal interfaces in the cellular structure, and scatters in all directions. A portion of the light is scattered back to the surface and leaves the apple. With the distance from the illuminated area, the light is scattered less and less. The soft apple has a larger image than the hard apple. To characterize the image for ripeness evaluation, the following image processing algorithms were carried out.

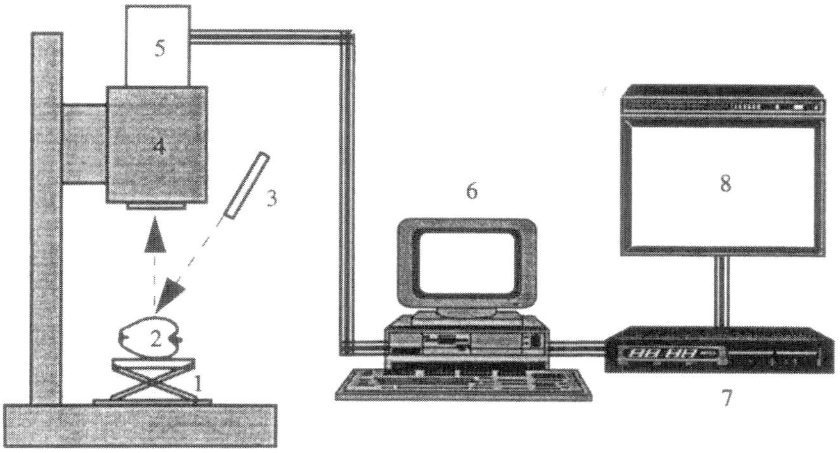

Fig.1 Image acquisition system. 1: apple holder, 2: apple, 3: laser light source, 4: stereomicroscope, 5: camera, 6: computer with A/D image board, 7: magnetoscope, 8: monitor

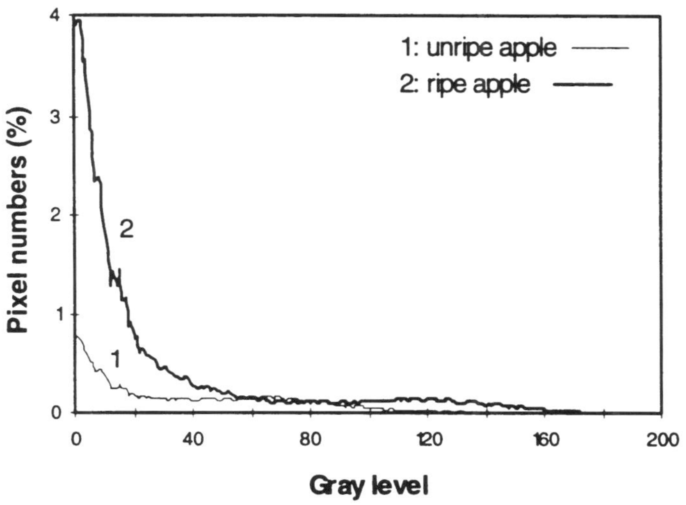

Fig.2 Gray level histogram of two apples

4.1 Smoothing the image

As we can see from Fig.3, it happens for both images at some pixels the sudden transitions in the gray levels. To eliminate or attenuate the sharp transitions, each pixel is averaged with its adjacent pixels (3x3, 5x5,...) for the gray levels. This is actually the discrete convolution, which is analogous to let the raw image pass through a low pass linear filter to eliminate high frequency components. Several filter matrix sizes of 3x3, 7x7, 11x11 and 15x15 were used in smoothing the images. It was found that bigger filters do not improve significantly the image for a better representation of the apple ripeness. Meanwhile bigger filters require much more calculation time for the smoothing procedure. A 3x3 filter was therefore selected in the following work.

4.2 Creating the binary image

A binary image is created by giving a gray level value of 255 (white) to all the pixels with gray levels equal to or above a constant threshold gray level, and giving a gray level value of 0 (black) to the rest pixels. The threshold gray level is an important factor in creating the binary image. From Fig.3, it can be noted that both images differ largely in the bottom part at lower gray levels. The upper part at higher gray levels differs much less with each other. Fig.2 shows that under the gray level of about 50, the difference of the pixels at the same gray level

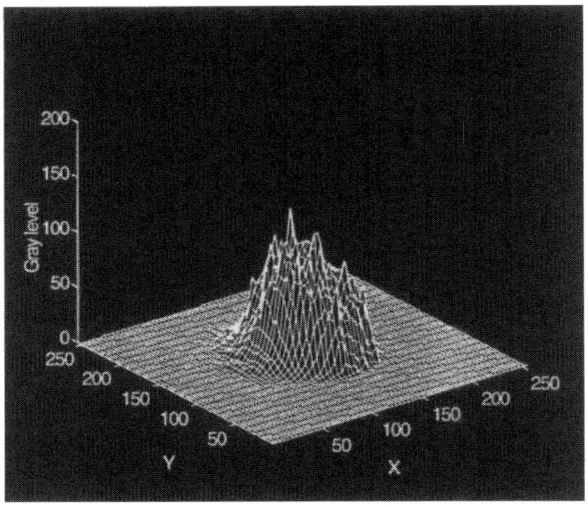

(a) an unripe apple

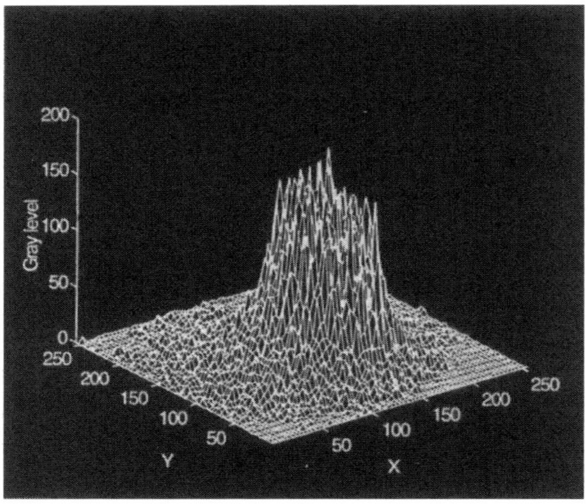

(b) a ripe apple

Fig.3 Three-dimensional images of an unripe apple (a) and a ripe apple (b)

between the ripe and unripe apple increases with decreasing gray level. To discriminate the image size for apple ripeness, the threshold should be set at a lower gray level. After some primary tests, a gray level of 15 was selected as the threshold. A threshold of less than 15 may cause the binary image to involve some noise from the background which is not eliminated by the smoothing procedure.

The created binary images corresponding to Fig.3 are illustrated in Fig.4. Both images look more like ellipses than circles. This is due to the incident angle of 30° of the laser beam onto the surface. The perimeter of the images is not smooth. Around the perimeter, there are some small single white spots which can be attributed to the effects of the apple surface.

4.3 Calculation of the image size

The normal way of calculating the image size is first to determine the perimeter of the image, and then to account the pixels within the perimeter. This procedure may exclude some white spots outside the perimeter, and therefore is more accurate. However this procedure may also increase the computer time. In the case of the apple images in Fig.4, the white spots outside the perimeter account only to a very small portion compared with the laser images, and they can be ignored in size calculation. The image size was therefore calculated by accounting all the pixels with the gray level of 255, and multiplying the pixel number by the pixel size (23.4 μm x 23.4 μm per pixel).

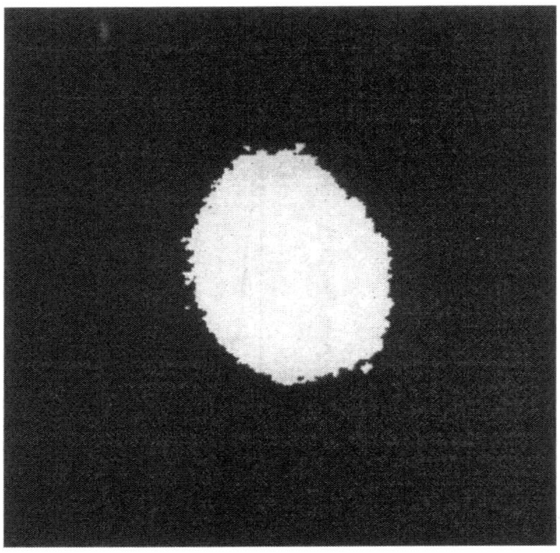

(a) an unripe apple

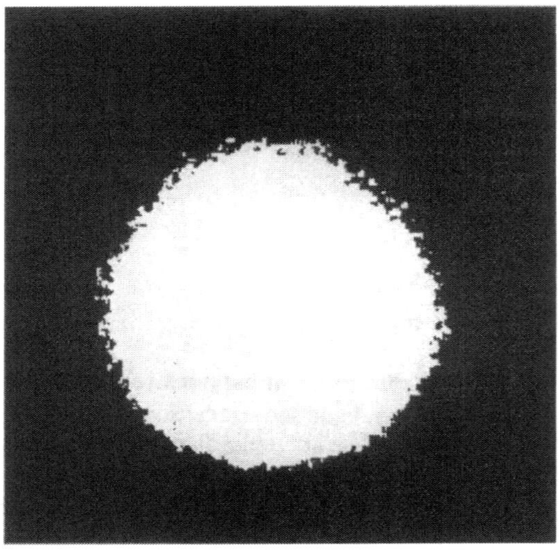

(b) a ripe apple

Fig.4 Binary images of two apples in Fig.3 at the threshold of 15

5. RELATION OF THE LASER IMAGE WITH THE APPLE RIPENESS

Firmness was used here as a ripeness index. To see how the laser image changes with the apple ripeness, 120 Golden apples were harvested from the orchard of our research station in September 1994 and stored for about one month in a cold air storage room under 3°C. Golden apples are characterized by relatively uniform color distribution as well as some lenticels over the surface. The apples were classified randomly into six groups and rested on shelf display at 20°C for accelerated ripening. In the determined shelf time intervals, six groups of the apples were measured in turn for the image and the firmness (Young's modulus). Each apple was measured at two locations with the maximum color difference and least black spots (lenticels) for the image. The average of the two measurements was used. The Young's modulus of each apple was measured at three equally distributed locations around the equator by using a penetro-tester.

Fig.5 shows the relation between the image size S and the Young's modulus E, where points represent the measurement results and the curve the fitted model. There is a negative relationship between the image size and the firmness. As apples mature, their image size tends to increase. This can be attributed to the loss of the apple chlorophyll during ripening. The apple texture with more chlorophyll shows stronger absorbance in the red light wavelength region, so that the reflectance is relatively lower (Sitkei, 1986). Correspondingly, the reflectance image has a relatively smaller size. With ripening of the apple, green color turns to yellow due to the loss of the chlorophyll, whereby the corresponding absorbance is lowered and the reflectance increases, causing a larger reflectance image. The relation between the image size and the Young's modulus can be expressed by the fitted model as:

$$E = 0.13\, S^2 - 1.76\, S + 7.54 \qquad (1)$$

where S is in mm^2 and E in MPa. This is a two order polynomial equation. From Fig.5, it can be seen that for the same amount of the change of the Young's modulus, soft apples change more in the image size than hard apples. Or, the image of the softer apples is more sensitive to the ripeness. The correlation coefficient r between S and E amounts to -0.84. It is evident that there exists a good correlation between S and E. The image size can be used as a nondestructive index to evaluate the apple firmness for ripeness monitoring. The correlation coefficient can possibly be improved by properly processing the image and developing better parameters to characterize the image.

Fig.6 shows the change of the image size during the shelf display under 20°C. The image size increases with the shelf time. For Golden apples, the average change of the image size with the shelf time under 20°C can be expressed by the linear equation as:

$$S = 3.55 + 0.10T \qquad (2)$$

where T is the shelf life in days. From Eq.2, it can be noted that one day of shelf life corresponds to 0.1 mm^2 of increase of the image size. The variation of the image size within group seems less at the beginning of the display and tends to increase following the shelf display, implying the increasing ripeness variation. This is caused by different ripeness rate for apples within group. The laser light image can be used to monitor the apple ripeness during storage and shelf display for the controlling of the apple quality. For example, once the image size exceeds some critical levels, the apples can not sustain further storage or shelf display. They should be processed, saled or consumed soon. Otherwise, considerable value loss of these apples may occur. For the critical levels of different varieties of apples, considerable measurements are required.

6. CONCLUSIONS

In this work, a machine vision system was introduced. A low power laser beam of 670 nm in wavelength and 2 mm in diameter was projected onto the apple surface, forming a ripeness dependent image caused by body reflectance and scattering. This image reflects more the texture property under the skin than the surface color.

The overall laser light reflectance image was captured and digitized into 256 pixels x 256 pixels digital image with each pixel having 256 gray levels. The raw digital image was then smoothed by using a linear low pass filter, converted into a binary image and calculated for the size. The matrix size of the filter and the gray level threshold were discussed and selected as 3x3 and 15 respectively.

Young's modulus of the apple flesh was used as a measure of the apple ripeness. 120 Golden apples were tested for both the image and the Young's modulus. There is a nonlinear negative relation between the image size and the Young's modulus. The image size increases with the decrease of the Young's modulus during ripening of the apple. The correlation coefficient (r) of 0.84 was obtained between them. This system is reliable for ripeness evaluation. The image size was monitored for its change with the ripening when apples were on shelf display under 20°C. It was found to increases linearly with time.

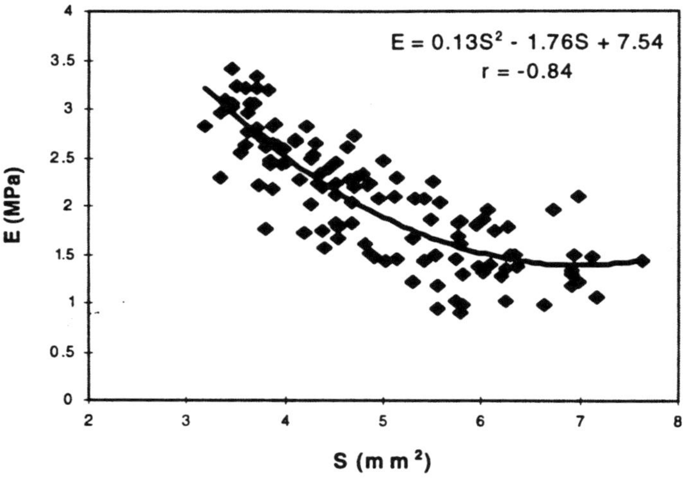

Fig.5 Relation between image size S and Young's modulus E

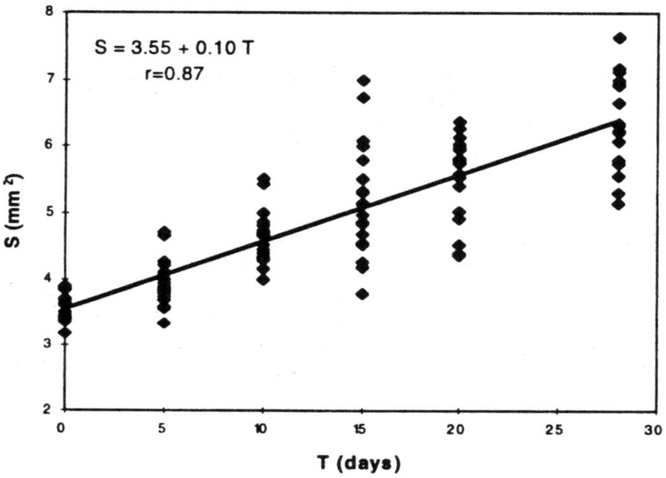

Fig.6 Change of the image size S during shelf display

REFERENCES

Batchelor, M.M., S.W. Searcy (1989). Computer vision determination of the stem/root joint on processing carrots. J. agric. Engng Res., **43**, 259-269.

Davenel, A., CH. Guizard, T. Labarre, F. Sevila (1988). Automatic detection of surface defects of fruit by using a vision system. J. agric. Engng Res., **41**, 1-9.

Delwiche, M.J., E.S. Drews (1992). Sensor mechanisms for improved peach pitting. Trans. of the ASAE, **35**, 903-908.

Graf, G.L., G.E. Rehkugler, W.F. Millier and J.A. Throop (1981). Automatic detection of surface flaw on apples using digital image processing. ASAE paper No. 81-3537, St. Joseph, MI.

Marchant, J.A., C.M. Onyango and M.J. Street (1988). High speed sorting of potatoes using computer vision. ASAE paper No.88-3540, St. Joseph. MI.

McClure, J.E., C.T. Morrow (1987). Computer vision sorting of potatoes. ASAE paper No.87-6501, St. Joseph, MI.

Miller, B.K., M.J. Delwiche (1989). A color vision system for peach grading. Trans. of the ASAE, **32**, 1484-1490.

Miller, B.K., M.J. Delwiche (1991). Peach defect detection with machine vision. Trans. of the ASAE, **34**, 2588-2597.

Nakahana, S., A. Maeda, Y. Nomura (1979). Automatic cucumber sorting using pattern recognition technique. IEEE Tokyo Proc. Denshi, **18**, 46-48.

Okamura, N.K., M.J. Delwiche, J.F. Thompson (1991). Raisin grading by machine vision. ASAP paper No.91-7011, St. Joseph, MI.

Pathare, N.P., P.H. Heihemann, S. Deck, C.T. Morrow (1993). Automatic inspection station for grading of potatoes. ASAE paper No.93-3092, St. Joseph, MI.

Rigney, M.P., G.H. Brusewitz, G.A. Kranzler (1992). Asparagus defect inspection with machine vision. Trans. of the ASAE, **35**, 1873-1878.

Rehkugler, G.E., J.A. Throop (1986). Apple sorting with machine vision. Trans. of the ASAE, **29**, 1388-1397.

Rehkugler, G.E., J.A. Throop (1989). Image processing algorithm for apple defect detection. Trans. of the ASAE, **32**, 267-272.

Sarkar, N., R.R. Wolfe (1985a). Feature extraction techniques for sorting tomatoes by computer vision. Trans. of the ASAE, **28**, 970-974.

Sarkar, N., R.R. Wolfe (1985b). Computer vision based system for quality separation of fresh market tomatoes. Trans. of the ASAE, **28**, 1714-1718.

Sites, P.W., M.J. Delwiche (1988). Computer vision to locate fruit on a tree. Trans. of the ASAE, **31**, 257-263, 272.

Sitkei, G (1986). Mechanics of agricultural materials. Akadémiai Kiado, Budapest, Hungary. ISBN 0-444-99523-4.

Storlie, C.A., A. Stepanek, G.E. Meyer (1989). Growth analysis of whole plants using video imagery. Trans. of the ASAE, **32**, 2185-2189.

Tao, Y., C.T. Morrow, P.H. Heiheman (1990). Automated machine vision inspection of potatoes. ASAE paper No. 90-3531, St. Joseph, MI.

Throop, J.A., G.E. Rehkugler, B.L. Upchurch (1989). Application of computer vision for detecting watercore in apples. Trans. of the ASAE, **32**, 2087-2092.

Throop, J.A., D.A. Aneshansley, B.L. Upchurch (1994). Camera system effects on detecting watercore in 'Red Delicious' apples. Trans. of the ASAE, **37**, 873-877.

Whittaker, A.D., G.E. Miles, O.R. Mitchell, L.D. Gaultney (1987). Fruit location in a partially occluded image. Trans. of the ASAE, **30**, 591-596.

Wolfe, R.R., W.E. Sandler (1985). An algorithm for stem detection using digital image analysis. Trans. of the ASAE, **28**, 641-644.

Wolfe, R.R., M. Swaminathan (1987). Determining orientation and shape of bell peppers by machine vision. Trans. of the ASAE, **30**, 1853-1856.

STUDY ON QUALITY EVALUATION OF ORANGE FRUIT USING IMAGE PROCESSING

Naoshi Kondo*, Haruhiko Murase**, Mitsuji Monta*,
Yasunori Shibano* and Kentaro Mohri*

*Faculty of Agriculture, Okayama University
**Faculty of Agriculture, University of Osaka Prefecture

Abstract: It is important to classify fruit after harvesting according to its quality, but many kinds of fruits are classified based on only their size or shape. Generally speaking, Iyokan orange fruit whose color is reddish, whose size is medium, whose height is short and whose surface is glossy is sweet, however, the criterion depends on worker for classification and on district in Japan. In this paper, investigation to evaluate Iyokan orange fruit using image processing objectively is described. From the results, it was observed that the features extracted from image had some relationship to the fruit quality.

Keywords: Evaluation, Image processing, Classification, Quality, Automation

1. INTRODUCTION

Quality evaluation is essential work for orange fruit classification in Japan. Many varieties of fruits are classified based on only their size or shape, however their inside qualities are different. It is desirable that the inside quality such as sugar content, acid etc. are evaluated non−destructively.

Generally speaking, orange fruit "Iyokan" whose color is reddish, whose size is medium, whose height is short and whose surface is glossy is sweet, however, the criterion depends on worker and on district. Most farmers, therefore, can distinguish sweet fruits based on their experiences.

In this study, possibility of the quality evaluation of Iyokan orange fruit using image processing technology was investigated to automatize classification system for orange fruit.

2. MATERIALS

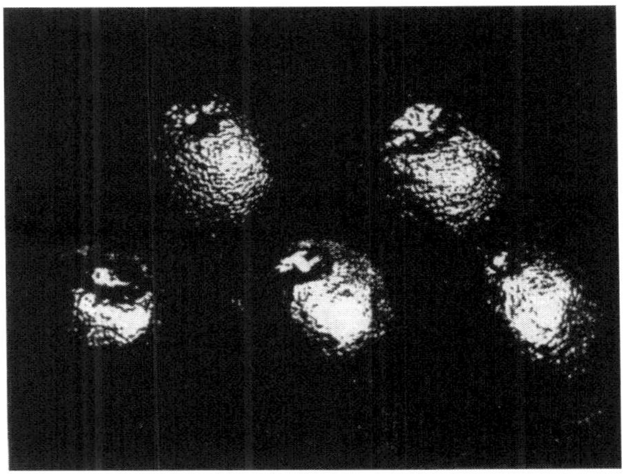

Fig.1. Iyokan orange fruits.

There are many varieties of citrus fruits in Japan. It is said that Iyokan orange and mandarin orange have much correlation between their appearance and taste and that especially, as for Iyokan orange, red, medium size, short height and glossy fruit is sweet.

In this experiment, 30 Iyokan orange fruits harvested in Ehime Prefecture where 90% of Iyokan orange is produced in Japan were used. Their sugar and acid content were measured after image input. As examples, Fig.1 shows the Iyokan oranges whose roughness of surface and shape are different. The degree of roughness was assigned to every fruit using number 1(glossy) to 5 (rough) by human eyes.

3. EXPERIMENTAL DEVICES AND METHODS

Fig.2 shows a blockdiagram of experimental apparatus and Fig.3 shows a scene of the image input. First, an image of the orange fruit was inputted to computer by TV camera through an image input board under the condition of illuminance 500 lx using lamps whose color temperature are 5500 K. The inputted image consisted of 256 × 256 pixels which have 256 grey-tone values.

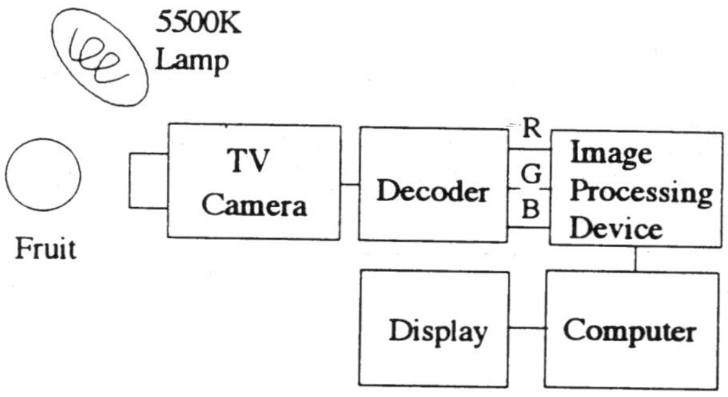

Fig.2. Blockdiagram of experimental apparatus.

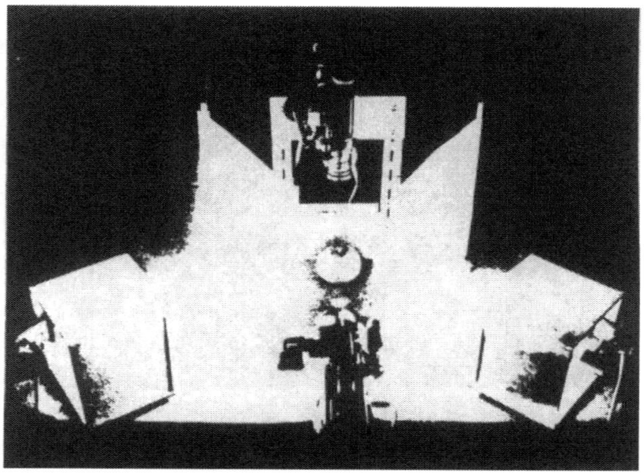

Fig.3. Method of image input.

The object color was decoded into red, green and blue color components, so it was easy to know the difference of fruit color using the ratio of color components such as combination of red and green.

Secondly, binary image could be made by using the one color image (red, green or blue). In this experiment, red image was used. The size and height of the fruit were extracted from the binary image as Feret's diameters. The shape of fruit could be expressed by Feret's diameter ratio R as shown Equation (1).

$$R = Dv/Dh \qquad (1)$$

Dv: Vertical Feret's diameter
Dh: Horizontal Feret's diameter

Thirdly, texture of fruit surface was investigated based on the 16 gray tone image through co-occurrence matrix. Angular Second Moment and Contrast (Robert, et al.,1973) were used as textural features in this study. The features are expressed by the following equations (2) and (3).

$$ASM = \Sigma\ \Sigma\ \{p(i,j)\}^2 \qquad (2)$$

$$Con = \Sigma\ \Sigma\ (i-j)^2 \cdot p(i,j) \qquad (3)$$

ASM: Angular Second Moment
Con: Contrast
p(i,j): (i,j)th entry in a normalized gray-tone spatial-dependence matrix

ASM means homogeneity on the fruit surface in the grey-tone image. The values of textural features depended on the distance and on angle between pixels. This experiment was conducted under the condition that the distance was changed from 1 to 16 and the angle was 0°

4. RESULTS AND DISCUSSION

Fig.4 shows an example of grey level on a horizontal line of light part of the fruit whose roughness is 1. The figures under the various condition were different as shown in Fig.5 to 7. From these figures, it was observed that the grey level depended on the condition.

Fig.8 shows relation between sugar content on Brix scale and red and green color ratio(R/G). It was observed that sugar content tended to increase when R/G became big, however correlation was not so high because the fruits whose shape and roughness were different were used as the materials. It was considered that correlation with color ratio of fruits which had similarroughness would become higher. The result of acid decreased conversely, when R/G became big.

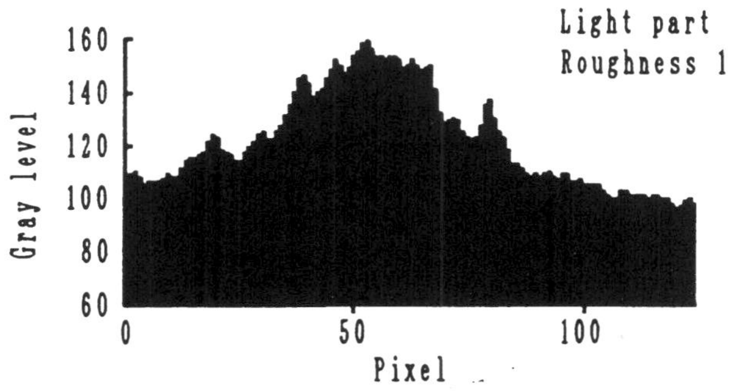

Fig.4. Grey level on horizontal line (1).

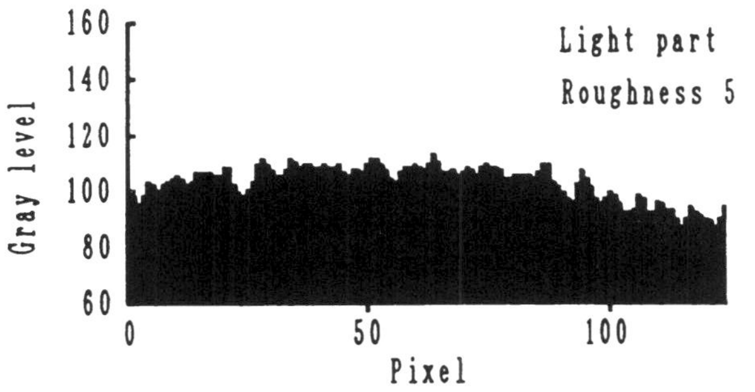

Fig.5. Grey level on horizontal line (2).

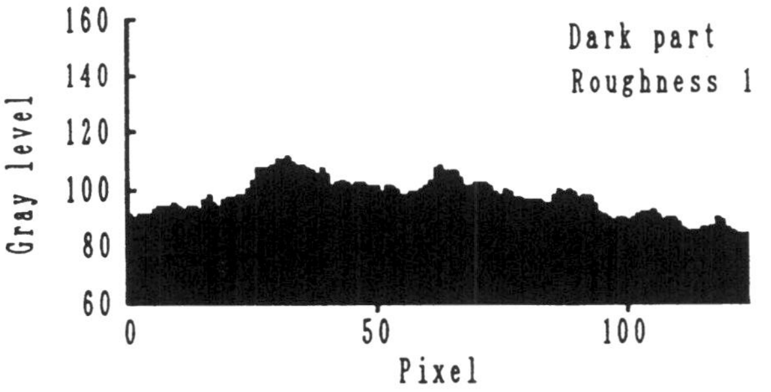

Fig.6. Grey level on horizontal line (3).

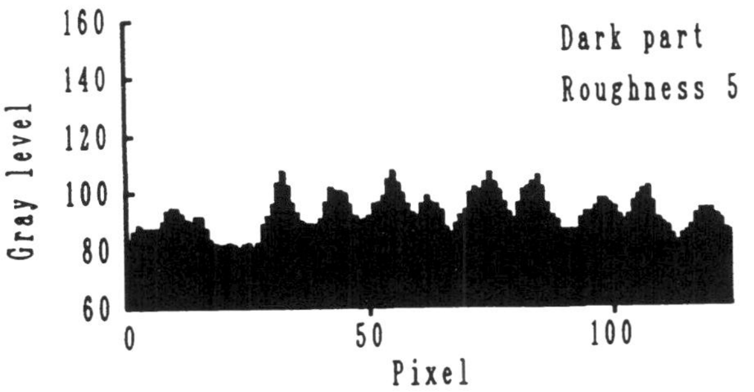

Fig.7. Grey level on horizontal line (4).

Fig.9 shows relation between sugar content and Feret's diameter ratio. From this figure, the reason that the fruit whose height is small is said to be sweet was understandable. Besides, it was observed that correlation with Feret's diameter ratio of fruits of similarroughness would become higher.

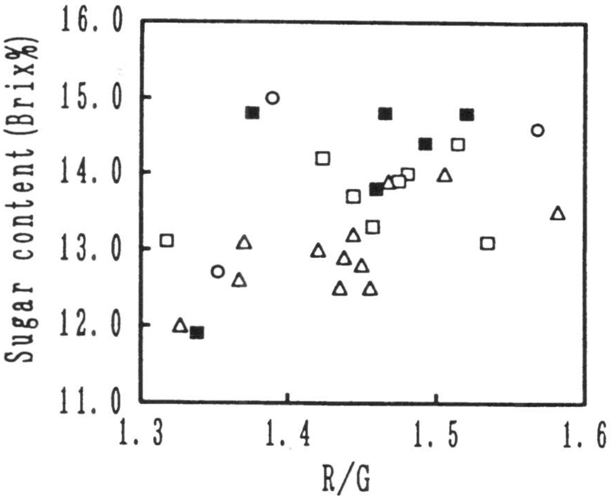

Fig.8. Relation betweeen color ratio and sugar content.

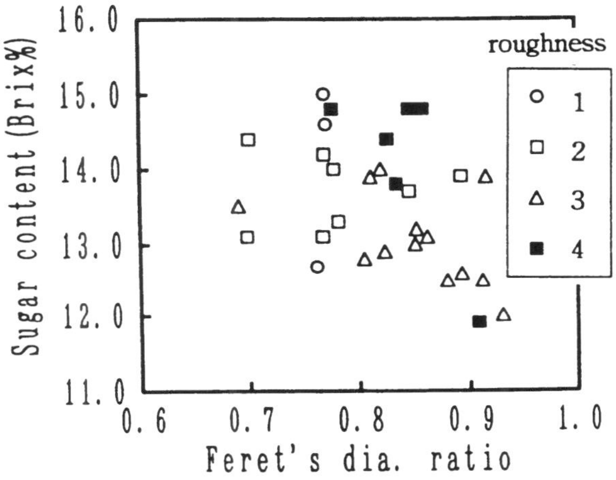

Fig.9. Relation betweeen Feret's diameter ratio and sugar content.

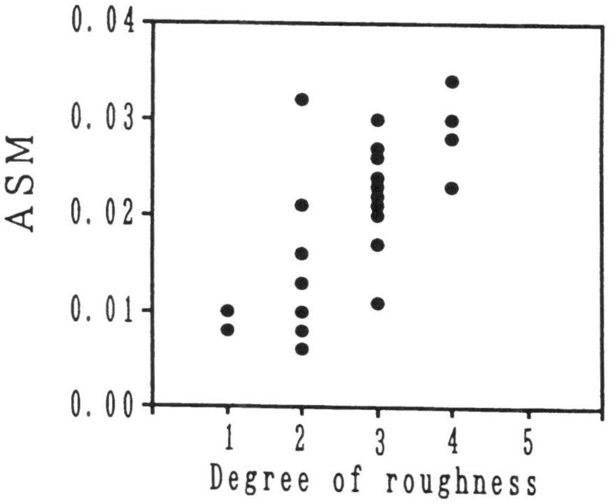

Fig.10. Relation betweeen degree of roughness and ASM (1).

Fig.10 shows relation between ASM and roughness

at light part of fruit. It was observed that ASM tended to increase when fruit surface became rough, since glossy surface had less homogeneity on grey–tone image than rough surface as shown Fig.4 and 5. On the contrary, ASM decreased at dark part of fruit as shown in Fig.11, because fluctuation of grey–tone on rough surface was bigger than that on glossy surface at dark part.

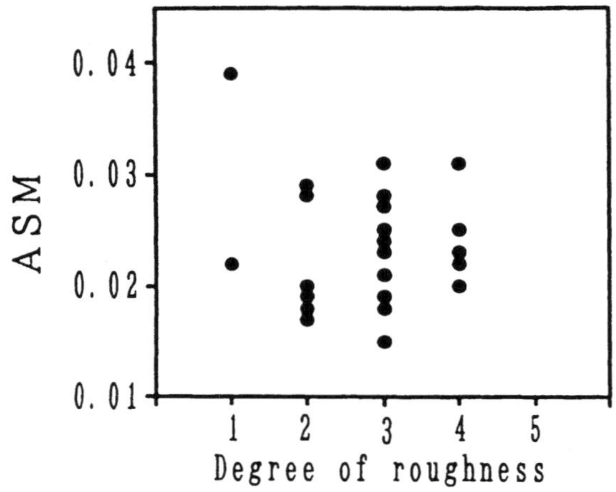

Fig.11. Relation betweeen degree of roughness and ASM (2).

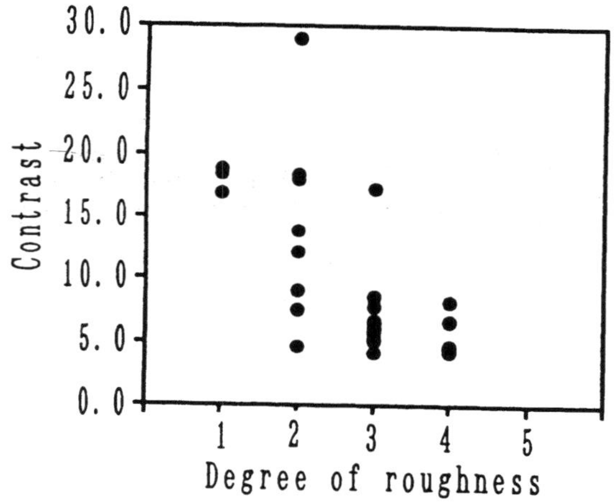

Fig.12. Relation betweeen degree of roughness and Contrast.

Fig.12 shows relation between Contrast and roughness at light part of fruit. Contrast tended to become small, when fruit surface became rough.

Fig.13 shows relation between sugar content and roughness. It was considered that rougher fruit had less sugar content as being said that glossy fruit is sweet except some fruits of roughness 1 and 4.

From the results, it was observed that the features extracted from image had some correlation to the fruit quality, however some of them had non–linear relation, since the features influenced on the quality complicatedly. Therefore, it was considered that the inside quality of fruit could be guessed by neural–networks or other methods so that the quality could be synthetically evaluated using the results obtained from image processing.

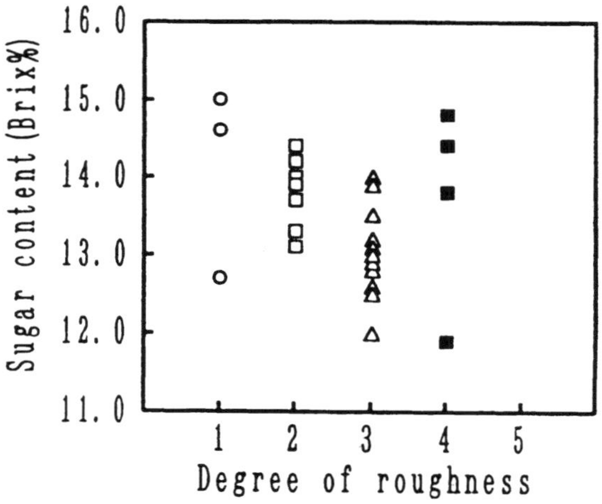

Fig.13. Relation betweeen degree of roughness and sugar content.

5. CONCLUSION

It was confirmed that common saying red, low height and glossy orange fruit is sweet by using the features extracted from image. However all features had not linear relationship with the inside quality, since the information extracted from image includes relative more errors than that of the system.

On the other hand, peach fruit grading system is already developed by using spectral photo reflectance in infrared region. The system can measure the sugar content of peach fruit accurately. However, the system may not be used for orange fruit, because skin of orange fruit is much thick than that of peach fruit.

It is, threfore, necessary to construct a non–destructive quality evaluation system for orange fruit by adding other features extracted from image or by using synthetical evaluation method.

REFERENCES

Haralick, R.M., K. Shanmugam and Its'hak Dinstein (1973). Textural features for image classification, IEEE Transactions on systems, man, and cybernetics, Vol.SMC–3, No.6, 610–621.

Recognition of Fruits by Image Processing
- Application of Template Matching -

S. Oke, M. Ookado and Y. Nakamura

Department of Agricultural Engineering
Ishikawa Agricultural Collage
1-308, Suematsu, Nonoichi Ishikawa, 921, JAPAN

Abstract: The object of the present study is to develop the algorism capable recognizing some sorts of particular fruits and vegetables, for the robots being operated in farm working, by using the images obtained from the visual sensors mounted thereon. In the researches reported up to the present, recognitions of fruits, by utilizing the difference of the reflectance spectroscopic characteristics between the fruits and plants' portions other than the fruits themselves in question, have been studied. In these cases, special types of filters and need, and moreover, the particular filters, best suited for each sort of the various harvests, are required. Therefore, in the present report, by means of the recognition algorithms using the image processing techniques not requiring special pre-processing procedures, recognition experiments of tomatoes were carried out.

Keywords: Image processing,Image recognition,FFT,Machine recognition

1. INTRODUCTION

The object of the present study is to develop the algorism capable recognizing some sorts of particular fruits and vegetables, for the robots being operated in farm working, by using the images obtained from the visual sensors mounted thereon. In the researches1,2 reported up to the present, recognitions of fruits, by utilizing the difference of the reflectance spectroscopic characteristics between the fruits and plants' portions other than the fruits themselves in question, have been studied. In these cases, special types of filters and need, and moreover, the particular filters, best suited for each sort of the various harvests, are required. Therefore, in the present report, by means of the recognition algorithms using the image processing techniques not requiring special pre-processing procedures, recognition experiments of tomatoes were carried out.

2. IMAGE RECOGNITION METHODS

As the recognition methods of fruits, we have adopted, among the families of techniques generally called as pattern matching, the one looked upon as a fundamental template matching3. The template matching is a technique of positioning a certain object by seeking the similarity for every points of the image with respect to the similarity for the overlapping portions, between the template image representing the object on one hand and the image in which the object is included on the other. As the similarity representing the degree of matching, we have adopted the cross correlation formula (1) written below.

$$m\ (u,v) = \iint_S f(x+u,y+v)\,t(x,y)\,dxdy \quad (1)$$

Since the left-hand term of (1) depends on the image including the object, we have also adopted, as the one representing similarity, the normalized cross correlation formula (2) written below

$$m^{*}(u,v) = \frac{\iint_S (f(x+u,y+v)-\bar{f})\left(t(x,y)-\bar{t}\right)dxdy}{\sqrt{\iint_S (f(x+u,y+v)-\bar{f})^2 dxdy \iint_S \left(t(x,y)-\bar{t}\right)^2 dxdy}} \quad (2)$$

where f and t are the are mean values, in the regions S of the images $f(x + u, y + v)$ and $t(x, y)$, respectively.

3. EXPERIMENTS ON POSITION DETECTIONS

We have adopted tomatoes as the objects to be recognized in the present study, and taken photographs, in the form of transparencies for film projector, of the tomatoes cultivated in a hot house of the experimental farm attached to our university, and after the development of the (transparency) film, prepared Photo CDs thereof. The image data, in which the objects in question are included, were transferred directly from the Photo CD to a personal computer, where the PICT data were transformed into general data and were adopted as the image data (384 x 256 pixels) to be processed in the calculation. To investigate the influences, concerning the regions' sizes corresponding to the tomatoes within the template image, on the tomatoes of various sizes included in the image, three sorts of the template images were selected, i.e., two templates consisting of circles (Red 98, Green 65, and Blue 34; each of which representing the optical density values) depicted within the squares of 32 x 32 pixels and 64 x 64 pixels, respectively, and a tomato's image (47 x 39 pixels) arbitrarily chosen from the Photo CD.

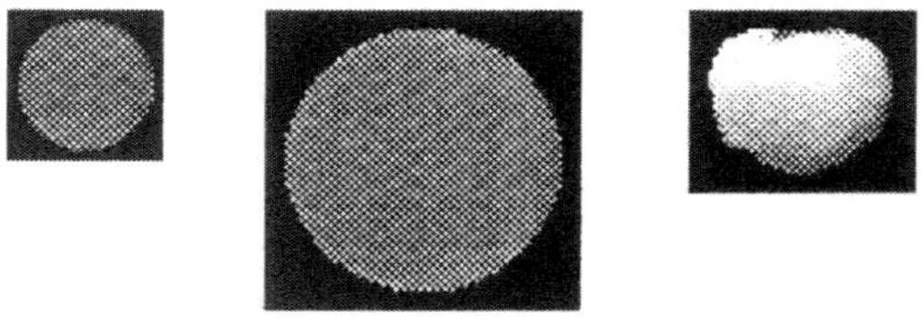

Circle 32x32 | Circle 64x64 | Tomato 47x39

Fig.1 Template Images

Figure 1 shows the three sorts of template images used in the recognition experiment. In addition, as to the color of the portions other than the circles and tomatoes in the templates, black (Red 0, Green 0, and Blue 0, representing the regions' optical density values) color was selected because it has no influence on the similarity calculations. Only the optical density value of red color was employed in the calculation of cross correlation because the objects to be recognized were red-colored tomatoes. To investigate the influences, concerning the regions' sizes corresponding to the tomatoes within the template image, the template size which was increased by 32x32 pixels to 131 pixels were selected too. As the images used in the present experiment, only those in which a single tomato is shown within were employed.

4. RESULTS OF EXPERIMENTS AND CONSIDERATIONS

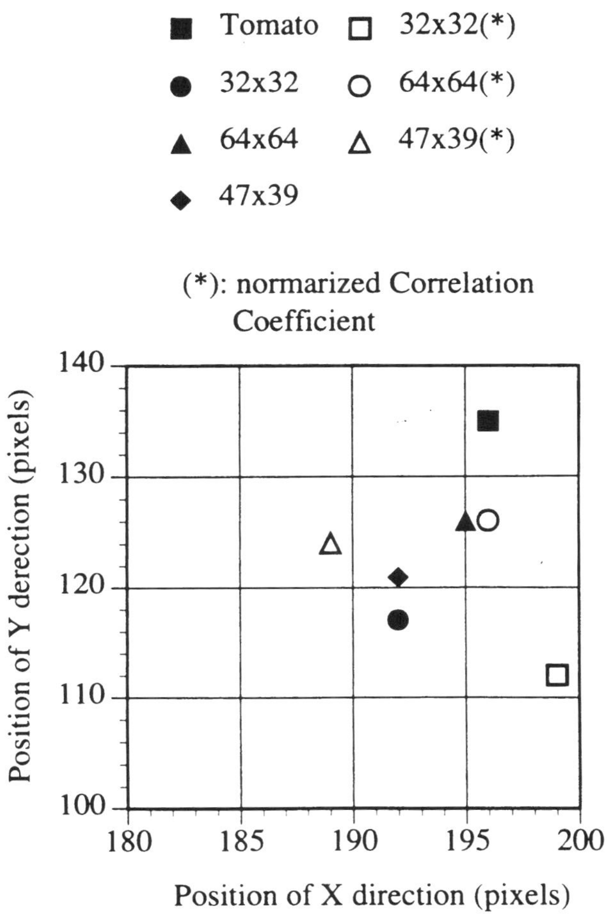

Fig.2 Results of detected position of image No.1

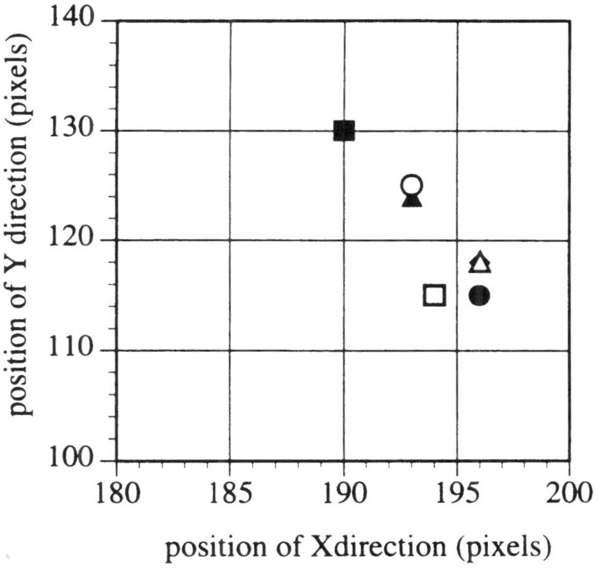

Fig.3 Results of detcted position of image No.2

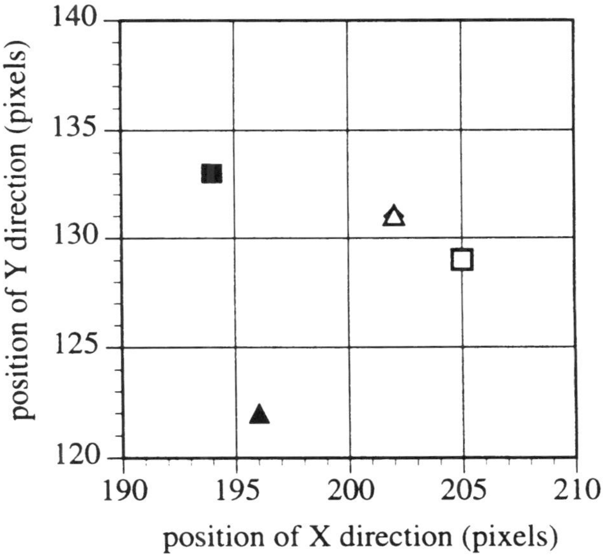

Fig.4 Results of detected position of image No.4

With respect to a single image, a combination of two cases of cross correlations and three cases of template images, totaling six calculation procedures, were carried out to detect the tomatoes' positions. Figure 2-5 show the results of the calculations for tomatoes' positionings.

Figure 2 shows the results of detected positions of image No. 1. It was possible to detect the tomatoes' positions in all of the above-mentioned combinations (for calculation). The detection errors were found to become larger according to the reverse order of templates' sizes, or 64 x 64 pixels, 47 x 39 pixels, and 32 x 32 pixels; the maximum errors thereof were 7 pixels in X axis direction and 23 pixels in Y axis direction, manifesting larger values in the latter case compared with the former case.

Figure 3 shows the results of detected position of image No. 2. It was possible to detect the tomato's position in all the above-mentioned combinations.

Similar to the case of images No. 1, the detection errors were found to become larger according to the reverse order of templates' sizes, or 64 x 64 pixels, 47 x 39 pixels, and 32 x 32 pixels; the maximum errors thereof were 6 pixels in X axis direction and 15 pixels in Y axis direction, manifesting larger value in the latter case compared with the former case. Moreover, the differences of the errors for cross correlation were small, which is conceivably ascribable to the influences of the templates' sizes on

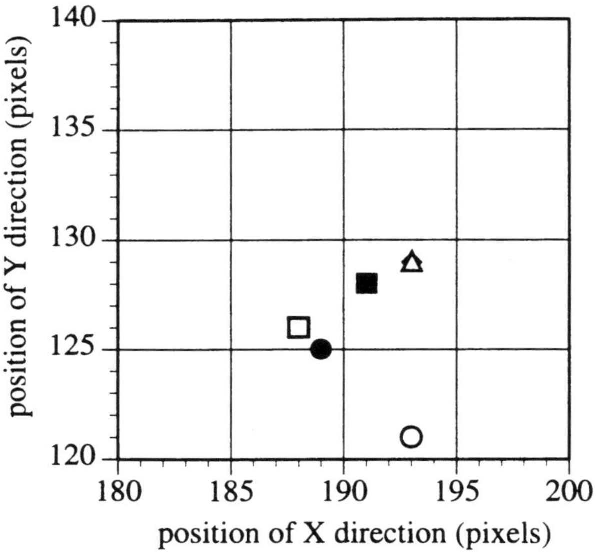

Fig.5 Results of detected position of image No.4

the errors.

Figure 4 shows the results of detected position of image No. 3. Except for the case in which the template's size is 64 x 64 pixels and besides the calculation is carried out using normalized cross correlation, it was possible to detect all of the tomatoes' positions. The detection errors were found to become larger according to the particular order of templates' sizes, or 47 x 39 pixels, 32 x 32 pixels, and 64 x 64 pixels, and the errors for the undetectable case were 159 pixels in X axis direction and 17 pixels in Y axis direction. The causes for the undetectable case are considered to be that the template image (64 x 64 pixels) was larger compared with the size of the tomato in the image to disable the latter's detection. Moreover, as to the difference of errors for cross correlation, in the case of the template's size of 64 x 64 pixels it was not possible to detect out by means of normalized cross correlation, but in the cases of the templates' sizes of 47 x 39 pixels and 32 x 32 pixels, no difference was found out. This is conceivably ascribable to, similar to the case of image No. 2, the influence of template size on the errors.

Figure 5 shows the results of detected position of image No. 4. Except for the case in which the template's size is 64 x 64 pixels and the calculation is carried out using unnormalized cross correlation, it was possible to detect all of the tomatoes' positions. The detection errors became larger according to the sizes of the template, or 47 x 39 pixels, 32 x 32 pixels, and 64 x 64 pixels, and the errors for the undetectable cases were 86 pixels in X axis direction and 94 pixels in Y axis direction. The cause for the undetectable case is considered to be, similar to the case of No. 3 image, that the template image (64 x 64 pixels) was larger compared with the size of the tomato in the image to disable the latter's detection.

Summarizing the above-mentioned results, the templates, sized 32 x 32 pixels, in which the circles' and tomatoes' images are depicted were smaller- or equal-sized compared with the tomatoes' size in the image, and it was possible to detect their positions regardless of the presence of the cross correlation. For the circle's template consisting of 64 x 64 pixels, the reason for the unsuccessful detection is considered to be ascribable to the larger tomatoes' size in the image compared with the size of the circle in the template.

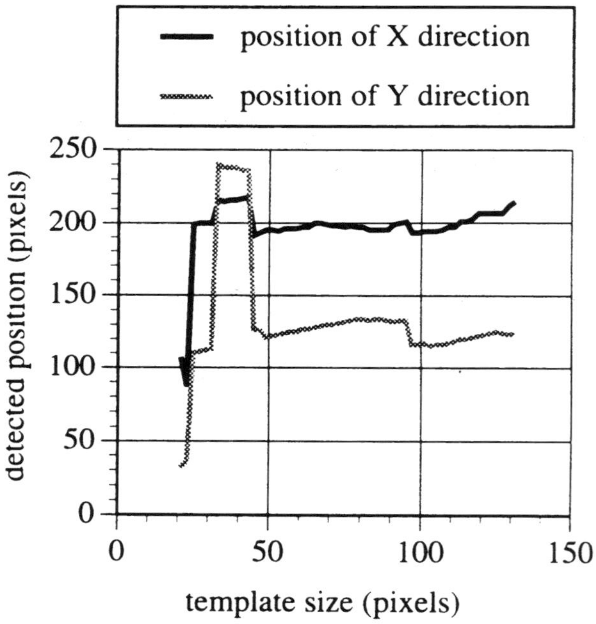

Fig.6 Results of detected posion of image No.1

Fig. 6 shows that the results of detected position of image No. 1. It was possible to detect all of the tomatoes' positions with the size of the circle by 25 to 31 and by 45 to 131 in the square template.

Fig. 7 shows that the results of detected position of image No. 2. It was possible to detect all of the tomatoes' positions with the size of the circle by 29

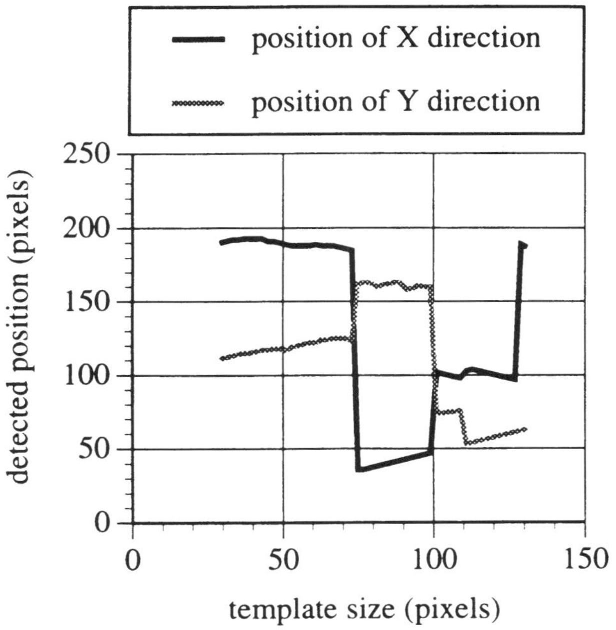

Fig.7 Results of detected posion of image No.2

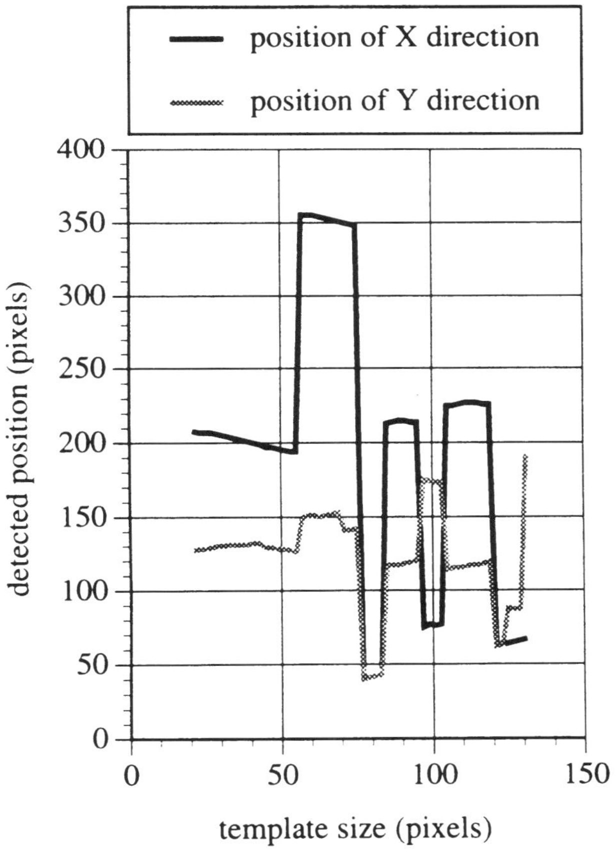

Fig.8 Results of detected posion of image No.3

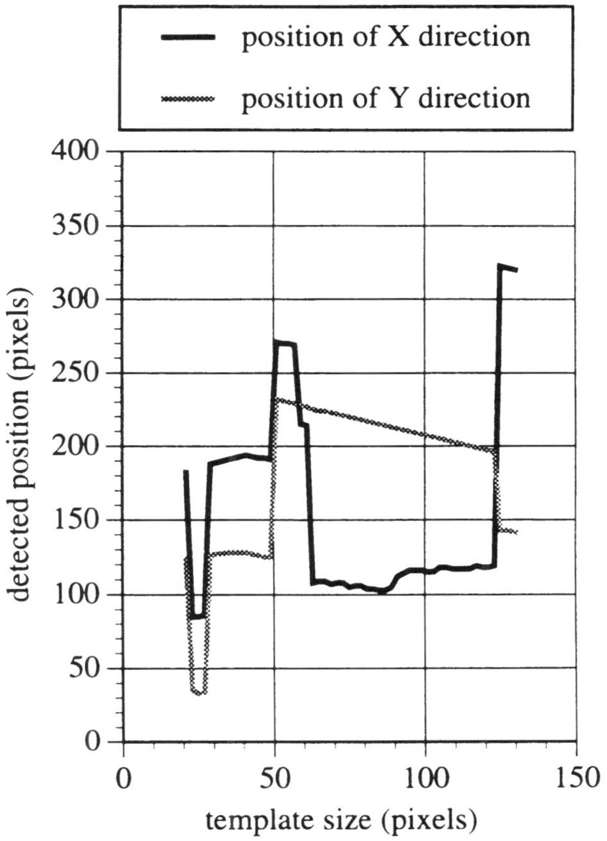

Fig.9 Results of detected posion of image No.4

to 73 in the square template.

Fig. 8 shows that the results of detected position of image No. 3. It was possible to detect all of the tomatoes' positions with the size of the circle by 21 to 31 , by 85 to 95 and by 105 to 119 in the square template.

Fig. 9 shows that the results of detected position of image No. 4. It was possible to detect all of the tomatoes' positions with the size of the circle by 29 to 49 in the square template.

5. SUMMARY

The results of the application of template matching technique to the image recognition of tomatoes can be summarized as follows.

1) It was clarified that there exists an optimum template image size against the size of the tomato in the image.
2) It was clarified that, if the size selection of the

template image is adequate, there is no significant difference among the correlation coefficients concerned.

REFERENCES

Naoshi Kondo et al., "Researches on the Visual Sensors for Recognition of Fruits", (Third Report) journal of the Institute of Agricultural Machinery, Vol. 50, No. 6, 83/89 (1988).

Makoto Dohi et al., "Researches on the Multi-function Robots for Grocery Harvesting", (First Report), Vol. 55, No. 6, 77/84 (1993).

Hideyuki Tamura: "Introduction to Computerized Image Processing", pp. 148-150. Soken press (1985).

MEASURING OF GRAINY MATERIALS COMPOSITION BY COMPUTER VISION

Dariusz Andrejko, Józef Grochowicz, Dariusz M.Stasiak

Agricultural University of Lublin, Dpt. Food Engineering & Technics

A qualitative and quantitative analysis of seed mixtures, for example lupin seeds, may be carried through by vision computer systems equipped by a software suitable for morphometric measurements. The measurement procedures carried out by image analysis are characterised by a fast and simply performance but need the relevant instruments and a strict keeping the methodical conditions.

Keywords: computer vision, image analysis, object recognition, agriculture, industry automation, process control

1. INTRODUCTION

A very fast population growth observed during last years, especially in the third world countries, causes new, sometimes very hard to solve, tasks and goals for the agriculture and the food processing industry.

One of the most important tasks is to search for new raw materials rich of proteins and to elaborate inexpensive processing technologies. A proper analysis of the performance of technological processes is depending on the velocity and the quality of measuring and evaluating the changes taking place in the processed raw materials. A very important parameter of the performance of almost all the technological processes is the estimated percentage of particular components of the raw material. For the purpose percentage composition estimating of the mixtures of biological grain substances the methods based on mask analysing and pneumatical separation are widely applied. Some methods using the effects of the electrostatic field are also utilised. It is also possible to combine various separation methods, ex. a method of mask-pneumatical separation. With the development of computer technologies, computer analysis vision systems has been successfully used in many industrial applications. Some applications of them include inspecting and sorting agriculture products, ex. classifying tomatoes (Sarkar and Wolfe, 1985), selective harvesting of asparagus (Humburg and Reid, 1991) and others.

For those reasons and others the purpose of researches carried through was to determine the conditions of utilising the vision system of a computer image analysis for a quick and contactless method of estimating the percentage of particular fractions in the mixtures.

2. MATERIALS

A wide range of seed raw materials has been analysed. Below some of chosen results of a quantitative analysis of a mixture of Sur and Aga lupin seed grades are presented. the choice of these two lupin grades was not accidental: Aga seed has an uniform light-creamy operculum whereas the operculum of Sur seed is dark-brown, almost black, with rare bleach spots on the surface. Such a composition of a raw material made it possible to examine the possibility of applying the vision system for a colour separation of the components.

Table 1 shows some chosen parameters characteristic the physical properties of the tested materials. They have small differences on the density, freeflow rate, granulation, and they differ on operculum colour.

Some preparations of mixtures of Aga percentage of 10, 15, 20, 25, 30 and 35% were prepared. Some random chosen samples of 100 g mass were then taken from the preparations to the vision analysis.

Table 1 Physical properties of lupin seeds

	Sur	Aga
bulk density [kg·m^{-3}]	777.3	763.2
specific density[kg·m^{-3}]	812.2	811.9
granulation [mm]	4.74	4.95
freeflow rate [deg]	13.5	13.9
mass of 1000 seeds [g]	121.3	125.2
porosity [%]	43.4	44.5
operculum colour	black-brown	light-creamy

3. DEVELOPMENT SYSTEM

The researches were performed using the SVIST v.1 computer image analysing system with SVISTMET module for morphometric measurements (Wikom, 1992). The system was installed in an AT486 computer. A CCD camera was used for image detection, the image was projected on a screen of a NEC colour monitor connected to a special vision card. The second monitor served the user's menu. Fig.1 shows the view of the apparatus used.

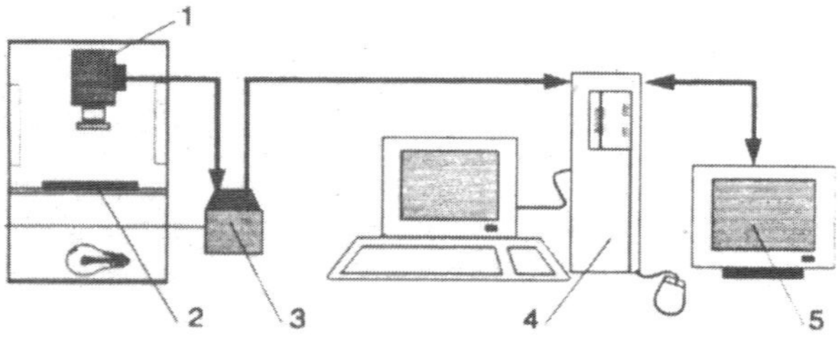

Fig.1. Scheme of the system (1 - CCD camera, 2 - lightproof chamber, 3 - image processing hardware, 4 - PC-system with software, 5 - truecolor monitor)

The applied system of image analysis has a multi level user's interface. The best operation conditions are by using a pulldown-menu system operated by a keyboard and a mouse. It is also possible to command the system by text orders typed on a keyboard. The system serves also the macrodefinitions compiled by the user and automating all the image recording and analysing process. Macrodefinitions eliminate errors, especially at analysis of histograms and determining of thresholds. System accepts C-modules too. It expands potential possibilities of presented image analysis system.

In order to get the uniform testing conditions the camera was fitted in a special lightproof chamber eliminating the influence of lumination fluctuations. A specimen of one-shift placed seeds was placed in a image field of the camera in a special square frame and illuminated by a dispersed light of an adjustable intensity. It have been recommended a monochrome sources of light, especially for differentiate similar operculum colour seeds.

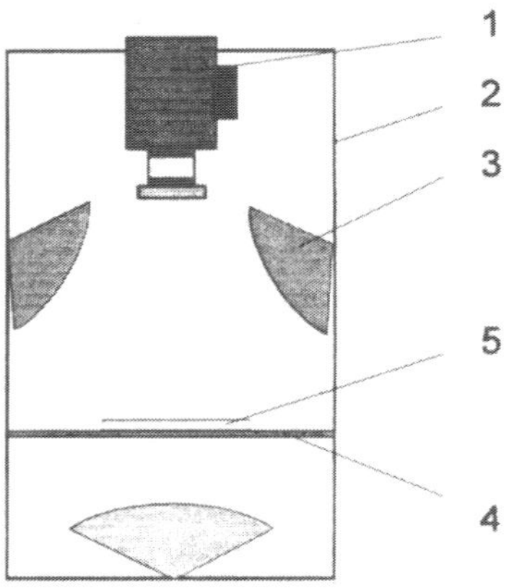

Fig.2. Scheme of the chamber (1 - CCD camera, 2 - walls of chamber, 3 - dispersed lights, 4 - colour filter, 5 - specimen frame)

3. PROCEDURE

The procedure of analysing the percentage composition of a lupin seed mixture was carried out on base of the algorithm shown on Fig.3.

Fig.3. Flowchart of analysis of fraction algorithm

The specimen image was digitalized by a CCD camera and then subjected to a graphic image processing in order to get the strictly determined conditions of image analysis (Ballard and Brown, 1982). Each time the histograms describing the brightness fluctuations of the image points of the mixture were elaborated. Based on that the binarization of the image was carried out by dividing it into three components: background, bright objects and dark objects. Then the procedures of SVISTMET software were activated in order to evaluate the percentage of particular fractions of the mixture basing on the component images.

4. RESULTS

In accordance with that procedure the images for Sur and Aga mixtures arose as presented on Fig.4. depending on the mixture composition. The presented image consists of a wide range of grey shades different however for particular components of the image.

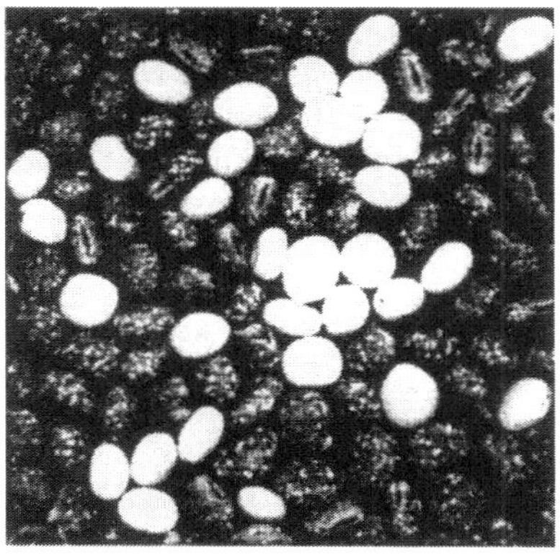

Fig.4. Input image of lupin mixture

The obtained histograms showed the range of variation of the brightness of analysed image points as shown on Fig.5.

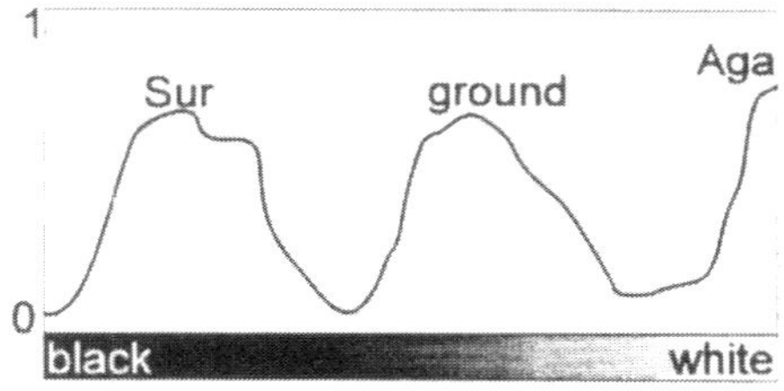

Fig.5. Histogram of image components intensity

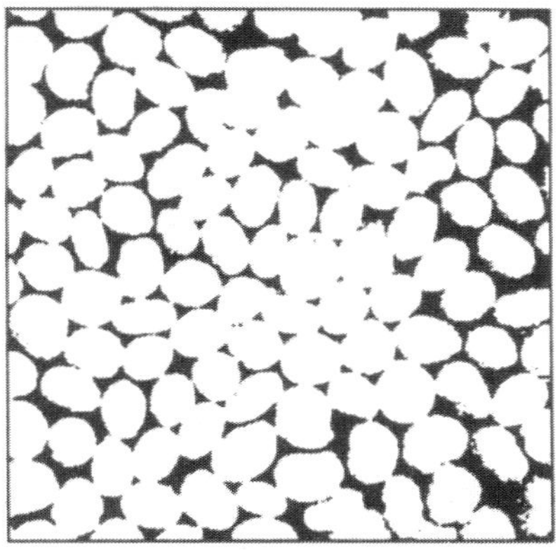

Fig.6. Binary images of lupin background

On base of those histograms the detailed condition of further image processing were determined and then the images were binarized (Fig.6,7) by using the graphic procedures of the vision system. For that purpose a double threshold function with values 27 and 150 was used obtaining the component images of bright fraction and background.

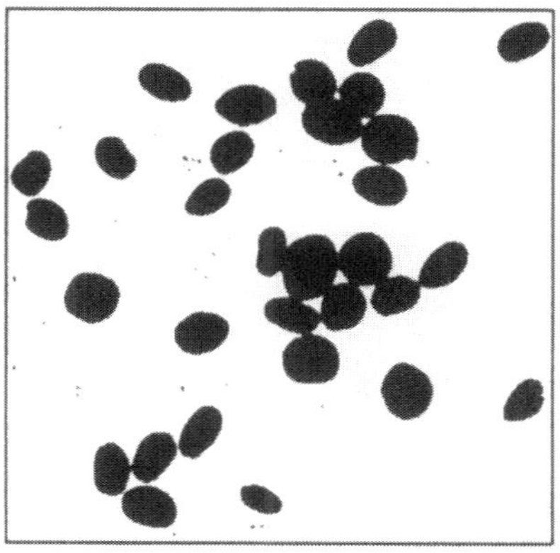

Fig.7. Binary image of a mixture's fraction

The images processed in this way were used for a fraction analysis of tested mixtures. For that purpose some procedures evaluating, among others, the percentages of component fields on the analysed image were activated.

Table 2 shows the results of the analysis of the fraction in 20 repetitions carried through for each mixture of Sur and Aga lupin seed.

Table 2 Results of image analysis

Real contents	white areas Mean	white areas SD	black areas Mean	black areas SD	Error
10 %	8.1	0.03	74.1	0.30	0.15
15 %	12.5	0.03	72.1	0.27	0.22
20 %	16.2	0.03	66.1	0.25	0.32
25 %	19.3	0.05	59.1	0.16	0.38
30 %	24.3	0.07	58.2	0.20	0.55
35 %	28.7	0.08	54.3	0.12	0.43

The above table shows the results of the analysis of the area of all component images of the seed, i.e. the bright (Aga) and the dark (Sur) fraction.

On the base of the results shown in Table 2 a calibration curve (Fig.8) for the mixture was compiled. In order to eliminate influence of the background areas, the shape of curve presents proportion of mixture's fractions in function of one component.

That curve enables a quick reading of the percentage of the mixture as a function of the ratio of mixture percentage of areas obtained from binary image of fractions. Fig. 8 shows both of curve: theoretical and calculated with empirycal points.

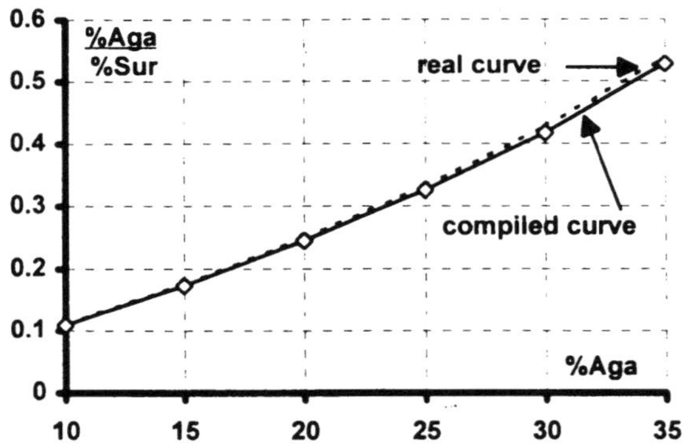

Fig.8. Aga-Sur calibration curve

5. SUMMARY AND CONCLUSIONS

The results of the presented measurements are generally charged by a maximal error of ±0.55% of the percentage of particular mixture component. This may be caused by an imperfection image analysis method, especially by analysing the histogram and choosing the limits values of image components. In order to eliminate that error the histogram analysis must be automatized. The researches must be however extended to achieve it.

The presented method of vision analysis of percentage composition of seeds mixtures enables to obtain the results charged by an error of ca. ±0.6% provided that the strictly determined test conditions are kept. In order to get a repeatability of the results the following conditions must be secured:

- proper conditions of preparing the representative specimen (sampling method),
- proper size of the specimen (filling the measure area),
- proper placing the specimen on a measure plane (single layer of smallest possible porosity),
- proper illumination of the specimen (adjustable light intensity, a monochromatic light recommended, else colour filters),
- proper parameters of specimen background (illuminated background the specimen is laid on, , a monochromatic light recommended, else colour filters),
- proper choice of threshold values of image components (threshold for white, grey, black)

The observance of above mentioned conditions will assure, that the results of measuring the percentage composition of seed mixtures will be repeatable.

The presented method is especially suitable for quick analysis the composition of mixtures consisting of particles of similar geometric forms or for particles similarly placed on the measuring plane.

The presented method may be substantially automated provided that the system is previously calibrated according to specific mixture, especially to forms and space orientation of their particles. For each particular mixture a system calibration is needed for taking the form and size fluctuations into account.

REFERENCES

Ballard, D.H. and C.M.Brown (1982). *Computer vision.* Prentice-Hall, Inc., NJ.

Sarkar, N. and R.R. Wolfe (1985). Computer vision based system for quality separation of fresh market tomatoes. *Transactions of the ASAE* **28(5)**, 1714-1718.

Humburg, D.H. and J.F.Reid (1991). Field Performance of Machine Vision for the Selective Harvest of Asparagus. *ASAE Paper* No. **911751**, ASAE, St.Joseph, MI 49085.

Wikom Ltd. (1992). *SVISTMET user's manual.* Wikom Ltd. Warsaw.

MACHINE VISION FOR THE AUTOMATIC MEASUREMENT OF BROKEN GRAIN FRACTIONS[1]

H. Georg, N. Guth and F.-J. Bockisch

Institute of Agricultural Building Research
Federal Research Centre for Agriculture, D-38116 Braunschweig, Germany

Abstract: An automatic measuring system for whole and broken kernels was developed. A digital image processing system was chosen as a sensor in combination with a computer controlled kernel positioning device. The machine vision system separated 6000 kernels per hour. Trial series with grain samples collected from different prototypes of combine harvesters are described. The machine vision system had a classification accuracy of 95.8 % for broken kernels. In the future it should be taken into consideration to use a modified machine vision system as a tool for a fast evaluation of the threshing equipment of combine harvesters.

Keywords: Automatic recognition, Image analysis, Evaluation, Quality control

1 INTRODUCTION

The development of image processing techniques for a wide range of applications in quality control is in progress. An analysis of literature showed the great potential of image processing for the evaluation of grain. Image analysis in cereal science is used to determine contents, varieties, seed properties and damage. Early studies on the classification of wheat varieties were made by Zayas, et al. (1985), using a canonical discriminant function. Canonical functions can be also used to measure the amount of broken grain fractions (Zayas, et al., 1990). The determination of the broken grain fraction is important as a quality factor concerning dockage and also for the evaluation of the working quality of threshing aggregates. Research on automation of measuring damaged grain fractions up to now is concentrated on optimised positioning and rapid feature detection. Casady and Paulsen (1988) developed an automatic device for kernel positioning and separation of grain fractions. Recently developed systems used a Fourier descriptor to distinguish between whole and broken kernels (Liao, et al., 1994)

The research presented here will show that a machine vision system could be developed to measure broken grain fractions automatically, helping the designing engineer to yield objective results on investigations of threshing aggregates. The study was carried out under practical conditions in co-operation with the research and development dept. of a manufacturer for combine harvesters.

Up to now two methods were inserted for the measurement of broken grain mainly: sievegrading with following manual determination and different mechanical instruments. The disadvantages of these methods are:

- the relatively small sample size,
- the subjectivity of manual determination and
- the high time required.

To facilitate the measurement with a Leitz - image processing system (CBA 8000), a particular

[1] This study was carried out at the Institute of Agricultural Engineering, University of Giessen, Germany

positioning system for grains was developed to fulfil the following specifications:

- the grains may not touch,
- the contrast between kernels and background should be maximum,
- the area of view should be as large as possible and
- the measurement should be automatically and rapid

The requirements of the producer of combine harvesters reflect upon the possible use of automatic broken grain measuring as an evaluation tool concerning the work quality of their machines. In discussion with the research and development department of the co-operating manufacturer the following demands were made:

- high capacity,
- automatic determination of broken grain fractions,
- higher precision than manual determination and
- availability on PC for field-scale tests.

2 MATERIAL AND METHODS

2.1 Developed equipment

Through the fulfilment of minimising touching kernels there is no algorithm for the separation of touching grains necessary. The maximum contrast of the images was reached through a black conveyor belt, on which the grains were placed individually through a vibrating plate as positioning system (Fig. 1). The positioning system was charged from a sample container, which had a capacity of approximately 300 g

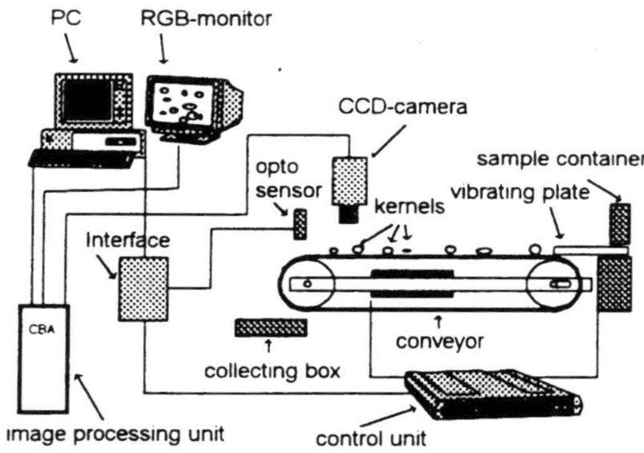

Fig. 1 Scheme of the automatic machine vision system for measuring broken grain fractions.

The kernels were moved individually under the camera with the help of a conveyor belt. The area of view was controlled by an optical sensor counting small reflecting plates on the surface of the belt. The conveyor was stopped while the image frame was processed. This was necessary because the camera had no shutter option. The image processing system measures approximately 50 - 70 grains per image. A larger area of view was not possible because of the limited resolution. The two lighting sources, which are not shown in Fig. 1, were mounted above the conveyor belt besides the camera. At the end of the belt the kernels fell into a collecting box.

The Leitz image processing system CBA 8000 measured with an image resolution of 750 x 582 Pixel. The system contained 2 signal processors ADSP 2100A, CPU 68000, 128 bit-image memory bus and VME-bus. The image processing unit was connected to the PC via RS 232. That means, all data-transfer of measured properties passed through this bottleneck and caused a higher processing time

The PC shown in Fig. 1 controlled both, the image processing unit and the positioning system. The positioning system consisted of a special vibrating plate with adjustable frequency, an opto-sensor, conveyor drive, interface and control unit. The interface was connected to the second serial port of the PC, so that all functions of the positioning system could be controlled together with the image processing system.

All controlling functions of the system were executed by a window-orientated program using the so called "SAA-standard". The programming language was Quick-Basic. The menu consisted of several submenus, covering handling functions like data storage, sample coding, calibrations, measuring options, filters, etc.

2.2 Data management

The measured data from the image processing unit was stored on the harddisk. Each record consisted of the following parameters:

- Area
- Perimeter
- Maximum feret
- Minimum feret
- Shape descriptor

For the separation of broken and whole grains a canonical discriminant function was calculated with the statistical package SPSS-PC+. This function based on a training data set, containing values of whole and broken kernels. The data analysis started after storage of a complete measured wheat sample by a batch process. The analysis included whole descriptive statistics for each parameter mentioned above and the quantity of whole and broken kernels.

2.3 Grain samples

For this study 93 wheat samples with an average weight of 100 g were measured. The material derived from different prototypes of combine harvesters, which had to be evaluated by measuring the broken grain fraction. The tested combine harvesters varied concerning the performance measured as output straw per hour (Fig. 2). The samples were also measured by the manufacturer of the combine harvesters with conventional methods, using a trieur and manual determination.

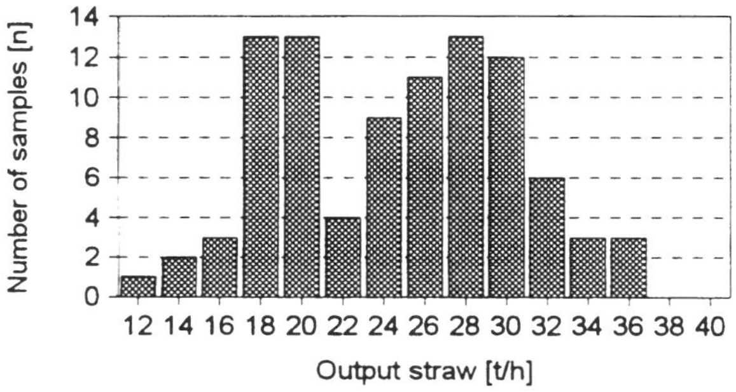

Fig. 2 Distribution of wheat samples on the base different straw output.

3 RESULTS

3.1 Feasibility of the developed system

The results of image analysed grain fractions were sent to the manufacturer as fast as possible to prepare further investigations on the threshing aggregates. The total duration of experiments was 3 weeks. The developed machine vision system processed a total number of 186000 (wheat-) kernels within that time. An average processing speed was found at 6000 kernels/h.

3.2 Data analysis

Concerning the accuracy of the presented machine vision system and manual determined broken kernels a comparison was made with three characteristic wheat samples. The broken kernel fraction of these samples was evaluated by machine vision, manually with a trieur and accurate by hand, named as control (Fig. 3). The results of broken kernel fractions evaluated by the machine vision system were transformed by a factor from quantity of broken kernels to percentage of weight. This is essential for a direct comparison between the manually determined broken grain fractions and the ones measured by machine vision, because the manual fractions were weighted in relation to the total weight of the sample. The transformation of data took place in accordance to results found in literature was done in accordance to Symons and Fulcher, (1988), which found a high correlation between projected area of kernels and the weight.

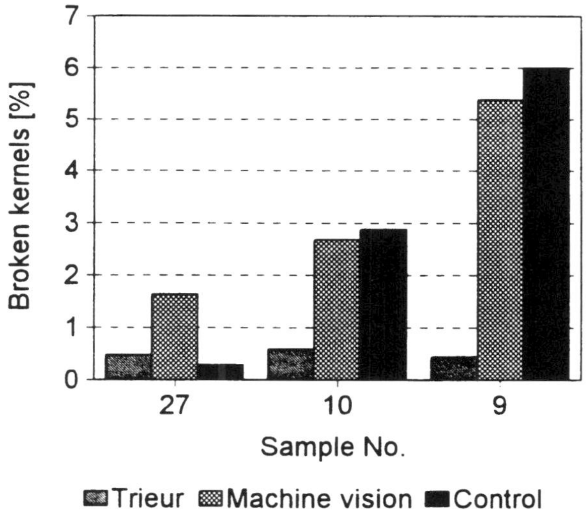

Fig. 3. Comparison between conventional and machine vision based determination of broken kernels (n=3).

Figure 3 shows that sample 10 and 9 were underestimated by the „trieur" system, whereas sample 27 was overestimated by the machine vision system. Considering the ranking among the three samples, machine vision and control variant are similar in contrast to the trieur-determined one, which indicated sample no. 10 to have more broken kernels.

Reflecting on the distribution of the complete trial series, it can be shown, that there were differences between the machine vision based method and the conventional method. Figure 4 indicates that values of the machine vision system covered a range from 0,5 to 5,5 per cent broken kernels. In contrast to this were the results of manually determined broken grain fractions, which covered only a small range from 0 to 1 % (Fig. 5).

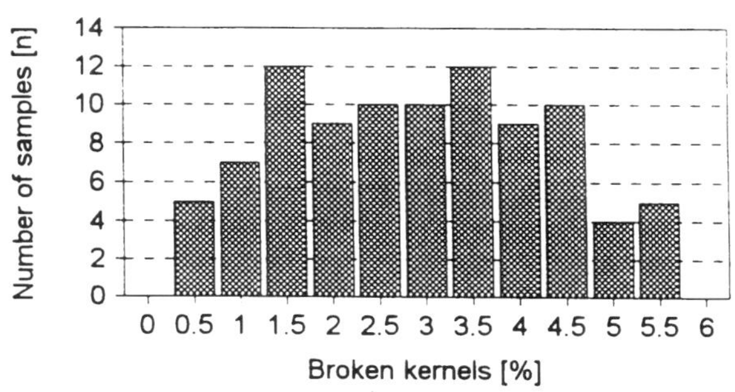

Fig. 4. Histogram of broken grain fractions evaluated by machine vision (n=93).

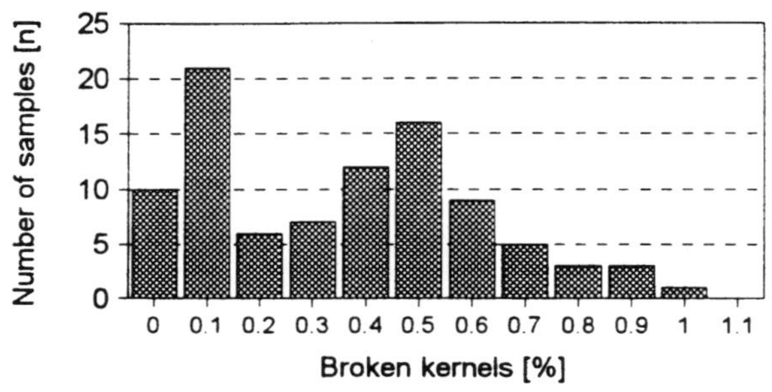

Fig. 5. Histogram conventional (trieur) measured broken grain fractions (n=93).

A correlation between output of the combine harvesters, measured as tons of straw per hour, and the percentage of broken kernels was tested. The measured correlation coefficient (r = 0.11) showed no relation between output (straw t/h) and broken grain fraction. Additional specific data of the combine harvesters, concerning threshing aggregates for example were not available.

The analysis of relations between the two methods showed a high correlation between automatic and manual determined broken kernels (Fig. 6). The correlation coefficient was 0.73; the number of values was n=93.

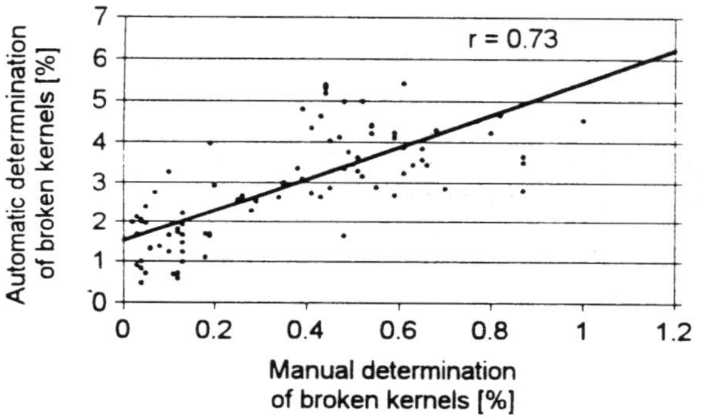

Fig. 6 Correlation between manual and automatic determination of broken kernels.

4 DISCUSSION AND CONCLUSIONS

The development of an automatic machine vision system for the measurement of broken grain fractions showed the feasibility and the practical use of such an application. With further on development of the system it should be possible to integrate most external devices used in the presented system within a PC. For example the interface and the image processing unit can build in as boards with direct access to the bus-system of the PC. An additional advantage of this integration would be the increasing processing speed, which was limited by the connection through the serial ports.

The ranking of the samples with respect to their content of broken kernels is important for the industrial producer of combine harvesters, because the evaluation of newly developed threshing aggregates depends on these results. For that it could be shown, that the manual determination of broken kernels is not exact (see Fig. 3). This result corresponds with earlier, not published comparisons between machine vision based and manual determination of broken kernels at the University of Giessen.

For a better result of determining broken kernels it should be taken into consideration to measure the 3D-properties of kernels. Specially with wheat there would be a higher percentage of correct classified broken kernels, because the longitudinal broken wheat kernels can not be mixed-up with whole wheat kernels. 3D-measurement is possible with structured lighting of the area of view, applications were demonstrated by Sakai and Yonekawa (1992) for soybeans. The usefulness of measuring 3D-properties has to be proved with regard to additional costs.

Reflecting on the great potential of machine vision systems, it should be a task for the future to develop a sensor for broken kernels, that will be mounted directly on the combine harvester. As a possible solution a bye-pass of the grain stream would be analysed by an automatic machine vision system, extracting the percentage of broken kernel as on-line signal for the controlling on-board computer of the combine harvester.

5 REFERENCES

Casady, W.W. and M.R. Paulsen (1988). An Automated Kernel Positioning Device for Computer Vision Analysis of Grain. ASEA-Paper No. 88-3051

Liao K., M.R. Paulsen and J.F. Reid (1994) Real-time detection of colour and surface defects of maize kernels using machine vision. Journal of Agricultural Engineering Research, **59**, 4, pp. 263-271

Symons, S.J. and R.G. Fulcher (1988) Determination of wheat kernel morphological variation by digital image analysis. Journal of Cereal Science, Vol. **8,** 3, pp. 211-231

Zayas, I., H. Converse and J. Steele (1990). Discrimination of Whole from Broken Corn Kernels with Image Analysis. Transactions of the ASAE, Vol. **33,** 5

Zayas, I., Y. Pomeranz and F.S. Lai, F.S. (1985). Discrimination between Arthur and Arkan wheats by image analysis. Cereal Chem., **62**, p. 478

Sakai, N. and S. Yonekawa (1992). 3-Dimensional Image Analysis of the Shape of Soybean Seed. J. Food Eng, vol. **15**, 3, p.221-234

Effect of different controlled atmospheres on processing quality of potatoes

O. S. Khanbari and A. K. Thompson

Cranfield University, Silsoe College, Silsoe, Bedford, MK45 4DT

Abstract: Four potato cultivars responded positively to a commercial reconditioning treatment after being stored for seven months at 4 °C and 8 °C under different controlled atmospheres. CA combinations of 9.7-12% CO_2 with 2.0-2.9% O_2 delayed sprouting in all the cultivars. It has been found that low CO_2 concentrations (0.7-6.3%) increased the number of sprouted eyes, and increased sprout growth in *Saturna* and *Panda* respectively. The highest weight loss after reconditioning was found in the potatoes which has previously been stored in the gas combination of 0.9% CO_2 and 21% O_2, while the gas combinations with high CO_2 showed the lowest weight loss after reconditioning. Results also indicated that the fry colour on the surface of crisps produced from all the cultivars for most of the CA combinations, at both temperatures, was lighter than that obtained after curing.

Keywords: *Solanum tuberosum*, crisps, oxygen, carbon dioxide

1. INTRODUCTION

Sprout control is very important when potato tubers must be stored for a long period. Storing potatoes at low (2-4 °C) temperatures, after the curing period, increases the accumulation of reducing sugar content which, depending on cultivar and growth circumstances, will in turn develop a dark, brown bitter product when processed into crisps (Harris, 1992). Isherwood (1976) and Iritani and Weller (1978) showed that reconditioning may remove most of the sugars developed during storage, but those accumulated as a result of senescent sweetening are irreversible. Storage at higher temperature (10 °C) reduces sugar accumulation, but tubers should be treated with a disease control chemical and a chemical sprout suppressant. Burton (1973) found that tubers reconditioned after storage temperature of 5 °C still maintain higher sugar levels than tubers stored at 10 °C.

Storage of potatoes in high CO_2 levels was reported to increase the sugar content of the tubers (Schouten, 1992; Smith, 1987, and Reust et al., 1984). Smith (1987) reported that tubers stored at higher CO_2 concentrations must be reconditioned before processing. Recently Khanbari & Thompson (1994), showed that potato tubers stored in high CO_2 levels responded to a commercial reconditioning treatment better than those stored in air.

In this study, four potato crisping cultivars were subjected to long term storage under different controlled atmospheres for the purpose of investigating their response to a commercial reconditioning treatment.

2. MATERIAL AND METHODS

Tubers of the cultivars *Record*, *Saturna*, *Hermes* and *Panda* grown in Nottingham, UK, were harvested on 28 September 1992 and cured for two weeks at 15 °C, and 90-95% relative humidity. Eighteen samples from each cultivar, each sample consisting of 4-5 medium size tubers (120-150g), were placed in perforated paper bags. The initial weight of the tubers in the bags was recorded. The eighteen paper bags from each cultivar were then placed randomly in each of eighteen air tight 75 litre polyethylene containers (Model C217, Mailbox International Ltd, Cheshire, UK). The containers were divided into two groups of nine and each group was stored, in air for 43 days, in a temperature controlled room adjusted to either 4 °C or 8 °C. Those containers were then connected to a gas distributor (Mercury, UK), by a PVC tubing coil of 6.5 mm internal diameter. The gas distributor was connected to a computer-controlled gas blender (850 series, Signal Instrument CO. Ltd, Surrey, UK). The CA system used in this trial was previously described by Khanbari & Thompson (1994). Eighteen CA combinations of CO_2 and O_2 were tested and details are given in Table 1 and Table 2.

After six months of CA storage, the tubers were removed from the containers, left at room temperature (about 20 °C) for two hours and weighed. Sprouts were cut off, and weighed before and after reconditioning the tubers for two weeks at 20 °C and 90-95% relative humidity. The tubers were then peeled, sliced and two middle slices from each tuber were washed under tap water, surface dried with paper tissues, and fried at 180-183 °C for 3.5 minutes.

The method for crisp colour measurement was by computer vision system, previously described by Khanbari and Thompson (1993 & 1994), which measures the proportion of light-ness or dark-ness (grey level) over the surface of each crisp, where 0=black and 256=white. The grey level, was measured for each of the eight crisps separately, and results were analysed statistically using Genstat 5 (Copyright 1990, Lawes Agricultural Trust, Rothamsted Experimental Station). The grey level of crisps, from tubers of the four cultivars, processed after curing was found 154, 153, 151 and 150.5 for *Hermes*, *Record*, *Saturna*, and *Panda* respectively.

Table 1 light-ness (grey level) of crisps from four potato cultivars when reconditioned for two weeks at 20 °C after 6 months of storage at 4 °C under different controlled atmospheres

Treatment CO_2%	O_2%	*Record*	*Saturna*	*Hermes*	*Panda*	Mean
0.7	3.7	160.7	158.8	160.4	160.6	160.6
1.8	3.9	157.9	158.4	158.9	158.4	158.4
5.3	3.9	156.4	157.1	160.3	157.8	157.9
10.2	3.8	159.1	158.3	160.5	157.7	158.9
0.7	2.4	158.3	157.1	161.8	158.5	158.9
1.6	2.1	159.1	158.8	159.5	160.9	159.6
4.8	2.0	158.4	158.2	157.7	159.4	158.4
9.7	2.0	158.3	158.6	160.5	160.0	159.4
0.9	21	158.9	159.0	156.6	159.5	158.5

LSD (p=0.05) for treatments = 2.16
S.e. = 0.78 CV% = 1.3
Cultivars = Not significant
Cultivars x treatment = Not significant

3. RESULTS

3.1 *Processing quality after reconditioning.*

All cultivars stored at 4 °C, showed a positive response to reconditioning, and light coloured crisps were obtained after frying. There were no statistically significant difference (p = 0.05) in fry colour between the cultivars (Table 1). At the same time none of the tubers in the CA combinations had a significantly lighter crisp colour than the controls (0.9% CO_2 and 21% O_2). There was no significant interaction between the cultivars and the gas combinations. However, the lightest crisps was obtained from tubers of *Hermes* followed by *Panda* and *Record*, but tubers of *Hermes* stored in the control treatment at 4 °C produced darker crisps compared with the other three cultivars.

Table 2 light-ness (grey level) of crisps from four potato cultivars when reconditioned for 2 weeks at 20 °C after 6 months of storage at 8 °C under different controlled atmospheres

Treatments CO_2%	O_2%	*Record*	*Saturna*	*Hermes*	*Panda*
12.3	5.3	157.7	160.6	156.5	161.2
10.3	10	156.7	160.5	159.3	160.2
14.8	3.0	156.3	161.6	157.4	163.0
12.0	2.9	158.6	159.9	158.8	161.8
9.3	2.8	158.7	160.2	159.6	162.2
6.3	3.0	159.0	159.8	157.3	161.0
3.7	3.2	155.8	160.6	158.3	160.8
0.8	2.8	158.2	161.2	158.2	162.2
0.9	21	156.1	160.8	159.8	160.9
Mean		157.4	160.6	158.4	161.5

LSD (p=0.05) for cultivars = 1.32
S.e. = 0.52 CV% = 2.7
For treatment = Not significant
Treatment x cultivar = Not significant

Tubers of *Panda* and *Saturna*, reconditioned after storage at 8 °C, had a significantly (p = 0.05) lighter crisp colour than tubers of *Hermes* and *Record* (Table 2). Most of the cultivars, stored in gas combinations of 0.8% CO_2 and 2.8% O_2, 9.3% CO_2 and 2.8% O_2, 12% CO_2 and 2.9% O_2, and of 14.3% CO_2 and 3.0% O_2, produced the lightest crisps after reconditioning. However, none of the treatments provided significantly lighter crisps than the controls. The quick response of the cultivars to the reconditioning treatment clearly indicated that tubers from the different treatments did not reach a stage of senescence sweetening even in those treatments (3.7% CO_2 and 3.2% O_2, 6.3% CO_2 and 3.0% O_2, and 0.8% CO_2 and 2.8% O_2) where tubers from *Panda* showed vigorous sprout growth after storage.

Furthermore, it has been observed that the reconditioned tubers from both storage temperatures, after total storage period of seven months, in air and the different CA combinations, produced crisps lighter than those directly produced after curing. The light-ness of crisps from tubers processed after curing showed an unusual dark colour. This could have been due to the unfavourable pre-harvest conditions, where the harvest time was delayed twice because of the wet and cold season (soil temperature at harvest was recorded as below 5 °C). Under these conditions potato tubers have been reported to develop an increased amount of reducing sugars and sucrose (Storey & Davies, 1992; Burton et al., 1992) sufficient to affect the light-ness of directly processed (not reconditioned) tubers.

3.2 *Sprout growth.*

At 4 °C, the tested cultivars appeared to behave differently, when stored in low CO_2 (0.7-1.8%), and low O_2 (2.0-3.9%) combinations, compared with the control (Table 3). In these gas combinations, tubers of *Panda* produced double the number of sprouted eyes than those of the control with similar results for *Record*. However, *Saturna* developed about the same number of sprouted eyes in all the gas combinations, with a similar trend for *Hermes* in the gas combinations of 0.7% CO_2 and 3.7% O_2, 0.7% CO_2 and 2.4% O_2, and of 1.8% CO_2 and 3.9% O_2. Tubers of this cultivar stored in an atmosphere of 1.6% CO_2 and 2.1% O_2, and 4.8% CO_2 and 2.0% O_2, produced half the number of sprouted eyes as those in the control.

Table 3 Sprout growth and number of sprouted eyes of four potato cultivars directly after storage at 4 °C under different controlled atmospheres

Treatments		*Record*		*Saturna*		*Hermes*		*Panda*	
CO_2	O_2 (%)	No. of sprouted eyes	Sprout Wt./tuber (g)	No. of sprouted eyes	Sprout Wt./tuber (g)	No. of sprouted eyes	Sprout Wt./tuber (g)	No.of sprouted eyes	Sprout Wt./tuber (g)
0.7	3.7	6.2	0.08	7.3	0.26	6.0	1.32	4.5	6.55
1.8	3.9	6.2	0.12	7.0	1.63	5.5	0.33	6.0	8.13
5.3	3.9	4.8	0.06	6.5	0.18	3.5	0.07	3.0	3.11
10.2	3.8	1.9	0.00	0.0	0.00	0.0	0.00	0.0	0.00
0.7	2.4	5.5	0.10	7.3	1.15	6.0	0.52	4.8	7.82
1.6	2.1	5.7	0.14	6.5	0.21	2.5	0.20	5.8	5.14
4.8	2.0	3.8	0.10	6.3	0.58	3.0	0.07	1.0	0.38
9.7	2.0	1.0	0.00	0.0	0.00	0.0	0.00	0.0	0.00
0.9	21	4.8	0.20	6.8	0.24	6.0	1.21	2.8	2.80

Table 4 Sprout growth[1] (g/tuber) of four potato cultivars when reconditioned for 2 weeks at 20 °C after storage at 4 °C under different controlled atmospheres

Treatments CO_2	O_2 (%)	*Record*	*Saturna*	*Hermes*	*Panda*
0.7	3.7	0.91	0.80	0.70	2.08
1.8	3.9	0.48	0.86	0.87	2.59
5.3	3.9	0.72	0.65	1.32	2.41
10.2	3.8	0.21	0.43	3.64	1.12
0.7	2.4	0.58	1.30	2.11	2.86
1.6	2.1	0.71	0.76	1.07	1.97
4.8	2.0	1.07	0.70	1.20	2.08
9.7	2.0	0.59	1.37	1.10	1.51
0.9	21	0.56	0.67	0.54	2.78

[1] Sprouts were removed prior reconditioning

High CO_2 (9.7% and 10.2%), and low O_2 (2.0% and 3.8% respectively) combinations, completely inhibited sprouting in tubers of *Saturna*, *Hermes* and *Panda*, while some tubers of *Record* showed eye opening (Table 3).

The highest weight of sprouts per tuber was found in *Panda* at the gas combinations of 0.7% CO_2 and 3.7% O_2, 1.8% CO_2 and 3.9% O_2, 0.7% CO_2 and 2.4% O_2, and of 1.6% CO_2 and 2.1% O_2. These accounted for twice or even three times that of the control. For *Saturna* the gas combinations of 1.8% CO_2 and 3.9% O_2, 0.7% CO_2 and 2.4% O_2, and of 4.8% CO_2 and 2.0% O_2 gave higher sprout growth than that of the control. In *Hermes* only the gas combination of 0.7% CO_2 and 3.7% O_2 gave about the same weight of sprouts per tuber as that produced by tubers stored in the control. Sprout dehydration in *Record* reduced the sprout weight per tuber for all the low CO_2 and low O_2 treatments (Table 3). Sprout dehydration in *Record* was also observed by Khanbari & Thompson (1994).

In all cultivars reconditioning of tubers enhanced growth of sprouts (Table 4). The highest sprout weight was found in tubers of *Panda*. Sprouting after reconditioning was highest in tubers of *Hermes* at the gas combination 10.2% CO_2 and 3.8% O_2. *Saturna*, *Panda*, and *Hermes* produced about the same weight of sprouts in the gas combination of 9.7% CO_2 and 2.0% O_2, but *Saturna* and *Record* stored in this gas combination produced twice or even three times the weight of those in the gas combination of 10.2% CO_2 and 3.8% O_2 respectively.

At 8 °C *Record*, *Saturna* and *Hermes* showed signs of sprout growth at the gas combination of 10.3% CO_2 and 10.0% O_2, 12.3% CO_2 and 5.3% O_2, and of 9.3% CO_2 and 2.8% O_2. Sprout growth was inhibited at 14.8% CO_2 and 3.0% O_2, and 12.0% CO_2 and 2.9% O_2. The four cultivars responded differently to the different CO_2 and O_2 combinations (Table 5 and Table 6). High CO_2 (9.3-15%) completely inhibited sprouting in *Panda*. *Saturna* and *Record* showed sprout initiation in the gas combination of 12.3% CO_2 and 5.3% O_2, 10.3% CO_2 and 10.0% O_2, and of 9.3% CO_2 and 2.8% O_2. With decreasing concentrations of CO_2 (0.8%-6.3%) *Panda* produced increasing sprout weight per tuber and an increased number of sprouted eyes (especially in the gas combinations of 3.7% CO_2 and 3.2% O_2, and of 0.8% CO_2 and 2.8% O_2) compared to the control in air. In the other three cultivars the number of sprouted eyes were the same as in the control. However, sprout weight per tuber was much less in *Hermes* than in the control. The gas combination of 6.3% CO_2 and 3% O_2 gave the highest sprout weight in *Saturna*, while less than 1 g per tuber was obtained in *Record* and *Hermes* after reconditioning (Table 6).

Table 5 Sprout growth and number of sprouted eyes of four potato cultivars directly after storage at 8 °C under different controlled atmospheres

Treatments		*Record*		*Saturna*		*Hermes*		*Panda*	
CO_2 (%)	O_2 (%)	No. of sprouted eyes/tuber	Sprout Wt./tuber (g)	No. of sprouted eyes/tuber	Sprout Wt./tuber (g)	No. of sprouted eyes/tuber	Sprout Wt./tuber (g)	No.of sprouted eyes/tuber	Sprout Wt./tuber (g)
12.3	5.3	0.5	0.00	2.4	0.00	2.0	0.02	0.0	0.00
10.3	10	2.2	0.09	4.0	0.22	6.0	1.08	0.0	0.00
14.8	3.0	-[1]	-	0.0	0.00	5.0	0.08	0.0	0.00
12.0	2.9	0.0	0.00	0.0	0.00	1.5	0.13	0.0	0.00
9.3	2.8	1.9	0.01	3.5	0.12	5.0	0.03	1.0	0.02
6.3	3.0	3.5	0.31	5.7	2.70	5.5	0.29	3.3	7.29
3.7	3.2	5.3	0.27	6.8	0.90	5.0	1.90	4.3	12.73
0.8	2.8	5.7	0.19	6.0	0.20	5.5	0.48	5.3	9.86
0.9	21	5.5	0.21	6.3	0.66	5.5	6.71	3.3	4.67

[1] All tubers were rotten

Table 6 Sprout weight[1] (g/tuber) of four potato cultivars when reconditioned for 2 weeks at 20 °C after storage at 8 °C under different controlled atmospheres

Treatments CO_2%	O_2%	*Record*	*Saturna*	*Hermes*	*Panda*
12.3	5.3	0.43	0.82	0.94	1.41
10.3	10	0.77	0.59	0.88	2.00
14.8	3.0	-[2]	0.86	0.61	1.00
12.0	2.9	0.70	0.82	0.20	1.73
9.3	2.8	0.60	1.02	0.77	1.95
6.3	3.0	0.90	0.66	0.62	1.21
3.7	3.2	0.82	0.84	1.32	1.88
0.8	2.8	0.55	0.56	2.25	2.51
0.9	21	0.77	0.80	1.75	2.75

[1] Sprouts were removed prior reconditioning
[2] All tubers were rotten

3.3 *Weight loss.*

All the cultivars at the different gas combinations had a weight loss within a range of 1.5% to 3% by the end of the CA storage, with *Record* showing the lowest loss. Those treatments which enhanced sprouting, 0.7% CO_2 and 3.7% O_2 in *Hermes*, and 1.8% CO_2 and 3.9% O_2, 0.7% CO_2 and 2.4% O_2, and 1.6% CO_2 and 2.1% O_2 in *Panda*, had weight losses of above 3%. High weight loss, also occured in tubers of *Panda* stored in the control. Reconditioning of tubers, previously stored in the control at 4 °C, appeared to have an increased weight loss in *Panda*, *Record* and *Hermes* (Table 7). The weight loss in these three cultivars was found to be higher than in *Saturna*. Tubers of *Saturna*, *Hermes* and *Panda*, had lower weight loss than *Record*, when reconditioned after removal from the gas combination of 10.2% CO_2 and 3.8% O_2, and of

9.7% CO_2 and 2.0% O_2. Those tubers which were removed from the above two gas combinations showed no sign of sprouting during storage.

Weight loss from tubers of the four cultivars immediately after storage at 8 °C showed an average range of 2-3%. After reconditioning, tubers which had been stored in the control produced the highest weight loss (Table 8). Those combinations which gave high sprout weight per tuber, in particular the low CO_2 (6.3% and below), had a higher percentage weight loss especially in *Panda*. At higher levels of CO_2 (9.3% and above) weight loss in tubers of *Record*, was found to be higher than that of *Saturna*, *Hermes* and *Panda*. There was an indication that increasing O_2 (5.3% and 10%) in the CA storage, at higher CO_2 levels, also increased the weight loss in *Record* and *Hermes*.

Table 7 Effect of different controlled atmospheres on percentage weight loss from four potato cultivars after 6 months of storage at 4 °C and after reconditioning for 2 weeks at 20 °C

Treatments CO_2%	O_2%	*Record*	*Saturna*	*Hermes*	*Panda*
0.7	3.7	3.4	2.9	3.2	11.1
1.8	3.9	3.4	4.5	3.5	11.8
5.3	3.9	2.6	2.7	3.1	5.9
10.2	3.8	3.8	1.9	2.3	2.4
0.7	2.4	4.6	4.6	3.9	11.9
1.6	2.1	4.3	3.2	2.7	8.7
4.8	2.0	4.8	3.3	2.2	3.6
9.7	2.0	4.9	2.3	2.5	2.6
0.9	21	6.5	3.4	4.8	7.8

4. DISCUSSION

Low temperature is an effective method of maintaining some of the quality characteristics in stored potato. The rate of sprout growth is reduced potentially allowing effective control with reduced chemical input. However, at low temperatures reducing sugar accumulation results in poor fry colour on processing. This may be improved by reconditioning which is achieved by holding tubers at an elevated temperature. Results of reconditioning after long term storage at low temperature showed differences between cultivar response. In previous work, Iritani and Weller (1978), and Storey and Shackley (1987) found that some cultivars required up to four weeks or more to bring down tuber reducing sugar to an acceptable level. Cunnington and Gerrish (1992) reported *Record* to be one of those cultivars that needs a minimum of 4 weeks to produce commercially acceptable fry colour. Results of this experiment showed the quick response to reconditioning of the CA stored tubers of all tested cultivars. Coffin et al. (1987), on long term storage at 5 and 10 °C, reported a deterioration in fry colour of potato chips after reconditioning of tubers at 20 °C. Parkin & Schwobe (1990) reported little colour improvement as a result of reconditioning tubers of two cultivars previously stored at 3 °C in air. results of this experiment also confirm our previous finding (Khanbari & Thompson, 1994), on the quick response of long term CA stored potatoes at 4 °C to reconditioning treatment.

Table 8 Effect of different controlled atmospheres on percentage weight loss from four potato cultivars, stored at 8 °C and then reconditioned for 2 weeks at 20 °C

Treatments CO_2%	O_2%	*Record*	*Saturna*	*Hermes*	*Panda*
12.3	5.3	5.3	2.1	3.3	2.6
10.3	10	6.3	2.7	3.7	3.0
14.8	3.0	-[1]	2.7	2.9	2.3
12.0	2.9	3.5	2.1	1.9	2.5
9.3	2.8	4.3	2.2	2.7	2.8
6.3	3.0	4.5	5.3	2.4	10.4
3.7	3.2	4.9	3.8	4.2	15.3
0.8	2.8	5.7	2.8	2.8	14.6
0.9	21	6.6	4.6	7.5	9.2

[1] Some tubers were rotten

The inhibitory effect of high CO_2 on sprout growth has been reported by several workers (Khanbari & Thompson, 1994; Schouten, 1992; Reust et al., 1984, and Burton, 1958). The results presented in this work confirm these findings at the CO_2 levels of 9-12% at both storage temperatures. However, as the level increased tuber rotting also increased on long term storage of potatoes and *Record* was found the most affected cultivar.

Results also indicate the extended effect of the high CO_2 combinations during reconditioning. Tubers of all tested cultivars, produced low weight loss at both temperatures compared with the controls. This could be the reason of maintaining a good skin quality in the CA stored potatoes.

REFERENCES

Coffin, R. H., K. L., Yada, B. G. Parkin and D. W. Stanley (1987). Effect of low temperature storage on sugar concentrations and chip colour of certain processing potato cultivars and selections. *J Food Sci*, **52,** 639-645

Cunnington, A. and S. Gerrish (1992). Potato Marketing Board, Annual review, Sutton Bridge Experimental Station 1991.

Burton, W. G. (1958). The effect of the concentration of carbon dioxide and oxygen in the storage atomsphere upon the sprouting of potatoes at 10 °C. *Eur. Potato J.*, **1** (2), 47-57.

Burton, W. G. (1973). Physiological and biochamical changes in the tubers as affected by storage conditions. Proc. 5th. trienn. Conf. Eur. Assn. *Potato Res*, Norwich, England, 1972, pp.63-81.

Burton, W. G., A. Van Es and K. J. Hartman (1992). The physics and physiology of storage. In: *The Potato Crop* (P. Harris, Ed.), pp. 608-709. Chapman & Hall, London.

Harris, P. (1992). *The potato crop*, 2nd eddition, pp. 640-643. Chapman & Hall, London.

Iritani, W. M. and L. Weller (1978). Factors influencing reconditioning of *Russet burbank* potatoes. *Amer. Potato J.*, **55**, 425-430.

Isherwood, F. A. (1976). Mechanism of starch-sugar interconversion in *Solanum tuberosum*. *Phytochemistry*, **15**, 33-41.

Khanbari, O.S. and A. K. Thompson (1993). Effect of amino acids and glucose on the fry colour of potato crisps. *Potato Res.*, **36**, 359-365.

Khanbari, O.S. and A. K. Thompson (1994). Effect of controlled atmosphere storage at 4 °C on colour , sprout growth, rotting and weight loss of potato tubers. *Potato Res.*, **37**, 273-282.

Parkin, K. L. and M. A. Schwobe (1990). Effect of low temperature and modified atomsphere on sugar accumulation and chip color in potatoes (Solanum tuberosum). *J. Food Sci.*, **55** (5), 1341-44.

Reust, W., A. SCHWARZ and J. Aerny (1984). Essai de conservation des pommes de terre en atomsphère contrôlée. *Potato Res.*, **27**, 75-87

Schouten, S. P. (1992). Possibilities of controlled atmosphere storage of ware potatoes. *Asp. Appl. Biol.*, **33**, 181-188.

Smith, O. (1987). Potato chips. In: *Potato processing* (W. F. Talburt and O. Smith, Eds.) 4th ed. Chap. 10, pp. 371-473. Avi Publishing CO, Westport, CT, USA.

Storey, R. M. J. and H. V. Davies (1992). Tuber quality. In: *The Potato Crop* (P. Harris, Ed.). pp. 507-569. Chapman & Hall, London.

Storey, R. M. J. and D. J. Shackley (1987). The effect of reconditioning on the processing quality of Pentland Dell, Maris Piper and Wilja. 10th Trienn. EAPR Conference, 1987, Aalborg. pp. 61-62.

THE DEVELOPMENT OF COMPUTER CONTROLLED ENVIRONMENT IN POTATO STORAGE

Felix Lamber B.Sc.

Fancom B.V. P.O. Box 7131
5980 AC Panningen

1. INTRODUCTION

The development of environmental control in potato storage has made a tremendous increase in the last decade. Manual controlled storages have made room for automatic control systems. The investment costs for electronic or computer controlled environmental equipment have decreased. The use of this equipment is becoming an economical necessity in modern storage.

After an inventory of the present control systems, the computer controlled systems are extensively analysed according:
- purpose
- environmental factors
- control functions
- management functions
- advantages

At the end a prediction is given of a control system in 2000.

2. INVENTORY OF SYSTEMS

The manual control of fans and louvers according to read outs of thermometers has been replaced by successively thermostats, differential thermostats, electronical air mixing controls. All controls only had a possibility to control temperature. Humidistats enabled to decrease weight losses by humidifying supplied air. Mechanical timers and hour counters provided poor programming and registration possibilities. Present electronic and computer controlled equipment offer extended possibilities.

2.1 Purpose controls

Purpose of using control equipment is:
- avoiding diseases, sprouting, damages by extreme temperatures, discolorisation and bruising or pressure spots
- minimise weight losses
- maintain quality (sugars, fry colour)
- management tool

2.2 Environmental factors

Since temperature control is not sufficient to store potatoes in the most optimal way, following climate factors are being controlled:
- pile temperature
- plenum temperature and humidity
- pile uniformity
- air flow and direction
- pile humidity
- above pile temperature and humidity
- CO_2 level

Control functions pile temperature
Pile temperature is controlled by using cheap outside air if suitable for cooling and heating. The computer measures both outside air temperature and humidity to determine heat content of the air according the psychrometric diagram and to compare it with the heat content of the ambient air in the pile. If outside conditions are suitable, the fan will be switched on.

Plenum temperature and humidity
Most computers control temperature difference between pile and plenum. Managers can choose for fixed or variable differences between pile and plenum.
If the store house is provided with a mechanical heating or cooling system, the computer decides to switch over from natural to mechanical heating or cooling.
If the outside air is too dry, the computer stops ventilation or humidifies the plenum air. Also mechanical dryers are seldomly used because of energy and investment costs.

Pile uniformity
To create pile uniformity following methods are used:
- recirculation based on cooling times
- recirculation based on temperature differences
- variable air speed based on pile temperature differences
- variable pile-plenum differences
- reversed recirculation

Air flow and direction
Depending on the phase of the storage, the computer is able to control a certain air flow. Most storages in Western Europa and the North of USA use single speed fans. The air flow can be controlled by switching on a number of fans, or modulating the running time. In Northern Europe where lower air flows are common, air flow is controlled by two speed or variable speed motors. In Germany inverse recirculation of fans is applied to equalise pile temperatures and humidities. Time clocks enable to use low energy rate periods, peak-shaving.

Pile humidity
In Denmark, Germany, Czechia / Slowakya and Holland humidity monitors have been developed. Some manufacturers use this information to decide for a manual switch-over from drying to wound-healing phase, while others choose for an automatic switch based on the measured resistance in the potatoes.

Above pile temperature and humidity
These climate factors are controlled to prevent condensation. Some manufacturers use timers to control circulation fans, other use circulation fans with electric heaters, while the most sophisticated systems use a sequency of circulation, ventilation and heating based on measured roof temperature and dewpoint of above pile air.

CO_2 level
The control of the CO_2 level has following functions:
- safety regulations
- prevent formation of reducing sugars

Programs
Climate control factors can be controlled in preset programs like: drying, woundhealing, cooling down, holding and reconditioning.

2.3 Management functions

Computer controlled environment offers following management facilities:
- alarms
- registration of climate, running times of fans, ventilation, internal circulation, suitable outside conditions

If a personal computer is connected to the controllers, all measured data can be stored, displayed in graphics or printed. It is possible to use the computer as a remote control. By means of modems it is possible for the manager to keep contact with the store house via the telephone line. Some suppliers of control equipment offer on-line support via modem.

2.4 Advantages

Computer controlled environment offers following advantages above electronic or manual controls:
- less weight loss
- saving of energy
- saving of labour costs
- management facilities
- better quality
- peak shaving

Investments in computer controllers can be paid back within some years. The exact returns on investment, depend on local potato sales prices, energy costs, interests and subvention possibilities.

3. LATEST DEVELOPMENTS

Computerized control systems are already used to dose anti-sprouting inhibitors into the store house. Because of efficiency of dosing it is necessary that the computer starts dosing in periods after outside air ventilation together with internal circulation. It blocks ventilation to enable the inhibitor to get effective. It is also possible to relate the amount of dosing to the temperatures of the pile.
Controlled Atmosphere (CA-control) is being researched, and computer control offers already control possibilities. Since controllers can be connected to weather forecast datalines it is possible to use computers for predictive control. Although technoloy offers already high-tech possibilities the biggest problem for the manufacturer is to keep it simple for the operator. The store house manager is no computer expert, but wants to use the computer as a management tool.

Storage control in 2000
If the same increase of development goes on, we might have fully computerized storage systems, that store all possible weather data during cropping, harvesting. Depending on variety, soil conditions and weather forecasts the drying and wound healing processes will be calculated by the computer. The type of application of the product (chips, fries etc.) determines settings of storage phases. A continuous monitoring of quality factors (sugars, fry colour) will influence the environmental regime.
The price level of the potato, however, might severely influence this positive look into the future.

ARTIFICIAL BANANA RIPENING BY FUZZY CONTROL

Y. SEO*, T. AMANO*, Y. KAWAGOE* and Y. SAGARA*

*Dept. of Agric. Engineering, Fac. of Agriculture, Univ. of Tokyo, Yayoi 1-1-1, Bunkyo-ku, Tokyo 113, JAPAN

Abstract: Fuzzy control was applied to 7-day artificial banana ripening with the fuzzy rules of three inputs and one output. The operating conditions for ventilation based on CO_2 concentrations and the deviation of CQEC from the target value were considered to reflect the effects of CO_2 and O_2 concentrations in the ripening chamber on the CO_2 evolution rate of bananas. Changes in the measured value of CQEC in ripening by fuzzy control closely followed changes in the target value. Ventilating operations were effectively conducted to improve the control. 6050 mg/kg of CQEC and 10.1 brix% of increase in sugar content were obtained at the end of the ripening against the target values of 6030 mg/kg and 11.8 brix%.

Keywords: Agriculture, Automatic control, Bio control, Fuzzy control, IF-THEN operators

1. INTRODUCTION

In past two years, Japan imported yearly over nine hundred thousand tons of bananas mainly from the Phillipines, Equador and Taiwan. Those imported bananas, so called green mature bananas, are mature but unripe with dark green peels and are not edible on their arrival in Japan because import of yellow and ripe bananas are prohibited by quarantine reasons. Therefore, all imported bananas are artificially ripened with ethylene gas before shipped to market. Commercial banana ripening still relies , to a large extent, on skills and experience of operators, although it is widely practiced under certain standard temperature schedules.

A previous study (Seo and Hosokawa, 1983a) on automation of banana ripening showed that increase in sugar content of bananas during artificial ripening was closely correlated with the cumulative quantity of evolved CO_2 (CQEC) by respiration. The correlation of two parameters was expressed by the regression equation below,

$$Y=-0.497 + 0.257\times10^{-2} X - 0.841\times10^{-7} X^2 \quad (1)$$

where, X: cumulative quantity of evolved CO_2 (CQEC), mg/kg

Y: increase in sugar content, brix%

Thus, a sugar content of bananas at any moment during ripening can simply be predicted by knowing the cumulative quantity of evolved CO_2 by bananas. Seo and Hosokawa(1983b) also showed that the CQEC at any moment of ripening process under a given temperature schedule could be calculated by CO_2 evolution rate curves at the temperatures of which the schedule consists, and conversely, it was possible to construct a ripening temperature schedule so as to achieve a given value of CQEC or a sugar content of bananas at the end of ripening. Amano et al.(1992) tried to manage CQEC by a conventional control algorithm to control banana ripening process, but the stable control of CQEC was not always carried out.

As for fuzzy control, the experience and the knowledge of skilled operators which are usually expressed qualitatively in words are applied to control of processes by putting them into logical forms as fuzzy control rules. Therefore, fuzzy control was

Table 1 Temperature schedule for 7-day ripening

Ripening day	1st day	2nd day	3rd day	4th day	5th day	6th day	7th day
Temp. °C	17.5	16.0	15.0	15.0	14.0	14.0	14.0

considered to be appropriate to apply for automation of banana ripening.
The objective of this study is to establish automatic banana ripening by fuzzy control.

2. MATERIALS AND METHODS

2.1 Bananas

Two cartons (12 kg banana/carton) of bananas from the Philippines were used for each ripening experiment. The variety of bananas was Cavendish.

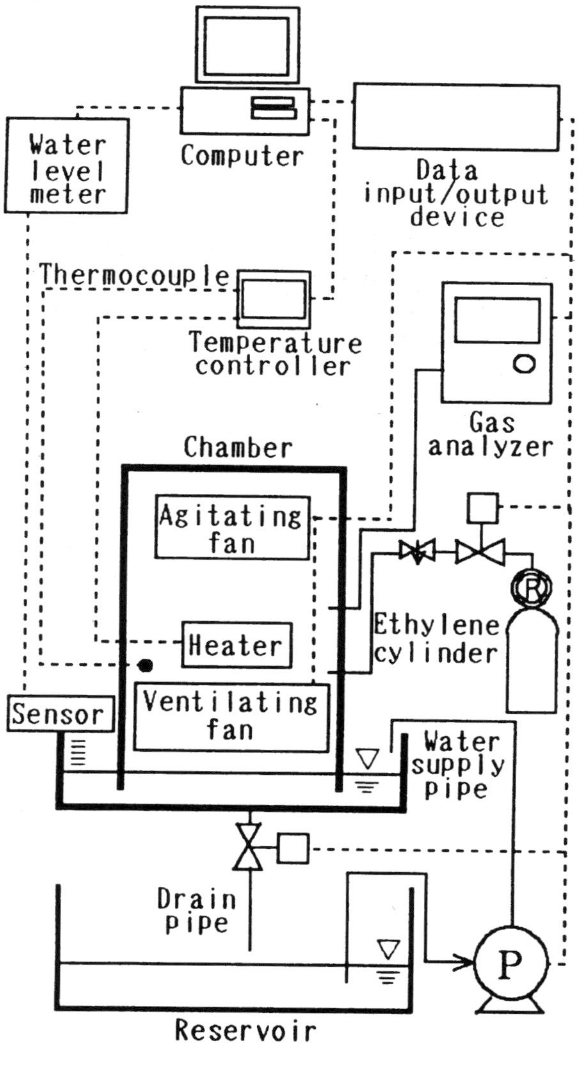

Fig.1 Arrangement of experimental apparatus

2.2 Ripening temperature schedule

Each banana ripening experiment by fuzzy control was carried out for seven days. The temperature schedule used for ripening is shown in Table1. This schedule which predicts about 6030 mg/kg of CQEC and 11.9 brix% of increase in sugar content by the end of ripening was proposed by Seo and Hosokawa(1983b) for 7-day medium banana ripening. The temperatures used for the schedule were set as standard temperatures for ripening by fuzzy control. The value of CQEC obtained by this temperature schedule was the target value for fuzzy control.

2.3 Experimental apparatus

The arrangement of the experimental apparatus used for the experiments is illustrated in Fig.1. The apparatus principally consists of two main components, namely a ripening chamber of 1050H×700W×550D mm with the volume of about 400 litres and peripheral instruments for control and measurements. The chamber is composed of an aluminum frame with a two-tier shelf to put bananas on, a plyvinyl sheet tent covering the frame and a base chamber. The chamber is equipped with a 500W electric heater, two cross-flow fans for agitation and ventilation and a ethylene cylinder. For confinement of ethylene gas and for accurate data acquisition of CQEC, the chamber must be airtight. Airtightness is achieved by immersing the skirt of the tent into water in the base chamber. The base chamber is equipped with a water level sensor, a water reservoir and a pump. The whole chamber was placed in a refrigerated room maintained at the temperature of about 10°C . Operations needed for the ripening experiment was fully conducted automatically with the apparatus.

2.4 Ethylene gas

At the beginning of ripening, ethylene gas was inserted to the chamber to stimulate ripening from the cylinder through the solenoid valve and was confined for a period of 24 hours. Ethylene concentration was maintained at the level of about 1000 ppm in the ripening chamber.

2.5 Ventilation

Ventilation of the ripening chamber was performed to remove the accumulated CO_2 and the confined ethylene gas from the chamber. Ventilation started with lowering the level of water in the base chamber by draining the water to the water reservoir under the chamber. After the water level was lowered so that the ripening chamber was open to the outside air, the ventilating fan started operating to introduce the outside air into the chamber. When CO_2 concentration became low enough, the ventilating fan

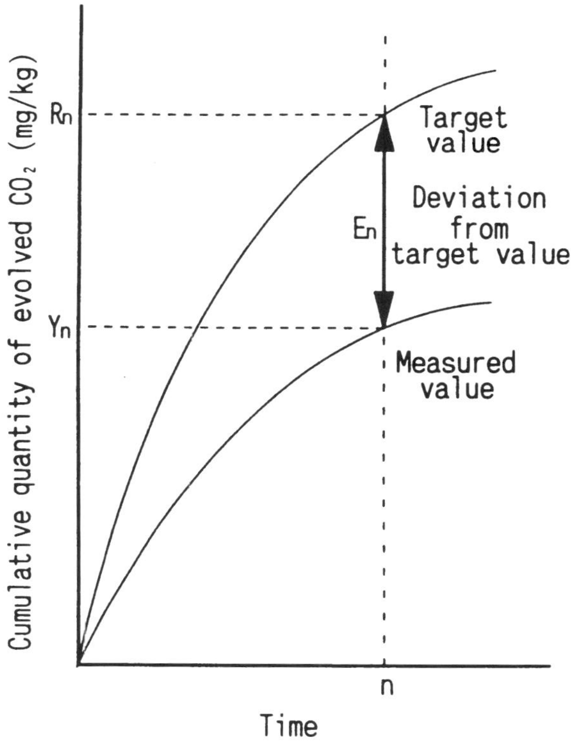

Fig.2 Explanatory illustration for E_n

stopped and at the same time, the water level was risen again by sending water back to the base chamber to keep airtightness of the chamber.

2.6 Fuzzy control

The control algorithm for fuzzy control is given in "if-then" expressions, such as

If x is small and y is big, then z is medium (2)

If x is big and y is medium, then z is big (3)

where, x and y are information about the condition of the process to be controlled, and z is the input for the process.

These expressions are called fuzzy control rules. In general, variables x and y are called the input and z the output. "Small" and "big" are fuzzy values for x and y (sometimes called fuzzy variables). Fuzzy controllers are constructed of groups of these fuzzy control rules, and when an actual input is given, the output is calculated by means of fuzzy inference (Terano et al. 1992). In this study, E_n, ΔE_n and C_n were selected for the input. As illustrated in Fig.2, E_n is the deviation of a measured value from a target value which is obtained by the temperature schedule in Table 1 and expressed by the equation below.

$$E_n = R_n - Y_n \qquad (4)$$

where R_n:target value at the time n

Y_n:measured value at the time n

ΔE_n is the change of E_n in a sampling interval for CO_2 analysis between the time n-1 and n, that is $\Delta E_n = E_n - E_{n-1}$. The CO_2 concentration was measured every ten minutes by the infrared gas analyzer. C_n is CO_2 concentration in the chamber at the time n. ΔU_n which is the change of U_n ($=U_n - U_{n-1}$) was chosen for the output. U_n is the difference between the standard temperature set by the temperature schedule and the temperature set by the control. Membership functions for E_n, ΔE_n, ΔU_n and C_n are illustrated in Fig.3 and Fig.4.

The domain from -300 to +300 (mg/kg) for E_n, from -18 to +18 (mg/kg/h) for ΔE_n and from -0.05 to +0.05 °C/10 min. for ΔU_n are standardized to the Fig. domain from -1 to + 1 for the membership function as shown in Fig.3.

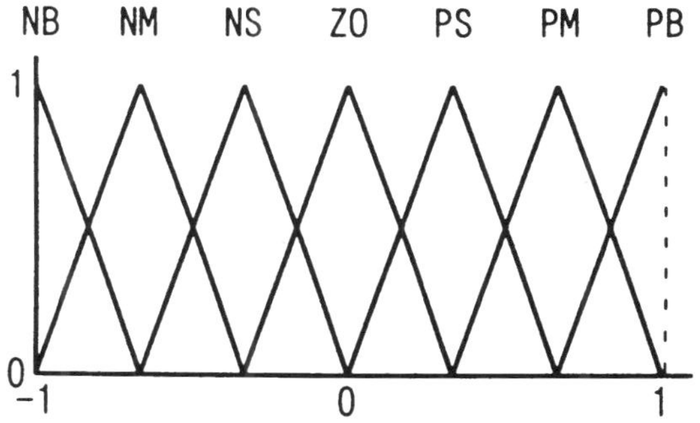

NB:Negative Big

NM:Negative Medium

NS:Negative Small

ZO:Zero

PS:Positive Small

PM:Positive Medium

PB:Positive Big

Fig. 3 Membership function for E_n, ΔE_n and ΔU_n

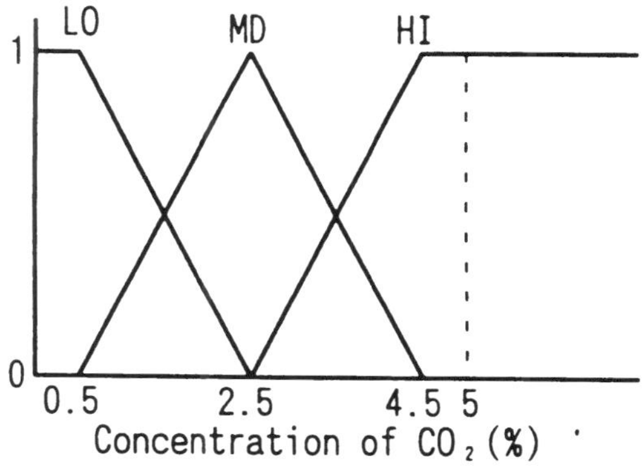

LO:Low concentration

MD:Medium concentration

HI:High concentration

Fig. 4 Membership function for C_n

3. RESULTS AND DISCUSSION

The first ripening experiment for 5-day ripening by fuzzy control carried out using two inputs, E_n and ΔE_n with simple fuzzy rules (Amano et al. 1993). However, it was found after several repetitions of ripening experiment that gaseous concentrations of CO_2 and O_2 considerably influenced ripening and played an important role in ripening control, in particular, for longer ripening. In the ripening chamber, CO_2 increases as much as O_2 decreases by respiration of bananas. Ripening was usually hastened by fresh air shortly after ventilation of the chamber, while ripening was suppressed in a certain period before ventilation due to high CO_2 concentration and low O_2 concentration. Besides E_n and ΔE_n, C_n was later added to the input. At the same time, fuzzy rules were tried to modify to improve the control. The fuzzy rules shown in Fig.5 is the latest ones obtained for the 7-day ripening so far after repetitions of experiment which was divided into three portions depending on the level of CO_2 concentration, low, medium and high. Fuzzy inference was conducted to calculate the output, ΔU_n by using the min-max method and the method for obtaining the central axis of the membership function for ΔU_n.

Case A : Cn for LO

		ΔEn						
		NB	NM	NS	ZO	PS	PM	PB
En	NB			NB				
	NM			NM				
	NS			NS	ZO			
	ZO	NM	NS	ZO	PS	PS	PS	PM
	PS	NS	ZO	PS				
	PM			PS				
	PB			PM				

Case B : Cn for MD

		ΔEn						
		NB	NM	NS	ZO	PS	PM	PB
En	NB	NB			NB			
	NM				NM			
	NS				NS	ZO		
	ZO	NB	NM	NS	ZO	PS	PS	PM
	PS	NM		ZO	PS			
	PM				PS			
	PB				PM			

Case C : Cn for HI

		ΔEn						
		NB	NM	NS	ZO	PS	PM	PB
En	NB	NB				NB		
	NM					NM		
	NS					NS	ZO	
	ZO	NB	NM	NM	NS	ZO	PS	PS
	PS	NM	NS		ZO	PS		
	PM					PS		
	PB					PM		

Fig. 5 Fuzzy rules used for banana ripening

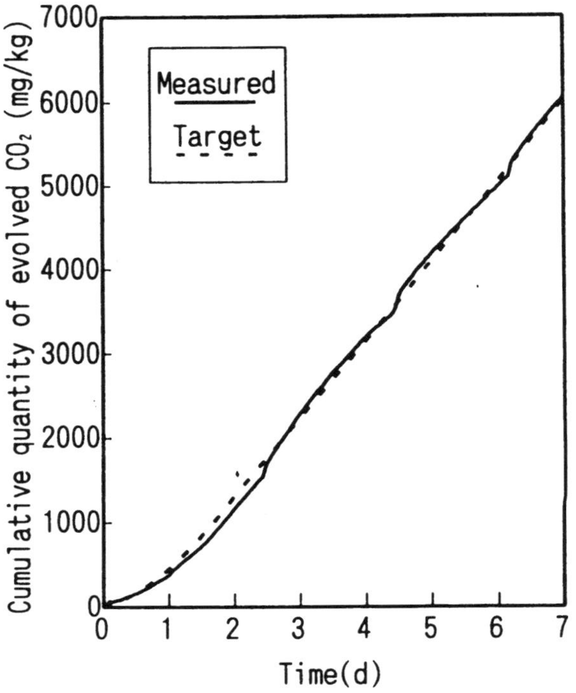

Fig. 6 Comparison of changes in measured value in fuzzy control with those in target value of the cumulative quantity of evolved CO_2

The conditions for ventilation below was applied to the experiment to reflect the effect of gaseous concentrations of CO_2 and O_2 on ripening.

$$「C_n > 8.5(\%)」$$
$$\text{or}$$
$$「C_n \geqq 5(\%) \text{ and } E_n \geqq 100(\text{mg/kg})」 \quad (5)$$

when these conditions were satisfied, ventilation was performed. Fig.6 shows the result of banana ripening by fuzzy control. Changes in the measured value of CQEC in fuzzy controlled ripening closely followed the target value all the way during ripening. In this experiment, 6050(mg/kg) of CQEC and 10.1(brix%) of increase in sugar content was obtained at the end of ripening against the target values of 6030(mg/kg) and 11.9(brix%) respectively. Based on the results, it can be concluded that banana ripening was carried out quite successfully by fuzzy control. Peakes of C_n in Fig.7 indicates that operations of ventilation were effectively performed for controlling ripening.

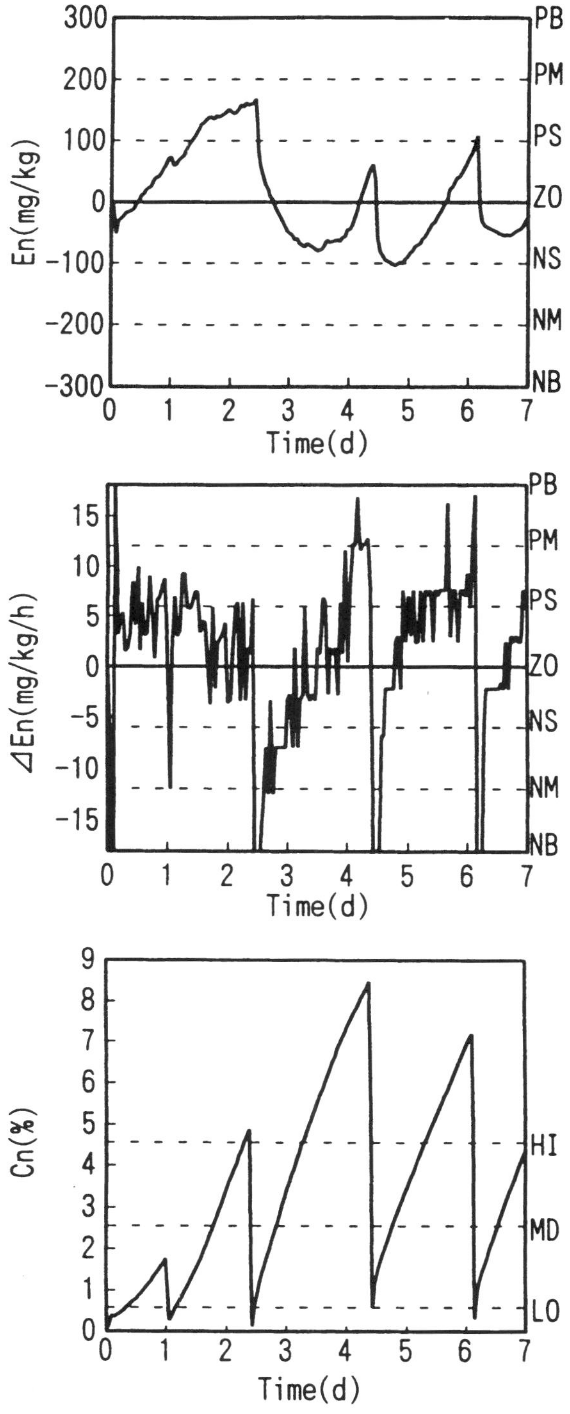

Fig. 7 Changes in E_n, $\varDelta E_n$ and C_n during ripening

REFERENCES

Amano, T., Y. Seo, H. Morishima, K. Imou and W.Yasui (1992). Development of automatic controlled system for ripening of fruits. *J. of The Japanese Society of Agricultural Machinery (JSAM)*, 54(2), 99-104 (in Japanese).

Amano, T., Y. Seo, H. Morishima and Y. Kawagoe (1993). Studies on ripening of fruits by fuzzy theory (Part I). *J. of JSAM*, 55(2), 137-142 (in Japanese).

Seo, Y. and A. Hosokawa (1983a). Relationship between evolved CO_2 and increase in sugar content during artificial banana ripening. *J. of JSAM*, 44(4), 633-638 (in Japanese).

Seo, Y. and A. Hoshokawa (1983b). Prediction of sugar content with temperature schedule in artificial banana ripening. *J. of JSAM*, 45(2), 229-234 (in Japanese).

Terano, T., K. Asai and M. Sugeno (1992). *Fuzzy systems theory and its applications*, Chap. 10, pp.159-160. Academic Press.

OPTIMIZATION OF THE CLIMATE FOR POTATO STOREHOUSES USING ADAPTIVE FUZZY-CONTROL METHODS

Klaus Gottschalk

Institut für Agrartechnik Bornim e.V. ATB
D-14496 Potsdam, Max-Eyth-Allee 100

Abstract: Keeping the quality of stored potatoes over a long storage time needs controlling an optimal climate within the potato bulk during this time. For a large number of big storehouses the potato bulk is air-conditioned simply by ventilation with outdoor air. A mathematical modelling allows to describe the heat and mass transfer processes and to investigate the behaviour of the potato bulk under varying conditions, with the aim to develop a controlling algorithm to optimize the climatization process. To simulate ventilation control algorithms, the mathematical model is included in the closed control loop. The condition rules and dependencies are implemented in the control algorithm by using fuzzy-logic methods. A genetic algorithm is used to vary relevant parameters to reach an optimal control, with the objectives to secure high potato quality, to save ventilation energy and to save costs for storing.

Keywords: Potato storage, Modelling, Automatic control, Fuzzy control, Genetic algorithms

1. INTRODUCTION

For big storehouses the bulk of potatoes is air-conditioned simply by ventilation with outdoor air, without any additional equipment for cooling and heating. The climatic process is therefore dependent on the actual outdoor climatic condition, i.e. on the condition whether the inlet air temperature and humidity are suitable for ventilation or not. Underfloor fans ventilate the bulk from bottom to top of it. The inlet air may be mixed with outlet air, controlled by a set of dampers. Mixed air can be suitable for certain climatic demands, e.g. to warm up or to humify the inlet air with 'used' outlet air. A totally recirculated air helps to equalize the temperature distribution within the bulk or to dry the potatoes from condensed water, adsorbed on the surface. The damper positions and the ventilation rate is controlled by an automatic climatic control equipment. To simulate the control algorithm, a model of the heat and mass transfer process between the stored potatoes and the flowing air is useful to calculate the air temperature, the potato temperature, the air humidity and the potato moisture, i.e. its mass (water) loss. The mathematical modelling allows to investigate the behaviour of the potato bulk under varying conditions. The model for the heat and mass transfer is embedded in the control loop to simulate the control process and to validate the control algorithm.

2. MODELLING THE HEAT AND MASS TRANSFER

The potato bulk is considered by the model of the heat and mass transfer as a stack of layers, each with the thickness of a typical potato diameter. Each layer has the dimensions of the storehouse ground area. The stack of the layers is ventilated homogenously through this area, disregarding the fact that in real storehouses the ventilation air is conducted through channels along the storing room, causing an inhomogeneous air flow distribution within the bulk. These circumstances are respected in another more advanced model, under investigation at the moment, but is not described here. The model simplified under these aspects is assumed as sufficient at the moment to allow to varify and modify the control algorithm. The mathematical model is a one-dimensional model of air stream lines passing from bottom to top of the bulk layer by layer, changing the air and potato conditions on each layer (fig. 1). The biological heat production, which causes a self-warming-up of the potatoes is also included in the model.

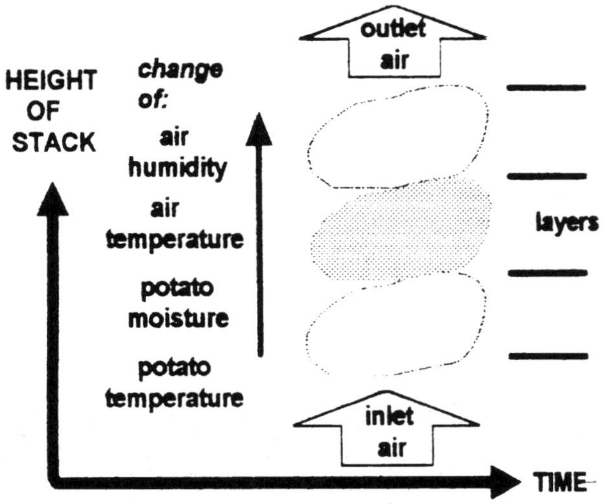

Fig. 1. Heat and mass transfer from layer to layer

With the model the air and potato temperature and the air humidity resp. the potato moisture can be calculated by time and heigth even for instationary inlet air conditions.

3. CONTROLLING THE CLIMATE

To simulate ventilation control algorithms, the mathematical model of the heat and mass transfer is included in the closed control loop. Hence, the model acts as a replacement for the real bulk. The climatic controller can be modified and tested to reach optimal results by varying relevant parameters. After validation, the controller is implemented in the real storehouse.

The climatic controller controls the damper positions and the ventilation rates. The ability to vary the ventilation rates depends on the equipment facilities. Generally, the fans can only be switched off and on, up to the maximum ventilation rate. More sophisticated equipments allow to vary the ventilation rate by varying the fan motor revolutions using frequency inverters. A continuous dosage of the ventilation rate has the advantage of a better adaptation facility to gain an optimal result. Simulations are made with both facilities.

First investigations are made with a conventional controller, i.e. a controller in which the control algorithm is implemented conventionally as a structured code of a programming language. Under the circumstances of Mid-European climatic conditions, the objectives to keep a good and an almost optimal climate within the storehouse leads to a set of 'rules' to control the climate, e.g.

- cool down the potato bulk as fast as possible
- not cool down too fast, to avoid too much mass loss
- dry the potato surface after storing from adsorbed water exactly to the point the surface is dry
- keep potato bulk cool during the whole storing time at a constant temperature
- minimize mass loss during the whole storage time
- avoid surface humidity on the potatoes to prevent rottenness (condensation effects may occur when temperature and air humidity are not suitable at the moment)
- minimize energy consumption

These and more 'rules' lead to a complex set of dependencies which are difficult to implement, to maintain and to test when using a conventional controller. Respecting these circumstances gave the idea to use a fuzzy control algorithm to profit from the ability to implement the 'rules' directly into the rulebase of a fuzzy controller.

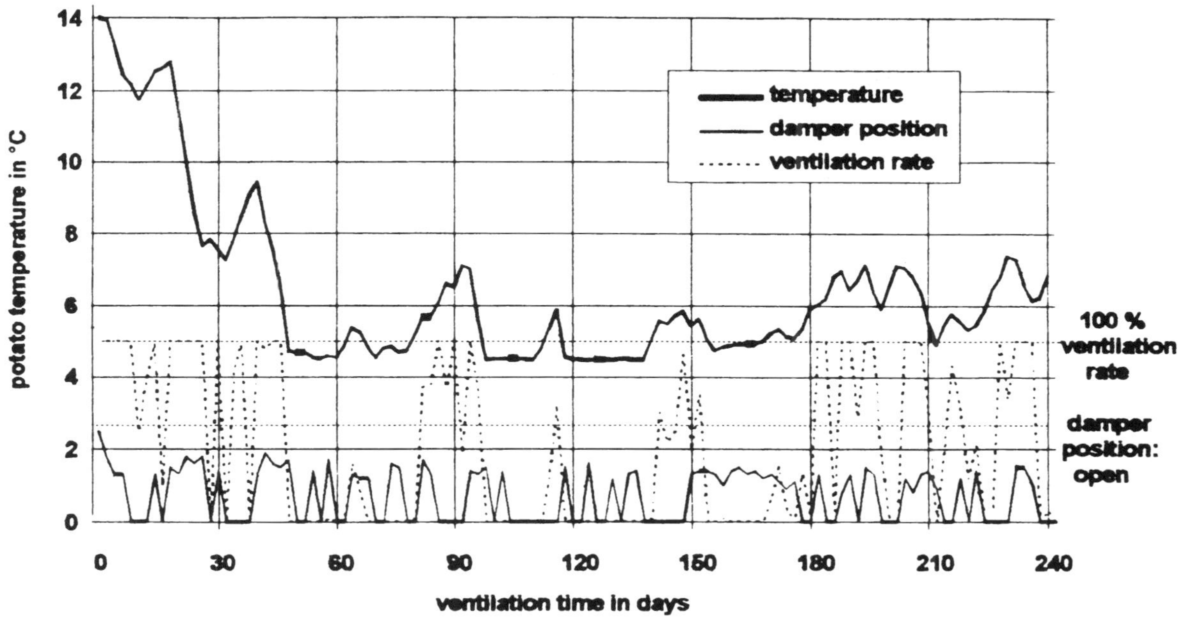

Fig. 2. Regulation of the climate of the potato bulk under real climatic outdoor conditions (Potsdam) from September to April with a fuzzy control algorithm. Simulation by using the mathematical model

The controller respects

- inlet air temperature
- inlet air humidity
- potato temperature within a depth of about 1 m from the bulk surface, as a typical temperature sensor is applied in a real storehouse.

More temperature sensors placed inside the bulk are difficult to handle. The reliability of humidity sensors placed in a soiled outlet air stream is not ensured. Therefore, the application of more temperature sensors and humidity sensors is avoided, except for one humidity sensor in the outdoor area, which is placed in a housing.

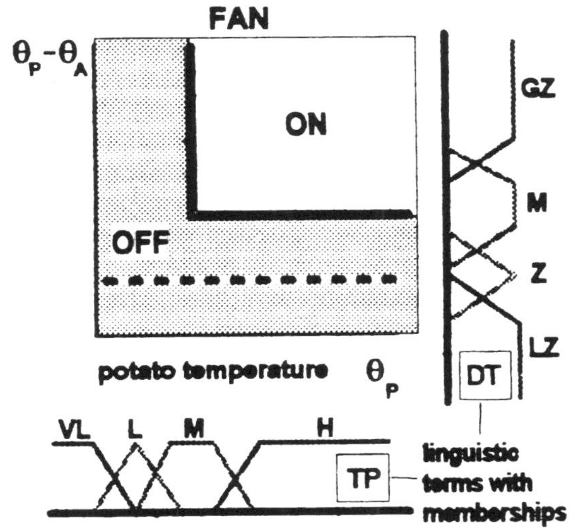

Fig. 3. Fuzzy controller response function and linguistic variables

But the model allows to calculate the temperature and humidity distribution inside the bulk, instead.

4. FUZZY CONTROL

The input values of the controller are

- temperature difference between potatoes (in 1 m depth from bulk surface) and inlet air θ_{pot}-θ_{air} (DT)
- potato temperature θ_{pot} (TP), in 1 m depth from the bulk surface
- outdoor air humidity; assumed as equal to the inlet air humidity; mixing with outlet air is respected, if applied.

The output value (control value) is the ventilation rate (continuous resp. discrete, i.e. 'on/off'). The ventilation rate is a result from the answer of the fuzzy-controller respecting the input values and the 'climatization rules'. The respond function of the fuzzy-control algorithm for the ventilation rate, dependent on the temperature difference and the potato temperature, reflects the rules (fig 3). The variables for the input values, e.g. of the temperature differences and the potato temperature are defined as linguistic variables DT resp. TP. For each of this linguistic variable, four fuzzy sets are defined. The definition of only these four fuzzy sets is assumed as sufficient. The delicate range to control the ventilation rate is about ±0 centigrades for the temperature difference and about 4 centigrades for the potato temperature (the optimal storage temperature). This means for example that the ventilation rate should be reduced, when the potato temperature reaches its optimal storage temperature or when the difference temperature is about 0 centigrades. To reduce the mass loss, additional rules are implemented to make use of a high inlet air humidity (resp. outdoor air humidity, if available). In this case a higher ventilation rate is allowed.

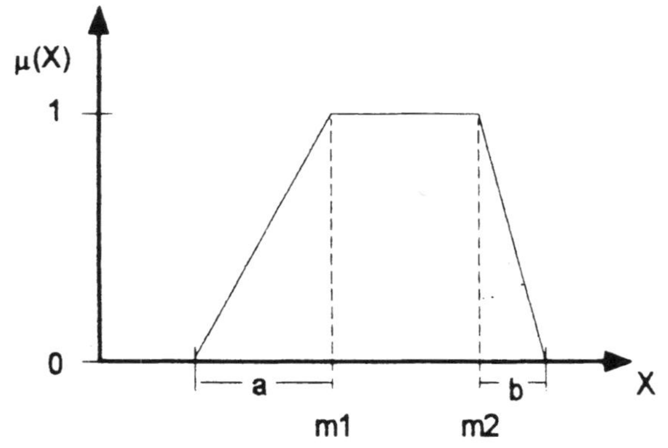

Fig 4 Member function of trapezoidal shape with parameters

When the inlet air is too dry, then the drying effect on the lower layers of the bulk must be reduced to avoid too much mass loss.

The most usefull shape for the member function of a fuzzy set

$$S = \{(x , \mu(x)) \mid x \in X\} \qquad (1)$$

(with X as DT, TP and so on) is the normalized trapezoidal function $\mu(x)$ with the parameters m1, m2, a and b (fig. 4).

Much more shapes are usable, but this function is chosen for its simplicity and for the assumption taken that the choice of the most suitable member function is not particularly relevant in this application. For all the fuzzy sets, defined for the controller, the parameters m1, m2, a and b are fixed heuristically first. The question is, whether it will be an optimal fixation related to a specific optimal valuation criteria. The objective is therefore to use an adaptive algorithm to fit the parameters to the criteria of minimizing total storing cost, for example. The genetic algorithm search procedure is a good technique to solve this problem.

5. ADAPTIVE CONTROLLING USING GENETIC ALGORITHMS

For each fuzzy set a number of members are chosen (e.g. 10 members), constituting a population. For the first member of each population, the parameters of the concerned fuzzy set are fixed heuristically. All further members of each population are derived from the first member and their parameters are varied randomly, so the parameters are all different among the population members. The algorithm starts its calculation with the first member of each population using the heat and mass transfer model to calculate the proceed of temperatures, humidities, mass loss, total energy consumption and total cost, respecting actual cost factors, for the whole bulk and the whole storing time. During the following steps of the optimizing procedure, for each population a pair of members are selected randomly to act as 'parents' to generate a 'child' for the next generation. The parameters of the child are a combination (crossover) of the parameters of the parents (fig. 5). The child 'heritates' the characteristics of the parents. The part of the characteristics (the parameters) of each parent is randomly selected (table 1). In the next step the parameters of the child are mutated with a low probability to avoid that the genetic procedure may stagnate. The probability to select parents with 'good' characteristics rises with their 'goodness', but parents with 'worse' characteristics are also selectable with a low probability. For the reproduced and (probably) mutated child, the calculation is made again, respecting the new parameters to try. The calculated total cost is regarded as a 'fitness' value for the child. The child replaces the member with the 'worst' characteristics within the population. The procedure continues like this until a (sub-) optimum is found (fig. 6).

The optimizing criteria may be:
- minimize energy consumption
- minimize mass loss
- minimize total cost.

The energy consumption is the consumption of electrical energy of the fans. The fan motor power consumption is proportional to the third power of the ventilation rate. Reducing the ventilation rate means to reduce energy consumption, to reduce mass loss but to rise storing time during cooling down, which means an increase in the total energy consumption. Hence, an optimal storing time with an optimal controlled ventilation rate is aspected. The quality of potatoes is directly related to their mass loss, but primarily, a mass loss means a weight loss, which results in a monetary loss. The primary aim to keep the quality of the potatoes, regardless of the energy consumption, fixes the optimizing criteria to minimize mass loss. In this simulation example, a total cost function c is defined

1) create origin population, each member with different parameters
2) select first member; calculate the model with this member to obtain validation
3) select 2 parents randomly
4) select crossover code randomly (table 1)
5) reproduct (crossover) 1 child from parents with probability 0.6 using selected code
6) mutate child with probability 0.1
7) calculate the model with child
8) replace 'worst' member in population by child ('live and die')
9) increase generation counter and repeat at 3)

Fig. 5. Genetic algorithm

Table 1 crossover code for the fuzzy set parameters; '0' means: change not the parameter from parent 1 but from parent 2 - '1' means: change the parameter from parent 1 but not from parent2

m1	m2	a	b	code
1	0	0	0	0
0	1	0	0	1
0	0	1	0	2
0	0	0	1	3
1	1	0	0	4
0	0	1	1	5
1	0	1	0	6
0	1	0	1	7

$$c = \Sigma_{\Delta t}\,(k_{TotalEnergy} \cdot Energy \cdot \Delta t) + k_{MassLoss} \cdot MassLoss \cdot TotalPotatoMassInStore \qquad (2)$$

as the validation criteria to be minimized (with constants k_T, k_M and fan-switch-on time-intervalls Δt; the energy is a function of the ventilation rate). The results show that the mass loss is reduced during the optimizing procedure, accompanied by a slight rise of energy consumption, while the (average) total cost is monotonously decreasing (fig. 6).

6. CONCLUSION

The concept of a fuzzy-controller has the advantage to implement the rules for an (almost) optimal climatization process for longtime stored potatoes into the rulebase of the fuzzy-controller. Varying and changing the rules and the parameters of the fuzzy controller is much easier, compared to a conventional controller, if the number of dependencies leads to a complexity difficult to handle. When respecting the actual outdoor air humidity during the control procedure, simulation results show that the mass loss of the potatoes can be reduced from approx. 2.7 % down to approx. 1.8 %. A genetic algorithm in conjunction with the model of the heat and mass transfer can vary the parameters of the fuzzy sets to reach an (sub-) optimum, on which the total storing cost (resulting from mass loss and energy consumption) can be reduced by approx. 8 %. The mass loss is reduced by approx. 0.2 %. Reducing the mass loss means reducing the loss of quality.

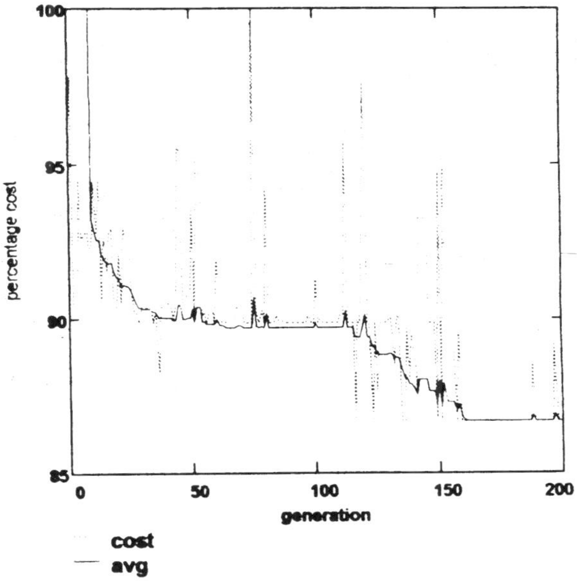

Fig 6 Proceed of total cost during optimization process

DYNAMIC CONTROL OF THE OXYGEN CONTENT DURING CA STORAGE OF FRUITS AND VEGETABLES

S.P. Schouten

ATO-DLO
Bornse Steeg 59
Post Box 17
6700 AA Wageningen
The Netherlands

1. ABSTRACT

The lowest possible oxygen content during CA storage of fruits and vegetables is considerably lower than what is applied in practice. Fermentation of the stored crops is avoided in this way. However, due to this safety margin, the maximum yield in product quality is not achieved. Application of the lowest possible oxygen concentration asks for a system in which the oxygen content can be watched dynamically and corrected immediatly when the stored product starts fermenting. This can be realized by making use of changes in the respiration behaviour of crops, monitored e.g. by the start of ethanol production. For the development of a Dynamic Control System (DCS) to obtain the optimal CA storage conditions with respect to produce quality, the tolerance of several crops for mild forms of temporarily fermentation was tested. Ethanol accumulation in apples and Brussels sprouts was studied. Both crops were stored in extremely low oxygen contents or in nitrogen. Quality development was measured by: firmness of apples, colour and leaf abscission of Brussels sprouts. A linear increase of the ethanol content was observed. The tolerance of Jonagold apples is very high, since accumulation of more than 100 mg/100 g apple tissue did not influence the firmness of the apples. In Brussels sprouts a decrease of ethylene sensitivity could be observed. The results will be discussed in relation to the development of a Dynamic Control System (DCS) for CA storage of fruits and vegetables.

Keywords: CA storage, dynamic control, apple, Brussels sprouts, ethanol, firmness, colour, abscission.

2. INTRODUCTION

Fruits and vegetables are stored in CA rooms in low oxygen and elevated carbon dioxide contents. Apples are stored nowadays in 1.2% oxygen and 0.7% - 5% CO_2, depending on the variety. Cabbage varieties are stored in 2.5%O_2 + 5%CO_2 and recent research showed, that the oxygen content could be much lower for oxheart cabbage (Schouten, 1994). Quality retention of these crops is much better under these circumstances than in cold rooms. For example colour loss of cabbage and firmness loss of apples are inhibited strongly.
Nevertheless, quality problems of stored fruits and vegetables still exist. Consumers complaint about too soft and mealy apples, and stored cabbages must be trimmed to remove rotten and yellow leaves before distribution. These problems are caused amongst others by sub-optimal storage conditions. Storage in very low oxygen contents in CA rooms introduces the danger of fermentation. The stored crop can no longer respire normally below a certain oxygen content. Respiration activity (CO_2 production) rises and special metabolites such as ethanol, lactate and ethyl acetate are synthesized (Ke et al, 1993).
Fermentation may destroy the stored crop completely for example apples develop brown colouration of the skin and fruit flesh. Almost every storage season, the stored crops in several rooms are lost by fermentation.

A better control of fruit and vegetable quality may be achieved by introduction of a dynamic feed-back system for oxygen in CA rooms. In that case the lowest possible oxygen content can be applied and maximal quality retention may be achieved. Early symptoms of fermentation, for example the start of ethanol production, must be used as a signal to adjust the oxygen content.
This seems to be possible since there is a stable ratio between the ethanol content in the tissue and the atmosphere around the crop (North & Cockburn, 1975). However, not much is known about the tolerance of

crops to repeated periods of anaerobic respiration, accompanied by accumulation of ethanol or other metabolites. Research with Jonagold apples and Brussels sprouts was concentrated on quality effects of long term and repeated exposure to anaerobic respiration to collect data for a feed-back system for the control of the oxygen content in CA rooms.

3. MATERIAL AND METHODS

Storage experiments with apples and Brussels sprouts were performed in two kinds of storage cabinets: flow-through and static CA containers. The first system consists of 65 liter stainless steel containers. The desired gases were flown through at any desired concentration. Pure nitrogen, oxygen and carbon dioxide were mixed using a mass-flow controller (5850 TR series, Brooks Instrument b.v., The Netherlands) per gas for every container. The gas concentrations were regularly measured by an O2 and CO2 analyser (Servomex analysers, series 1400). The entering gas (flow rate 200 ml.min) was humidified by passing through a 500 ml water flask.
The static storage system consists of polypropylene containers of 600 liters. CA conditions were hold constant by a KOH scrubber and CO2 injection for the CO2 content. The oxygen content is kept constant by the respiration of the stored crop and by nitrogen and air injections. The containers were equipped with small ventilators, placed in a duct outside the containers. The system is equipped with a computer controlled sample system, which allows measurement and adjustment of the gas content every hour.
Jonagold and Elstar apples were picked in local orchards on the optimal date for long term storage e.g. on 8th september (Elstar) and 23rd september (Jonagold) in 1993. The fruits were put directly after harvest into static CA containers until july 1994 in standard CA conditions: 1.2%O2 + 2.5%CO2 (Elstar) and 1.2%O2 + 4.5%CO2 (Jonagold). Conditions were kept constant untill performance of the experiments in the period between december and may 1994. Stems of Brussels sprouts of the variety Khundry were harvested 19th november 1993 at a grower's in the South-West of the Netherlands. The stems were put into static CA containers at 0-1°C in recommended CA conditions: 1%O2 + 5%CO2 until perfromance of experiments in january 1994. The experiments were conducted with Brussels sprouts harvested from the stored stems.

Quality parameters for apples were: colour and firmness. In one experiment respiration and ethylene production were also measured . Quality of Brussels sprouts was measured by assessment of colour and abscission of leaves. Ground colour of apples was measured by reflection analysis (Minolta Chromameter). Every individual apple was measured on two sides of the fruit: the green (shadow) and red (sunside). Green/yellow colour assessment of Brussels sprouts was done visually of individual sprouts, using a 4 category system: 0 = no yellowing, 1 = a few yellowing, 2 = moderate yellowing and 3 = severe yellowing. Firmness of apples was measured by pentrating a 9 mm plunjer (Bishop Fruittester FT 327) into two sides (shadow and sunside) of the fruit. The skin was removed with a knife before penetration.
Abscission of leaves was assessed by weighing the sprouts and the abscised leaves.
Ethanol content was measured enzymatically (Boehringer Mannheim Cat. No. 176 290). Pieces of each individual fruit were collected and stored at -20°C. These samples were homogenized in a Waring blendor before analysis.
Respiration and ethylene production were analyzed by measuring the accumulation of CO2 and ethylene during a certain time in the flow-through system. Ethylene was measured by a Chrompack gas chromatograph.

4. RESULTS

4.1 Apples

Jonagold apples were stored in flow through containers in 0.2%O2 at 1-2°C. Ethanol content and firmness were monitored regularly. Samples were removed and placed at 15°C in air to study changes in alcohol content. The results concerning firmness and ethanol content are given in figure 1.
The increase of the ethanol content during storage in 0.2%O2 is almost linear between 2 and 7 weeks of storage.

Firmness is neither influenced during 7 weeks storage at 1-2°C nor during the extra periods of 1 week at a higher temperature. Disorders on the skin or in the flesh were absent during the whole storage period.

Jonagold and Elstar apples were stored in at 1-2°C in 0.2, 0.4, 0.6 and 1.0%O2 during 2 weeks. Respiration, ethylene production and ethanol content were measured twice during this period. The results are given in table 1.

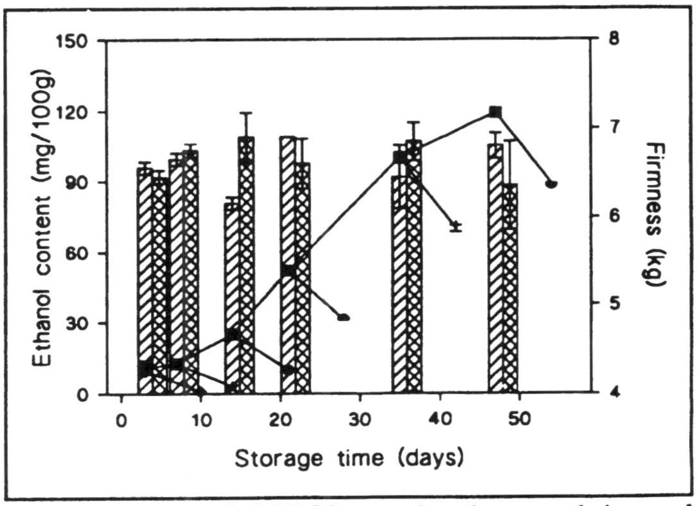

Fig 1: Influence of 0.2%O2 on ethanol accumulation and firmness of Jonagold apples during storage at 1-2°C, followed by 1 week at 15°C in air.

Table 1: Influence of the oxygen content on respiration, ethylene production and ethanol content of Jonagold and Elstar apples during storage at 1-2°C.

Oxygen Content (%)	Respiration Activity (mgCO2.kg.hr)		Ethylene Production (µl.kg.hr)		Ethanol Content (mg.100g)	
	Jona.	Elstar	Jona.	Elstar	Jona.	Elstar
1.0	0.89	2.89	4.37	1.35	0.8	1.0
0.6	1.35	2.61	4.13	1.75	0.5	1.3
0.4	0.84	2.52	2.44	1.23	0.8	3.8
0.2	1.39	3.63	0.72	0.54	6.5	77.5

Respiration of Jonagold is hardly influenced by these very low oxygen contents, although a little increase in ethanol content can be observed in 0.2% O2. The variety Elstar is more sensitive to the very low oxygen contents. In 0.2%O2 a rise in carbon dioxide production is observed. Elstar shows anaerobic respiration, since there is also a sharp increase in the ethanol content. Ethylene production is depressed in the lowest oxygen content for both varieties.

Jonagold apples were stored in static containers at 1-2°C in 1%O2 + <0.5%CO2 during 10 months. Storage was interrupted once, twice and four times for a stay of 3 days at 1-2°C in nitrogen to induce an ethanol content of 10-15 mg.100g. Quality assessment was performed after 8 and 10 months of storage. The results concerning firmness are given in the figures 2 and 3.

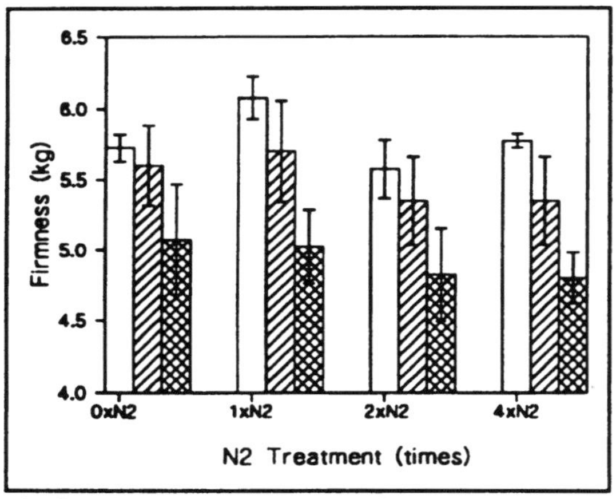

Fig 2: Influence of repeated N2 treatments on the firmness of Jonagold apples after 8 months at 1-2°C in 1%O2, followed by 1 and 2 weeks at 15°C in air (2nd and 3rd column respectively).

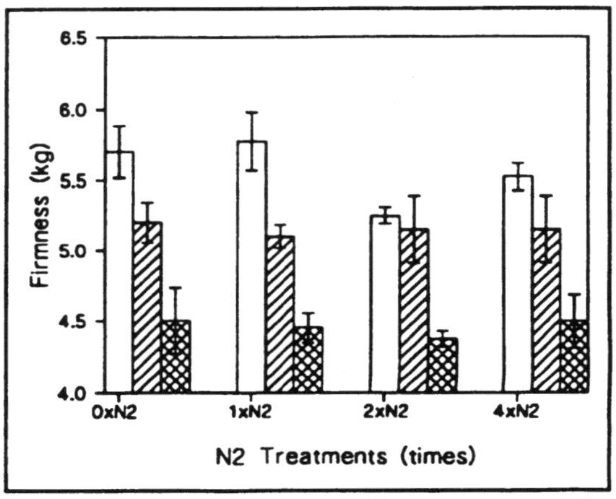

Fig 3: Influence of repeated N2 treatments on the firmness op Jonagold apples after 10 months storage at 1-2°C in 1%O2, followed by 1 and 2 weeks at 15°C in air.

The N2 treatments did not have significant influence on firmness after 8 months ULO storage (fig 2). This was somewhat different after 10 months (fig 3). A difference then was present among the nitrogen treatments. 0xN2 and 1xN2 showed a higher firmnesss than 2xN2 and 4xN2. This difference however, was absent after the extra storage periods of 1 and 2 weeks at 15°C.
Big differences in firmness are present between the measurements directly after ULO storage and storage at 15°C in air after the ULO storage period.

4.2 Brussels sprouts

Brussels sprouts were stored at 0-1°C in air and nitrogen during 4, 8 and 15 days. After this period the sprouts were stored at 15°C in air with 10 ppm ethylene to test their sensitivity to ethylene.

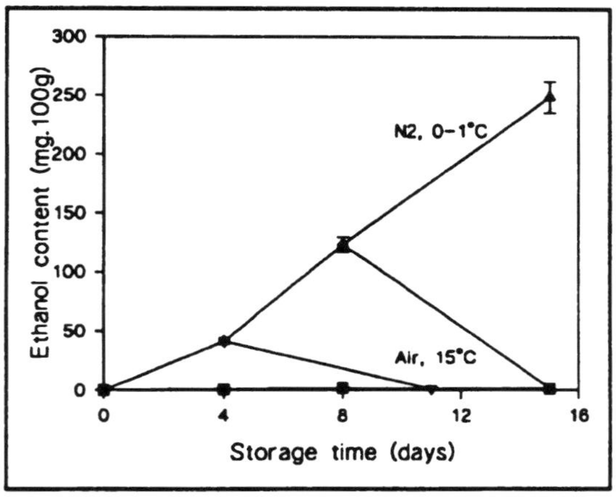

Fig 4: Influence of nitrogen on the ethanol content of Brussels sprouts during storage at 0-1°C, followed by storage in air at 15°C.

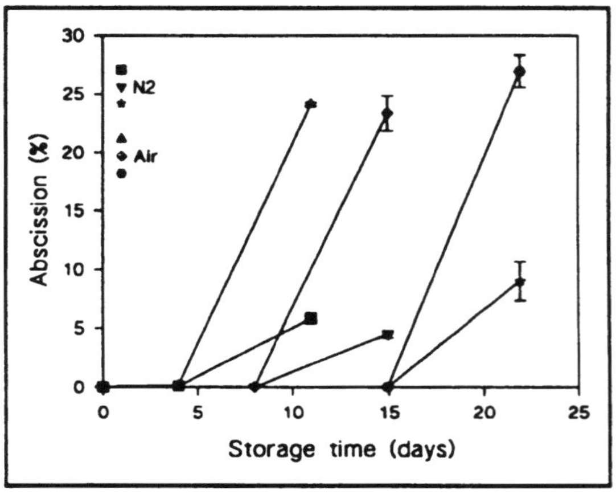

Fig 5: Influence of nitrogen on leaf abscission of Brussels sprouts during storage at 0-1°C, followed by starage in air at 15°C.

The accumulation of ethanol and leaf abscission (typical ethylene reaction of cabbage) are shown in

the figures 4 and 5. The accumulation of ethanol (fig 4) shows a linear increase and there is a quick disappearance of ethanol from the tissues during storage afterwards at 15°C. A great difference was present between abscission of leaves of Brussels sprouts stored in air and in nitrogen. The typical ethylene response for cabbage was much stronger after storage in air than in nitrogen (fig 5), indicating that nitrogen reduced the sensitivity of Brussels sprouts to ethylene.
Also the colour of the Brussels sprout was influenced by the N2 treatments; the results are shown in table 2.

Table 2: Influence of nitrogen on the colour* of Brussels sprouts after storage at 0-1°C, followed by storage at 15°C in air with appels.

Storage time (days)	Storage at 0-1°C	Storage afterwards in air at 15°C - ethylene	at 15°C + ethylene
4	air	0.12	2.16
	nitrogen	0.14	1.50
8	air	0.36	2.38
	nitrogen	0.98	2.02
15	air	0.80	2.22
	nitrogen	1.50	2.25

* visual colour assessment (0-3)

Conclusions from this research are as follows. The storage in nitrogen caused more colour loss than storage in air, especially after 8 and 15 days. On the other hand Brussels sprouts were less sensitive to colour loss during the extra period at 15°C if ethylene was present.

5. DISCUSSION

Some conditions must be fulfilled for the development of an interactive system for the automatic control of the lowest possible oxygen content in CA rooms for fruits and vegetables. At first a metabolite should be measured, preferentially in the gas phase of the storage room. This seems possible by making use of changes in respiration rate or ethylene production of the stored crop. CO2 production rises below the lowest possible oxygen content, ethylene production decreases and ethanol is produced. We observed an almost linear increase in ethanol content of Jonagold and Elstar apples and in Brussels sprouts (figures 1 and 4 and table 1). The changes in respiration behaviour and ethylene production are smaller than the rise in ethanol production (table 1), which makes ethanol more attractive. A further advantage of ethanol for this purpose is its specificity. Anaerobic respiration is always accompanied by ethanol accumulation, although some other metabolites may also be synthesized, for example lactate and ethylacetate (Ke et al, 1993). Involvement of these compounds is not excluded in further research.
A second condition is, that the volatile should be measured in low concentrations. A rather stable relationship between the ethanol content in apple tissue and in air was found by North & Cockburn (1975). The ratio between the two contents is about 3000 according to these authors. This means roughly, that an alcohol content of for instance 10 mg/100g equals about 0.003 mg/L. These concentrations and lower (less than 1 ppm) can easily be measured with a gas chromatograph. It may also be, that development of an ethanol sensor will prove to be necessary.
A third condition is, that ethanol should disappear from the tissue and have no consequences for quality. Ethanol was lost from apple tissue rather quickly in air at 15° (fig 1), but this speed was much lower in 1% oxygen at 1-2°C (data not shown). Ethanol was disappearing very quickly from Brussels sprouts in air at high temperature (fig 4). The periods of anaerobic respiration therefore should last as short as possible in an interactive system for the lowest possible oxygen content.
The storage of apples and Brussels sprouts in an atmosphere causing fermentation did not have consequences

for the quality. Neither firmness of apples nor abscission of Brussels sprouts were increased as a consequence of fermentation. Apples were stored for weeks in 0.2%O2 and the firmness was not adversely affected. Brussels sprouts were even less sensitive to the adverse effect of ethylene after a period of fermentation. This effect was also observed by Saltveit & Sharaf (1992). They observed an inhibiting effect of ripening of tomatoes after exposing the fruits to ethanol. Exposing Brussels sprouts to fermentative conditions may have some advantage, although colour loss was promoted. However in this research the period of 4 days in nitrogen may have been too long.
Also repeated short periods of fermentation did not show quality disadvantages for apples.

6. CONCLUSION

Development of a Dynamic Control System (DCS) for the oxygen content in CA rooms is possible on the basis of the presented results. Ethanol production, indicating fermentation of the stored crops, might be used as a signal to control the lowest possible oxygen content. This is possible, because a stable relationship exists between ethanol in the tissues of the crop and the atmoshere around it. Measurement of ethanol can be performed with a gas chromatograhp, although development of an ethanol sensor may prove to be necessary. The tolerance of crops, including apples and Brussels sprouts, towards ethanol was rather high and gives support to the concept of ethanol as signal for the control of the lowest possible oxygen content in CA rooms.
Future prospects of DCS are as follows. The measuring device for ethanol can be built in in existing measuring and controlling systems for oxygen. Nitrogen flushing and ventilation systems will be controlled on the basis of ethanol measurements. Optimal quality retention is expected from DCS, because the average oxygen content in CA rooms will be lowered. The oxygen content in CA rooms wil be set on the basis of the properties of the stored crop and no longer on the basis of general recommendations.
Patent for DCS system is applied for by ATO-DLO.

7. REFERENCES

Ke D. M. Mateos and A.A. Kader. Regulation of Fermentative Metabolism in Fruits and Vegetables by Copntrolled Atmospheres. Proc. 6th Int. Contr. Atm. Res. Conf. Ithaca, N.Y. 63-77 (1993).

North C.J. & J.T. Cockburn. Ethyl Alcohol Levels in Apples after Deprivation of Oxygen and the Detection of Alcohol Vapour in Controlled Atmosphere Stores using Indicator Tubes. J. Sci. Fd Agric. 26, 155-1161, (1975).

Saltveit M.E. & A.R. Sharaf. Ethanol inhibits ripening of Tomato Fruit harvested at various Degrees of Ripeness without affectingsubsequent Quality. J. Am. Soc. Hort. Sci. 117 (5): 793-798 (1992).

Schouten S.P. Koelen kan met nog minder zuurstof. Groenten en Fruit/Vollegrondsgroenten 38, 13 (1994).

NEURAL NETWORK CLASSIFIERS IN APPLE BLEMISH INSPECTION

Qingsheng Yang and Robin D. Tillett

Silsoe Research Institute, Wrest Park, Silsoe, Bedford MK45 4HS, UK

Abstract: When automatically grading non-orientated apples with machine vision, stalk and calyx areas look very similar to blemishes in images and a classification process is often required to distinguish between them after initial image segmentation. In this work, multilayer feed-forward neural networks are used to perform the classification. The segmentation is accomplished with a flooding algorithm. Structured light is applied to obtain the 3D geometric shape information of apples. The input features of the neural networks are extracted from both the image under normal diffuse light and the image with structured light. The classifiers were tested with sample apples and good classification accuracies were achieved.

Keywords: Image analysis, neural networks, pattern classification.

1. INTRODUCTION

In high speed automatic inspection of apples with machine vision, fruits may be presented under a camera in arbitrary stalk-calyx orientation. In images stalks and calyxes may show up. Stalks, calyxes and the shadows around them appear in images as small dark areas that are very similar to discoloured patch-like blemishes. An example apple image with a blemish and a stalk area is shown in Fig. 1. When detecting blemishes with 2D image analysis techniques, they can be segmented out along with blemishes. In order to send the correct grading signal to the controller of the handling mechanism, a classification step after the initial segmentation is required to distinguish between blemishes and stalks/calyxes.

Because of the similarity between stalks/calyxes and blemishes, it is generally difficult to differentiate them by their 2D shape properties or their intensity properties. such as mean, variance and gradient. However, many blemishes, e.g. bruises and abrasions, occur on the flat or convex surfaces of a fruit while the surface around a stalk or calyx is concave. The 3D surface shape information can be utilized for the differentiation.

A structured light technique (Ballard and Brown, 1982) has been used to obtain the information of 3D surface shape (Yang, 1993a, b). A set of parallel light stripes are simultaneously projected onto an apple surface and an image with the stripes, in addition to the apple image, is grabbed for each view of the fruit. Fig. 2 shows an example image with light stripes on the same apple as in Fig. 1. The shapes of the stripes on the normal convex surfaces of an apple are different from those on the surfaces related to stalks and calyxes. Therefore, the 3D information can be obtained by the analysis of the different shape patterns of the curved stripes.

Thus, in this work we treat the distinction between stalks/calyxes and blemishes as a pattern classification problem. The pattern classification system is shown in Fig. 3. Multilayer feed-forward neural networks are used to perform the classification. Dark patches are initially segmented out from the apple image. For each patch, the input features of the neural networks are extracted from both the image under normal diffuse light and the image with structured light. The following sections explain the details of the system. Test results with sample apples are presented.

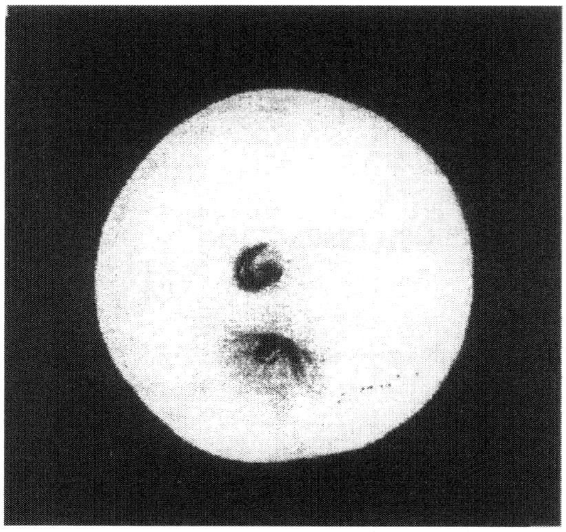

Fig. 1. An example apple image with a blemish and a stalk area.

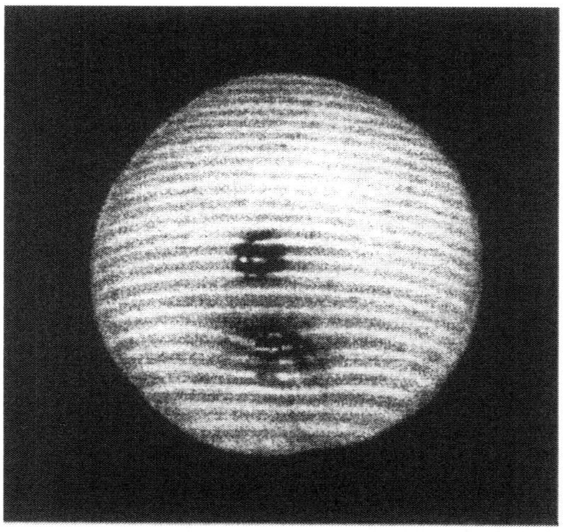

Fig. 2. An example image with light stripes on the same apple as shown in Fig. 1.

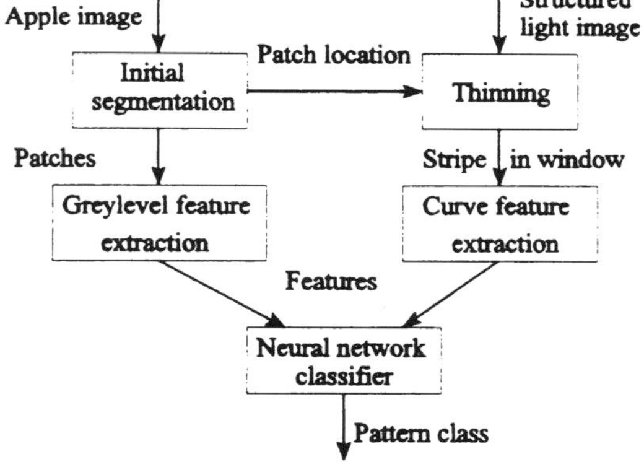

Fig. 3. The pattern classification system

2. THE PATTERN CLASSIFICATION SYSTEM

Pattern classification, a subfield of pattern recognition, is concerned with determining whether the pattern in an input image belongs to one of the predefined classes. Developing pattern classification systems is usually a two-stage process. First, the characteristics of patterns should be examined carefully, which involves lengthy trial-and-errors. The result is a set of features chosen to represent the original image pattern. Second, a suitable classifier should be chosen from a range of techniques to classify the pattern represented by the features.

2.1 Image patterns

For each view of an apple, two images are acquired, one is the apple image under ordinary diffuse light and the other is the same apple image viewed with structured light. In the apple image, both blemishes and stalks/calyxes are generally darker than their surrounding surfaces, and in image grey-level landscapes they usually appear as concavities. With the concept of topographic representation of images, they can be modelled as water catchment basins in the grey-level landscapes. A flooding algorithm has been developed to detect the basins, i.e. the patches (Yang, 1994). The detection result for the example image in Fig. 1 is shown in Fig. 4. The detected dark patches are highlighted with their bright borders.

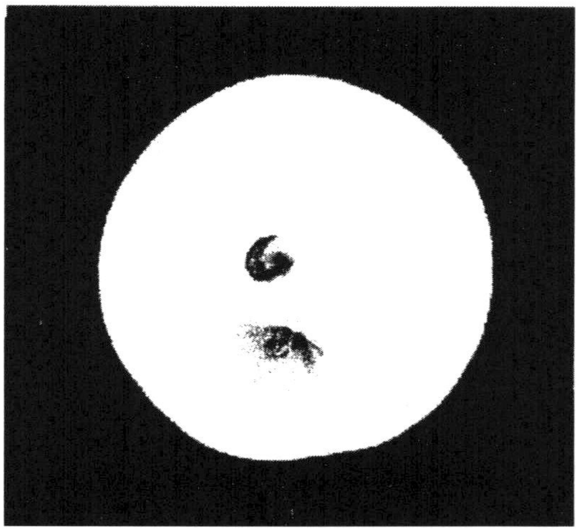

Fig. 4. Results of patch segmentation for the image in Fig. 1. Detected patches are highlighted by bright borders.

For each detected patch in the apple image under normal diffused light there is a corresponding area at the same location in the stripe image. Therefore, a rectangular window around each corresponding area is placed in the stripe image. The image patch and the stripe image region within the window form a pattern. Blemish and stalk/calyx are two pattern classes. By classifying each pattern into one of the two classes, blemishes and stalks/

calyxes can be distinguished.

2.2 *Feature extraction*

Feature extraction is a process that converts the input image pattern into a set of quantitative values. Because of the extensive data reduction in the extraction, some information will inevitably be lost. Therefore, an important thing is to preserve the discrimination information. Features should be chosen such that patterns from the same class should end up with numerically close values for the features whilst patterns from different classes should have quite different quantitative values. Since blemishes and stalks/calyxes can be anywhere in fruit images, it is preferable that the features used are insensitive to rotation and translation transformations in the input image pattern.

In stripe images a principal parameter for describing the shape of stripes is curvature, which is invariant to rotation and translation. To reduce the noise caused by taking the second-order derivatives in the usual calculation of curvature, we use the following practical method to compute the estimator of point curvature:

$$k_\psi = \psi(t+1) - \psi(t) \quad (1)$$

where

$$\psi(t) = \tan^{-1}\frac{\dot{Y}(t)}{\dot{X}(t)}, \quad (2)$$

and

$$\dot{X}(t,\sigma) = x(t) * \frac{\partial g(t,\sigma)}{\partial t}$$

$$= \int_{-3\sigma}^{3\sigma} x(u)\frac{-t}{\sigma^3\sqrt{2\pi}} e^{-(t-u)^2/2\sigma^2} du, \quad (3)$$

$$\dot{Y}(t,\sigma) = y(t) * \frac{\partial g(t,\sigma)}{\partial t}$$

$$= \int_{-3\sigma}^{3\sigma} y(u)\frac{-t}{\sigma^3\sqrt{2\pi}} e^{-(t-u)^2/2\sigma^2} du, \quad (4)$$

where $x(t)$ and $y(t)$ represent a curve in parametric form, $g(t, \sigma)$ is a Gaussian kernel of standard deviation σ. Note that whilst the angle function $\psi(t)$ is multiple-valued, under practical constraints on continuity of the derivative, it is uniquely defined. The actual features used are the statistics of curvature. Before the curvature calculation, the stripes are binarized by a local thresholding process and then thinned to their one pixel wide skeleton curves by a shape preserving thinning method (Gonzalez and Woods, 1992). So the stripes are actually represented by their skeleton curves, from which their curvatures are computed. The thinning result of Fig. 2 is shown in Fig. 5. Other features from stripe images, including a similarity measurement (correlation coefficient) of curvatures between curves within the window, are also extracted.

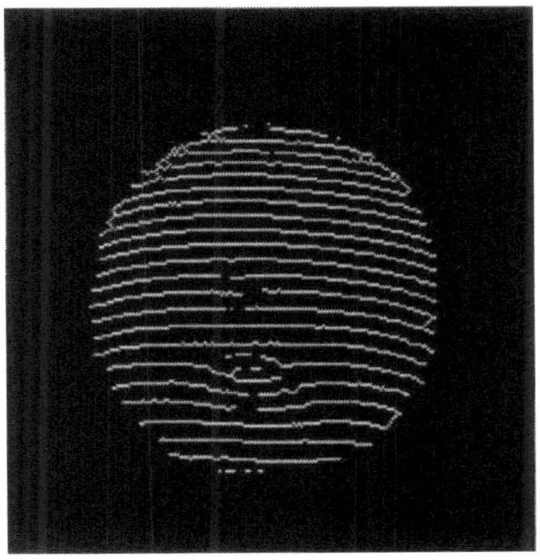

Fig. 5. The thinning result of Fig. 2.

In apple images, once a small patch is segmented out from an apple image, some features are already available as the results of the flooding algorithm, which include the lowest and the highest grey-level, the area, the boundary length and the water volume V of the water catchment basin. Other geometric parameters of the basin are also extracted. A total of 18 features are computed for each pattern and may be listed as follows:

From the stripe image
-- average positive curvature,
-- average negative curvature,
-- average absolute curvature,
-- frequency of points with positive curvature,
-- frequency of points with negative curvature,
-- frequency of points with zero curvature,
-- average correlation coefficient of curvature between curves,
-- average curve length,
-- percentage of short curve segments,

From the fruit image
-- average grey level intensity,
-- variance of grey level intensity,
-- lowest grey level of the patch,
-- highest grey level of the patch,
-- area of the patch,

-- compactness of the area,
-- water volume of the lake,
-- conical angle of the volume,
-- average depth of the volume.

The selection of the features is intuitive and provides reasonable basic coverage of the feature space. As can be seen, these features are not independent. But in this study, we do not address the issue of salient feature selection. Since the dependencies are not easily modelled and the features are not difficult to compute, we retain all the features for the following classification.

2.3 *Neural network classifiers*

Having extracted the feature vector for each pattern, the features are fed into a classifier to classify the pattern as either blemish or stalk/calyx, which are the two output classes. As the input feature data to the classifier possess a non-linear relationship to its output class, a non-linear classifier is required.

The current implementation of the system employs a multilayer feed-forward network for the classifier block. Such a network has a layered structure and only connection between neurons at subsequent layers are permitted. The number of neurons in the input layer is fixed and equal to the number of extracted features. Since the output neurons are organised such that each neuron represents a class, the number of neurons in the output layer is fixed to two. Hence, given the input and output layer size, the number of hidden layers and the number of neurons in each hidden layer should be decided. The neurons in output layer and hidden layer have a non-linear activation function of sigmoidal shape.

The network is trained with a variety of back-propagation learning algorithm (Chan and Fallside, 1987). The algorithm accelerated the learning by adapting the two coefficients used in the conventional back-propagation algorithm, learning rate and momentum factor. During the training the weights are updated after each pass through all the training samples.

3. TEST RESULTS

The proposed system was tested for Golden Delicious and Granny Smith apples with a laboratory imaging setup. Apples were placed in the middle of the field of view in a lighting chamber, which provided diffuse illumination. The orientation of the sample fruits was made arbitrary. The light stripes were formed by projecting light through an optical grid of thin transparent lines. The background of the viewing field was black. All images were stored in the memory in the size of 256 x 256 pixels.

Two images were acquired in sequence for each view of an apple. One was under diffuse illumination so that dark patches can be segmented out, another was under structured light. It is worth pointing out that if the structured light is from an appropriate narrow spectral range, the two images can be grabbed simultaneously by two camera sensors (Crowe and Delwiche, 1994). For each pattern, its 18 features were extracted from both images. The type of the patch, i.e. blemish or stalk/calyx, was recorded together with the features, thus to form a labelled pattern sample. As shown in Table 1, a total of 274 samples were collected and arranged in 6 sets. Two of them were training samples, the rest were used to test the performance of the trained neural network classifiers. There were no training samples included in the test sample sets. Because of different value ranges, the values of all the features were normalized to the range of [0, 1.0].

Table 1. Sample patterns

Apple variety	Samples	Blemish	Stalk/ calyx	Total	Total
Golden Delicious	Training1	47	37	84	
	Test 1	23	12	35	274
	Test 2	21	13	34	
Granny Smith	Training2	51	15	66	
	Test 3	21	7	28	
	Test 4	20	7	27	

The neural networks were simulated on a transputer-based system. The training and performance test of the neural networks are organized according to the samples.
In all the trainings the initial learning rate and momentum factor were fixed to 0.1 and 0.5 as the training algorithm employed was not very sensitive to these initial values. The weights of the neural networks were all initialized with random numbers in the range of [0, 1.0]. We experimented three network configurations as shown in Table 2, although no attempt was made to find optimal network configuration. The third configuration had two hidden layers. A few different sets of initial weights were used for each configuration. All trainings were converged rapidly, from about 100 to 1200 iterations.

In the performance evaluation for trained neural networks, the classification accuracy was computed by dividing the number of correctly classified samples by the number of samples in the test group. The classification results on testing data are listed in Table 2. It can be seen that the difference in classification accuracy is small for the three network configurations. The equal values of classification accuracy for test sample groups 1 and 4 indicate that one or two specific sample groups 1 and 4 indicate that one or two specific pattern samples in the groups have feature values in the range of the ones for the other class. In these cases persistent errors would happen. On the whole, good classification accuracies have been achieved.

Table 2. Classification accuracy of trained neural networks

Network configuration	Classification accuracy (%)			
	Test 1	Test 2	Test 3	Test 4
18:10:2	94.3	97.1	92.9	96.3
18:27:2	94.3	97.1	96.4	96.3
18:10:6:2	94.3	100.0	92.9	96.3

4. CONCLUSIONS

A neural-network-based pattern classification system has been proposed for the distinction of blemishes from stalks/calyxes in apple blemish inspection with machine vision. Dark patches which may include both blemishes and stalks/ calyxes were first segmented out with a flooding algorithm. Structured light was used to provide necessary 3D shape information about apple geometric surfaces. For each pattern corresponding to a segmented image patch, the qualitative features of the stripes are extracted from the stripe image and the 2D intensity properties are extracted from the apple image under normal diffused light. By feeding the features into the neural network classifiers, each patch was classified as either blemish or stalk/calyx. The proposed technique was tested with sample apples. The results demonstrate that the system can effectively detect discoloured blemish patches which occur on convex or flat surfaces and distinguish them from stalks and calyxes which are surrounded by concave surfaces.

ACKNOWLEDGEMENT

This work was partially funded by the EC CAMAR program (project 8001-CT91-0206).

REFERENCES

Ballard, D.H. and Brown, C.M. (1982). *Computer vision*, pp. 52-54. Prentice-Hall, Inc., Englewood Cliffs, New Jersey 0732.

Chan, L. W. and Fallside, F. (1987). An adaptive training algorithm for back propagation networks. *Computer Speech and Language*, **2**: 205-218.

Crowe, T. and Delwiche, M. (1994). Real-time defect detection in fruit. *International Conference on Agricultural Engineering (AgEng'94)*, Milano, Italy, 29 Aug.-1st Sept. 1994, Report No. 94-G-027.

Gonzalez, R.C. and Woods, R.E. (1992). *Digital image processing*, pp. 492-494. Addison-Wesley Publishing Company, Inc.

Yang, Q. (1993a). Finding stalk and calyx of apples using structured lighting. *Computers and Electronics in Agriculture*, **8**:31-42.

Yang, Q. (1993b). Classification of apple surface features using machine vision and neural networks. *Computers and Electronics in Agriculture*, **9**: 1-12.

Yang, Q. (1994). An approach to apple surface feature detection by machine vision. *Computers and Electronics in Agriculture*, **11**: 249-264.

ENVIRONMENTAL CONTROL STRATEGIES BASED ON PLANT RESPONSES USING INTELLIGENT MACHINE VISION TECHNIQUE

H. Murase, Y. Nishiura, K. Mitani

Lab. of Bioinstrumentation, Control and Systems Engineering
Department of Regional Environmental Science
University of Osaka Prefecture
1-1, Gakuen, Sakai, 593, JAPAN

Abstract: The finite element method was utilized in image feature extraction. Non-linear relationships between plant moisture content and the finite element image features were identified by using a neural network in order to predict water status of plant from digital image of a wilting plant. In this paper, bio-response feedback control strategies for developing an intelligent water management system in plant production were discussed. A non-invasive sensing technique for monitoring plant moisture conditions were proposed.

Keywords: Computer vision, Finite element method, Neural networks, Kalman filter, Image compression

1. INTRODUCTION

In a protected plant production system such as a plant factory, the control applications have been limited to its environmental controls. The feedback control technology for greenhouse environmental factors such as temperature, humidity, radiation intensity, carbon dioxide concentration and so forth has been developed and successfully implemented (Hashimoto and Nonami, 1992). The development of bio-response feedback control system has been a challenging task for plant production engineers and scientists. The plant growth can be optimized or controlled by adjusting the environmental factors. Plants respond to the change of environmental parameters. For example, the stomata activity is sensitive to ambient humidity and CO_2 concentration. The plant tissue rigidity is affected by the availability of water at its root zone. Those environmental factors should be controlled based upon the plant status responding to the environmental conditions. Water constitutes 80-90% of the living plant body and has a large heat capacity. Changes in water status drastically affect the growth and metabolism of plants.

Effective water management is an essential procedure in production of high quality marketable crops and vegetables. For example, precise water care should control the size of plant to facilitate such as small size packaging, vigorous grafting, and shipping adjustment for better market price. Figure 1 illustrates an example of bio-response feed back control system including an intelligent non-invasive sensing system for the plant moisture conditions in the feed back path.

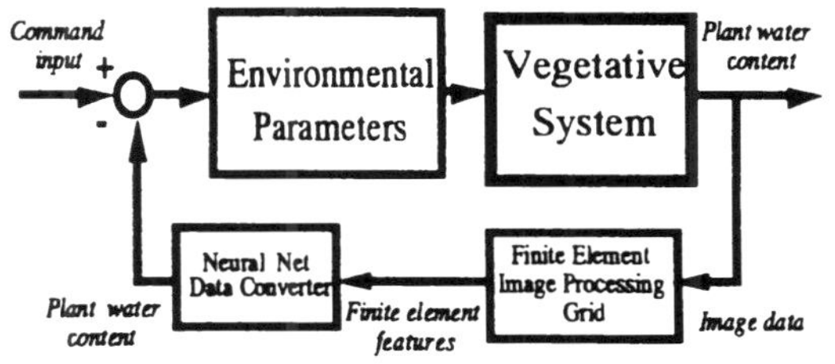

Fig. 1 Bio-response feedback control system block diagram.

Plant moisture conditions in terms of plant water potential level can be measured by attaching a thermocouple psychrometric transducer directly to the reverse side of a leaf where the stomata are located or extracting a small sample tissue from a leaf, which should be placed in a chamber of a psychrometric transducer. The latter method is a destructive technique. Even a slight contact of foreign material with the plant tissue disturbs physiological activity of the plant. The disturbance may be negligible over the short time period needed for the measurement, say, for a few minutes. The practice of non-invasive measurements against plants is essential for the bio-information feedback and/or feed-forward control system of a plant factory. No non-invasive method for direct determination of plant moisture conditions is currently available.

One of alternatives is the use of indirect measurement techniques. For instance, those who are skilled in raising plants can sense whether their plants are under adequate water conditions or not from minor changes in the appearance of their plants through their color and tone before the plants wilt. It may be possible to recognize changes in some kind of indices that describe the plant moisture conditions from the appearance of wilting plants by machine vision.

Changes in plant moisture conditions reflect tonal variations over the plant canopy. The tonal variation can be transformed into pictorial information electronically in retrieval form. The some kind of image features can be related to the tonal characteristics of the plant canopy that also substantially reflect the plant moisture conditions. The textural analysis can be considered as one of applicable techniques for extracting image features (Murase *et al.*, 1994; Haralick *et al.* 1973). Some of problems in implementing the textural analysis are that there is too much flexibility to construct the co-occurrence matrix and the construction of the co-occurrence matrix requires impractical long calculation time. Broadly speaking, image features are any extractable measurement of use. Examples of low-level image features are pixel intensities or geometric distances between pixels. Features may also result from applying a feature extraction algorithm or operator to the image data. In this research work, the finite element algorithm is use to extract image features (FE features;finite element features).

In order to achieve the aim of developing such a bio-response feedback control system, the primary concern should be to develop a practical non invasive technique for monitoring plant moisture conditions. In this paper, non-linear relationships between the plant moisture content and the finite element features of the pictorial information of a plant (tomato seedling) were identified by using a neural network in order to predict plant moisture content from the pictorial data.

2.KALMAN FILTER NEURON TRAINING

In this research work, three-layered neural networks were used. The learning performance of neural network is of extreme importance for the users especially in such a case that the training data contain significant amount of noise in measurement. The kalman filter can be used as a learning algorithm for the neural network. Figure 2 illustrates an example of a three layered neural network structure. The mathematical explanation for the process of signal transfer in the network may help understand the mechanism of layered neural networks (Murase *et al.*,1991).

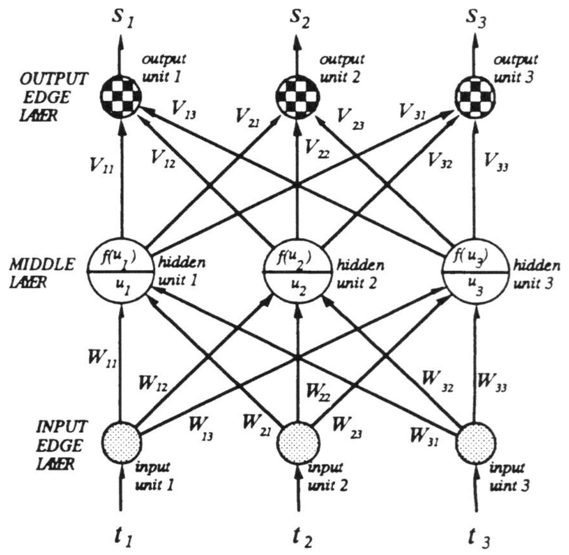

Fig. 2. Three-layered neural network structure.

The input T can be expressed in a vector form as $T=\{t_1,t_2,....,t_n\}$. In Fig.4, n is equal to 3. I-th component of the inputs T, i.e., t_i that comes out from the input unit i is transferred to a hidden unit j through the synapse weight W_{ij}. Since each hidden unit has a summation function operating on inputs, the total input u_j received by the hidden unit becomes

$$u_j = \sum_{i=1}^{n} W_{ij} t_i \quad (1)$$

The hidden unit i has also a transfer function that performs nonlinear transformation on the total input u_i and then gives an output which becomes the next input fed into the output unit j, which has also a summation function, through another synapse weight V_{ij}. The total input received by the output unit j becomes directly its output s_j expressed as

$$s_j = \sum_{i=1}^{n} V_{ij} f(u_i) \quad (2)$$

The outputs can be given in a vector form as $S=\{s_1,s_2,.....,s_m\}$. After all, what this neural network does is to perform a nonlinear transformation on T as expressed in the following equation.

$$S = F(T) \quad (3)$$

Once those nonlinear functions (transfer functions) of hidden units are specified, the behavior of the network can be identified by determining all synapse weights contained in the network. The sigmoid function is often employed for the transfer function. The learning of neural network is a procedure to determine optimal values of synapse weights by adjusting them step by step using known input data and their associated output data called training data. The most common algorithm for this learning procedure is the back propagation. The Kalman filter can be also used as a training algorithm.

In this study, the Kalman filter algorithm was employed due to its quicker converging characteristics than any other algorithm (Murase, 1991).

The state equation and the observation equation for the neural network can be described by the following forms:

1) State equation

$$\{x\}_{k+1} = [I]\{x\}_k \quad (4)$$

$\{x\}$:synapse weights to be determined

$[I]$:unit matrix

k :discrete time

2) Observation equation

$$\{y\}_k = \{h(\{x\})\}_k \quad (5)$$

$\{y\}$:unit values

$[h]$:observation matrix

The discrete time can be taken as the iterative step, i.e., learning iteration. $[I]$ in Eq.4 is due to that synapse weight in this problem are independent upon time. Equation 5 is considered as a nonlinear observation equation that can be expressed in simpler form using the sensitivity matrix $[H]$ as given by Eqs.6,7, and 8.

$$\{p\}_k = [H]_k \{x\}_k \quad (6)$$

where,

$$\{p\}_k = \{q\} - \{F(\chi,t)\}_k + [H(\chi,t)]_k \{x\}_k \quad (7)$$

$$[H_{ij}] = \partial F_i / \partial x_j \quad (8)$$

$\{q\}$:training data

χ :synapse weights estimated at one step prior to the present iteration.

The observation vector $\{p\}$ and matrix $[H]$ can be evaluated using a priori information.

3. FINITE ELEMENT FEATURES

Figure 3 shows a schematic representation of the finite element image processing grid that converts pictorial image into non-parametric image features numerically. Each of input nodes serves as a photosensitive receptor. In practice, for instance, signals transferred from sensing elements of CCD area array should be given to the input nodes. Nodal values of the output nodes become the finite element features. Other nodes are boundary nodes on which boundary conditions are specified.

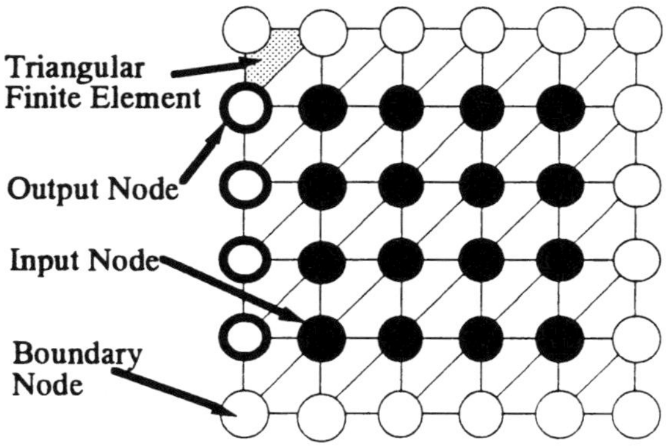

Fig. 3. Finite element image processing grid.

The algorithm to relate input and output signals of the finite element image processing grid can be a linear mapping as described by a linear finite element equation. In this research work, 2-D Poisson's equation was utilized as a governing equation given by Eq.(9). The finite element equation used here is expressed as Eq.(10). The basic mechanism of finite element image processing grid for generating image features is the conversion of incident light intensity distribution projected over the area comprising of finite element input nodes into a vector form of image features distributed over the output nodes.

$$K_x \frac{\partial \phi^2}{\partial^2 x} + K_y \frac{\partial \phi^2}{\partial^2 y} = Q \qquad (9)$$

K_x:information conductivity in x direction
K_y:information conductivity in y direction
ϕ :potential
Q :constant

$$[K]^{-1} \begin{Bmatrix} \vdots \\ A \\ \vdots \\ B \\ C \\ \vdots \end{Bmatrix} = \begin{Bmatrix} 1 \\ \vdots \\ 2 \\ \vdots \\ 3 \\ \vdots \end{Bmatrix} \qquad (10)$$

$[K]^{-1}$: inverse matrix of stiffness matrix
{A} : input vector
{1} :output vector (image features)

4. NEURAL NETWORK CONTROL SYSTEM

Figure 4 shows the neural network control system with a bio-feed back loop containing another neural network which converts the pictorial information into the plant moisture content. The vegetative system is the controlled system of which controlled variable is the plant moisture content. The control element of this system is the environment surrounding the plant with which the control inputs are temperature, humidity, CO_2 concentration, light intensity and water potential of root zone.

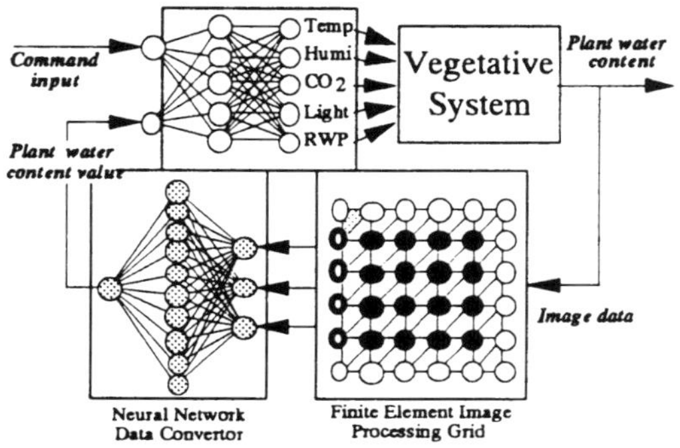

Fig. 4. Neural network controller.

5. NON-INVASIVE MEASUREMENT FOR PLANT MOISTURE CONTENT

5.1 Finite Element Image Processing Grid

A tomato seedling (30 days after seeding) was used to obtain input data of a finite element image processing grid. A single tomato seedling was placed in a growth chamber to provide proper temperature, CO_2 and light conditions as illustrated by Fig.5. Water supply to the seedling was stopped to observe change in appearance of wilting plant. The water loss from the plant was measured continuously by the electronic balance on which the plant with water pot was set.

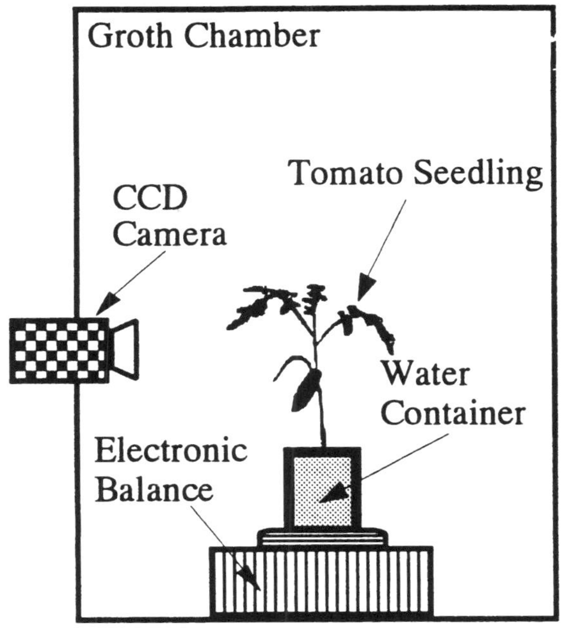

Fig.5 EXperimental setup.

Movement of leaves and stems of the wilting plant due to the loss of turgidity of tissues was recorded on a video tape through a CCD color camera. An example of binary image of the wilting plant (30, 60 and 90 minutes after stopping water supply) is shown in Fig. 6. For extraction of finite element features, brightness distribution in 256 gray scale over captured image frame of the wilting plant was fully used as input data.

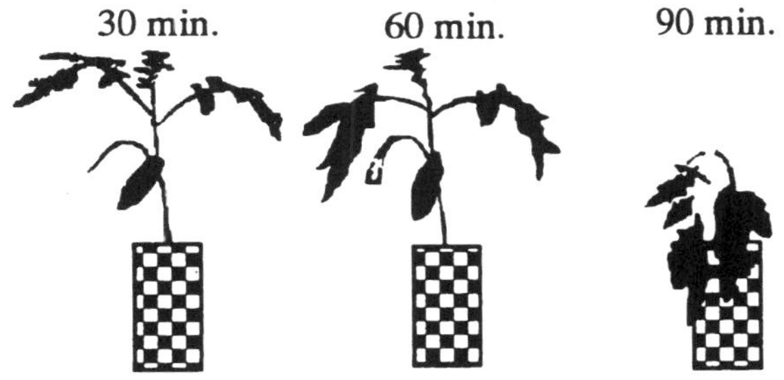

Fig.6. Change in appearance of wilting plant.

Figure 7 illustrates the procedure to prepare input data for the finite element image processing grid. The captured still image frame was divided into 12 x 12 segments. The image dimensions of each segment were 25 x 30 pixels. The average value of brightness (Y signal) of each segment was calculated. The calculate average brightness values made 144 input data which were fed into the finite element image processing grid. Figure 8 shows the finite element image processing gird with 144 input nodes corresponding to the input data obtained from the image frame. Six nodes were selected for output nodes as shown in Fig.8.

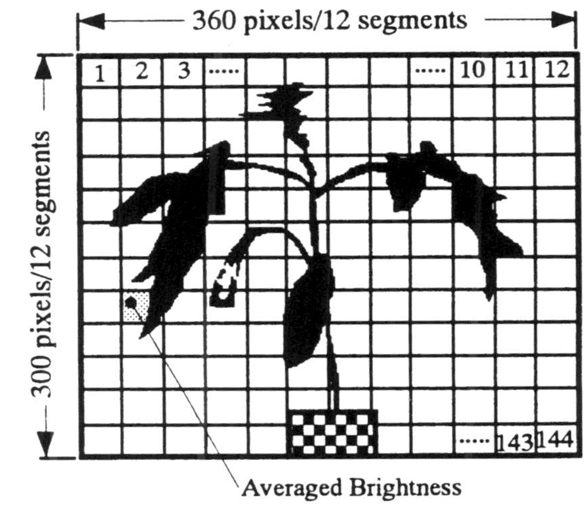

Fig. 7. Digital image frame consisting of 144 segments.

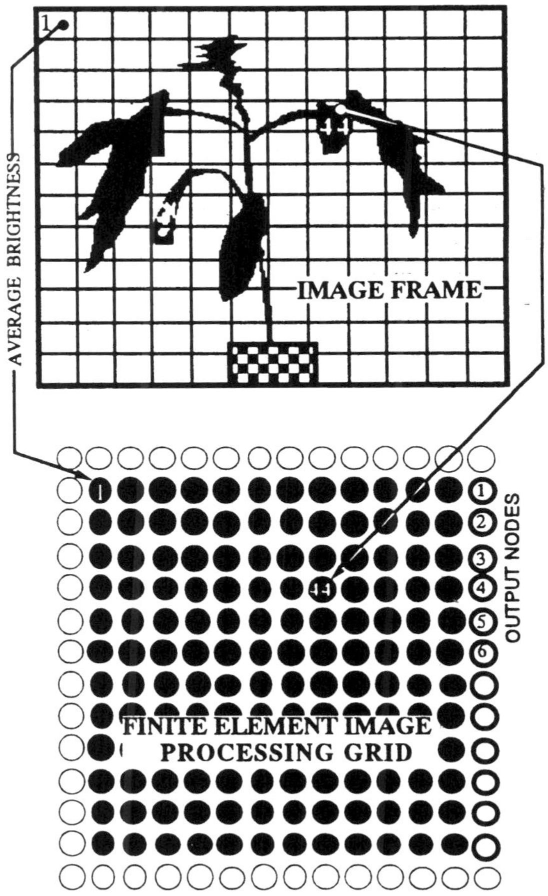

Fig. 8. Each of segments of the image frame is coincident with the finite element input node in order.

The variations of the finite element features with time after wilting started were plotted in Fig.9. Tne output nodes closer to the center of the output node array are the more sensitive to the change in plant appearance in the image frame. The relationship between water content of the plant and the value of the finite element feature obtained at the node no. 6 is indicated in Fig.10 as an example.

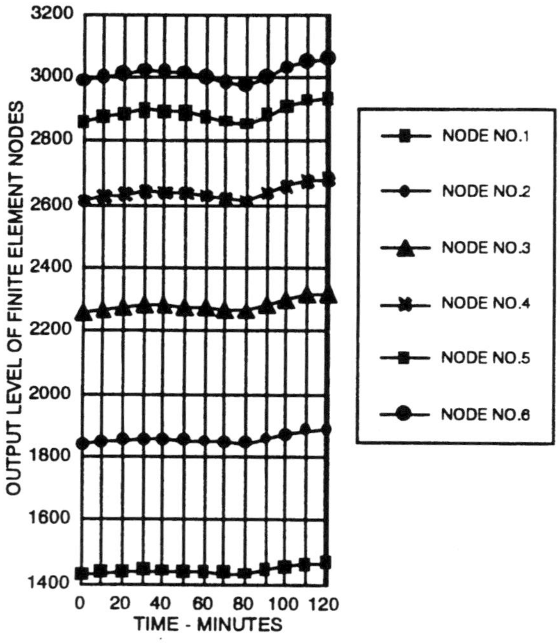

Fig. 9. Variation of the finite element features with time after wilting started.

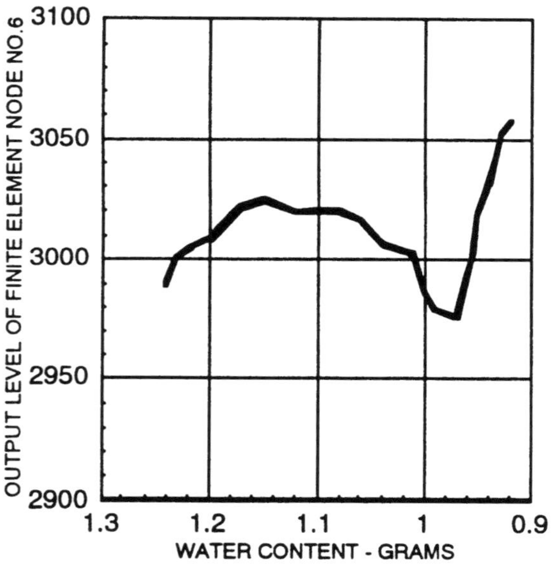

Fig. 10. The value of one of finite element features changes with the wate r content of the wilting plant.

5.2 Neural Network Model

A three-layered neural network was used to model the non linear relationship between the moisture content and the finite element features. Three finite element features obtained at the most sensitive output nodes (Nodes 6,5 and 4) were designated as input parameters of the neural network model. The three-layered neural network with the sigmoid non-linearity contains 10 units in the middle layer that makes the total number of units 14 as shown in the feed back path of Fig. 4. A single output unit was provided for the plant moisture content.

This neural network serves as a data convertor that transforms the three entities of finite element features extracted from the pictorial image of plants into moisture content value. The Kalman neuron training algorithm was applied for this problem.

5.3 Training Data

During the 2-hour video taping, 25 image frames from 0 to 120 minutes at 5- minute intervals and 24 readings of plant water content were taken. Two data at 30 min. and 90 min. were left to use for inspection data. The remaining 23 data sets were used for the neuron training. Table 1 shows some of the training data sets.

6. RESULTS AND DISCUSSION

The learning process was terminated when the root mean squared error of outputs reached 8.11×10^{-3} after 500 times of iterative calculations. Table 2 indicates satisfactory agreement of the estimated plant water content with the actual measured data. As shown in Fig.10, a strong non-linearity exists in the relationship between water content of the plant and the value of the finite element feature. Inspire of that, the result indicated that the neural network model proposed here was able to correlate two naturally different quantities. It is also clear that the finite element features can be use effectively to quantify a variation of pictorial image.

TABLE 1 Som of training data.

#	OUTPUT DATA (WATER CONTENT)	INPUT DATA NODE 4	NODE 5	NODE 6
1	1.24	2615	2863	2990
2	1.22	2628	2877	3005
3	1.23	2625	2873	3000
4	1.20	2631	2881	3009
5	[illegible]	[illegible]	2885	3013
			2894	[illegible]
[illegible]	0.95	2651		[illegible]
21	0.94	2661	[illegible]	3032
22	0.93	2670	2918	3041
23	0.93	2679	2928	3052
24	0.92	2683	2933	3056
25	0.92	2686	2835	3059

TABLE 2 Estimated water content of the wilting plant..

WATER CONTENT		INPUT DATA		
Measured	Estimated	NODE 4	NODE 5	NODE 6
1.15	1.133	2645	2896	3025
0.95	0.947	2639	2883	3004

7.ACKNOWLEDGEMENT

A research plant for extending this study has been accepted for a grant of the Ministry of Education, Science and Culture of Japan by a Grant-in-Aid for Developmental Scientific Research (Project No.06556042).

8.REFERENCES

Haralick, R. M., K. Shaanmugan, and I. Dinstein (1973). Textural features for image classification. *IEEE Trans. on Systems, Man, and Cybernetics,* Vol. SMC-3,6,610-621.

Hashimoto, Y. and H. Nonami (1992). Measurement and control in transplant production systems. In:Transplant Production Systems. (K.Kurata and T. Kozai, Ed.) pp117-136. Kluwer Academic Publishers, Dordrecht, The Netherlands.

Murase, H., S. Koyama, N. Honami and T. Kuwabara (1991). Kalman filter neuron training. Bull. Univ. Osaka Pref. Ser. B. 43,91-101.

Murase, H., N. Honami and Y. Nishiura (1994). Image information feedback using textural features for plant growth control. Proc. of the First Asian Control Conference, Tokyo, July 27-30, 1994. Vol.3, pp.17-20.

NEW APPROACH TO TOTAL PRODUCTION SYSTEMS BASED ON AN INTELLIGENT CONTROL

Y. Hashimoto, T. Morimoto and J. De Baerdemaeker

Dept. of Biomechanical Systems, Ehime University, Tarumi, Matsuyama, Japan

Dept. of Agric. Engineering, Katholieke Universiteit, Leuven, Belgium

Abstract: Total plant production process consisted of cultivating process and storage process is quite complex and uncertain. A new intelligent control technique, including an expert system and a hybrid system combined with neural networks and genetic algorithms, is proposed for the optimization of total plant production process. An expert system is used for normal (rough) control through the whole growth stage. A hybrid system is used for optimization (careful control) during the important growth stages. For the optimization, plant responses, as affected by environmental factors, are first identified using neural networks and then the optimal setpoints of environmental factors are searched for through simulation of the identified model using genetic algorithms. This intelligent control technique was shown to be very useful to the optimization (or optimal control) of such complex systems as plant production processes.

Keywords: Total plant production, optimization, intelligent control, expert system, neural network, genetic algorithm

1. INTRODUCTION

In recent years, advanced agricultural production systems, such as plant factories and protected cultivation, have been extensively developed (Hashimoto, 1991). These systems equip many computers and control devices and consume much electrical energy. In these systems, therefore, optimization techniques to obtain higher yields and better quality of plants are severely required from the viewpoint of cost-performance. Until now, studies on such optimal control have been performed (Van Henten and Bontsema, 1991; Tantau, 1991).

For optimization of plant production processes, the monitoring of plant (or fruit) responses and its effective uses for the control algorithms seem to be essential. We call this concept "speaking plant approach (SPA) or speaking fruit approach (SFA)" (Hashimoto, 1993). In general, however, it is not easy to realize the optimization of plant production processes because the physiological responses are characterized by complexity and uncertainty.

Recently, intelligent control techniques have been markedly developed for the control of really complex systems to which conventional mathematical approaches cannot easily applied. There has been considerable interests in expert systems, neural networks and genetic algorithms for the intelligent control. Expert systems are useful for the control of really complex systems in which a skilled human operators knowledge plays an important role (Hatou *et al.*,1991; Hashimoto and Hatou, 1992; Martin-Clouaire *et al.*, 1993). Neural networks have capability to identify the nonlinear behavior of a complex system with their own learning ability (Chen *et al.*, 1990; Hunt *et al.*, 1992). Genetic algorithms are one of the combinatorial optimization techniques. It searches for an optimal value from a complex objective function by simulating the evolutionary process based on a crossover and mutation in genetics (Goldberg, 1989). Active applications of neural networks and genetic algorithms to the optimization of agricultural production processes are observed in recent years (Morimoto *et al.*, 1993 and 1995).

The present works are attempts to apply a new intelligent control technique based on expert system, neural networks and genetic algorithms to the optimization of total plant production process consisted of cultivating process and storage process. In the cultivating process, the optimal control of net CO_2 uptake as affected by intermittent drainage of the solution and the optimization of plant growth as affected by nutrient concentration in hydroponics were performed. In the storage process, on the other hand, the optimal control of absolute humidity in the storage house as affected by ventilation was carried out through simulation.

2. OPTIMIZATION PROBLEMS

In the present study, we dealt with three types of optimization problems for total plant production.

Optimization problem I; The nutrient concentration of the solution is one of the most important manipulating factors for the growth optimization of plants in hydroponics (Ehret and Ho, 1986). It is usually increased along with the growth of plants. For the cultivation of fruit vegetables, it is important to promote the reproductive growth rather than the vegetative growth because both growth are competitive with each other. The control problem here is to determine the adequate setpoints of nutrient concentration to promote the reproductive growth through all growth stage using an expert system. Experiments were carried out in a computer controlled greenhouse. Tomato plants (*Lycopersicon esculentum* Mill. cv. "Kyoryoku-Kyokko") grown in deep hydroponic culture were used.

Optimization problem II; The control of net photosynthetic rate of leaves is essential for the growth optimization. In hydroponics, active oxygen supply to the roots makes it possible to accelerate the net photosynthetic rate. The oxygen supply is usually controlled by the drainage and supply of the solution. The optimization problem here is to determine the 4-step drainage and supply times which maximize the net photosynthetic rate using a hybrid system. The experiment was carried out in the growth cabinet.

Optimization problem III; During the storage process of a citrus fruits, because of the continuous evapotranspiration, the humidity in the storage house increases excessively. The excessive high humidity induces many diseases caused by mold fungi. A ventilation is simple and effective way for reducing the humidity in the storage house. The optimization problem here is to determine the 6-step operational on-off times of ventilation to maintain the relative humidity at the optimal setpoint (85%RH). Iyokan fruits (*Citrus Iyo Hort.* Tanaka) were used for the experiments. The storage house is 19.8m^2 area × 3m height and is equipped with one ventilator whose ventilation rate is 4m^3/min.

3. COMPUTER CONTROL SYSTEM FOR TOTAL PRODUCTION

To treat the complexity and the large scale of plant production process, a computer network system may be useful. Figure 1 shows the schematic diagram of a decentralized computer control system for total production of plants, which has several types of computers, (c): computer I for process control of greenhouse and storage house, (a): computer II for identification and optimal control of plant responses and (b): computer III for planning, diagnosis and analysis of physiological status of plants. All computers are connected with a local area network (LAN) and communicates with each other. Base on this computer control system, the optimization of complex plant production processes are performed. Here, control techniques in (a) and (b) are discussed.

3.1 Control system for optimization

For realizing the optimization of plant production

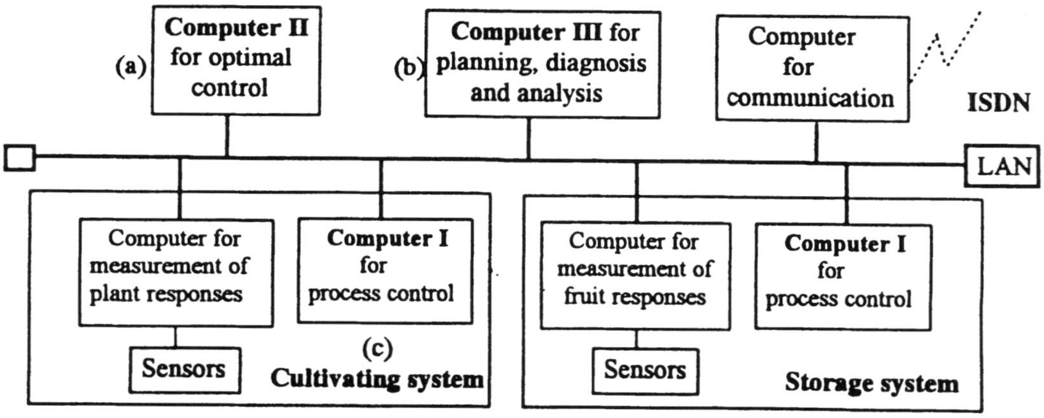

Fig. 1 Schematic diagram of a decentralized typed computer control system for total production

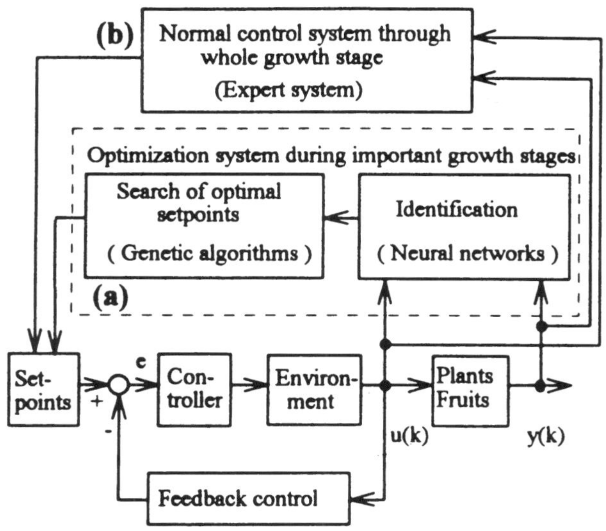

Fig. 2. Control system for realizing the optimization of total plant production process.

processes, the effective uses of plant responses, which reflect the current physiological status of plants, are essential. This concept is called "SPA". Figure 2 shows a block diagram of SPA-based control system for optimization which corresponds to the parts (a) and (b) in Fig.1 (Hashimoto, 1991; Morimoto *et al.*, 1993). It is mainly consisted of two control systems; an expert system (b) for normal (or rough) control through all the growth stage and a hybrid system (a), which is combined with neural networks and genetic algorithms, for optimal (or careful) control during the important growth stages. Plants are usually controlled and managed with the instructions of an expert system. However, the optimization is carried out using a hybrid system when the plants reached important growth stages, where plant responses, as affected by environmental factors, is first identified using neural networks, and then the best setpoints of environmental factors is searched for through simulation of the identified model using genetic algorithms.

3.2 Expert system

An expert system, which is placed in the part (b) in Fig.1, was used for determining the rough setpoints of environmental factors through all the growth stages. It is mainly composed of an user interface, an inference engine and a knowledge base. Here, optimal setpoints of nutrient concentration were determined. At first, the present state of plant growth is evaluated. The vegetative growth is evaluated from growth data (stem diameter, leaf number, leaf shape and leaf color), solution data (nutrient concentration and pH), cultivation data (transplanting day and days after transplanting) and climate data (amounts of temperature and solar radiation). The reproductive growth is evaluated from growth data (fruit diameter, fruit number in 1st, 2nd or 3rd cluster and fruit colors), solution data, cultivation data and climate data. Furthermore, the present conditions of the nutrient solution are also evaluated from the solution and cultivation data. The future states of plant growth, nutrient solution and climate are then predicted. Finally, the appropriate setpoints of nutrient concentration are determined based on these prediction data.

3.3 A hybrid system based on neural network and genetic algorithm

A hybrid system combined with neural network and genetic algorithm, which is placed in the part (a) in Fig.1, was used to determine the optimal drainage and supply times in hydroponic (optimization problem II) and the optimal on-off ventilation times during the storage process (optimization problem III).

Neural network for identification; Two types of three-layer neural networks were used for the identification. For dynamic model identification, (n+1)th time series of input variable {u(k), ..., u(k-n)) and nth past time series of output variable {y(k-1), ..., y(k-n)) are applied to input layer and the current output y(k) is applied to the output layer as a training signal (u(k): input, y(k): output, n: number of system parameters). Here, net photosynthetic to drainage and supply operations and relative humidity to ventilation were identified. For static model identification, on the other hand, input variables are applied to the input layer and output variables are applied to the output layer. Here, the relation between evapotranspiration of fruits and temperature was identified. The learning method is error back-propagation (Rumelhart *et al.*, 1986).

Genetic algorithm for optimization; We defined an individual which means the decision variable to be obtained. An individual is often expressed by finite-length binary strings. Individuals in the optimization II and III are defined as follows:

(optimization II) (t_1, t_3: off time; t_2, t_4: on time):
Individual= 4-step drainage and supply times
$=(t_1, t_2, t_3, t_4)$
=(000111, 101010, 110011, 011001)

(optimization III) ($t_{init,on}$: initial on time; t_{on}: on time; t_{off}: off time):
Individual= 6-step ventilation on-off times
$= (t_{init,on}, t_{off}, t_{on}, t_{off}, t_{on}, t_{off})$
=(011010, 111100, 000100, 110011, 001100, 011000)

The set of individuals is called "population". Also, one character, 0 or 1, in each parameter is called "gene". At first, initial population P(0) consists of N_i sorts of individuals are generated at random. Second, a crossover and mutation are applied to those individuals. Third, the fitness of all individuals are

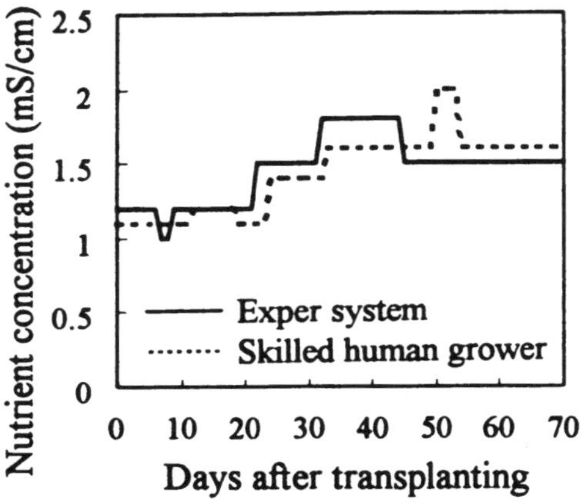

Fig. 3. Daily setpoint of nutrient concentration of the solution in hydroponic tomato cultivation.

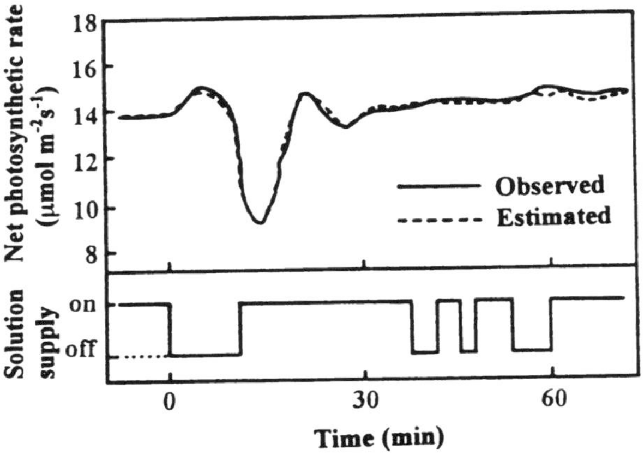

Fig. 4. Dynamic change in net photosynthetic rate as affected by drainage and supply of the solution and its identification using the neural network.

calculated. Here, the fitness are obtained from the model simulation. Finally, better individuals with lower (or higher) fitness are retained for next generation P(t+1). The optimal values can be obtained by repeating these procedures. For the crossover, two individuals (e.g., 000101 and 001111) are first mated at random and then cut at the center position (3-bit position). Then, two new individuals (e.g., 000111 and 001101) are generated by swapping all binary characters from 1-bit to 3-bit position. For the mutation, one individual (e.g., 000011) is first selected at random, and then a new individual (010011) is created by inverting one character from 1 to 0 or vice versa.

4. IDENTIFICATION AND CONTROL RESULTS

4.1 Control result of nutrient concentration using an expert system through whole growth stage

The adequate setpoints of nutrient concentration through the entire growth stages were determined using an expert system. Figure 3 shows the daily setpoints of nutrient concentration in the hydroponic tomato cultivation. The most severe disorder during the tomato cultivation is blossom-end rot (BER) of fruits. It occurs when the plant cells suffer from a calcium deficiency. Poor cultivating management induces BER. It can be seen that the daily setpoints from an expert system are similar to that from a skilled grower. In both experiments, the occurrence of BER was only slightly observed. This shows that the expert system used here worked well for growth optimization of tomato plants.

4.2 Identification and control results of net photosynthetic rate using a hybrid system

Figure 4 shows the dynamic change in the net photosynthetic rate as affected by the drainage and supply of the solution and its identification using a neural network, where four drainage operations were conducted. From the response to first drainage (12-minute drainage, which is the longest time of the four drainage cycles), a slight increase in the net photosynthetic rate was observed during the first 5 minutes after the drainage. This gradually decreased later with the progression of water stress. Significant reduction was then observed when the solution was supplied. In the latter three responses, there were few significant increases. These results show that optimal drainage and supply times for maximizing the net photosynthetic rate of leaves do exist. The dotted line represents the estimated response computed by the neural network model. The system parameter number and hidden neuron number used here are 5 and 2, respectively. The estimated response is in good agreement with the observed response. Thus, a successful model was obtained for computing the fitness (net photosynthetic rate at stage four) for each individual (4-step drainage and supply times).

The search process of an optimal value using genetic algorithms is analogous to a natural evolutionary process. Figure 5 shows search evolutionary curves of an optimal value for three mutation rates. The maximum value could be obtained at 5 generations when the crossover and mutation rates had high values (P_c=0.8 and P_m=0.8). The evolutionary rate was slightly delayed with a decrease in the mutation rate, but the optimal value could be successfully obtained in all cases, without falling into local optima. This result suggests that the high diversity of population is essential for faster evolution.

Figure 6 shows the optimal control performance of the net photosynthetic rate, obtained from the neural-network model. The lower figure shows the optimal 4-step drainage and supply times, which were determined to be t_1=4-minute drainage, t_2=8-minute supply, t_3=4-minute drainage and t_4=2-minute supply. It is clear that the optimal 4-step drainage

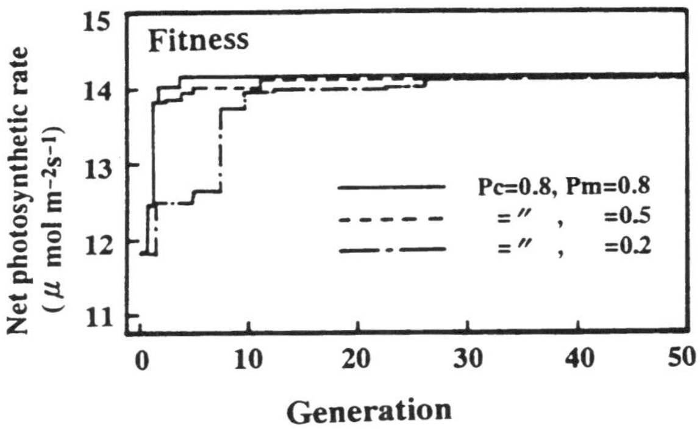

Fig. 5. Search evolution curves of the optimal 4-step drainage and supply times for mutation rates.

and supply operations obtained here markedly increased the net photosynthetic rate. Thus, the optimal value in the biological control problem can be successfully obtained by using this intelligent technique.

4.3. Identification and control results of fruit responses during the storage process

Figure 7 shows daily changes in evapotranspiration of fruits under three temperature conditions, 5, 10 and 20 ℃ during the storage process. The evapotranspiration in all cases increased almost linearly with time. However, the relationship between the temperature and the evapotranspiration appears to be nonlinear. These data are identified using the neural network and we'll obtain a computational model for calculating the evapotranspiration of fruits. Figure 8 shows estimated relationships between temperature and absolute humidity in the storage house under suitable amount of fruits (0.8, 1.0, 1.2, 1.4, 1.6 and 1.8 ton). The closed circles on those curves represent the observed values. All estimated values are closely related to the observed values. This means that we could obtain a successful model for calculating the absolute humidity in the storage house.

Figure 9 shows dynamic changes in absolute humidity in the storage house as affected by ventilation. Here, air temperature was kept at 15 degree C. Also, the outside absolute humidity was 5 (g/m^3) and the ventilation rate was 4 (m^3/min). The estimated response is in good coincident with the observed response. This means that we could obtain a successful model for calculating the absolute humidity as affected by any combination of the 6-step operational times of ventilation.

Figure 10 shows the optimal control performance of the absolute humidity as affected by optimal 6-step ventilation. From simulation of the identified model in Fig.8, the absolute humidity was estimated to be 9.8 g/m^3. The optimal control aim is to reduce the

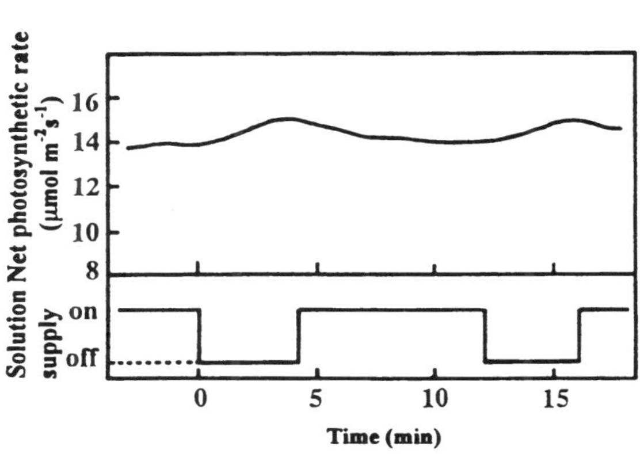

Fig. 6. Optimal control performance of the net photosynthetic rate as affected by the drainage and supply of the solution.

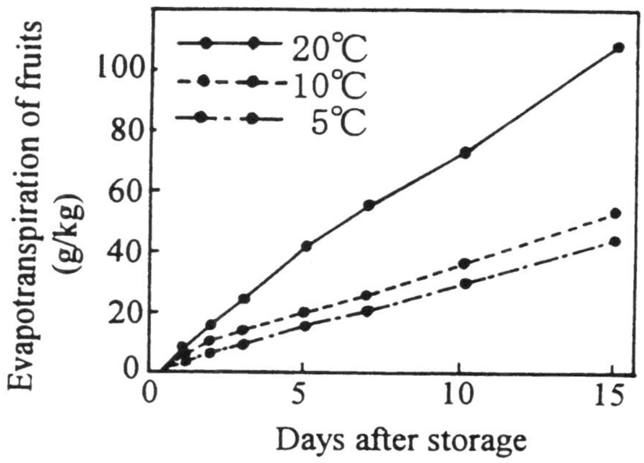

Fig. 7. Observed daily changes in evapotranspiration of fruits after storage.

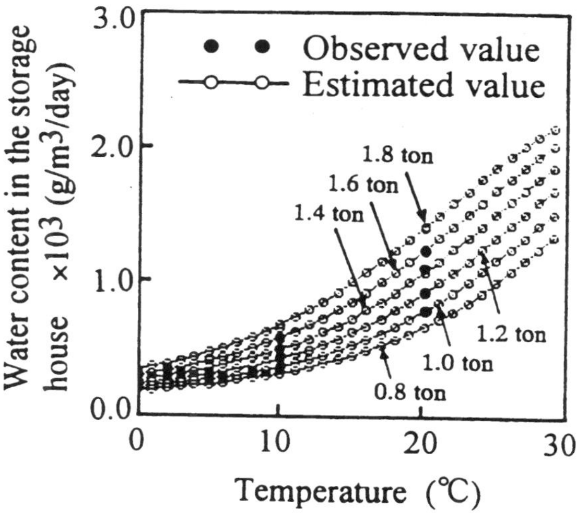

Fig. 8. Estimated relation between temperature and evapotranspiration of fruits in the storage house.

absolute humidity in the storage house and then keep it at the optimal level (9.0g/m^3). This can be realized by minimizing the error between the current values and desired values of humidity. The optimal value is $t_{init,on}$= 28-min on, t_{off}= 2-min off, t_{on}= 58-min on, t_{off}= 2-min off, t_{on}= 58-min on, t_{off}= 2-min off. The optimal control performance obtained here looks like best under this limited condition.

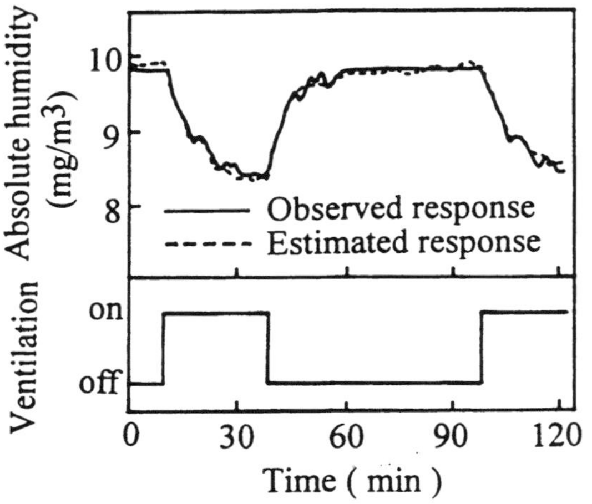

Fig. 9. Response of absolute humidity in the storage house as affected by ventilation.

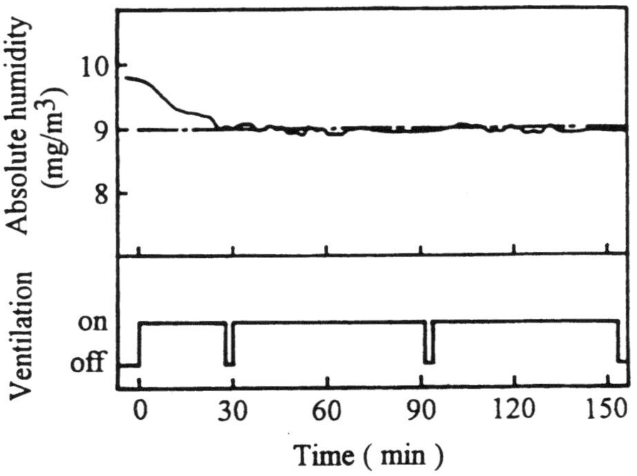

Fig. 10. Optimal control performance of the absolute humidity in the storage house.

5. CONCLUSION

In this study, for realizing the optimization of plant production process, we proposed a SFA-based intelligent control technique composed of an expert system and a hybrid system based on neural networks and genetic algorithms. By using an expert system, we could obtain the same strategy of nutrient concentration as a skilled grower's plan. In a hybrid system, furthermore, a three-layer neural network was effective for the identification of plant and fruit responses as affected by environmental factors. Next, the genetic algorithm was used for the search of an optimal value through simulation of the identified model. It allowed an optimal value to be quickly sought under high crossover and mutation rates. Thus, this intelligent control technique was shown to be quite useful for the optimization of such complex processes as plant production processes.

REFERENCES

Chen, S., S.A. Billings and P.M. Grant (1990). Non-linear system identification using neural network. *International Journal of Control,* **51(6)**, 1191-1214.

Ehret, D.L. and L.C. Ho (1986). Effects of osmotic potential in nutrient solution on diurnal growth of tomato fruit. *J. Exp. Bot.,* **37(182)**, 1294-1302.

Goldberg, D. (1989). *Genetic algorithms in search, optimization and machine learning,* Addison-Wesley, Reading, Massachusetts.

Hashimoto, Y. (1991). Computer integrated plant growth factory for agriculture and horticulture. *Proc. of IFAC/ISHS Workshop on Mathematical and Control Applications in Agric. and Hort.*, Pergamon Press, Oxford, pp.105-110.

Hashimoto, Y. and K. Hatou (1992). Knowledge-based computer integrated plant factory. *Proc. IFAC Workshop on Expert System in Agric.*, 9-12.

Hashimoto, Y., (1993). Computer integrated system for the cultivating process in agriculture and horticulture. In *The Computerized Greenhouse* (Hashimoto eds.). pp.175-196, Academic Press.

Hatou, K., Y. Kamio and Y. Hashimoto (1991). Computer integrated plant factory based on AI. *Proc. 9th IFAC Symp. on Identification and System Parameter Estimation, Vol.1,* pp.616-621.

Hunt K.J., D. Sbarbaro, R. Zbikowski and P.J. Gawthrop (1992). Neural networks for control systems - Survey. *Automatica,* **28(6)**, pp.1083-1112.

Martin-Clouaire R, T. Bouland and B. Jeannequin (1993). Using empirical knowledge for the determination of climatic setpoints: An artificial intelligence approach. In: *The Computerized Greenhouse* (Hashimoto eds.), 197-224, Academic Press.

Morimoto, T., T. Takeuchi and Y. Hashimoto (1993). Growth optimization of plant by means of the hybrid system of genetic algorithm and neural network. *Proc. of Inter. Joint Conference on Neural Networks,* pp.2979-2982.

Morimoto, T., J. De Baerdemaeker and Y. Hashimoto (1995). Optimization of storage system of fruits using neural networks and genetic algorithms. *Proc. of 4th IEEE International Conference on Fuzzy Systems,* Vol.1, pp.289-294.

Rumelhart, D.E., G.E. Hinton and R.J. Williams (1986). Learning representation by backpropagation error. *Nature,* **323(9)**, 533-536.

Tantau, H. -J (1991). Optimal control for plant production in greenhouse. *Proc. of IFAC/ISHS 1st Workshop on Math. and Control Applications in Agric. and Hort.*, Pergamon Press, Oxford,

Van Henten, E. J. and J. Bontsema. (1991). Optimal control of greenhouse climate. *Proc. of IFAC/ ISHS 1st Workshop on Mathematical and Control Applications in Agriculture and Horticulture,* Pergamon Press, Oxford, pp.27-32.

QUANTIFICATION OF STRUCTURE AND QUALITY OF CHOPPED CROPS BY AN AUTOMATED IMAGE PROCESSING[1]

N. Guth, H. Georg and F.-J. Bockisch

Institute of Agricultural Building Research
Federal Research Centre for Agriculture, D-38116 Braunschweig, Germany

Abstract: Classification methods for the structure of forage, utilising physical properties, are not able to make an exact classification of all cutting structures. With image analysis each particle is measured exactly with its real physical sizes. In order to reduce measurement time, a full-automated image analysis system was developed. First task was to realise the separation of the particles which was achieved with three connected conveyers. The second task was to write a control system program. The measuring time could be diminished by 92.5 % compared to hand-laying. A 25 g dry-matter sample could then be analysed in 3.2 h.

Keywords: Automatic, Food processing, Classification, Image analysis

1. INTRODUCTION[1]

The judgement of the physical structure of forage is accomplished mainly by mechanical classifiers (sieving and air flow apparatus). The problems with all sieving analyses are the many parameters, which influence the passage of particles through the sieving hole, for example the vibrating length and frequency, the sample height and the kind of the sieves with round, squared or oblong holes (Wessel, 1968; Orth and Peter, 1975; Baader, 1961). The diversity of the enumerated influential factors make a direct application of the results from one device to another and the repeatability even at one same device difficult. For this reason, the ASAE (1976) tried to standardise apparatus construction and parameter assessment methods (ASAE Standard S319).

The generally used sorting apparatus have no capability to indicate the surface of a particle, but can give only an approximate reference point for the diameter or the length of a particle, which emerges from the sieve hole diameter.

Through hand-measurement (planimetry) the assessment of particle surface can be undertaken, but generally, it is not feasible due to the large number of particles to be measured (50000 particles or more per 100 g dry matter).

The image analysis is suitable for the classification of the structure of forage, since in contrast to the sieve analysis the physical dimensions of each food particle are measured individually and the exact counts are recorded and furthermore the surface is also indicated.

For the measurement of the particles by means of the image analysis, it is necessary, that the single particles should not be in contact with one another in the image frame. At hand-imposition and following image analysis, measurement of 43.5 g sample material required about 250 hours. From this, the quantity of random sample must be very small which makes the sample less representable. Consequently

[1] This study was carried out at the Institute of Agricultural Engineering, University of Giessen, Germany

the application of this method for agricultural purposes is not conceivable.

In this study the quality and reliability of the judgement of the physical structure of forage using the sieve analysis and the image analysis are examined.

Considering the very high time required for the implementation of image analysis measurements, it was necessary to study and develop the technology and methodology for the measurement of the forage surface in laboratory.

2. MATERIAL AND METHODS

The feed structure of two maize silage sets was analysed. The silage was prepared on one hand by means of a conventional precision chopper (CC), fixed 10 tangentially ordered knives on the cylinder, which does exact cutting. In this system the stalks are cut off in a cross section pattern, so that the particles are displayed in a spherical-cubical form. The theoretical cutting length in this case was 6.8 mm. On the other hand a multiknife chopper (MC) was used. The cylinder was here fitted with 2x24 knives in appropriate radial form. Thus a hitting tearing cut is executed causing the stalks and husks to be macerated in longitudinal direction.

For the determination of the food structure was accomplished, using an image analysis apparatus from Leitz Company (CBA 8000). The images with a resolution of 750 x 582 pixels were taken with a CCD-camera. The parameters recorded were projected surface, perimeter, length (maximum Feret) and width (minimum Feret) of each particle. The arithmetic mean value and the sum of each parameter were calculated.

An important requirement for image analysis measurements is the appropriateness of feed sampling. The deep-frozen food materials were thawed, thoroughly mixed, and then decomposed in four parts from which two opposite parts were taken. These two parts were furthermore mixed and the procedure repeated till a desired volume was achieved. After the sample reduction the samples were dried with 105 °C and then subsequently measured by image analysis. The sample weights used in CC- and MC-variants were 14.6 g and 18.9 g respectively.

For the comparison of sieve and image analysis, a sieve apparatus with three hole openings (2 mm, 1 mm and 0.5 mm) was used. The sieve boxes had a base of 0.5 x 0.5 m^2, which were ordered vertically one above the other. The sieve comprised of square grid. The tray oscillated horizontally; the sieving time totalled 5 minutes. Each fraction was thus evaluated separately by the image analysis. For the total sample therefore the recorded data of the single fractions were aggregated.

The precision of the automated edition was examined. Through which the image analysis of straw particles using three sieve fractions (1-2 mm, 2-5 mm and 5-10 mm) of two sets was done both over the automated pattern as well as over the manual corrected pattern. At the manual correction, touching particles were separated from each other. Set 1 was treated with normal flail chopper (chopping aggregate at the combine), whereas in set 2 the treatment was done with a more intense flail chopping system (spliced).

3. RESULTS

3.1 Comparison of sieve and image analyses

The average length of the particles between any two adjacent sieves does not correspond necessarily to the sieve hole sizes. However, due to vibration and raising of the longer particles with smaller diameter, as a rule these particles passed in a vertical position through the narrow sieves (Table 1). For instance, the average length of the particles of the CC-variant in the fraction 1-2 mm did not correspond to the mean of the adjacent sieve openings but 2.59 mm. Thus, the measurement of the actual average length of the particles was not exact.

Furthermore, the average scores within the classes of both variants did not agree. For example, the average length of the particles from the sieve class greater than 2 mm at the CC-variant totalled to 7.38 mm, while for MC-variant this was only 4.11 mm. Accordingly, there was doubling of particle counts per unit g dry matter in the case of MC compared to CC (Table 1).

Table 1 Results of image analysis measurements of maize silage, sieved in several fractions

Sieve opening	<0.5 mm	0.5-1 mm	1-2 mm	>2 mm
Number of particles per unit g dry matter				
CC-variant	45456	8010	2002	143
MC-variant	36164	8025	2546	280
MC/CC (%)	97.5	100.2	127.1	195.8
Average length of the particles				
CC-variant	0.41	1.42	2.59	7.38
MC-variant	0.47	1.47	2.41	4.11
MC/CC (%)	114.6	103.5	93.1	55.7

3.2 Development of a full-automated image analysis system for measurement of forage.

The results of the work, which was carried out in accomplishing an image analysis system for the judgement of food structure are summarised here below.

Automation of the particle separation. The requirements on an automatic particle separation technique were executed as follows:

1. The particles to be measured may not touch (Fig. 1). But none the less, a high particle density is expected.

2. The time required for the separation of the particles should be reduced as much as possible in comparison to manual hand-laying of the particles on the measurement surface.

3. The equipment must sample sufficient particles and have adequate reserve.

4. The structure of the particles must remain mechanically unchanged as much as possible.

5. The equipment must surely function and be operable through a computer program.

Fig. 1. Sprinkled chopped particles (grain silage).

To find a suitable separation technique, different alternatives were studied. For instance a mutual repulsion of particles through high-resistance loading separation and vibration techniques were used. However, these two procedures delivered no satisfying results. As a third alternative, a system was engineered involving three mini-conveyers (Fig. 2).

In this method (Fig. 2), the first conveyer served as reserve and supplier of the sample at very slow speed to the second belt. The second belt was characterised by a variable slope and diagonally ordered drivers. The particles fell from the second belt to the third one through a 20 - 35 cm long free case and were well distributed on a larger plane. The separation of particles was mainly achieved through the differential speed of the belts.

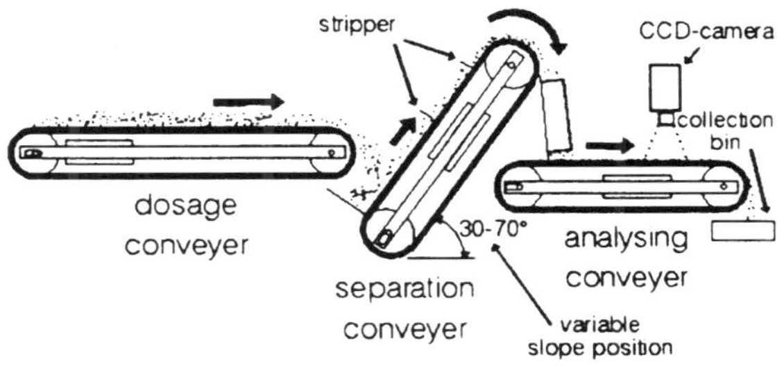

Fig. 2. A sketch of the separation technique for chopped forage.

Examination of the sum of the surface area revealed that an error due to automation occurred of less than 2 % in comparison to manual application. At the same time, a difference in medium particle length of 2-8 % for the two application methods was observed. This perhaps was due to complete separation of particles and/or faulty image detection.

Table 2 Comparison of the particle number by two chopped straw samples in three sieve fractions between automatically and manually corrected edition

sieve opening (mm)	weight (g)	auto-mated	manual corrected	autom./ manual (%)
Set 1				
1-2	0.87	2074	2195	94.5
2-5	1.17	748	792	94.4
5-10	5.28	331	327	101.2
Set 2				
1-2	0.66	773	843	91.7
2-5	1.71	525	579	90.7
5-10	2.31	259	270	95.9

Control system program. As a prerequisite an automated separation technique was required for the image analysis under this full automation system. Although the time for the measurement was considerably reduced, the separation technique still required some set program for its regulation in the different measurement steps. In full automation manpower is needed only for setting and control tasks. For the computer control system in this full automation, a program (programming language Quick Basic) was written, which operated the following functions:

1. Control the separation system: During the measurement of forage the three belts must be set in motion. To facilitate good quality and avoid unsharpness of the images over the CCD - camera, the conveyers were stopped until the image grabbing, analysis and saving of the results had been completed. This procedure was repeated after every other image had been processed. The positioning of the belt is accomplished by an optosensor. To the edge of the belt reflecting marker plates are fastened.

2. Control of the image processing: After the image grabbing, digitisation and the processing of the image results followed. In the initialising phase the parameters like thresholds and minimal pixel size were determined. The corresponding image analysis functions (erosion, dilatation, closing) may be executed automatically during the measurement, if necessary. Finally the detected objects were defined on the basis of projected surface, perimeter, length and width.

3. Control the saving of results: The characteristics of the sample and the image analysis assessment data were saved in an info file to allow a repetition of the measurement. After measurement, the recorded data may be compressed automatically (about 80 % space-saving). A 50 g dry matter sample requires without compressing 6 to 8 MB.

With the device for forage samples, as it is to be seen in figure 3, were measured without constant monitoring up to now about 50 tests in respectively 3 to 17 hours. This corresponded to a performance of 2.4 - 7.8 g sample material/h or 6100 to 9900 particles/h.

Fig. 3. Image analysis system for the measurement of forage.

Interpretation of the measured parametric data. The data files are transferable into statistical programs (here SPSSPC+). For forage structure characterisation, descriptive analyses for the determination of mean values, standard deviations, sums, minima, maximums, kurtosis and skewness ,may be of interest. Besides, a distribution may be displayed with arbitrarily fine grouping (Fig. 4). The statistical comparison between two tests are undertaken with the Kruskal-Wallis-test, because of the non-parametric distributions.

The time required for the statistical analyses for one sample (25 g dry matter) totalled at an 486-PC with 33 MHz up to 3-5 h. This included the calculation of the descriptive statistics for the parameters surface, perimeter, length and width for the whole sample and additionally the distribution of each parameter.

3.3 Results of image analysis measurements.

The total surface area of the MC-maize silage sample was with 0.87 opposite to 0.56 m^2/100 g dry matter at the CC-test around 55 % larger. The sum values of the other features perimeter, length and width are to be recognised at the MC-test approximately doubly as high as at the CC-test (Table 3). At the mean values a larger difference is to be ascertained however only at the surface.

Table 3 Comparison of recorded features with image analysis by two chopping variants of maize silage, converted on 100 g dry matter

	arithmetic mean		
feature/chopper	CC	MC	MC/CC (%)
particle number			
surface (mm)	4.49	3.44	76.6
perimeter (mm)	6.01	5.90	98.2
length (mm)	2.70	2.63	97.4
width (mm)	1.21	1.18	97.5
	sum		
particle number	123574	250692	202.9
surface (m^2)	0.55	0.86	156.4
perimeter (m)	742	1479	199.3
length (m)	333	658	197.6
width (m)	150	296	197.3

In figure 4 it is to be seen, that the distribution of the sum of the surface in size classes displays a asymptotic course, with a high surface share at the small particles. One must note, that up to the 16 mm^2-class was undertaken a finer division, so that with an exclusive 8 mm^2-class-division the asymptotic course would be still more distinctive.

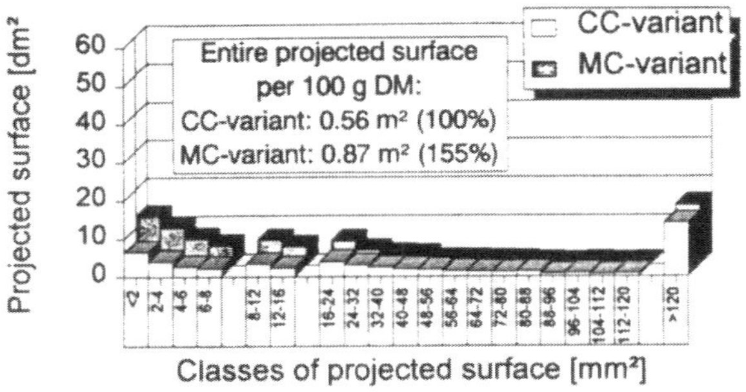

Fig. 4. Distribution of the projected particle surface of maize silage per 100 g dry matter in dependence of the chopping system (CC, MC).

The MC-tests display relating to this a still steeper curve, what means, that the amount of the small particles is very large. The smallest recognised particles had a surface area of 0.05 mm^2, a perimeter of 0.94 mm as well as a length and width of 0.23 mm.

4. DISCUSSION AND CONCLUSIONS

The problems described in the literature in the application of sieve analysis for the judgement of the forage structure was also estimated in our investigations. It was shown, that the methodical principle in the sieve analysis to allocate the mean chop length from the sieve hole sizes, is not exact, since the longer particles with small diameter also fall through the sieve after longer vibration time. On the other hand the material located in a sieve fraction does not agree in the composition with different chopping structures, as had been observed with the CC and MC-silage (Table 1).

The image analysis is suitable for the exact appraisal of the structure of forage, since each particle is measured with its real physical sizes. Additionally, beside the length some other physical properties sizes like surface, perimeter, width, form shape and colour for the characterisation can be recorded. Also spatial representation of the particles are possible. Furthermore the image analysis delivers an arbitrar distribution with fine gradation for all features (Fig. 4). These advantages face however to a high apparatus expense.

The degree of separation between the particles by the use of the three conveyers was on the higher side with more than 90 % (Table 2). The value of 1.2 % more particle at the automated edition of test 2 (5-10 mm-class) resulted into a very high score. This perhaps was due to complete separation of particles or faulty image detection.

A certain error is still occurring with the presented equipment, because larger particles, which are situated in two adjacent images are taken as two particles. There was no big difference in the sum of the surface area, but the average size is underestimated. Remedial measures could involve joining of the two successive images and following interpretation or through the use of a line camera.

Through the development of an automated separation technique and the control system program, a high separation share (90 - 95 %) of a time reduction of around 92 % compared to hand-laying culd be reached. The program enabled automation of the regulation of the different steps in the measurement avoiding error and minimising manpower requirement. A 25 g dry-matter sample could then be analysed in only 3.2 h. The time required for the measuring is accordingly more influenced by the number than of the weight of the particles. It is assumed, that this is found mainly through the very slow serial data transfer the used image analysis system.

A further acceleration should be possible with more powerful computers and image analysis boards. Another improvement could be the three-dimensional description of the particles and the indication of integrity of stalk superficial cuticle. It thus recommended, that the measurement of the structure of forage using image analysis in laboratory is possible.

REFERENCES

ASAE Subcommittee of the American Feed Manufactures (1976). Method of determining and expressing fineness of feed materials by sieving - ASAE Standard S319. Agricultural Engineers Yearbook. 410-411

Baader, W. (1961). Das Verhalten eines Schüttgutes auf schwingenden Siebrosten. - The performance of a bulk freight on swinging sieves. Grundl. Landtechnik 11, 13, 13-20

Orth, W. and H. Peters (1975). Ein Verfahren zur Bestimmung der Halmlänge mit Hilfe eines Schwingsiebes. - A procedure for the determination the stalk length by sieve analysis. Grundl. Landtechnik 25, 6, 187-188

Wessel, J. (1968). Verfahren des Siebens und Windsichtens. - Proceed of sieving and air flow. Grundl. Landtechnik 18, 4, 151-157

METHODOLOGY AND INSTRUMENTATION FOR CONTAMINANT DETECTION IN WINES, FRUITS AND VEGETABLES USING BIO & CHEMILUMINESCENCE (BCL)

V. Issert, P. Grenier, L. Dusserre-Bresson, V. Bellon-Maurel (1)
B. Gouteyron (2)

(1) Cemagref, BP 5095, 34033 Montpellier cedex 1
(2) SEEE Automation, 43750 Vals-Près-Le-Puy

Abstract: The aim of this work is to valid and improve rapid pesticide immunodetection in food, especially in fruits, vegetables and beverages, using BCL detection. A commercialized kit has been chosen to valid procymidone immunodetection on different food matrices (wine, grappe and lettuce) and to test advantages and drawbacks of a BCL detection. Sensitivity of BCL detection measured on a portable luminometer has been comparable to colorimetric detection measured with an efficient laboratory spectrophotometer. The assay logarithmic response was linear from 10 to 200 ppb procymidone in wine matrix. But test procedure must be improved to allow quantitative assays on matrices other than wine. We are studying the possibility of increasing sensitivity of the method by improving the opto-electronics of the luminometer.

Keywords: Immunoassay, Procymidone, Bio&Chemiluminescence, Colorimetry

1. INTRODUCTION

Immunoassays are specific methods, sensitive, easy to implement, and rapid. For pesticides residues control, they make today a very interesting alternative to chromatography, and many kits based on competitive ELISA test are already on sale.

Today, the limit of detection of immunoassays associated with a colorimetric detection reaches generally the low residue levels imposed by present regulations (Bushway et al, 1991, 1992; Feng *et al*, 1990; Thurman E. M. *et al*, 1990, Bushway *et al*, 1991, 1992; Van Emon J.M. and Lopez-Avila, 1992). Because we may expect a progressive decrease of these maximal admissible concentration, it is important to anticipate this evolution and investigate new methods for improving detection limits (Seiber J.N., 1991; Wittmann C. and Cook B., 1993; Ulrich J., 1993). Considering the performances of BCL in the medical area for traces detection (Van Dyke K. and Van Dyke R., 1990), this detection method should be of interest in food processing (Evrard J.-C., 1993).

2. OBJECTIVES

The research presented in this paper emphasizes the feasibility of the BCL detection of pesticides residues in matrices as different as beverages, fruits, vegetables, with respectively wine, grappes, and lettuce as examples. Procymidone has been chosen because of its wide-spread use. The goal of this research is to set advantages and drawbacks of a BCL detection compared to a colorimetric detection when combined to immunoassays. For each matrix, a calibration curve should be set up for evaluating the concentration range of a quantitative assay and for estimating the assay sensitivity. The results should be compared to reference chromatographic analysis.

As the commercial kit chosen for our research is designed for procymidone in wines with a colorimetric detection, the chromogene reagent has to be replaced by a luminescent one. Kinetics of the luminescent reaction have to be considered.

Finally, instrumentation modifications can be looked at for improving the overall BCL immunoassay method. The volume of the sample and its dilution influence the light emission. As the measurement is luminescence intensity, light collection has to be optimized. The quality of the overall method must be an optimization of the combination between sample volume, dilution, and light collection through a multidisciplinary approach. The optics of the luminometer is improved by insertion of a fiber optic bundle inside of the luminometer for remote sensing. This remote sensing allows modifications of the reading chamber so that the apparatus constraint be softened and the interaction between sample and light collection be optimized.

In this paper, only results about the biochemistry aspects of the immunoassay are presented. The multidisciplinary approach will be developped only after respective further advances in biochemistry and optics and it will be the topic of a future paper.

3. MATERIALS AND METHODS

3.1. Materials

The reagents used were Millipore kits for procymidone detection in wine, luminol-H2O2 (Amersham) diluted to 1/5 in Tris buffer (Trizma Base Sigma) 0,1 M pH 8,5, procymidone (1 g/l in acetone), acetone and Pestipur methanol. A BCL-Book^{TM} luminometer with its supervision software BCL-View^{TM} from SEEE Automation Corporation, and a Milton Roy spectrophotometer were used for measurements. The analysed samples were red wine certified procymidone free, table grapes and lettuces both treated in controled conditions by various amounts of procymidone.

3.2. Principle of immunoassay

An ELISA immunoassay has been used. It implies three kinds of analites: polyclonal antibodies able to bind specifically to procymidone, procymidone molecules linked to an enzyme marker, Raifort peroxydase, and free procymidone molecules. The enzyme labeled by the peroxydase is called the conjugate. For each assay, antibodies and conjugate concentrations are constant, and only free procymidone concentration can vary according to the sample contamination which has to be determined.

The antibodies are bound to the tube walls and can complex free procymidone as well as the conjugate. There is competition between both. For measuring only the antigene-antibodies complexes fixed on the tube wall, the excess of non complexed procymidone is washed out.

Finally, the antibodies-conjugate complexes are detected by colorimetry or by luminescence and the signal intensity is inversely proportional to the procymidone concentration.

3.3. Methodology

3.3.1. Test kit procedure for procymidone detection

The kit implies the following sequence of operations: dilution to 1/3 of wine with a specific dilutant and incubation during 10 mn (500 µl of wine for 1 ml of dilutant); addition of 200 µl of sample and 200 µl of conjugate in tubes coated with antibodies; gentle agitation during 10 to 15 seconds; incubation without agitation during 10 mn; emptying and rinsing with distilled water several times and drying of the tubes with absorbant paper in between.

For colorimetric detection, 160 µl of substrate and 80 µl of chromogene are added. A 10 mn incubation is ended by 40 µl stop solution and following by a dilution with 500 µl of distilled water. The absorbance reading is measured with a spectrophotometer at 450 nm.

When detecting by luminescence, 300 to 400 µl of luminescent reagent is added and immediate reading is measured with a luminometer.

The ELISA procedure was performed according the manufacturer's intructions for wine analysis. For other matrices (grapes and lettuces), a sample preparation was needed and is described below.

3.3.2. Sample preparation

Because grape and lettuce are solid matrices, procymidone has to be extracted before analysis. In usual procedures, solid matrices are first blended. Then a portion of the sample is mixed with the extraction solvants and the mixture is filtered. The filtrate has then to be diluted with an aquaous phase before analysis so that the final concentration of solvant should be low enough for not affecting significantly the characteristics of the ELISA. Because procymidone is not a systemic pesticide (most of this coumpound is located on the fruit or vegetable surface), we have tested an extraction by only washing the product with solvant before the filtration. Also these two extraction procedures were used as following.

For sample preparation of grapes, 300 g of grapes were blended and 100 g were frozen, while 200 g were simply washed with 100 ml of methanol. After defreezing, 50 g of blended sample were mixed with 100 ml of methanol. Blended grapes and methanol mixture as well as methanolic solution for washing were filtered under vacuum conditions through a funnel containing a Whatman N°1 filter paper. 100 µl of methanolic filtrate were 10 or 100 fold diluted with distilled water. Total dilution of the matrix resulted in 1/20 to 1/200 for blended grapes, and 1/5 to 1/50 for washed grapes.

This procedure was closely followed for lettuces, with 350 g blended and frozen, and 350 g simply washed with 200 ml of methanol. Final dilutions resulted in 1/20 to 1/200 for blended lettuces, and 1/17.5 to 1/175 for washed lettuces.

3.3.3. Calibration curves for each matrix

Calibration curves have been established with samples of the three matrices (wine, grape and lettuce) as well as distilled water, each one of these four matrices spiked with different procymidone concentrations, from 0 to 100 ppb. For grape and lettuce, we have spiked the methanolic filtrate from negative control sample which has been submitted to a blending before extraction.

For comparing colorimetry to luminescence detection, results were expressed as % control calculated as follows : the measurement (Optical Densities, DO or Relative Light Units, RLU) of the sample containing procymidone was divided by the measurement of the negative control times 100.

4. RESULTS AND DISCUSSION

4.1. Modification of test kit procedure from colorimetric to luminescent detection

For samples containing 0, 1, 5, 10, 50 and 100 µg/l (ppb) of procymidone, kinetics were studied during 5 or 15 mn. Each concentration was tested in 2, 3 or 4 replicates. The luminescent reagent volume was 400 µl and the reading chamber temperature of the luminometer was maintained at 33 °C. Integration time of the luminometer was 1 second. Measurement of luminescent signal during 15 minutes with samples containing 0, 1, 10 and 100 ppb of procymidone have showed that the signal was maximal at 1 minute for each sample and the maximum was maintained during all the period of measurement.

After measurement of all samples, a significant variability was observed for the procymidone free sample and for the 1 ppb sample in spite of temperature control. Above 5 ppb, the relative error of the calibration curve is below 15 %. The curve can be best used from 10 to 100 ppb (Table 1 and Figure 1). This upper bound should be precised by testing samples containing higher concentrations of procymidone.

Tab.1 BCL measurements of wine samples fortified with various concentrations of procymidone.

[Procy-midone] in ppb	Replicates	RLU *10-3 average	% Rel.Err.	%RLU
0	3	410.7	38.1	100
1	3	312.8	33.2	76.2
5	4	325.1	13.1	79.2
10	4	322.8	14.9	78.6
50	3	160.5	18.1	39.1
100	3	116.1	4.5	28.3

RLU: Relative Light Units. % Rel. Err : Relative Error calculated as : $\frac{\text{RLU max} - \text{RLU min.}}{\text{RLUaverage}} * 100$

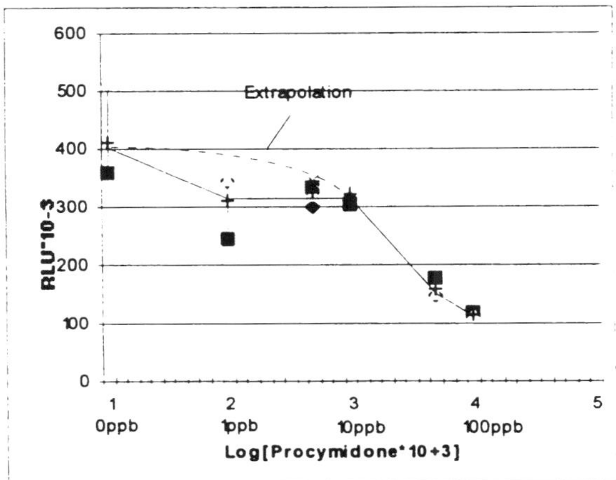

Fig.1 Standard curve obtained with wine sample fortified with various concentrations of procymidone.

These results obtained with a luminescent detection were compared to results published by Wynn *et al*

(1993) where a colorimetric detection was used. Table 2 and Figure 2 sum up the results. The results of Wynn *et al* (1993) in colorimetry are better than the ones obtained here in luminescence detection for samples containing less than 10 ppb of procymidone. This difference could be explained either by the luminescent detection itself or by difference in immunoassay procedure in two laboratories.

Tab.2 Comparison of luminescent and colorimetric detection with wine samples fortified with procymidone. SD: Standard deviation.

	Colorimetry		Luminometry	
[Procymidone] in ppb	Replicates	%DO (average ± SD)	Replicates	%RLU (average ± SD)
0	4	100 ± 3.1	3	100 ± 22
1	3	93.4 ± 4.0	3	76.2 ± 14.3
2	3	84.6 ± 4.4	-	-
5	4	73.5 ± 4.0	4	79.2 ± 3.9
10	4	61.5 ± 2.7	3	78.6 ± 6.1
20	4	51.3 ± 4.2	-	-
50	-	-	3	39.1 ± 6.1
100	-	-	3	28.3 ± 0.7

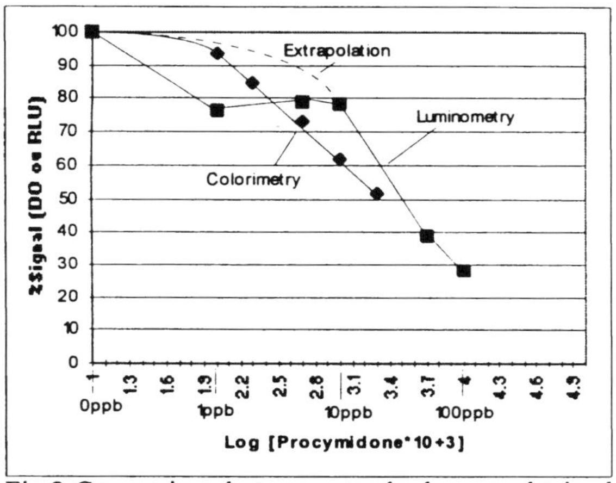

Fig.2 Comparison between standard curve obtained with luminescent and colorimetric detection. Wine samples were fortified with procymidone.

In order to compare sensitivity of only the two detection systems, assays were realized using the conjugate solution of the Millipore procymidone kit. After establishing a range of various dilution samples with the conjugate, the enzyme was analyzed by colorimetry and by luminescence. For the intensity of the signals to be in the same order of magnitude as the ones obtained with immunoassays, the choice of the dilution samples set was oriented by preleminary tests aimed at optimizing the luminescent reactive addition for procymidone immunoassay detection. For colorimetry as well as for luminescence, each sample was measured in duplicates.

Table 3 and Figure 3 indicate that the sensibility of both detection systems are very comparable. Any difference that might appear from one experimentation to another might consequently be due to different success in the immunoassay reaction prior to detection or to the physical differences between both detection methods.

Tab.3 Comparison of luminescent and colorimetric detection using free conjugate solution.

Volume of conjugate (µl)	%DO (average ± SD)	%RLU (average ± SD)
0.2	37.38 ± 0.15	31.46 ± 0.26
0.24	44.43 ± 0.75	36.71 ± 0.91
0.3	56.57 ± 1.26	49.70 ± 0.30
0.4	69.42 ± 0.39	61.85 ± 1.01
0.6	100.00 ± 1.21	100.00 ± 3.58

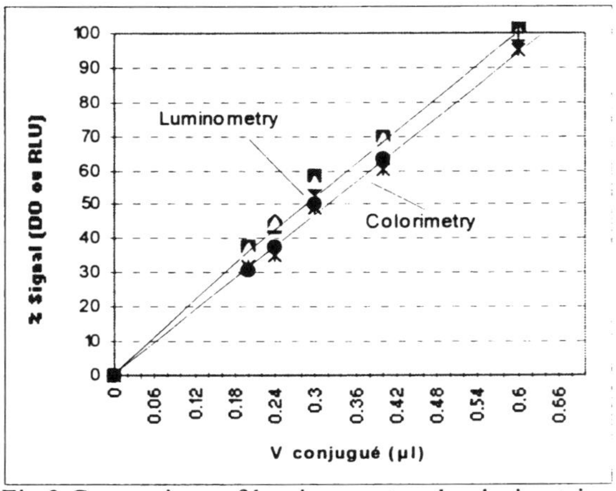

Fig.3 Comparison of luminescent and colorimetric detection using free conjugate solution.

4.2. Matrix effect

Washed samples and blended samples of grapes and lettuces were used. It seemed that a better reapitability was obtained when blending was used for sample preparation. Consequently, the following results were obtained with this preparation mode.

The calibration curve of the BCL detection of procymidone is presented at Figure 4, with concentration of procymidone of 0, 10 and 100 ppb, respectively in distilled water, grapes and lettuces. It means clearly that increasing procymidone concentration competes with conjugate-antibodies complexations and yield lower detection signals. From one matrix to another, similar decreases of signal are observed.

Considering the uncertainty of BCL immunoassay which can be observed at Figure 4a, it can be concluded that distilled water and grapes yield similar calibration curves. But with a lettuce extract matrix, there is a higher uncertainty and a more difficult distinction between the 0 and the 10 ppb samples.

This shows that matrix effects can be avoided when comparing the quite similar results of grapes and distilled water, that these effects should be specifically worked out for being avoided when comparing lettuce and grapes matrices.

The calibration curve of colorimetry detection presented at Figure 4b suggests similar conclusions. It reinforces the idea that matrix effects are more important for lettuces than for grapes compared to distilled water. It confirms also that these matrix effects influence the immunoassay by itself and not the detection, whatever method is used.

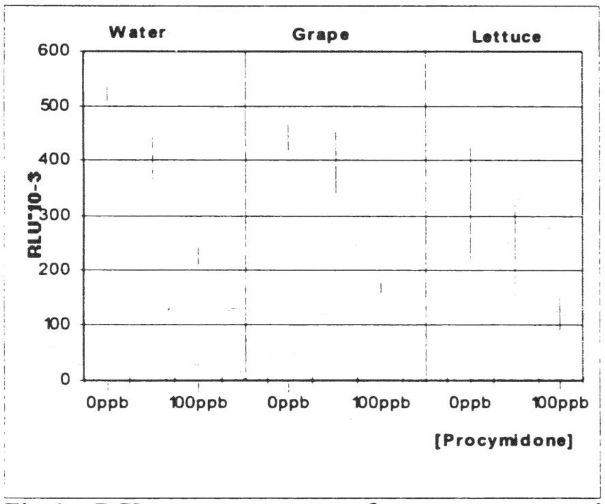

Fig.4a BCL measurements of water, grape and lettuce samples fortified with procymidone at 10 and 100 ppb and with 3 replicates for each concentration.

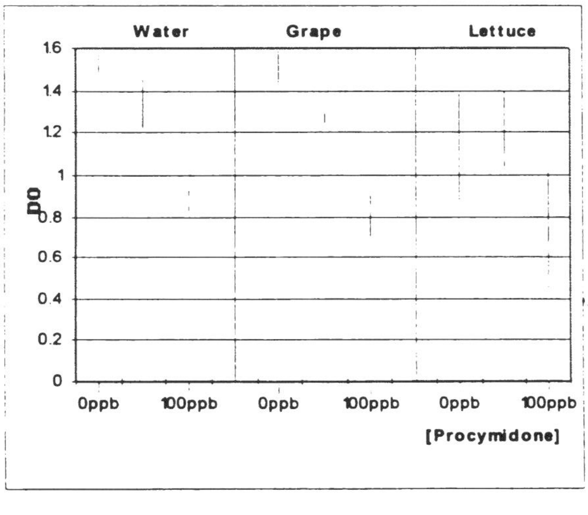

Fig.4b DO measurements of water, grape and lettuce samples fortified with procymidone at 10 and 100 ppb and with 3 replicats for each concentration.

4.3 Comparison of BCL and colorimetric detection

The presented results have shown that sensitivity of the two detection system are similar. But the detection test procedure with luminometry is more simple and more rapid than with colorimetry. Effectively, colorimetry detection requires successive additions of substrate and chromogene. Then a period of incubation of 10 mn is required so that color intensity reaches detection level. As reaction kinetics are not stabilized after 10 mn, it is necessary to stop them with an appropriate solution. In our procedure, the mixture has to be diluted with water so that the volume be large enough to be read with a classic spectrophotometer. In the case of luminometry, after adding the luminescent reagent in one step, reaction kinetics are stable 1mn after and intensity of the signal is maintained during at least 15 mn. This could be an advantage for the development of an automated immunoassay.

Colorimetry implies the measurement of light absorbtion by an homogenous solution and precision of measurement is limited by signal intensity and performances of the spectrophotometer. Luminometry implies measurement of a light emission. Conjugate molecules are punctual light sources which location can effect the precision of the measurement. Heterogeneity of measurements has been studied. Successive measurements were made after rotation of the sample tube by a quater of circle. It can be observed that the BCL signal varies regularly according to some geometry with a maximum variability of 10 %. This confirms the punctual aspect of the luminescence due to antibodies-conjugate complexations on the surface of tube walls as well as the physical characteristics of the tube. All the luminescence light coming out of the many punctual sources happens to cross the tube which releases this light with an intensity varying according to the heterogeneities of the tube, but stable with time in a period including minutes 4 to 9. This point encourages to work out solutions for getting rid of tube effects and gaining more reapitability through improved light collection.

5. CONCLUSIONS AND PERSPECTIVES

This work shows the feasibility of turning a colorimetry-detection immunoassay of procymidone into a luminescence-detection immunoassay just by combining commercial reagents.

The reapitability of sample preparation should not be overlooked. But the limits encountered here in the procymidone analysis are so far due to the optimization of the antibodies-antigenes complexations including ionic force of the preparation, temperature, agitation, and way of emptying and washing the tubes prior to the detection step. More work is necessary in order to be able to achieve successfully this essential step and to explain how these parameters have been managed. The matrix effects concern this complexation step and here the grapes matrix has been handled better than the lettuces matrix. It seems logical, as the initial kit was designed for wine and grapes have more in common with wine than with lettuce.

BCL detection and colorimetry seem quite similar in efficiency, excepted some time saving in the case of BCL because it does not need incubation at the detection step. But the tube wall heterogeneities suggest some improvements by working out light collection.

Instrumentation improvement includes investigating light collection within the luminometer as it is now for increased reapitability and sensitivity, but also remote sensing for benefiting from an « open space » and designing optimal light collection systems. Interactions between samples volumes, dilutions, and light collection should be discussed through implementions. This may lead to a full redesigning of the luminometer, and it is only after achieving this investigation that an effective comparaison between colorimetry and luminescence detection will be set up.

REFERENCES

Bushway *et al*, Comparison of Enzyme Linked Immunosorbent Assay and High-Performance Liquid Chromatography for the Analysis of Atrazine in Water from Czechoslovakia, *Arch. Environ. Contam. Toxicol.*, 21, 365-370, 1991,

Bushway R.J. *et al*, Determination of Methyl 2-Benzymidazole Carbamate in Blueberries by Competitive Inhibition Enzyme Immunoassay, *J. AOAC int.*, 75(2), 323-327, 1992

Evrard J.-C., Les Auto-Contrôles, Dossier "Nouvelles Réglementations", *L'Arboriculture Fruitière*, 459, 26-29, 1993

Feng *et al*, Development of an Enzyme Linked Immunosorbent Assay for Alachlor and its Application to the Analysis of Environmental Water Samples, *J. Agric. Food Chem.*, 38, 159-163, 1990

Seiber J.N., New trends in Analytical Methods for Pesticide Residues in Foods, in "Pesticide Residues and Food Safety", Chp.14, American Chemical Society, 1991.

Thurman E.M. *et al*, Enzyme Linked Immunosorbent Assay compared with Gas Chromatography/Mass Spectrometry for the Determination of Triazine Herbicides in Water, *Anal. Chem.*, 62, 2043-2048, 1990

Van Emon J.M. and Lopez-Avila V., Immunochemicals Methods for Environmental Methods for environmental monitoring, *Anal. Chem.*, 64(2), 79-88,1992

Van Dyke K. and Van Dyke R (Eds), Luminescence Immunoassays and Molecular Applications, CRC Press, Boca Raton, 1990

Wynn N.R. *et al*, Détermination de Résidus de Procymidone dans le Vin : Méthodes Immunoenzymatiques et Couplage Chromatogrphie en Phase Gazeuse-Spectrométrie de Masse, *Bull. OIV*, 749-750, 551-557, 1993

SENSOR FUSION FOR QUALITY GRADING OF MELONS

V. Steinmetz *, G. Rabatel *, M. Crochon *
T. Talou **, B. Bourrounet **

**CEMAGREF - Agricultural Equipment and Food Process Engineering Division*
BP 5095 Montpellier 34033 Cédex 1- France Tel:(33) 67 04 63 00 Fax:(33) 67 63 57 95
Email: vincent.steinmetz@cemagref.fr
***Ecole Nationale Supérieure de Chimie. Laboratoire de Chimie Agro-Industrielle. 118, route de Narbonne. 31077 Toulouse. France. Tel:(33) 61 17 57 24. Fax:(33) 61 17 57 30*

Abstract: Various destructive and non-destructive sensors are applied to melon in order to assess its quality components. Destructive measurements (firmness and sugar content) made on the fruits are used as references, and are compared to the non-destructive sensors measurements (vision system, electronic sniffer). Classes of melons are made based on an expert, but also on unsupervised classification (Kohonen network). It is shown that the classification provided by the expert is not completely related to the destructive measurements, and that a satisfying artificial classifier can be built based on the electronic sniffer.

Keywords: Sensor fusion, quality control, melons

1. INTRODUCTION

Melon is a highly perishable fruit, and about 2 millions of tons were harvested in Europe in 1992, mainly in Spain, Italy and France. Melons are mainly field produced and grading melons is a labour intensive task. Melons are mostly graded based on their weight or size with mechanical systems, and based on colour. The latter characteristic is so far performed by human graders in charge of removing the non-mature or advanced-mature fruits based upon subjective visual determination (colour or changes in the physical exterior aspect such as the appearance of a slip near the stem).

Research for non-destructive maturity evaluation of melons has already been conducted on different aspects of melon maturity. A sensor system, based on detection of concentrations of headspace gases, has been developed and tested on several cultivars of melons (Benady *et al.*, 1992). The ability of this sensor for discriminating between ripeness stages was higher than all over destructive and non-destructive applied techniques. External melon firmness was also studied (Aubert *et al.*, 1977).

Sensor fusion is analogous to the ongoing cognitive process used by humans to integrate data continually from their senses to make inferences about the external world. Though few sensor fusion models for classification are available in the literature, some research has already been conducted on architecture selection for sensor fusion (Steinmetz *et al.*, 1994), and on the application of sensor fusion to tomato classification (Edan *et al.*, 1994). Promising results in this area leads our study of sensor fusion techniques applied to melon quality assessment.

The goals of this study were to combine the sensors in order to match the classes made by the expert, to compare the performance of the classifiers using different sensors, and to compare different fusion techniques.

2. MATERIAL

Five different sensors were used: a CCD camera, a near-infra-red sensor (Bellon, 1994), electronic sniffer (all non-destructive), and a refractometer and a manual penetrometer (both destructive).

2.1 Vision system

Inspection chamber: The inspection chamber includes consideration of melon position and size, lighting design and image resolution. Melon colour and size are one of the major elements in designing the inspection chamber. Background was selected to be black because it provided the best contrast. To provide uniform and diffuse lighting above the melon, an illumination chamber providing day light with ten 58W fluorescent lamps was used. These lamps were placed circularly around the inspection chamber. Views of the inspection chamber are shown in Figure 1. The chamber is a half-sphere with a 2 meters radius. Using a 15 mm focal length, the vertical and horizontal resolutions were respectively 0.91 mm/pixel and 0.65 mm/pixel. The aspect ratio was 1.4:1. Distance of the camera to the melon was set to 0.6 m.

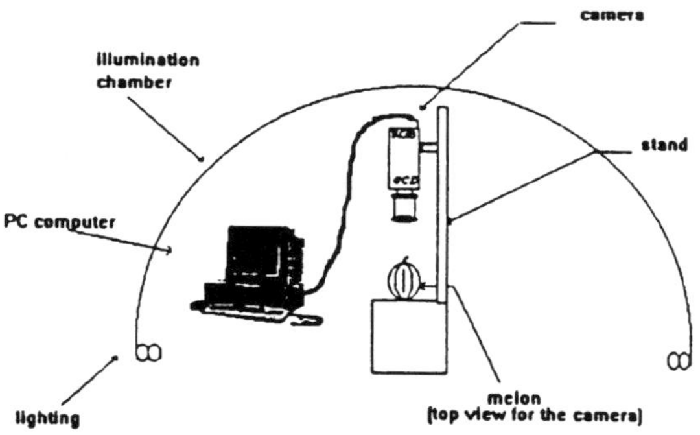

Figure 1. Inspection chamber (not to scale).

Image acquisition: The image processing hardware includes a colour CCD camera, a microcomputer and acquisition board necessary to acquire and perform basic image processing operations. The camera was centred at the middle of the background and balanced by adjusting the colour gains. The aperture of the lens was set to 16 to allow the segmentation of the melon from the background and colour segmentation.

Video signals from the camera were digitized with a colour frame grabber (Model Bytech) mounted in a personal microcomputer (PC 386). The camera was calibrated before each experiment with a colour reference frame. Average R, G and B values were computed for this reference frame in order to check the camera calibration.

2.2 Firmness

Firmness was measured with a Magness-Taylor apparatus. Melons were cut in two halves and firmness was measured on the same half on two opposite points (Figure 2). The average of these measurements was used as an indicator of the fruit firmness. Since the same distance from the skin was always used, it may induce a bias in the measurement. Indeed smaller melon in size will appear softer than large melon, because the flesh is softer close to the heart of the melon.

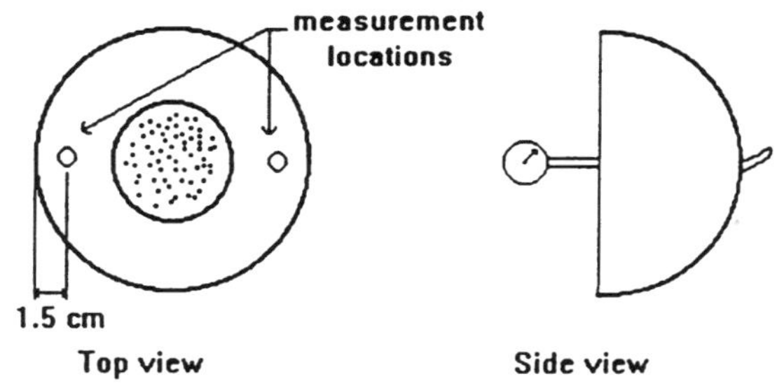

Figure 2. Firmness measurement on melons.

2.3 Sugar content

Sugar content was measured with a refractometer providing Brix degrees. Measurements were made with the same method than firmness, i.e. two measurements on each opposite side of a half-cut melon. Data were average and used as an indicator of the fruit sugar content.

2.4 Aromas

Aromas were measured with an electronic sniffer. The sensors used are of M.O.S. (Taguchi, 1971) type (Meta-Oxide Sensors) and were composed of sinfered tin dioxide deposit on an alumina ceramic tube with an heater coil inside. In presence of volatile compounds, the electrical resistance of the sensor was modified due to volatile adsorption on its surface. The adsorption kinetics is several seconds, is reversible at ambient temperature and proportional to volatiles concentration in the atmosphere. The sensor is coupled with a data acquisition hardware and the analog output signal is digitised and stored in the computer. The measurement system consists of a glass flask (0.6 l capacity) in which are placed five commercial MOS. This flask is linked to a sample air input port and an air output port, a control unit which provides both constant tension of 5 V to the sensor circuit voltage and variable tension, (4 to 6 V) to the heater circuit, and a personal computer allowing continuous monitoring and measurement of all electrical signals via five A/D converters.

3. METHODS

3.1 Image analysis

Two views were acquired for each melon: the first view is a top view as presented in Figure 1. A 90 degrees rotation was applied to the melon in order to acquire the side view. Each melon image was 256 x 256 pixels and covered 23.2 cm x 16.0 cm area. Each pixel was assigned three different values: R, G and B on an 8-bit scale. Background was almost completely black.

Figure 3 shows a description of the different steps performed during image analysis. Global segmentation involves segmentation of the whole melon from the background. Local segmentation involves segmentation of the greenest part of the melon. The main extracted characteristics were the diameters, the histograms of the top view for each colour component, and the average blue component of the top view.

Feature extraction includes size and colour considerations. Three parameters were extracted from the image, i.e. two diameters from the top view and one diameter from the side view. The variables derived for colour are based on the red, green and blue histograms. Since the global colour of the melon is used by the human grader, the histogram was computed based on the top view of the melon. However, the melons present some greenest part which are placed symmetrically around it. These parts are schematically represented by the black lines on the melon of Figure 1. These greenest parts may present some noise for the histogram colour, that is why they were eliminated before the computation of the histogram. Furthermore, the area around the stem of the melon is also important for the maturity evaluation, especially for mature melons. Skin of mature melons often splits apart around the stem, allowing the flesh to be seen. Therefore, the colour histogram around the stem was computed within a window which size was a function of the size of the melon.

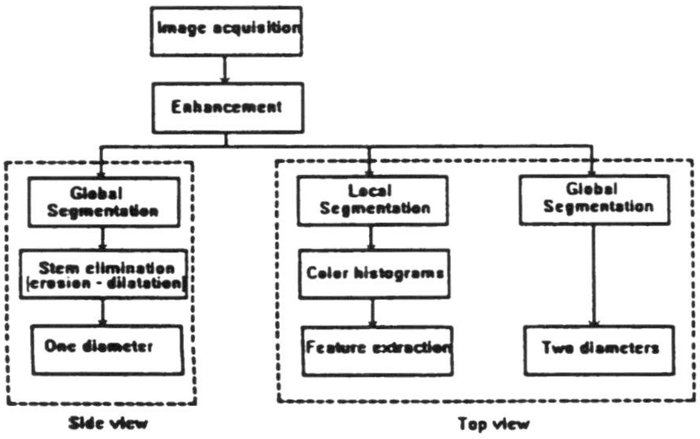

Figure 3. Image analysis for the two views of the melon.

3.2 Sensor fusion

Multisensor fusion is an evolving technology concerned with the problem of how to combine data from multiple sensors in order to make inferences about a physical event, activity, or situation (Hall, 1992). The ultimate goal of multisensor fusion is to find a combined declaration that is more specific and more accurate than any of the declarations based on each individual sensor. The approach relies on the hypothesis that fusion of signals from several sensors (similar or disparate sensors) will give better results than the signal from one single sensor (Holmbom, 1989).

The fusion process was applied to the features extracted from each individual sensor. Discriminant analysis and neural network approaches are proposed for testing the combination of the different sensors for class prediction. These classifiers were applied for separating the melons into the four categories commonly used by the expert ('green' (G), 'light green' (G1), 'light yellow' (Y1) and 'yellow' (Y)). Different classifiers were built by associating different kind of sensors, and performance of the different classifiers were compared. Since aromas sensor was applied only on a part of the samples, principal component analysis is also used for including the aromas sensor within the sensor fusion process description, because this method requires less numerous samples.

Features extracted from the histogram are calculated based on the pixels values which lead to the best discrimination among the classes (Ros *et al.*, 1990). This provided five different colour features extracted from the histograms.

3.3 Artificial classifiers

Features used by the artificial classifiers are the features extracted from the different sensors. Tested classifiers include discriminant analysis (Duda, 1973) followed by a nearest cluster center classifier, and a neural network. Some research has been conducted on the application of neural networks technology for inspection of fruits, vegetables and horticultural products. The properties of reflectance spectra for tomatoes have been studied (Thai *et al.*, 1992), and recent research on flower inspection includes work on grading cyclamens using an artificial expert (Brons *et al.*, 1991) and roses inspection (Steinmetz *et al.*, 1993).

Grouping of the samples into classes was made based on Kohonen's algorithm (Kohonen, 1990). This algorithm (KA) generates a topologically ordered mapping based on a competitive scheme.

Topological ordering means that neurons that are close to each other represent similar input patterns. The results of this algorithm is a four 'natural' grouping of the samples.

4. EXPERIMENTAL PROCEDURES

One variety (Galoubet) representing a total of 176 samples were collected in early June 1994 at a grower's place located in the Camargue area (France). The samples were classified by the same expert in the 4 different classes previously described.

The samples were collected at two different harvesting dates. Each sample was processed successively through the individual sensors for non-destructive measurements in the following order: weight, image acquisition, electronic sniffer (43 samples only), near-infra-red measurements. Destructive measurements, i.e. firmness and sugar content, were then processed. Melons were stored in a plastic bag during a 12 hours period before being sensed by the electronic sniffer. Atmosphere was sampled from the bag with a 50 cm^3 syringe, and then introduced into the gas flask.

5. RESULTS AND DISCUSSION

5.1 Weight prediction based on image analysis

The purpose was to predict the weight based on image processing, i.e. the three diameters of the melon . Results are presented in Figure 4. This model was obtained with a linear regression between the weight and the product of the three diameters of the melon measured by image analysis.
This linear regression relies on a constant density for the melons, which is obviously not correct. Indeed, melon density ranges from 0.90 to more than 1, depending on the variety. The fairly good results presented here can be explained by the fact that all the melons used during the experiment belong to the same variety, and all of them come from the same growing place and the same field. The results would certainly not be as good if many different varieties of melons had been used. For example, the variety 'Canari' presents a more ellipsoid shape, and a different calibration would be necessary for this kind of variety.

5.2 Colour histograms

Typical colour histograms from the top view of the melons are presented in Figure 5. Figure 5 is an illustration of the discrimination between the different classes around the pixel value 80.

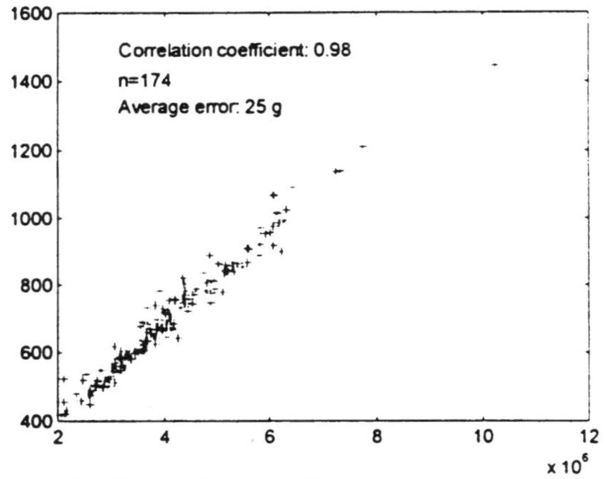

Figure 4. Observed weight (grams) versus the product of the three diameters.

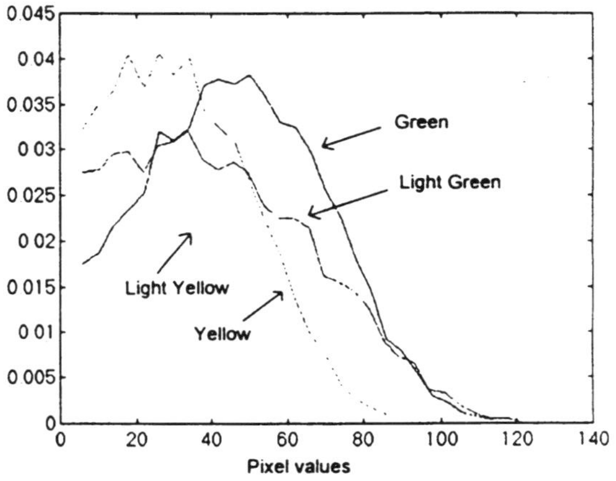

Figure 5. Histogram of blue components for different samples of the four classes.

5.3 Sugar content and near-infra-red sensor

No significant relationship was found between the sugar content measured by the refractometer and the spectra measured by the near-infra-red sensor. The main reason is due to the fact that the measurements was made on the skin of the melons, since sugar content was measured in the flesh. The near-infra-red sensor should then be equipped with a special device able to go through the skin until reaching the flesh. An other approach could be checking relationship between the sugar content of the skin and the sugar content of the flesh.

5.4 Sugar content and firmness

Relationship between grader classification and fruit maturity was studied by comparing means and variances for each day and each class (Tables 1 & 2).

Table 1. Firmness (kgs/0.5cm²) for the classes made by the expert (mean - stand. deviation)

	Y	Y1	G1	G
Day 1	1.68	1.69	1.76	2.50
	(0.37)	(0.48)	(0.27)	(1.23)
Day 2	1.86	1.77	1.85	3.57
	(0.51)	(0.43)	(0.35)	(1.39)

In Table 1, for both days, mean of class G is different (at a 5% significant level) from the means of the other classes.

Table 2. Sugar content (degree Brix) for the classes made by the expert (mean - stand. deviation)

	Y	Y1	G1	G
Day 1	12.23	12.3	12.29	10.84
	(1.79)	(1.05)	(0.80)	(1.15)
Day 2	12.3	10.76	12.05	10.18
	(1.25)	(0.98)	(1.07)	(1.84)

In Table 2, for day 1, mean of class G is different (at a 5% significant level) from the means of the other classes. For day 2, means of the class G and mean of class Y1 are not different, and means of the class G1 and mean of class Y are not different.

Only the 'Green' class is highly related to the destructive measurements. These results tends to prove that the expert classification, mainly based on colour, is not really related to melon maturity, except for the 'Green' class. It has to be noticed that the mean values of the second day are not consistent with the mean values of the first day. This can be explained by the dates of sample collection: the first day was a Monday, that is to say, the day where the grader had the largest choice among melons for building the classes, because melons are not harvested during the week-end. On the other side, the second collecting day was a Wednesday where a smaller set of melons was available for building the classes. In other words, it is clear that the expert did not build the classes for the first day in the same way than he did it during the second day. These results are very similar to the ones exposed in (Guedalia, 1994), where it is shown the difficulty to correctly classify the components that form the complete manual classification.

Based on these results, the Kohonen's algorithm was applied in order to build quality classes based on the destructive measurements. Results are presented in Table 3. Four classes are provided by the algorithm. Class 1 represents sweet and soft melons, while class 4 represents non-sweet and firm melons. Classes 3 and 4 represent intermediary quality classes (sweet and firm for class 2, and non-sweet and soft for class 3).

Table 3. Sugar content (degree Brix) for the classes provided by Kohonen's algorithm (mean - stand. deviation)

Class number	1	2	3	4
Sample number	89	9	27	51
Sugar content	12.78	12.40	10.73	10.23
	(0.79)	(0.34)	(0.71)	(1.25)
Firmness	1.69	2.69	1.47	2.86
	(0.35)	(0.32)	(0.27)	(1.24)

5.5 Principal component analysis

Principal component analysis (PCA) was performed with 43 samples, i.e. the 43 samples that were measured with the electronic sniffer. Measurements used for the PCA were weight, size, sugar content, firmness, aromas and colour features extracted from the image analysis process. Figure 6 represents the correlation circle.

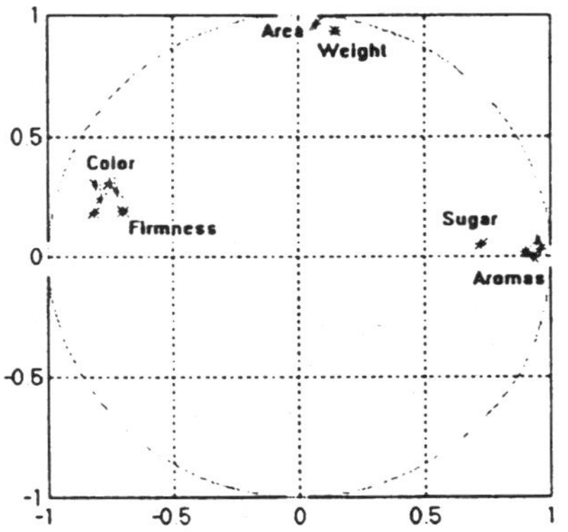

Figure 6. Correlation circle based on principal component analysis.

Based on the PCA, the two axis represents 72 % of variance. The first axis (48 % of variance) represents the opposition between firmness and sugar content. The second axis (23% of variance) is more related to the weight.
PCA was also processed with the 176 samples without including aromas measurements. Results were found to be similar to the previous PCA with aromas (52% of variance on the first axis, and 24% on the second axis).

5.6 Neural network classifier

Results obtained with a neural network classifier are presented in Table 4. These results were obtained with the following features as inputs for the network: vision system, the sugar content and firmness feature. The average error rate is 35%, and the error rate within one class is 12%. Additional networks were built by combining some features as inputs of the network. Table 5 presents the results by combining the different sensors. Colour is one of the most important feature for the classes built by the experts. Including sugar content or firmness do not show any important improvement in the classification process. This is a confirmation of what has been shown in paragraph 5.4, where destructive measurements were shown to be poorly related to the classification of the expert.

Table 4. Classification of the neural network based on vision sensor, refractometer and penetrometer.

		Classes made by the network			
Classes		Y	Yl	Gl	G
made	Y	33	1	2	8
by	Yl	2	33	12	1
the	Gl	1	11	27	4
expert	G	12	3	6	23

Table 5. Error classification of the neural network depending on the sensors.

Sensor combination	Error classification rate
Colour	37%
Color and refractometer	38%
Color and penetrometer	38%
Color, refractometer and penetrometer	35%
Refractometer and penetrometer	62%

A backpropagation neural network was also trained based on the classes provided by Kohonen's algorithm. The five aromas components were used as inputs. Table 6 shows that the classification is greatly improved , the error rate being about 8%.

Table 6. Neural netword classification for aromas

		Classification made by the neural network			
		1	2	3	4
KA	1	3	0	0	0
Classes	2	0	24	3	1
	3	0	0	6	0
	4	0	0	0	6

5.7 Discriminant analysis

Results from the discriminant analysis applied on the features from the vision sensor, refractometer and manual penetrometer are presented in Figure 7, where the two first axis are sketched. It is visible that the two extreme classes, i.e. the greenest class and the more yellow class are fairly well separated one from the other. However, the two intermediary classes ('light yellow' and 'light green') are poorly separated. This is not surprising because these two classes are rather similar one to the other from the colour point of view, and from the maturity point of view as well. Results from the cluster-minimum distance classifier applied after the discriminant analysis are presented in Table 7. The average error rate is 45% and the error rate within one class is 10%. The average error rate is larger compared to the results from the neural network. It has to be noticed in Table 7 that the 'Green' class was rather well separated from the other classes.
Different combinations of sensors were also tested with the discriminant analysis. They led to similar conclusions than the ones obtained from the neural network classifier.

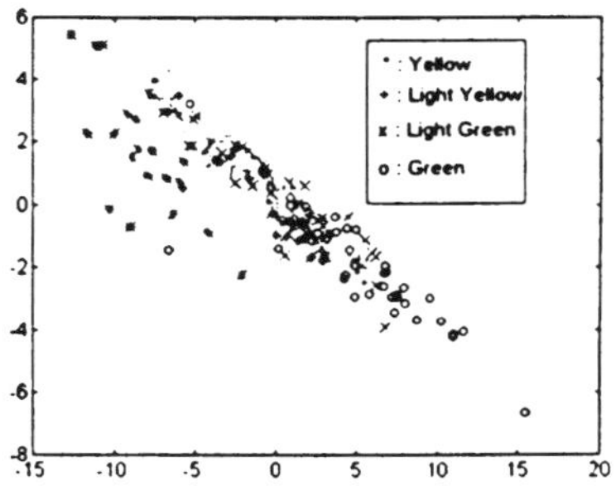

Figure 7. First two axis from the discriminant analysis

Table 7. Classification of the discriminant analysis

		Classes made based on discriminant analysis			
		Y	Yl	Gl	G
Classes	Y	31	4	6	3
made by	Yl	11	19	7	11
the expert	Gl	7	10	15	11
	G	1	0	12	31

6. CONCLUSIONS

Sensor fusion was studied in order to improve the classification performance of the sensors with respect to the expert classification. Features extracted from each individual sensors were used and two different classifiers were compared: a neural network classifier, and a minimal distance classifier to cluster center associated with a discriminant analysis. These classifiers provided the classification of melons among the four classes 'green', 'light green' and 'light yellow' and 'yellow'.

Weight of melons could be easily computed based on vision system with an average error of 25 grams. It was shown that the vision sensor is the most related to expert classification, though this classification is not completely related to melon maturity, except for the 'green' class. Classes made by the expert were shown to be highly subjective, based on colour histogram analysis and on maturity parameters as well. Aromas was shown with the PCA to be related to sugar content. Near-infra-red sensor was not measuring sugar content of the flesh. Neural networks classifier was shown to have a higher performance than discriminant analysis classifier, with 35% versus 45% classification error rate, respectively. However, this error rate was still high.

By removing the classes provided by the expert, and by using an unsupervised classification, it was shown that the classification could be improved by using aromas only. However, it must not be forgotten that only 43 samples were used by the electronic sniffer, and all samples belonged to the same variety. This is why further research should be done in this area, with a larger number of sample, including different varieties of melons.

7. ACKNOWLEDGEMENTS

We express our appreciation to the CEHM growers in Marsillargues, Gard, France, and particularly Mr. Yard, for providing the experimental samples.

REFERENCES

Aubert, S. and R. Dumas de Vaulx. (1977). Undestructive firmness test for the maturity control of melons. *Ann. Technol. Agric.* 27(**3**):243-254.

Bellon, V. (1994). Infrared and near-infrared technology for the food industry and agricultural uses. On line applications. *Food Control* 5:(**1**)

Benady, M., J.E. Simon, D.J. Charles and G.E. Miles. (1992). Determining melon ripeness by analyzing headspace gas emission. ASAE International Summer Meeting. Charlotte, North Carolina. *ASAE paper* 92-6055.

Brons, A., G. Rabatel, F. Sevila and C. Touzet (1991). Neural network techniques to simulate human judgement on pot plant quality: an artificial expert for beauty evaluation. *ASAE paper* No. 91-9893.

Chen, H., J. De Baerdemaeker and V. Bellon. (1993). Finite element study of the melon for nondestructive sensing of melon quality. ASAE International Winter Meeting. Chicago, Illinois. *ASAE paper* 93-6599.

Crochon, M. Eating quality of melons. (1986). EEC Conference. Aix-en-Provence, France.

Duda, R. and P. Hart (1973). *Pattern Classification and Scene Analysis*. John Wiley & Sons.

Edan, Y., H. Pasternack, D. Guedalia, N. Ozer, I. Shmulevitch, D. Rachnmani, E. Fallik and S. Grinberg (1994). Multi-sensor quality classification of tomatoes. ASAE International Summer Meeting, Kansas City, Missouri. *ASAE paper* 94-6032.

Guedalia, D., and Y. Edan (1994). A dynamic artificial neural network for coding and classification of multi-sensor quality information. ASAE International Summer Meeting, Kansas City, Missouri. *ASAE paper* 94-3053.

Hall, L. D. (1992). *Mathematical techniques in multi-sensor data fusion*. Artech House. Boston

Holmbom, P., O. Pedersen, B. Sandell and A. Lauber (1989). Fusing sensor systems: promises and problems. *Sensor review.* 9(**3**):143-152.

Kohonen, T. (1990) The self-organizing map. *Proceedings of the IEEE.* (78)**9**:1464-1480.

Ros, F., S. Guillaume, G. Rabatel, F. Sevila, and D. Bertrand. (1994). The characterisation of granular product populations as a pattern recognition problem: application to meal products classification. Accepted to Journal of Chemometrics.

Steinmetz, V. and M.J. Delwiche. (1993). Neural network analysis of rose straightness. *Proceedings of the Artificial Intelligence for Food and Agriculture Conference*, Nimes, France.

Steinmetz, V., M. Crochon, V. Bellon-Maurel, J. L. Garcia Fernandez, P. Barreiro Elorza and L. Verstreken (1994). Sensors for fruit firmness assessment: comparison and fusion. Submitted to Journal of Agricultural Engineering Research.

Steinmetz, V., M. Crochon, T. Talou and B. Bourrounet (1995). Sensor fusion for fruit quality assessment: application to melon. Proceedings of the International Conference ASAE 'Harvest and post-harvest technologies for fresh fruits and vegetables'. Mexico.

Taguchi, N. (1971). Method for making a gas sensing element. US Patent 3,625,756.

Thai, N.C., R.L. Shewfelt. and J.G. Latimer (1992). Neural network analysis of tomato reflectance spectra. *ASAE Paper* No. 92-7057.

INFLUENCE OF STOCHASTIC AIR TEMPERATURE DISTURBANCES ON THE COLD STORAGE OF FRUITS

B.M. NICOLAI and J. DE BAERDEMAEKER*

**Katholieke Universiteit Leuven, Kardinaal Mercierlaan 92, B-3001 Heverlee, Belgium. Email: bart.nicolai@agr.kuleuven.ac.be; Tel: +32(16)321456; Fax: +32(16)321994*

Abstract. A steady state variance propagation algorithm is derived to investigate the effect of stochastic air temperature disturbances on the variability of the centre and surface temperature of pineapples during cold storage conditions. The disturbances are modelled by continuous-time autoregressive processes of variable order. The algorithm is based on the finite element formulation of the heat conduction equation and involves the numerical solution of matrix equations of the Lyapunov and Sylvester type. It is shown that disturbances of different order but with the same scale of fluctuation result in comparable centre and surface temperature variances.

Key Words. Fruits, cold storage, temperature disturbances, heat transfer, stochastic analysis, finite element method

1. INTRODUCTION

Many fruits are susceptible to low temperature injury during cold storage if unappropriate temperatures are maintained in the storage facilities. The injury manifests itself by a variety of symptoms, including surface lesions (e.g. cucumbers), browning of flesh tissue (e.g. black heart in pineapples) and development of water–soaked areas (cucumber). The extent of the injury depends on several factors such as the chilling temperature, the duration of the exposure, and the sensitivity of the species involved.

The occurrence of chilling injury can be prevented by careful monitoring and control of the temperature in the cold storage facilities. However, due to unpredictable disturbances, such as the opening of doors or introduction of fresh produce, it is possible that the storage temperature temporarily drops into the region of increased chilling injury risk. This will affect the product centre temperature in such a way that, because of the thermal inertia of the product, high frequency fluctuations are smoothed. On the other hand, low frequency fluctuations will be phase shifted but not attenuated.

In the case of pineapples, the chilling injury starts at the core of the product, and the frequency of the storage temperature fluctuations will therefore affect the development of chilling injury to a large extent. A quantitative insight in the propagation of temperature fluctuations inside the fruits may guide towards the establishment of temperature control performance criteria which minimize the occurrence of chilling injury.

In this paper a variance propagation algorithm will be outlined to calculate the mean value and the variability of the temperature in steady state conditions at an arbitrary position inside a complex shaped cooled/heated object corresponding to a stochastically fluctuating ambient temperature. The algorithm is an extension of the algorithm which was described in Nicolaï and De Baerdemaeker (1994).

2. COMPUTATION OF HEAT CONDUCTION

The propagation of heat in linear conductive media is governed by the Fourier equation (Incropera and De Witt, 1990)

$$k\nabla^2 T = \rho c \frac{\partial T}{\partial t} \tag{1}$$

where

- T : temperature [°C]
- t : time [s]
- k : thermal conductivity [W/m°C]
- ρ : density [kg/m^3]
- c : heat capacity [J/kg°C]

Internal heat generation by respiration is neglected in the following stochastic analysis. The initial condition is defined as

$$T = T_0 \qquad \text{at} \quad t = t_0 \tag{2}$$

During precooling and cold storage, convection heat transfer occurs at the boundaries of the product

$$k\frac{\partial}{\partial n_\perp}T = h(T_\infty - T) \qquad \text{on } \Gamma \qquad (3)$$

$n_\perp$: outward normal to surface
h : surface heat transfer coefficient [W/m²°C]
T_∞ : ambient temperature [°C]
Γ : exposed surface

Because of the often complicated shape of the chilled produce, in general no analytical solutions of (1) subject to (2) and (3) are available and numerical solution becomes mandatory.

The finite element method is widely used for the simulation of conductive heat transfer. In this method the continuum is subdivided in (finite) elements of variable size and shape which are interconnected in a finite number n_{nod} of nodal points. In every element the unknown temperature can be approximated by a low order interpolating polynomial in such a way that the temperature is uniquely defined in terms of the (approximate) temperatures $u_i(t)$ at the nodes. It is possible to show (Segerlind, 1984) that the unknown nodal temperature vector $\mathbf{u} = [\ u_1\ \ldots u_{n_{nod}}\ ]^T$ may be found by solving the following linear differential system

$$\mathbf{C}\dot{\mathbf{u}} + \mathbf{K}\mathbf{u} = \mathbf{f} \qquad (4)$$

$$\mathbf{u} = \mathbf{u}_0 \quad \text{at} \quad t = t_0 \qquad (5)$$

with $\mathbf{C}$ the capacitance matrix and $\mathbf{K}$ the stiffness matrix, both $n_{nod} \times n_{nod}$ matrices and $\mathbf{f}$ a $n_{nod} \times 1$ vector. The superscript ^{T} denotes the transpose of a vector of matrix. $\mathbf{C}$ involves ρ and c, $\mathbf{K}$ involves k and h, $\mathbf{f}$ involves T_∞ and h. This system is usually discretized with a finite difference scheme.

3. AUTOREGRESSIVE RANDOM PROCESSES

Random processes provide a convenient way to describe phenomena which fluctuate in an unpredictable way as a function of time. The mean value $\overline{x}$ and the (auto)covariance $V_{x,x}$ of the (stationary) random process x are defined by

$$\begin{aligned} \overline{x}(t) &= \mathcal{E}(x) \\ &\triangleq \int_{-\infty}^{\infty} f(x,t)dx \\ V_{x,x}(\Delta t) &= \mathcal{E}\{[x(t) - \overline{x}][x(t+\Delta t)) - \overline{x})]\} \end{aligned}$$

If $\Delta t = 0$ then $V_{x,x}(0) \triangleq \sigma_x^2$, the variance of the process. The covariance function describes how much the current value of the random function will affect its future values. The correlation function $R_{x,x}$ is found by normalization of the covariance function:

$$R_{x,x}(\Delta t) = V_{x,x}(\Delta t)/\sigma_x^2 \qquad (6)$$

Autoregressive processes are a special class of random processes. A continuous time *autoregressive* process $x(t)$ of order m is described an the mth order linear differential equation (Priestley, 1981)

$$\frac{d^m}{dt^m}x(t) + a_1\frac{d^{m-1}}{dt^{m-1}}x(t) + \cdots + a_m x(t) = w(t) \qquad (7)$$

where $a_1, a_2, \ldots, a_m$ are constants, and $m \geq 1$. w is a zero mean stationary white noise process with covariance

$$V_{w,w}(\Delta t) = \sigma_w^2 \delta(\Delta t) \qquad (8)$$

where δ is the Dirac delta. The notation AR(m) will be used further on to refer to an autoregressive process of order m.

The autoregressive process (7) can be written in the following *state space form* form (Brogan, 1985):

$$\frac{d}{dt}\mathbf{x} = \mathbf{A}\mathbf{x} + \mathbf{B}w(t) \qquad (9)$$

where

$$\mathbf{x} = \begin{bmatrix} x \\ dx/dt \\ \vdots \\ d^{m-2}x/dt^{m-2} \\ d^{m-1}x/dt^{m-1} \end{bmatrix} \quad \mathbf{B} = \begin{bmatrix} 0 \\ 0 \\ \vdots \\ 0 \\ 1 \end{bmatrix}$$

$$\mathbf{A} = \begin{bmatrix} 0 & 1 & \cdots & 0 \\ \vdots & & \ddots & \vdots \\ 0 & 0 & \cdots & 1 \\ -a_m & -a_{m-1} & \cdots & -a_1 \end{bmatrix}$$

The matrix $\mathbf{A}$ is called the *companion matrix*. The dimension of $\mathbf{A}$ is equal to $m \times m$; $\mathbf{B}$ and $\mathbf{x}$ are of dimension $m \times 1$.

In order to describe the smoothness of a random process by means of a single measure, Vanmarcke (1983) introduced the concept *scale of fluctuation* which is defined as

$$\theta = \int_{-\infty}^{+\infty} R_{x,x}(\Delta t)d\Delta t$$

It gives an indication of the time beyond which a future value of a random process will not be affected anymore by its current value. In Fig. 1 some correlation functions and corresponding realizations of an AR(2)–process with different scales of fluctuation are compared. If $\theta \rightarrow 0$ then the process approximates a white noise process.

Table 1 Autocovariance function and scale of fluctuation of AR(1) and AR(2) processes

AR(1)	$V_{x,x}(\tau) = \sigma_x^2 e^{-a_1\|\tau\|}$ $\theta = 2/a_1$
AR(2) Real roots	$V_{x,x}(\tau) = \sigma_x^2(\xi_2 e^{\xi_1\|\tau\|} - \xi_1 e^{\xi_2\|\tau\|})[(\xi_2 - \xi_1)]^{-1}$ with ξ_1 and ξ_2 the roots of $\xi^2 + a_1\xi + a_2 = 0$ $\theta = 2a_1/a_2$
AR(2) Complex roots	$V_{x,x}(\tau) = \sigma_x^2 e^{-\frac{1}{2}a_1\|\tau\|}\left[\cos(p\|\tau\|) + \frac{a_1}{2p}\sin(p\|\tau\|)\right]$ with $p = (a_2 - a_1^2/4)^{1/2}$ $\theta = 2a_1/a_2$

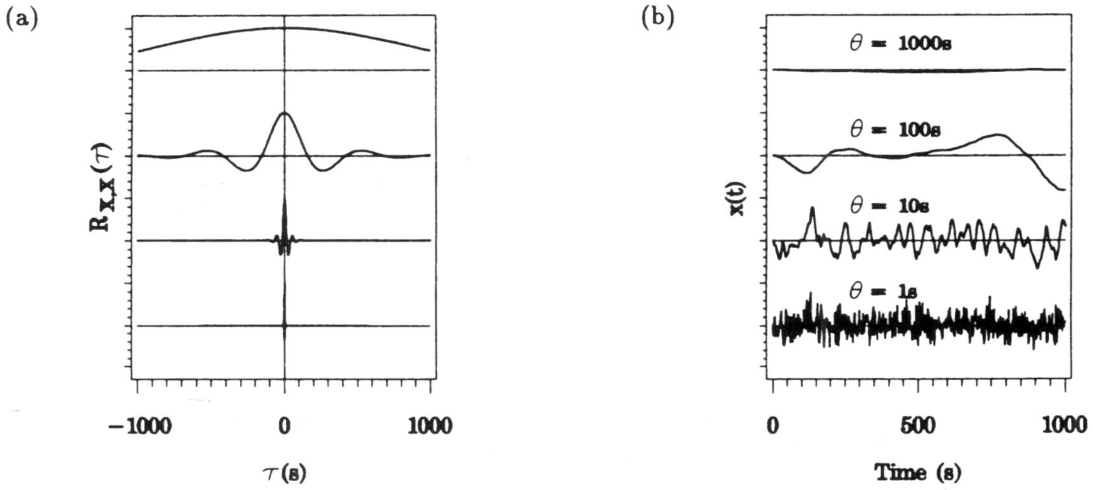

Fig. 1. Correlation functions (a) and corresponding realizations (b) of a AR(2)-process with different scales of fluctuation

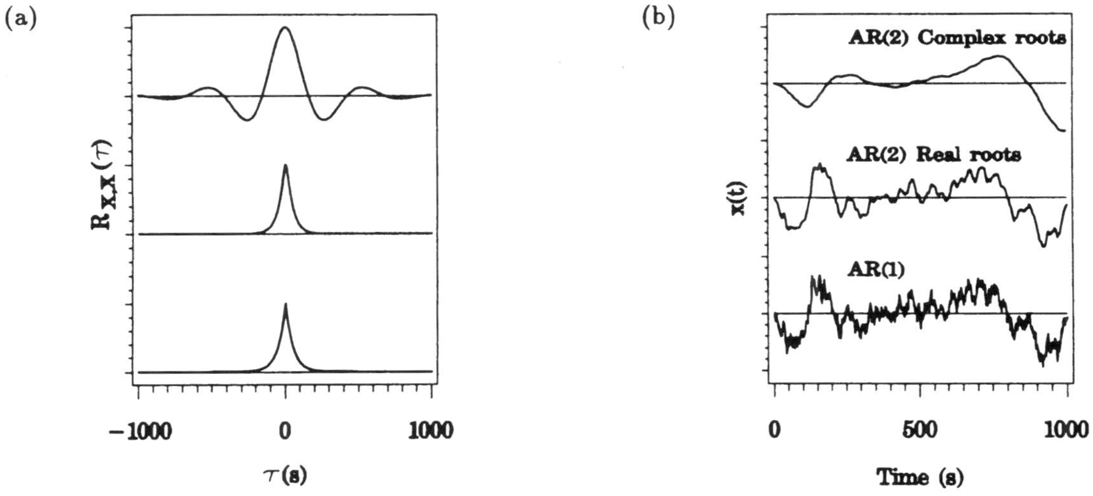

Fig. 2. Correlation functions (a) and corresponding realizations (b) of several types of autoregressive processes with the same scale of fluctuation $\theta = 100s$

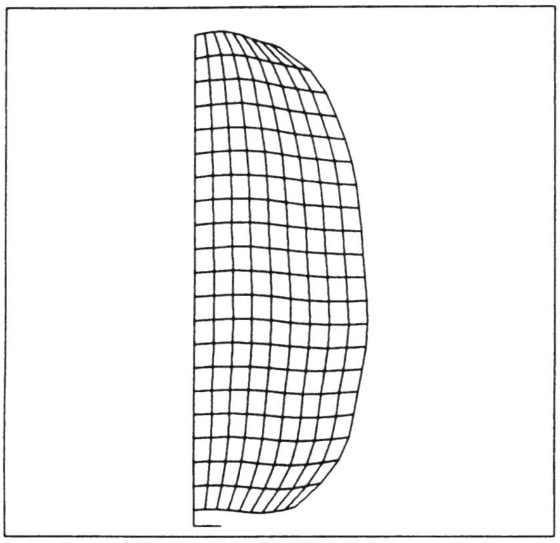

Fig. 3. Finite element grid of a pineapple

On the other hand, if $\theta \longrightarrow +\infty$ then the values of the realization at arbitrary points are completely correlated. In this case the random process concept is far too sophisticated to describe the physical quantity since all the meaningful probabilistic features of the quantity can be captured by a simple random variable model.

In Table 1 the variance, the autocovariance function and the scale of fluctuation are given for AR(1) and AR(2) processes. For the latter, the characteristic polynomial

$$\xi^m + a_1\xi^{m-1} + \cdots + a_m = 0 \tag{10}$$

has two real or two complex conjugate roots, resulting in non–oscillating or oscillating correlation functions, respectively. The correlation functions and corresponding realizations of different types of random processes are shown in Fig. 2. Clearly the scale of fluctuation is a measure of how frequent the process wiggles around the mean–axis, irrespective of the type of the process.

4. A VARIANCE PROPAGATION ALGORITHM

In the analysis below it is assumed that the ambient temperature T_∞ can be modelled by means of an autoregressive random process with the following state space representation

$$\frac{d}{dt}\mathbf{x}_{T_\infty} = \mathbf{A}\mathbf{x}_{T_\infty} + \mathbf{B}w(t) \tag{11}$$

Equation (4) and (11) can be combined in the following stochastic differential system

$$\frac{d}{dt}\mathbf{y}(t) = \mathbf{g}[\mathbf{y}(t)] + \mathbf{h}w(t) \tag{12}$$

where

$$\mathbf{y} = \begin{bmatrix} \mathbf{u} \\ \mathbf{x}_{T_\infty} \end{bmatrix} \tag{13}$$

$$\mathbf{g}[\mathbf{y}(t)] = \begin{bmatrix} \mathbf{C}^{-1}\mathbf{f} - \mathbf{C}^{-1}\mathbf{K}\mathbf{u} \\ \mathbf{A}\mathbf{x}_{T_\infty} \end{bmatrix} \tag{14}$$

$$\mathbf{h} = \begin{bmatrix} \mathbf{0} \\ \mathbf{B} \end{bmatrix} \tag{15}$$

A known result from stochastic systems theory is the following ***steady state variance propagation algorithm*** for the mean value and the covariance matrix of $\mathbf{y}$ as described by equation (12)

$$\mathbf{0} = \mathbf{g}[\overline{\mathbf{y}}(t)] \tag{16}$$

$$\mathbf{0} = \frac{\partial \mathbf{g}[\overline{\mathbf{y}}(t)]}{\partial \overline{\mathbf{y}}(t)} \mathbf{V}_{\mathbf{y},\mathbf{y}}(t) + \mathbf{V}_{\mathbf{y},\mathbf{y}}(t) \times \left(\frac{\partial \mathbf{g}[\overline{\mathbf{y}}(t)]}{\partial \overline{\mathbf{y}}(t)}\right)^T + \mathbf{h}\sigma_w^2(t)\mathbf{h}^T \tag{17}$$

Substitution of (13)–(15) in (16) and (17) yields the following system

$$\mathbf{K}\overline{\mathbf{u}} = \overline{\mathbf{f}} \tag{18}$$

$$\mathbf{0} = \mathbf{C}^{-1}\left(\frac{\partial \mathbf{f}}{\partial T_\infty}\mathbf{V}_{\mathbf{u},T_\infty}^T - \mathbf{K}\mathbf{V}_{\mathbf{u},\mathbf{u}}\right) + \left(\frac{\partial \mathbf{f}}{\partial T_\infty}\mathbf{V}_{\mathbf{u},T_\infty}^T - \mathbf{K}\mathbf{V}_{\mathbf{u},\mathbf{u}}\right)^T \mathbf{C}^{-T} \tag{19}$$

$$\mathbf{0} = \overline{\mathbf{C}}^{-1}\left(-\mathbf{K}\mathbf{V}_{\mathbf{u},\mathbf{x}_{T_\infty}} + \frac{\partial \mathbf{f}}{\partial T_\infty} \times \left[\, \mathbf{V}_{T_\infty,T_\infty} \,\middle|\, \cdots \,\middle|\, \mathbf{V}_{T_\infty,T_\infty^{(m-1)}} \,\right]\right) + \mathbf{V}_{\mathbf{u},\mathbf{x}_{T_\infty}}\mathbf{B}^T \tag{20}$$

with

$$T_\infty^{(i-1)} \triangleq \frac{d^{i-1}}{dt^{i-1}}T_\infty$$

The Jacobians must be evaluated using the mean parameter values and the dependency on the time is omitted for readability reasons.

Equation (19) is of the general form

$$\mathbf{A}(t)\mathbf{V}(t) + \mathbf{V}(t)\mathbf{A}(t)^T + \mathbf{C}(t) = \mathbf{0}$$

with $\mathbf{V}$, $\mathbf{A}$, and $\mathbf{C}$ square matrices of equal dimension, and is called a *Lyapunov* equation. Equations (20) is of the general form

$$\mathbf{A}(t)\mathbf{V}(t) + \mathbf{V}(t)\mathbf{B}(t) + \mathbf{C}(t) = \mathbf{0}$$

with $\mathbf{A}$, $\mathbf{B}$ square matrices, and $\mathbf{V}$ and $\mathbf{C}$ matrices which are in general not square. Equation (21) is called a *Sylvester* equation. Methods for the numerical solution of Lyapunov and Sylvester matrix differential equations are discussed in Bartels and Steward (1972).

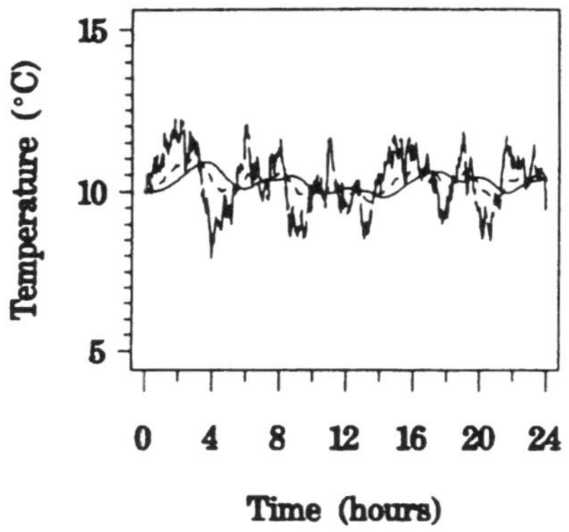

Fig. 4. Ambient (— —), surface (- - -) and centre temperature (———) in a pineapple; $\theta = 2$ h

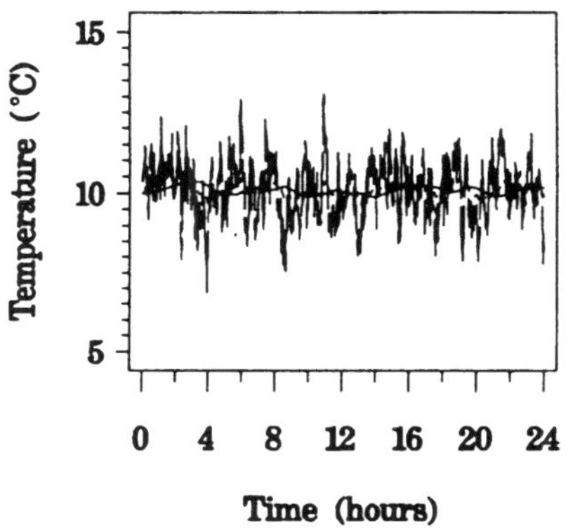

Fig. 5. Ambient (— —), surface (- - -) and centre temperature (———) in a pineapple; $\theta = 0.2$ h

5. STOCHASTIC ANALYSIS OF STEADY STATE HEAT TRANSFER IN PINEAPPLE

The variance propagation algorithm has been applied to investigate the effect of random storage temperature fluctuations on the centre and surface temperature of a pineapple. Hereto the pineapple has been subdivided into an axisymmetric finite element grid using the grid generator described in Nicolaï *et al.* (1994) (Fig. 3). The grid consisted of 180 quadrilateral elements with 4 nodes per element, yielding a total of 210 nodes. The following values have been used for the thermophysical parameters (Sweat, 1974): ρ=1010 kg/m^{3}°C; c=3686 J/kg°C; k=0.549 W/m°C. The surface heat transfer coefficient has been set equal to 10 W/m^{2}°C, respectively. For the stochastic ambient temperature, $\overline{T}_0 = 10$°C and $\sigma_{T_\infty} = 1$°C.

The actual computations were done on a HP/720 UNIX workstation using a modified version of the heat transfer finite element code DOT (Polivka and Wilson, 1976; Nicolaï, 1994). The time step for the transient Monte Carlo analysis was equal to 60 s. The total period considered was equal to 24 h. Preliminary simulations indicated a good agreement between the results obtained with the Monte Carlo method (100 runs) and the variance propagation algorithm. However, because the CPU time required by the variance propagation algorithm was several orders of magnitude smaller than that required by the Monte Carlo analysis, the former was used uniquely for the further analysis. Samples of a stochastically varying storage temperature and the corresponding surface and centre temperature in the pineapple are shown in Fig. 4 ($\theta = 2$ h) and Fig. 5 ($\theta = 0.2$ h). The samples were generated by straightforward evaluation of the time series which was obtained after temporal discretization of the continuous-time autoregressive processes by means of implicit Euler finite differencing. Gaussian white noise was generated by means of NAG routine G05DDF.

It is clear that for $\theta = 0.2$h the fast fluctuations are smoothed out as a consequence of the thermal inertia of the pineapple. Also, the surface temperature is more sensitive to storage temperature fluctuations.

In Fig. 6 the influence of the scale of fluctuation on the centre temperature variance corresponding to autoregressive chilling temperature fluctuations of different order is shown. While for the AR(1) process and the AR(2) process with non-oscillating correlation function the variance is almost equal for equal scale of fluctuation, for the AR(2) process with oscillating correlation function the variance is slightly larger.

In Fig. 7 the the mean and confidence interval of the centre temperature are shown as a function of the scale of fluctuation for AR(1) ambient temperature disturbances. For $\theta \rightarrow +\infty$ the 95% confidence interval becomes equal to 10±2°C, or equal to the confidence interval of the ambient temperature. In this case the random process degenerates and becomes a random variable.

6. CONCLUSIONS

Random temperature fluctuations can cause the centre temperature of chilled produce to decrease temporarily below the threshold level beyond which cold injury may develop. A variance propagation algorithm was derived to calculate the steady state temperature variance at an arbitrary position inside the pineapple. Simulations revealed that in the case of autoregressive ambient temperature disturbances the centre temperature variance is dependent on the scale of fluctuation, and, to a much lesser extent, on the order of the autoregressive processes. This indicates that knowledge of the scale of fluctuation of the disturbances might be sufficient to predict the variability of the internal temperature. As a consequence, stochastic estimation of the coefficients of the au-

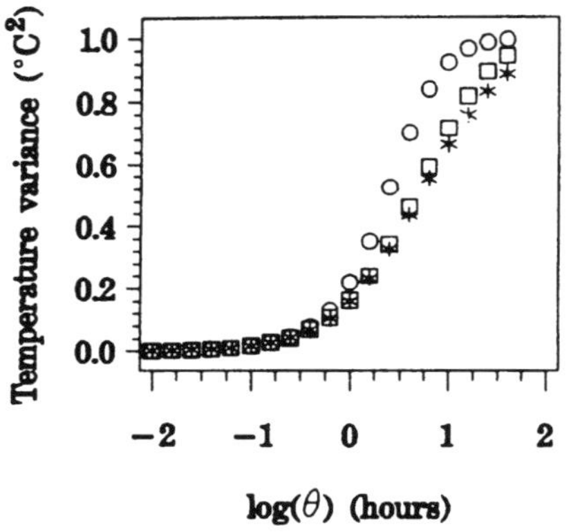

Fig. 6. Centre temperature variance as a function of the scale of fluctuation for autoregressive processes of different order. Legend: ⋆: AR(1); □: AR(2), real roots; o: AR(2), complex roots.

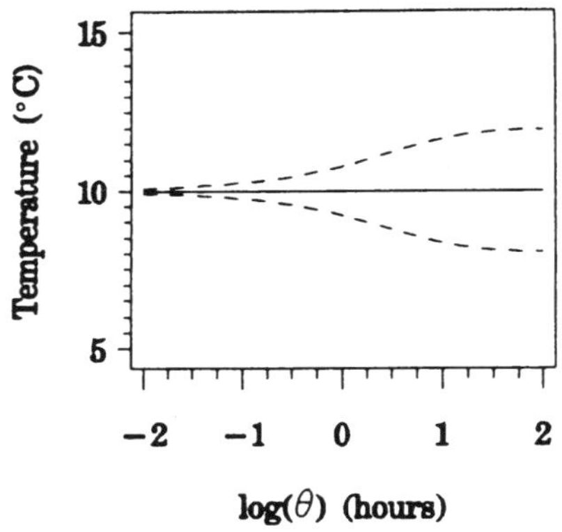

Fig. 7. Mean (——) and 95% confidence margins (- - -) of the core temperature in a pineapple as a function of the scale of fluctuation of a autoregressive ambient temperature of order 2 (real roots)

toregressive processes would be unnecessary.

7. ACKNOWLEDGEMENTS

The European Communities (FLAIR project AGRF-CT91-0047 and AIR project AIR2-CT93-1519) are gratefully acknowledged for financial support.

8. NOMENCLATURE

c	: heat capacity
$\mathbf{C}$	: finite element capacity matrix
$\mathcal{E}$	: mean value operator
$\mathbf{f}$	: finite element thermal load vector
h	: surface heat transfer coefficient
	: [W/m²°C]
k	: thermal conductivity [W/m°C]
$\mathbf{K}$	: finite element stiffness matrix
n_{nod}	: number of nodes
$n_{\perp}$	: outward normal
$\mathbf{0}$	: zero matrix
t	: time [s] or [h]
T	: temperature [°C]
T_0	: initial temperature [°C]
T_{∞}	: ambient temperature [°C]
$\mathbf{u}$	: nodal temperature vector
$\mathbf{V}$	: covariance matrix
w	: white noise process
Γ	: convection surface
θ	: scale of fluctuation [s] or [h]
ρ	: density [kg/m³]
σ	: standard deviation

9. REFERENCES

Brogan, W.L. (1985). *Modern Control Theory.* 2nd ed.. Prentice-Hall, Inc. Englewood Cliffs, USA.

Incropera, F.P. and D. De Witt (1990). *Fundamentals of Heat and Mass Transfer.* 3th ed.. John Wiley and Sons. New York, Chichester, Brisbane.

Nicolaï, B.M. (1994). Modeling and Uncertainty Propagation Analysis of Thermal Food Processes. PhD thesis. K.U.Leuven. Leuven, Belgium.

Nicolaï, B.M. and J. De Baerdemaeker (1994). Stochastic analysis of heat transfer in stored produce related to chilling injury risks. In: *International Symposium on New Applications of Refrigeration to Fruit and Vegetable Processing.* International Institute of Refrigeration. Istanbul, Turkey.

Nicolaï, B.M., W. Obbels, M. Schellekens, B. Verlinden, T. Martens and J. De Baerdemaeker (1994). Computational aspects of a computer aided design package for the preparation of cook-chill foods. In: *Proceedings of the Food Processing and Automation Conference III.* ASAE. Orlando, U.S.A.. pp. 190–199.

Polivka, R.M. and E.L. Wilson (1976). Finite element analysis of nonlinear heat transfer problems - DOT user's manual. Technical Report UC SESM 76-2. University of California. Berkeley, U.S.A.

Segerlind, L. (1984). *Applied Finite Element Analysis.* 2nd ed.. John Wiley and sons. New York.

Sweat, V.E. (1974). Experimental values of thermal conductivity of selected fruits and vegetables. *Journal of Food Science* **39**, 1080–1083.

Vanmarcke, E. (1983). *Random Fields — Analysis and Synthesis.* The MIT Press. Cambridge, Massachusetts.

NEURAL BRUISE PREDICTION MODELS FOR FRUIT HANDLING AND MACHINERY EVALUATION

Barreiro P.[*], Steinmetz V.[#], Ruiz-Altisent M.[*]

[*] *ETSIA. Rural Engineering Dept. Avda. Complutense s/n. 28040 Madrid*
Phone: 34-1-3365856 Fax:34-1-3365845
E-mail: mruiz@ccupm.upm.es
[#] *CEMAGREF. Agricultural Equipment and Food Process Engineering Division.*
361 Rue Jean Francois Breton. 34033 Montpellier
Phone: 33-67046300 Fax:33-67635795
E-mail: vincent.steinmetz@cemagref.fr

Abstract: Some neural bruise prediction models have been implemented in the laboratory, for the most traded fruit species and varieties, allowing the prediction of the acceptability or rejectability for damages, with respect to the EC Standards. Different models have been built for both quasi-static (compression) and dynamic (impact) loads covering the whole commercial ripening period of fruits.
A simulation process has been developed gathering the information on laboratory bruise models and load sensor calibrations for different electronic devices (IS-100 and DEA-1, for impact and compression loads respectively).
Some evaluation methodology has been designed gathering the information on the mechanical properties of fruits and the loading records of electronic devices. The evaluation system allows to determine the current stage of fruit handling process and machinery.

Keywords: Load sensors, network models, bruise simulation, decision support system.

1. INTRODUCTION

The current stage of fresh fruit retail market has been worsening. Fresh fruit consumption may decrease due to the lack of quality in the markets. Recent research results (Kamp and Pedersen, 1990) have shown that less than a 10% of the fruits titled Class I followed the EC Standards mainly due to the presence of mechanical damages.

Fruit handling processes and mechanical equipment get more and more complex. The need for an evaluation standarization is clear, in order to objectively compare differences between the machineries. Within this aim the electronic products have been developed (Halderson and Skrobacki, 1986; Brown et al., 1990). However, mechanical properties of fruits have also a great effect on the mechanical damage susceptibility of fruits (Chen and Zongnan, 1981; Chen et al., 1987; García and Ruiz, 1988; Rodriguez et al., 1990; Ruiz Altisent, 1990). So gathering the information on the mechanical properties of fruits is also needed for the evaluation of processes and machinery.

Until now some comparisons between the electronic devices measurements and the mechanical properties of fruits have been made (Brown et al., 1989; Schulte et al., 1992 & 1993; Sober et al., 1990). Day by day a wider range of fruit varieties and storage treatments are used through the same machinery, therefore a wider link between the electronic devices and the mechanical properties of fruit is needed.

Mechanical properties of fruits have been studied

(Barreiro and Ruiz, 1993) and a statistical approach to bruise modelization has been used in order to build up the bruise prediction models and evaluation methodology. Neural network process (Robert et al., 1989; Ros et al., 1993; Steinmetz and Delwiche, 1993) leads to consider whether this type of models would be adapted to bruise modelization or not.

Within this context, the objectives for the current study were:

1) to search for the optimal bruise prediction models,

2) to gather the needed information on the mechanical properties of fruits and the electronic products to simulate bruise appearance, and

3) to design an evaluation methodology for determining the current stage of fruit handling process and machinery with respect to the EC legislation.

2. MATERIAL AND METHODS

All the data employed for the current study were the consequence of a two seasons' work (Barreiro and Ruiz, 1993) where an exhaustive research on fruit mechanical properties and bruise susceptibility was carried out. Both the most common cold storage treatments and load types (compression and impact) were studied; the most exported pome and stone species: apple, pear, peach and apricot were tested. A trading study was made to select the most important varieties:

- apple: "Golden Delicious"
- pear: "Conference" and "Doyenne du Comice"
- peach: "Maycrest" and "Springtime"
- apricot: "Búlida" and "Canino"

2.1. Selection of variables for bruise modelization

The first step for developing the bruise prediction models was to select the variable which would best represent bruise damage. The EC Legislation gives the maximum surface damage allowable for each quality class: 1 cm^2 and 0.5 cm^2 for pome and stone fruits respectively. However, the external surface damage depends on the size of the contact surface between two bodies' during loading, also a function of both bodies' curvature radius (Horsfield et al., 1972). On the other hand, the smaller the curvature radius, the deeper the bruise is. Due to all these facts it was possible to find bruises having an equivalent bruise volume but different bruised external surface, which would consequently have different quality evaluation. Therefore, a new parameter, "bruise section", was created to include these effects. This way, the units for the bruised section is the same as for the bruised external surface (mm^2).

$$b.s. = \pi \frac{w}{2} d\,(mm^2) \qquad (1)$$

Bruise section is defined in Eq. 1, where w is bruise width (mm) and d is bruise depth (mm); it represents twice the internal bruised section area.

As for the selection of the explicative variables (*), it was decided to apply the parsimony principle on the machinery and on the data to be employed, so that the process and the machinery evaluation could be carried out easily. On the basis of previous studies, (Chen et al., 1987; Rodriguez et al., 1990; Ruiz Altisent, 1990; Barreiro and Ruiz, 1993) it was decided to perform a different bruise model for each variety and loading type (compression or impact), including in all of them as explicative variables the following magnitudes: fruit firmness (FR, 1), loading level (HI for impact or FZC for compression, 2) and storage treatment (ST, 3).

Measuring fruit firmness. The Magness-Taylor penetration test is the most common test used for fruit firmness assessment. This testing procedure has two main handicaps: it is a destructive test (so it can only be carried out on samples of fruits), and its variation coefficient is high, especially for stone fruits. Therefore, through the latest years a new testing machine for fruit firmness sensing has been developed (Chen and Zongnan, 1981; García and Ruiz, 1988; Ruiz Altisent, 1990; Jarén et al., 1992; Correa et al., 1992). This device uses low energy impacts for firmness sensing. It allows to measure a great number of variables related to the deceleration suffered by the impactor, including the impact duration. Previous research results (Barreiro and Ruiz, 1994; Steinmetz et al., 1995) have shown the clear relationship between the ratio of maximum impact force over maximum impact deformation (FDR) with the maximum force of the Magness-Taylor penetration test.

The impact tester has the advantage of being much more repetitive (usually 10% of coefficient of variation aproximately), than Magness-Taylor which may surpass 30% for stone fruits (Barreiro and Ruiz, 1992). It can also be performed non-destructively for a low drop height (Chen and Ruiz, 1994). For all these reasons the selected variable for firmness was the FDR in impact.

* explicative variables are all correlatively numbered.

Loading level variable. Research results carried out previously (Barreiro and Ruiz, 1993) included laboratory and field studies. During the laboratory testing different loading levels were applied for each variety, cold treatment and loading type (compression or impact) in order to achieve bruise damages below and above the EC Standards limits. For impact testing a constant mass was dropped upon the fruits and the impact loading level was measured as a function of the impacter mass dropping height. The compression loading level was fixed as a maximum compression force. The loading levels were included for each compression and impact bruise prediction model in this same way; the relationship between these laboratory load parameters and the electronic devices is shown in Paragraph 4. The electronic devices used during the current research have been an IS-100 (Halderson et al., 1986) measuring impact loads, and a DEA-1 measuring compression loads (Various authors, 1995).

Cold storage variable. Three storage treatment modalities were employed: no cold storage, cold storage with 80% of R.H. and with plastic covering; 1°C for pome fruits and 4°C for stone fruits. As cold storage had shown its influence on bruise susceptibility, it was decided to include this variable as a discrete parameter and to test afterwards whether it was significant or not for each bruise prediction model.

For the stone fruits, the 80% R.H. cold storage data were not used for bruise modelization, as the treatment showed to be non-commercial: more than 20% of fruit weight losses were produced.

Remarks on apricot models. Apricot varieties present a great variation in their mechanical properties during the harvesting season. This fact is due not only to their biological properties but also to the commercial competition made by the earliest varieties of peaches. This market demand determines the presence of immature fruits in the stores.

Non-mature apricots show great differences in their mechanical properties when compared to mature fruits. A new discrete variable, named "maturity" (M, 4), with two modalities was created. Both non-mature and mature fruits coming out of three different harvesting dates were tested.

A "minimal distance classifier" (Judez, 1989) based on normalized data was used for the maturity assessment of apricots. Inputs to the classifier were all the variables described in previous paragraphs of section 2.

2.2. Models evaluation methodology

As a first step, an analysis of variance (ANOVA) at 5% level was carried out to check whether the selected variables were significant for bruise prediction. Afterwards, different neural approaches for bruise prediction were carried out.

The bruise prediction models were compared using the correlation coefficient between bruise prediction and observation. Another way for evaluating bruise prediction models, was the classification under the EC Standards. An error below the EC tolerance (10% for I Class) was used also as a criterion for model acceptation.

Comparison between fruit classification under the EC Standards for predicted and observed damage is presented in Fig. 1, where

A is the percentage of unclassified fruits,
B is the percentage of fruits well classified,
C is the percentage of non detected damage, and
D is the percentage of overdetected damage; both "C" and "D" should be below 10% for the model to be acceptable.

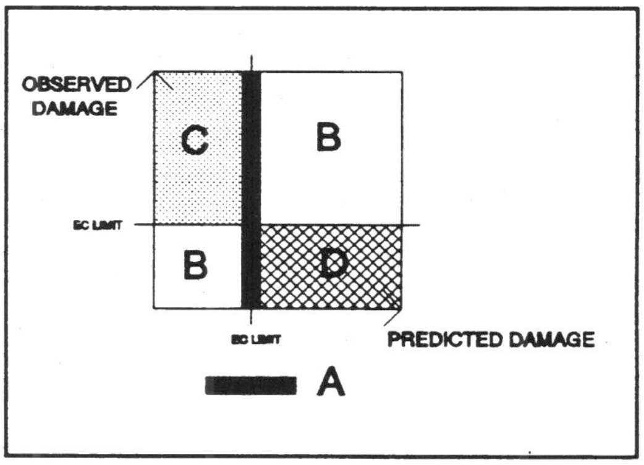

Figure 1. Comparison between fruit classification for predicted and observed damages.

Each group was defined using the whole confidence interval for each prediction (5% of significance level). One prediction had to show the same classification as the observed damage for the whole confidence interval to be considered well classified. The percentage of well classified increased with decreasing confidence interval, that is, increasing model adjustment.
Whenever the confidence interval was between acceptable or rejectable (below <u>and</u> above the EC limit) the fruit was considered unclassified, except if the observed damage was inside the band 5% plus or minus the EC limit. In this case the fruit would be considered well classified, as the observed damage would also be inside the threshold band.

2.3. Neural network approach

Neural networks are based on the emulation of human brain. A net is built with many neurons or processing elements (PE) which are connected. Normally, the neurons inside the brain are connected in a hierarchic way, as it makes the information easier to be gathered. On this basis the feedforward neural networks have been created (Schalkoff, 1992).

The PEs placed at the begining of the network (inputs) were the explicative variables and the PE placed at the end (output) was the bruise section. There was a PE called BIAS equivalent to the offset in the statistical procedures.

The transformation function (Eq. 2) can be selected depending on the characteristics of the inputs and the outputs. For bruise prediction, a sigmoid function was selected. This function was obtained initially from a biological model and allows to find non linear relationships.

$$f(x) = \frac{1}{1+e^{-\lambda x}} \qquad (2)$$

The number of iteractions (weight corrections) was fixed empyrically in 40.000 for pome fruits and 20.000 for stone fruits. These numbers correspond to the number of iteractions required until the convergence of the learning process is achieved. There were about 30% more individuals for the pome than for the stone varieties.

For neural approaches, a training as well as a testing set were used. The testing set contained about half the number of samples of the training set; this testing set was made out of the second repetition of the 1993 season's tests. The testing set was not included for the models adjustment, allowing to determine the models robustness by comparing the coefficient of determination for the training and the testing sets. The evaluation of the samples classification under the EC Standards was used only for the training set as a low number of samples may lead to a biased result.

Whenever it was not possible to achieve a better "generalist" solution, that is using all data for each model, for neural network approaches it was decided to check whether a sample selection or a "specialist" net training could give better results or not. Two approaches were studied:

1) the first one was sample selection. As the training set contained a large number of individuals, the presence of high number of individuals of certain characteristics could determine an overtraining learning for the generalist network. Therefore, a training sample was created following the Ward aggregation criterion (Judez et al., 1989). An inertia threshold was chosen bellow which every group of individuals would be selected. The net training was carried out on this selected sample and afterwards retrained for the whole sample; the learning coefficients for the second approach were lower, to avoid oscillation.

2) the second one was specialist network. For this solution, it was necessary to find out the areas where learning was low and to build a network covering that range of data. Once the area had been chosen, a "minimal distance classifier" was used to know under which network unknown individuals should be tested.

3. RESULTS FOR BRUISE PREDICTION MODELS

3.1. Results for the initial neural approaches

Neural approaches were carried out using the Neuralworks Professional II plus software. The first net architecture is shown in Table 1. Each net is identified by three digits corresponding to the number of inputs, PEs in the hidden layer and outputs.

Table 1. First neural architecture.

	Model	
Variety	Compression	Impact
"Conference"	3-1-1	3-1-1
"Doyenne"	3-1-1	3-0-1
"Golden"	3-1-1	3-1-1
"Maycrest"	3-1-1	3-1-1
"Springtime"	3-1-1	3-1-1
"Búlida"	4-1-1	4-1-1
"Canino"	4-1-1	4-1-1

The results obtained with neural processing did not achieve the desired error range ("C" and "D" below 10%, see Fig. 1) in most cases; even increasing the number of hidden PEs did not improve the overall results. Therefore, it was decided to check whether a sample selection or a "specialist" network could lead to better results, as it has been explained in section 2.5. The method showing best results was the specialist training, though the results did not satisfy the EC tolerance restrictions.

3.2. New variables for model optimization

As an alternative solution it was decided to search for new significant variables for bruise prediction. However, this step was difficult as the variables initialy were selected with the parsimony critera as main restriction.

At least an impact tester and an electronic device (IS-100 or DEA-1) would be necessary to evaluate handling procedures and machinery. It was decided, as far as possible, not to add new equipment. Therefore, it was decided to use the variables gathered by the impact tester in order to check whether the models could be improved with them. A bibliography review leads to some recommended variables (Chen and Zongnan, 1981; García and Ruiz, 1988; Ruiz Altisent, 1990; Jarén et al., 1992; Correa et al., 1992).

- DFI (5), maximum deformation in impact
- TPI (6), impact duration
- FZI (maximum force in impact)/TPI (7)
- FZI* TPI (8)
- DFI/TPI (9)
- FZI*DFI (10)
- FZI^2/TPI (11)
- FZI/DFI^2 (12)

Also new variables were created on a theoretical basis.

- FZI*TPI^2 (13); (energy dimension)
- **FZC**/FR (14); (deformation dimension)
- **HI**/FR (15); (searching for the impact analogy with 14)
- (FZI*DFI)/**FZC** (16); (deformation dimension)
- **FZC***DFI (17); (energy dimension)
- **FZC^2**/FR (18); (energy dimension)
- FR*DFI^2 (19); (energy dimension)

FZC and HI have been related to the electronic devices records (IS-100 and DEA-1) in Paragraph 4.

To the whole number of variables, 1 to 19, a stepwise regression analysis was applied in order to find out the significance level for bruise prediction. The variables selected (Table 2) always showed a significance level below a 5%. Therefore, the results could not be improved for both compression and impact models of "Golden" or the "Doyenne du Comice" impact models.

With the variables shown in Table 2, new generalist and specialist networks were built (Table 3), searching for a non-linear optimization. The generalist networks covered the whole commercial maturity stage of fruits. As for the specialist solutions, in most cases, it was appreciated that there was a bad learning for an observed damage above two times the EC limit. As this value was quite high, it was decided to train a specialist net only for an observed damage size below that amount. Any unknown individual to be classified would be submitted to a "minimum distance classifier". The decision would be based on the observed damage of the nearest neighbour.

Table 2. New impact variables selected through a stepwise regression analysis. The numbers refer to the variables defined before.

	Model	
Variety	Compression	Impact
"Conference"	1,2,3,7,11,12,14	1,2,3,10,11,15
"Doyenne "	1,2,3,5,14,16	2,3
"Golden"	1,2,3	1,2,3
"Maycrest"	1,2,3,13,14	1,2,3,6,15
"Springtime"	1,2,3,5,9,12,14	1,2,3,5,6,15,19
"Búlida"	1,2,4,7,10,14	1,2,4,7
"Canino"	1,2,4,6,10	1,2,4,10,15

Only for the compression and impact models of "Maycrest" it was shown that the bad learning area was placed around the EC acceptation threshold. This case was more difficult as the threshold area was the most important one. Therefore, it was decided to train a generalist network. Afterwards, a minimal distance evaluation would be made in order to know whether the individuals could belong to the bad trained area or not. In case of belonging to this area, a specialist network would be made to arrive to a final response. The final results for neural approaches are shown in Table 3.

4. RESULTS FOR THE BRUISE SIMULATION PROCESS

As it has been already mentioned in paragraph 2.1, it was necessary to relate the laboratory loading levels with those commonly applied to fruits by handling processes and grading machinery. Lately, Simulated Electronic Products (SEP) have been developed in order to do so (Halderson and Skrobacki, 1986; Brown et al., 1990). Therefore a calibration test was carried out for an IS-100 impact sensig device, as well as for a DEA-1 compression sensing device. In both cases a calibration equation was calculated between the laboratory loads and the electronic devices parameters.

For impact, a calibration equation (Eq.3) was

determined where HI is the impacter drop height (cm) of equal energy of the drop height of the IS-100, and AC_{IS-100} is the electronic device acceleration (g). The results showed a correlation coefficient of 0.98 for 127 samples (calibration limits 30-160g).

$$HI=0.347AC_{IS-100}-6.968 \quad \textbf{(3)}$$

Table 3. Final results for the neural approaches. Where: r is the correlation coefficient between predicted and observed damage, n is the number of samples, and "C" and "D" the classification errors of the models; "C" and "D" should be below 10% to accept the model as shown in Paragraph 2.2. The range of values obtained for the correlation coefficient are due to the wide conditions modelized which cover the whole commercial ripening period of fruits.

	Model	
Variety	Compression	Impact
"Conference"	*Specialist*	*Generalist*
	r=0.78;n=480	r=0.82;n=480
	C=5.4%	C=7.5%
	D=6.9%	D=2.3%
"Doyenne"	*Specialist*	*Generalist*
	r=0.72;n=400	r=0.79;n=400
	C=8.5%	C=6.8%
	D=4.8%	D=5.2%
"Golden"	*Generalist*	*Generalist*
	r=0.82;n=456	r=0.88;n=456
	C=7.5%	C=0.7%
	D=4.8%	D=7.7%
"Maycrest"	*Specialist*	*Specialist*
	r=0.79;n=228	r=0.71;n=228
	C=3.9%	C=7.9%
	D=2.7%	D=3.5%
"Springtime"	*Generalist*	*Generalist*
	r=0.81;n=238	r=0.80;n=238
	C=1.7%	C=3.4%
	D=9.2%	D=9.7%
"Búlida"	*Specialist*	*Specialist*
	r=0.64;n=184	r=0.64;n=184
	C=8.2%	C=4.9%
	D=2.2%	D=2.7%
"Canino"	*Specialist*	*Specialist*
	r=0.67;n=139	r=0.68;n=139
	C=7.2%	C=8.6%
	D=2.2%	D=0.0%

For compression, a calibration equation (Eq.4) was determined where FZC is the compression maximum force (N) and R_{DEA-1} is the electrical resistance registered by the electronic device (Kohm). The results showed a correlation coefficient of 0.94 for 252 samples (calibration limits 20-0.4 Kohm).

$$FZC=e^{(5.783-1.338\ln(R))} \quad \textbf{(4)}$$

Finally a simulation process was carried out (Fig. 2-5), were the load threshold necessary to cause damage above the I Class EC limit is 100mm^2 and 50mm^2 for pome and stone fruits respectively; the load limit is measured through the maximum acceleration for impact simulation and through the electrical resistance for compression simulation.

The main results obtained for the bruise simulation process can be summarized as follows:

- there is a great effect of fruit firmness evolution during storage on bruise susceptibility for "Conference" pears (Fig. 2) as well as for peaches and apricots. Firmness evolution did not show this effect for "Golden" apples (Fig. 3) or "Decana pears",

- the effect of fruit firmness evolution on bruise susceptibility is always higher for compression than for impact loads,

- the reference impact loading threshold, damage above the EC limit, for pome fruits reaches 70g while it decreases to 50g when working with stone fruits. These reference values change along the commercial ripening period depending on variety as shown in Fig.s 2 to 5, and

- the reference compression loading threshold, damage above the EC limit, for pome fruits reaches 5Kohmn. Stone fruits show higher compression susceptibility being their loading threshold 20 Kohm (note that the electrical resistance decreases with increasing loading level).

5. RESULTS FOR HANDLING PROCESS AND MACHINERY EVALUATION

Once an acceptable solution (classification error below the EC tolerance level) had been found for each bruise prediction model, an evaluation methodology for handling equipment was designed. This evaluation system consisted of a decision system to aid growers on the knowledge of their procedures and machinery qualities by gathering the information on fruit firmness and storage treatments and relating it with the records obtained with the electronic devices, IS-100 and DEA-1, for bruise prediction.

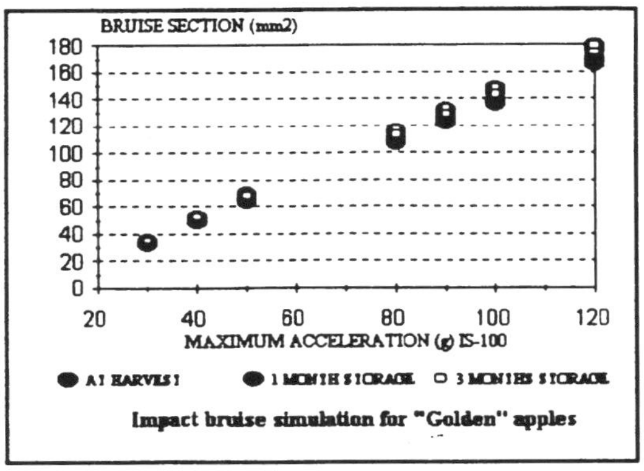

Fig. 2. Impact bruise simulation for "Golden" apples. A low effect of firmness evolution on bruise susceptibility is shown. An acceleration level of 70g leads to damages about the sizes of the EC limit for I Class (100 mm²).

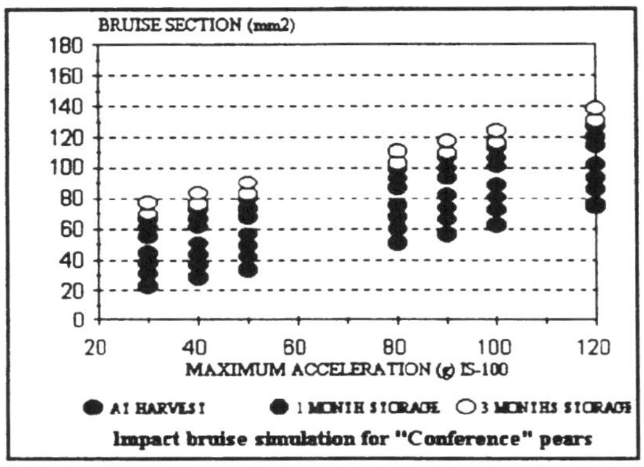

Fig. 3. Impact bruise simulation for "Conference" pears. A great effect of firmness evolution on bruise susceptibility is shown. An acceleration level of 70g leads to damages about the sizes of the EC limit for I Class (100 mm²); higher level of impact would be needed at harvest to reach the EC limit of damage.

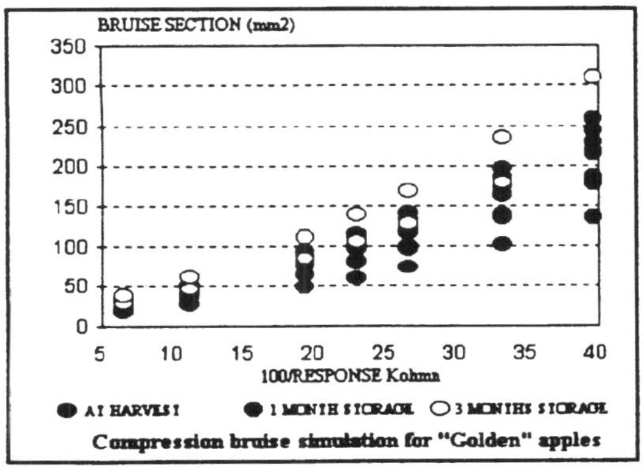

Fig. 4. Compression bruise simulation for "Golden" apples. An electrical resistance level of 5 Kohm (20*100/R) leads to damages about the sizes of the EC limit for I Class (100 mm²); 3.3 Kohmn (30*100/R) are needed at harvest.

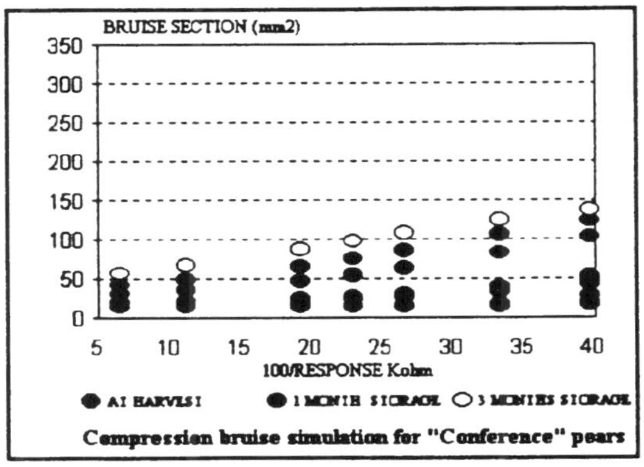

Fig. 5. Compression bruise simulation for "Conference" pears. A great effect of firmness evolution on bruise susceptibility is detected. After 1 month of cold storage an electrical resistance level of 3.3 Kohm (20*100/R) leads to damages about the sizes of the EC limit for I Class (100 mm²); note that there are no bruise predictions over 200mm² due to the bruise range covered by the model, though pears after 3 months storage may show greater bruise values.

Four classes of load levels were created for evaluation:

- **Acceptable loads** whenever predicted damage is between 0-80% of the EC limit,
- **Low risk loads** for predicted damages between 81-90% of the EC limit,
- **High risk loads** for predicted damages between 91-99% of the EC limit, and
- **Rejectable loads** for predicted damages equal or higher than the EC limit.

Once the classes were defined, it was necessary to design the evaluation methodology. As it has already been mentioned, both devices (impact tester and electronic sensors) are used for the evaluation. The electronic devices record the information on the loading levels. However, the number of loading levels registered can be very high, as well as the time needed to process this information. Using the maximum loading level, the process gets faster but the amount of information is reduced. Therefore, it was decided to employ the four maximum loads (L_i) corresponding to each quartil (q_i) of the whole load histogram as a loading selection criteria.

As for the impact tester, the evaluation is made using the information of a mechanical database in order to standardize the comparison between different procedures or machineries.

The number of processed samples is equal to the

number of fruits times four (four loads per fruit). The bruise prediction is made on these data obtaining as many damage predictions as processed samples; the whole methodology is described in Fig. 6. Over these results two different evaluations were designed: general evaluation and deficient process or machinery elements assessment.

5.1. General evaluation

One fruit needs only one damage above the limit to be rejected. For this reason it was decided to choose the worst damage prediction per fruit to find the potential rejectability of fruits. As there are four different classes of loads defined the results will also be divided in four categories

- -**AC** as the percentage of acceptable fruits,
- -**LR** as the percentage of fruits under low rejectability risk,
- -**HR** as the percentage of fruits under high rejectability risk, and
- -**RJ** as the percentage of rejectable fruits.

The Evaluation Criterium was defined acording to the EC tolerance (Eq.5), 10% in number or weight of fruits for I Class. Whenever the percentage of fruits classified as rejectable or under high rejectability risk is above a 10% a revision or replacement of the concerned process or machinery element is necessary. Also if the percentage of fruits under any rejectability risk is above 20% a revision will be made.

If $HR+RJ \geq 10\%$ or $LR+HR+RJ \geq 20\%$ then revision of the machinery is needed (5)

5.2. Deficient process or machinery element assessment

Each of the four selected loadings were decided to be studied through the different percentages of AC, LR, HR and RJ fruits obtained.

A decision system for deficient process or machinery element assessment was created as follows:

1) if the four L_i lead to a further revision (Eq. 5), all the records of the electronic devices above the minimum L_i are identified as needing an improvement. Once the element improvement is made, the evaluation procedure is repeated,

2) if any of the four L_i is within the tolerance criteria, an interpolation for the threshold loading level determination will be made. As before, all the records of the electronic devices above that threshold are identified. After improving the elements, the evaluation procedure is repeated,

3) if all the four L_i are within the tolerance level the final evaluation is positive. Therefore the handling process or mechanical equipment is acceptable with respect to the EC Standards.

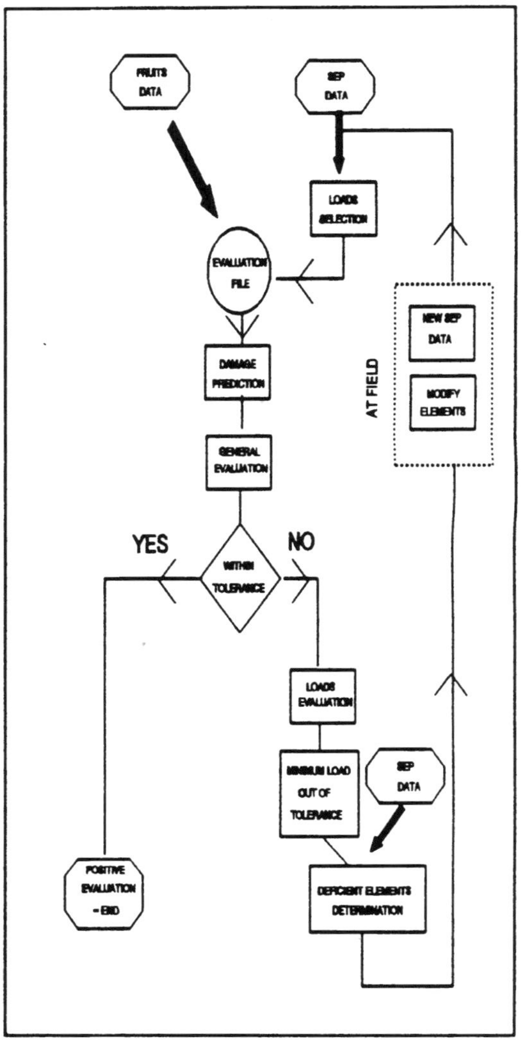

Fig. 6. Evaluation algorithm for fruit handling processes and mechanical equipment. The information on fruit firmness, storage treatment and on electronic devices (SEPs) is put together.

6. CONCLUSIONS

Bruise prediction models have been built, using linear and non-linear regressions as well as feed forward neural networks, for the main traded species and varieties: apple ("Golden Delicious"), pear ("Conference" & "Doyenne du Comice"), peach ("Maycrest" & "Springtime") and apricot ("Búlida" & "Canino").

Those bruise prediction models allow to classify mechanical damages (as acceptable or rejectable) under the EC Standards with errors (undetected or overdetected damages) within the EC tolerance threshold, i.e. 10%.

The prediction models gather the information about bruise susceptibility evolution of fruits at harvest, cold storage and subsequent ripenning at room temperature; the whole commercial ripeness range is covered within the models.

The prediction models integrate also information about different loading types (compression or impact) allowing to simulate the effect of loading under different bruise susceptibility conditions of fruits.

A calibration testing was performed to introduce Electronic Products information (IS-100, DEA-1) within the bruise prediction models.

A simulation process was carried out gathering the information on bruise models and impact and compression sensing devices (IS-100, DEA-1).

An evaluation of handling processes and machinery equipment as well as a decision system was built in order to determine current stage of fruit handling processes in relation to fruit physical quality.

7. ACKNOWLEDGEMENTS

To CAMAR Project n° 8001CT91-0206: "Quality of fruits: engineering research for improving the quality preservation during pre and post harvest operations".

To the CEBAS (Murcia) and specially to Dr. Fernando Riquelme for their aid in apricot research.

To the Fruit Cooperative of Montoliu (Lérida) and specially to Mr. Jaume Roque Calzada for their collaboration in pome fruits research.

to ANECOOP (Valencia) for its cooperation in peach research as well as handling process and machinery evaluation.

8. REFERENCES

Barreiro P. and M. Ruiz-Altisent. 1993. *Bruise susceptibility in pome fruits under different loading and storage conditions*.IV International Symposium on Fruit, Nut and Vegetable Production Engineering Mechanization, Valencia/Zaragoza. Paper n° 935.22

Brown G.K.; C.L. Burton; S.A. Sargent; N.L. Schulte-Pason; E. J. Timm and D.E. Marshall. 1989. *Assessment of apple damage on packing lines*. Applied Engineering in Agriculture **Vol.5(4)** pp:475-484

Brown G.K.; N.L. Schulte-Pason and E. J. Timm. 1990. *Impact classification using the Instrumented Sphere*. ASAE paper n° 90-6001

Correa P.; M. Ruiz-Altisent Altisent and J. L. De la Plaza. 1992. *Physical parameters in relation to physiological changes of avocado during ripening (20°c) and cold storage (6°c) in different conditions*. International Conference on Agricultural Mechanization, Uppsala. 1992 Paper n° 9211-16

Chen P. and S. Zongnan. 1981. *Impact parameters related to bruise injury in apples*. ASAE paper n°81-3041

Chen P.; M. Ruiz-Altisent; F. Lu and A.A. Kader. 1987. *Study of impact and compression damage on Asian Pears*. Transactions of the ASAE **Vol. 30(4)**:1193-1197.

Chen P. and M. Ruiz Altisent. 1994. *Effect of impacting mass on firmness sensing of fruits*. ASAE paper n°93-6542

Garcia C. and M. Ruiz-Altisent. 1988. *Impact parameters related to bruising in selected fruits*. ASAE paper n°88-6027 (Rapid City)

Halderson J.L. and A. Skrobacki. 1986. *Dynamic performance of an impact telemetry system*. ASAE paper n°86-3030

Horsfiled B.C.; R.B. Fridley and L. Claypool. 1972. *Application of theory of elasticity to the design of fruit harvesting and handling equipment for minimum bruising*. Transactions of the ASAE **vol15(4)**pp:746-750

Jarén C.; M. Ruiz-Altisent and R. Perez de Rueda. 1992. *Sensing physical stage of fruits by their response to non-destructive impacts*. International Conference on Agricultural Mechanization, Uppsala 1992 paper n° 9111-113

Judez L. 1989. *Técnicas de análisis de datos multidimensionales*. Ed. Ministerio de Agricultura, Pesca y Alimentación, pp:301

Kamp J. and J. Pedersen. 1990. *Quality of imported and domestic fruits and vegetables in the Danish retail trade with special reference to mechanical damage*. Workshop on impact damage of fruits and vegetables. Zaragoza. **Vol.II**pp:9-16

Robert P.; D. Bertrand; M. Crochon and J. Sabino. 1989. *A new mathematical procedure for NIR analysis: the lattice technique. Application to the prediction of sugar content of apples*. Applied Spectroscopy 43(6) 1576 GAL 76

Rodriguez L.; M. Ruiz-Altisent and M.R. De Felipe. 1990. *Differences in the structural response of "Granny-Smith" apples under mechanical impact and compression"*. Journal of Texture Studies 21(1990) 155-164

Ros F.; A. Brons; F. Sevila; G. Rabatel and C. Touzet. 1993. *Combination of neural network and statistical methods for sensory evaluation of biological products: on-line beauty selection of flowers*. Proceedings of the International Workshop on Artificial Neural Networks, Spain 1993 pp:726-731

Ruiz-Altisent M. 1990. *Impact prameters in relation to bruising and other fruit properties*. Workshop on impact damage of fruits and vegetables. Zaragoza. **Vol.II**pp:27-31

Schalkoff R. 1992. *Pattern recognition. Statistical, structural and neural approaches*. Ed. John Wiley & Sons, pp:364.

Schulte N.L.; G.K. Brown and E.J. Timm. 1992. *Apple impact damage thresholds*. Applied Engineering in Agriculture **Vol.8(1)** pp:55-60

Schulte N.; E. Timm and G. Brown. 1993. *"Red Heaven" peach impact damage thresholds*. IV International Symphosium on Fruit, Nut and Vegetable Production Engineering Mechanization, Valencia/Zaragoza. Paper n° 935.18

Sober S.S.; H.R. Zapp and G.K. Brown. 1990. *Simulated packing line impacts for apple bruise prediction*. Transactions of the ASAE **Vol.33(2)** pp:629-636

Steinmetz V. and J. Delwiche. 1993. *Neural network analysis of rose straightness*. International Conference for Agriculture and Food Equipment and Process control, Nimes pp:253-263.

Steinmetz V.; Crochon M.; Bellon V.; Barreiro P.; García J.L.;1995. *Sensors for fruit firmness assessment: comparison and fusion*. accepted by the Journal of Agricultural Engineering Research.

Various authors. 1995. *CAMAR Project n° 8001CT91-0206:" Quality of fruits". Final Report.*

MEASUREMENT OF PRODUCE RESPONSES ON CLIMATIC IMPACTS IN POSTHARVEST

M. Schreiner*, M. Linke*, S. Huyskens**

**Institute of Agricultural Engineering Bornim, Germany*

***Humboldt-University Berlin, Institute of Horticultural Crop Science, Germany*

Abstract: Numerous factors that affect quality in horticultural produce have to be taken into consideration in order to minimize losses in quality of harvested produce. The present study will investigate the effects of climatic impacts on harvested lettuce. The produce response, measured as CO_2 and H_2O gas exchange, is used as an indicator of quality changes. The gas exchange is related to the overall climatic effects and also to quality changes of the produce. To maintain produce quality a knowledge database will be developed as the basis for decisions on the configuration of single process phases during postharvest.

Keywords: quality, climatic impact, produce response, postharvest, lettuce

1. INTRODUCTION

Numerous factors that affect both internal and external quality in horticultural produce and their marketing capability have to be taken into consideration in order to minimize losses in quality of harvested produce (e.g. Shewfelt and Prussia, 1993; Kader, 1992). The present study will investigate the effects of climatic impacts on harvested vegetables such as lettuce. The methods conventionally used in determining produce quality, e.g. the analysis of constituents, are associated with the destruction of the material. However, the investigation of changes in quality related to cumulative climatic influences during

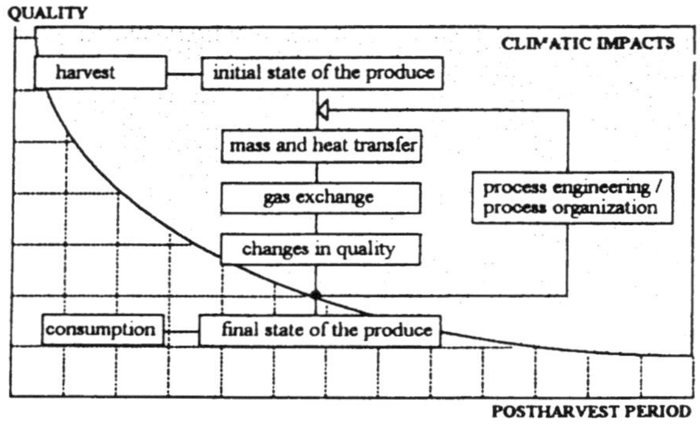

Fig. 1. Quality changes in postharvest period due to climatic impacts

several phases of the postharvest period requires continuous information on produce quality. As harvested plants respond to changes in environmental conditions by altering the CO_2 and water vapour gas exchange, gas exchange is used as an indicator of quality changes (Kader 1987; Ben-Yehoshua 1987). The relationship between climatic impacts and CO_2 and water vapour gas exchange on one hand and between CO_2 and water vapour gas exchange and changes in quality on the other hand are used to derive measures for produce handling during climatic controlled and uncontrolled phases of the postharvest period (Figure 1). To maintain quality of perishable horticultural produce in postharvest a knowledge database will be developed as the basis for decisions on the configuration of single process phases during the postharvest period (Linke, et al., 1994). This knowledge database might be used to develop special air-conditioning controls for short-term storage in cold storage rooms with the objective of carefully handling produces.

2. MATERIAL AND METHODS

2.1 Experimental equipment

The experimental equipment consists of a cuvette system with six measuring and one reference cuvette for the simultaneous investigation of six lettuces, integrated CO_2 and water vapour gas analysers combined with a multiplexer system for switching gas channels, and a computer-aided datalogging system with test-related sensors as described by Linke and Schreiner (1995). Individual lettuces in the measuring cuvettes are exposed to a defined climate (air temperature, relative humidity, air velocity, flow conditions, atmospheric pressure, atmospheric composition). Climatic parameters and plant responses are determined. Dark respiration, transpiration, and in some cases net photosynthesis are measured using differential infrared gas analysers for CO_2 and water vapour in an open system.

Internal quality parameters, such as cell wall carbohydrates and storage carbohydrates are determined spectrophotometrically and using high pressure liquid chromatography, respectively.

2.2 Produce material

Field grown lettuces of the day-neutral cultivar 'Libusa' are used in the investigation. The lettuce is harvested in various stages of development, i.e. from the rosette stage to beginning of shoot elongation, under controlled environmental conditions (air temperature, relative humidity, photosynthetic active irradiation, wind speed).

2.3 Methodical approach for determination of quality changes

The CO_2 and water vapour gas exchange are used as indicators for quality changes of the produce as a result of climatic impacts. Climatic impacts are expressed as mass and heat transfer mechanisms between the produce and the ambient air revealing different intensities during postharvest. Mass and heat transfer are due to varying potential differences (differences in partial pressure of water vapour, differences in temperature). As a consequence, both water loss and changes of produce temperatures occur. On one hand, the produce responds to water loss inducing internal water redistribution due to differences in water potential (Kays, 1991). Moreover, the produce responds with reversible stomatal closure, when a produce specific turgor threshold is passed (Raschke, 1975). On the other hand, changes in the internal produce temperature effect enzymatic processes (e.g., respiration, cell wall decomposition). All these produce responses cause changes in constituents attributed to internal and external quality losses.

Water loss leads to wilting symptoms as an important quality change in lettuce during postharvest. Lettuce shows a locally differentiated wilting behaviour beginning at the outside leaves. To determine locally varying wilting the following equation for the transpiration rate (E) is used:

$$E = \frac{x_{leaf} - x_{air}}{r_b + r_s} \qquad (1)$$

where

x_{leaf} is the saturated air within the leaf

x_{air} is the water content of the ambient air unaffected by the produce

r_b is the boundary layer resistance

r_s is the stomatal resistance.

Equation (1) is valid for leaves or leaf sections, which are unaffected by other leaves. Therefore, the application of this equation to an entire harvested lettuce requires a more specific consideration of single parameters participating in mass transfer. Due to the varying degree of overlapping leaves all single

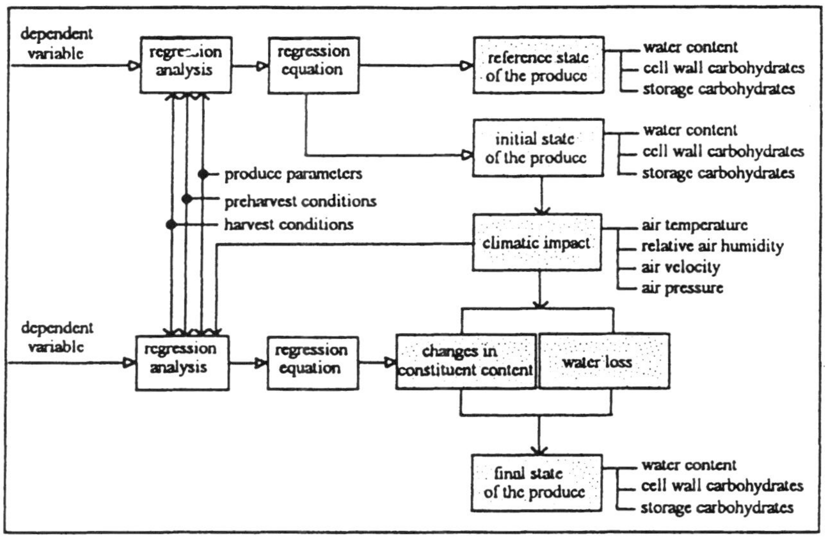

Fig. 2. Methodical approach for determination of external and internal quality changes.

parameters, effecting the transpiration rate, are locally differentiated. Therefore, the leaves are divided in several leaf sections. The transpiration rate per leaf section is calculated by the water content distribution before and after the climatic impact as described below. Due to the environmental conditions the leaves are often covered by the open water surface. In order to measure the transpiration rate, which is unaffected by the open water surface, it has to be corrected. This correction of the transpiration rate is carried out comparing climatic stressed lettuce with reference states of lettuce.

It is assumed, that independent variables such as produce parameters, preharvest conditions, and harvest conditions might be used in order to estimate the reference state of a dependent variable via a multiple regression analysis (Figure 2). Dependent variables are parameters effecting the external and internal quality of the produce, such as the water content distribution, the content of cell wall carbohydrates and of storage carbohydrates. The regression analysis is based on data of lettuces harvested under defined environmental conditions. This relationship is used for climatic stressed lettuce to calculate the initial state of the dependent variables. After the climatic treatment the final state of the dependent variables is measured. Loss of water and changes of other constituents are calculated from the difference between the initial and final state of the dependent variable. They are related to the measured CO_2 and water vapour gas exchange in the experiments. The methods conventionally used in determining produce quality, e.g. the analysis of constituents, are associated with the destruction of the material. However, the investigation of changes of quality related to cumulative climatic influences during several phases of postharvest needs continuous information on produce quality. This requires knowledge of the correlation between CO_2 and water vapour gas exchange and quality parameters. Consequently, transpiration, dark respiration, and in some cases net photosynthesis are used in the present study as indicators of external and internal changes of quality.

Cell wall carbohydrates (pectic substances, cellulose) are chosen as quality parameters indicating textural properties, such as crispness and freshness of the produce. In order to determine the internal quality storage carbohydrates (glucose, fructose, sucrose) are measured. Both, cell wall carbohydrates and storage carbohydrates are analysed using separately samples of the outside and the inside leaves.

3. RESULTS AND DISCUSSION

3.1 Correlation between climate conditions and gas exchange

Emphasis of the investigation is placed on the measurement of produce responses in respect to the overall climatic effects. The harvested lettuce responds to climatic effects by both the internal redistribution of water and the change of stomatal aperture. The varying degree of overlapping lettuce leaves results in locally and temporally differentiated produce responses. These responses are covered by phenotypic effects, causing locally differentiated partial differences in water vapour pressure. Consequently, measurements related to the entire lettuce in order to determine produce responses are not completely suitable. Measurements related to leaf sections are possible, but they are only valid for special measured locations. According to equation (1) these statements will be explained. Figure 3 shows the relation between transpiration rate, the resistances in the water vapour pathway, and the difference in water content between produce and ambient air. The range of variation of the stomatal resistance of harvested lettuce is between 15 and 25 s cm^{-1} at an irradiation below the saturation level, measured at the top of the outside leaves (Linke and Schreiner, 1995). The boundary layer resistance essentially depends on the flow conditions. According to Monteith (1975) and Larcher (1994) a maximum of approximately 1 s cm^{-1} is reported. The boundary layer resistance is assumed to be constant at a level of 1 s cm^{-1} , because the boundary layer is relative low in comparison with the stomatal resistance. The border lines describe the lowest and highest difference in

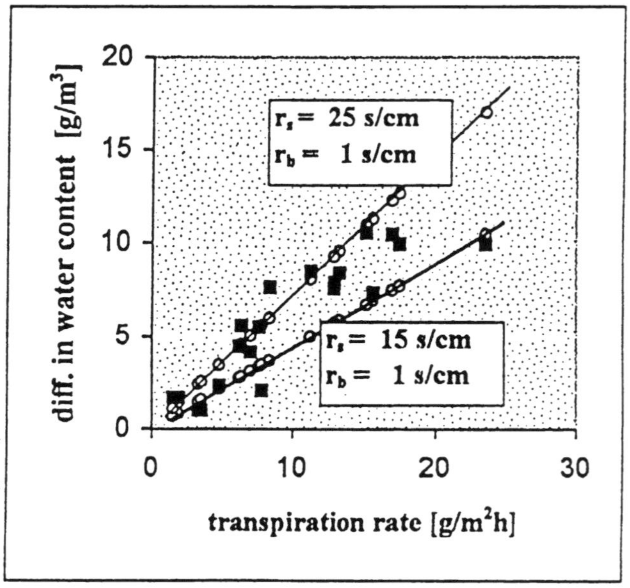

Fig. 3. Correlation between transpiration rate and difference in water content in dependence of total resistance

water content assuming total resistances of 16 s cm^{-1} and 26 s cm^{-1}, respectively. The measured values of the differences in water content between produce and ambient air taken from realized investigations vary greatly due to stomatal movements. The stomatal response via the stomatal H_2O feedback system is caused either by water loss of the guard cells or by low relative humidity of the ambient air. Differences in water content outside the border lines indicate methodical problems, e.g. determination of the open water surface. In further investigations the stomatal movements of the leaf sections have to be calculated by determining correlation, including stomatal closure and water potential differences between the ambient air and the leaf.

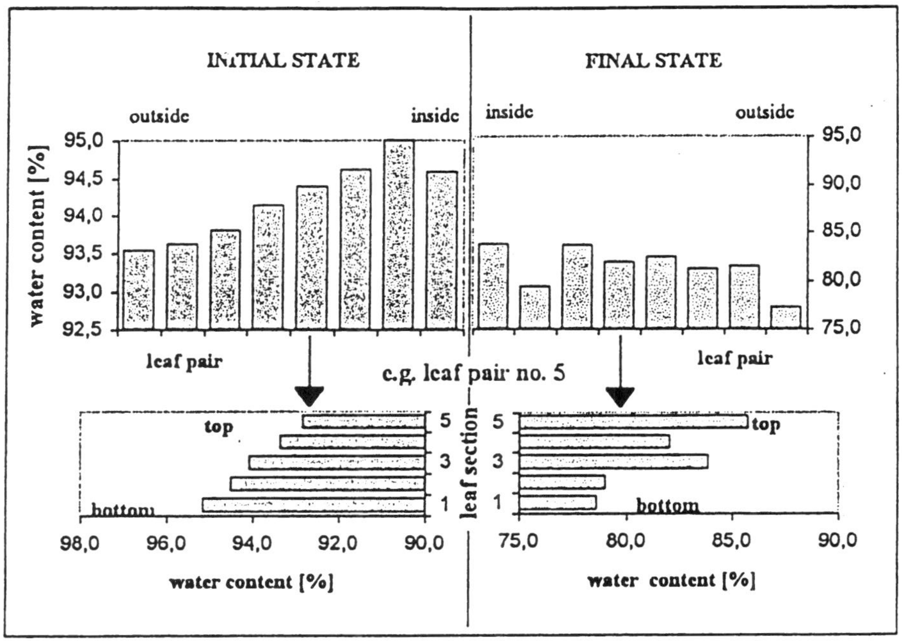

Fig. 4. Differences in water content of leaves and leaf sections (5. leaf section) between initial state and final state of an individual lettuce

The second produce response associated with transpiration is the internal water redistribution mechanism after cutting. Exemplary figure 4 shows the distribution of water content of an entire lettuce at the initial and final state. It is evident that water redistributions occur in two directions: from the inside covered leaves to the outside leaves and from covered basis towards the top of the outside leaf. In order to determine the transpiration rate of single leaf sections via the initial and final state of water content, it is necessary to consider the water redistribution in the entire lettuce.

The application of a multiple regression analysis is used as another possibility to find correlation between the overall climatic parameters and the CO_2 and water vapour gas exchange of the entire lettuce. This analysis reveals that the transpiration rate is

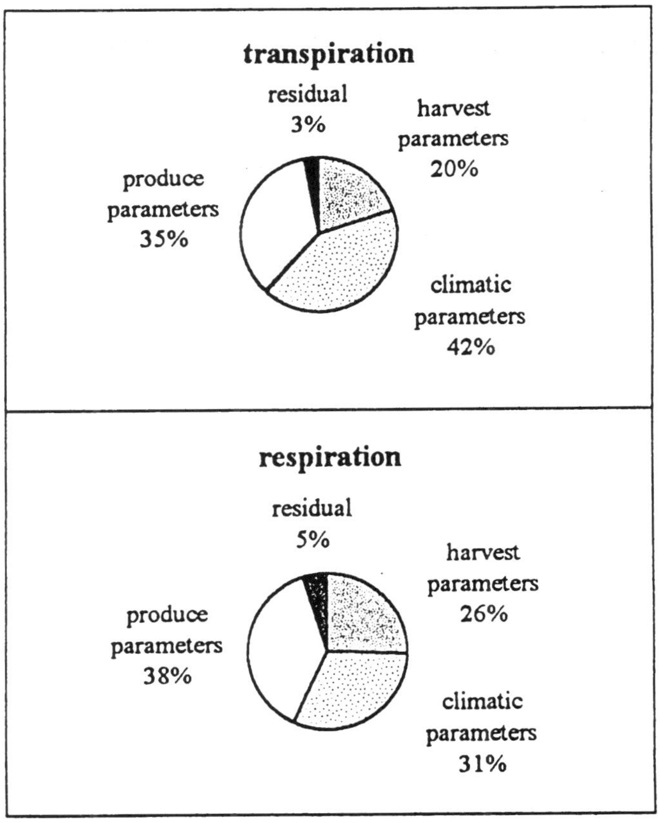

Fig. 5. Parameters effecting transpiration and respiration

greater influenced by climatic parameters and produce parameters, still harvest conditions have also to be taken into consideration (Figure 5). This is attributed to the different physiological states of the lettuce, depending on harvest conditions. During the diurnal course, the water content of lettuce, harvested at different hours decreases leading to varying transpiration rates in postharvest.

Compared to the transpiration rate, climatic parameters and produce parameters mainly effect the respiration rate. Harvest parameters additionally influence the intensity of the respiration (Figure 5). The respiration depends on the water status of the produce at harvest (Kays, 1991). Therefore, the harvest parameters have a comparably high effect on respiration. Moreover, the respiration as an enzymatic process is strongly influenced by temperature. Therefore, the respiration is mainly affected by heat transfer mechanisms. With increasing air temperature and produce temperature the respiration increases till the temperature optimum of the respiration is reached. In order to determine changes in internal quality parameters via gas exchange, respiration and produce temperature have to be considered.

3.2 Correlation between gas exchange and quality characteristics

Gas exchange, i.e. the rise in respiration and transpiration was found to be closely correlated with changes in the texture of the cell wall during produce development and ripening (Kader, 1987; Dick and Labavitch, 1989; Sonego, et al., 1995). Depending on the physiological stage of a produce break down and transformation of insoluble pectic substances to water-soluble pectin occur, leading to quality loss. Furthermore, the alteration of the cell wall structure due to the catabolism of pectic substances is strongly associated with changes in organic cell wall components, such as carbohydrate content and composition.

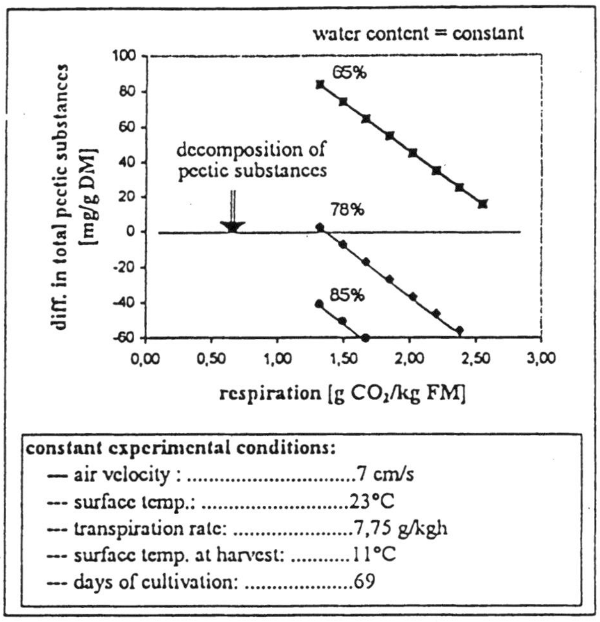

Fig. 6. Relation between respiration and differences in total pectic substances (example)

The strong relationship between the change of total pectic substances during treatment, respiration, and water content related to the initial state of lettuce is presented in figure 6. According to figure 2 the data presented are estimated by a regression analysis, among other parameters respiration rate and water content are used as independent variables. Therefore, the relation presented is only valid for the constant values of the independent variables described in figure 6. It is shown that respiration, expressed as total CO_2 efflux and content of total pectic substances depend on the water status of the produce, i.e. a rise of respiration coincides with a decrease of pectin at a low water content. Drought stress might have led to an enhanced metabolism, thus to an enhanced respiration, where monosaccharides attached to pectin molecules might serve as additional respiration substrate, indicating the break down of insoluble pectic substances. At a high water status the pectin content almost remains constant. The negative values of differences in total pectic sub-

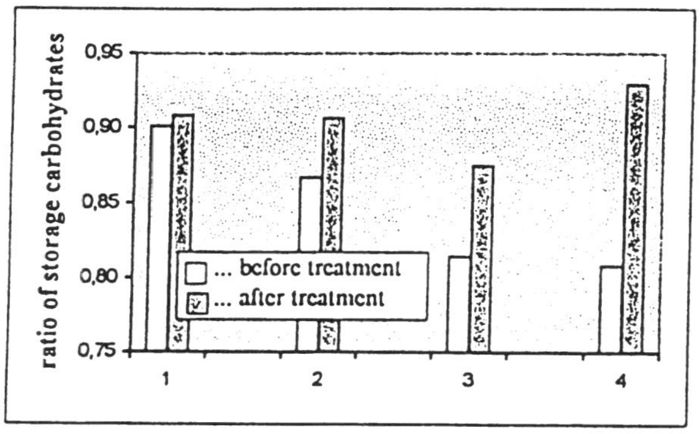

Fig. 7. Ratio of storage carbohydrates of the inside leaves to the entire lettuce

stances per gram dry matter at higher water status indicating the preferred respiration of storage carbohydrates. In further investigations the composition of the total pectic substances needs to be analysed regarding the different fractions, e.g. soluble and insoluble pectic substances in order to gain more detailed information on quality changes.

At harvest the content of storage carbohydrates (glucose, fructose, sucrose) is 10 fold higher in the inside leaves than in the outside leaves. According to the results obtained on water distribution, changes in the storage carbohydrate ratio indicate transport of storage carbohydrates, i.e. changes in source-sink relation. Figure 7 represents the ratio of storage carbohydrates, defined as the ratio of the carbohydrate content of the inside leaves to the carbohydrate content of the entire lettuce of four investigation series. The high ratio of storage carbohydrate after the climatic treatment emphasis the reversion of the source-sink relation. Further investigations are necessary to determine the effect of these transport mechanisms on the postharvest behaviour of lettuce.

4. CONCLUSION

The results presented show the possibility to interpret changes of quality parameters via the climatic conditions in postharvest. The CO_2 and water vapour gas exchange are suitable as indicators of quality changes regarding to climatic impacts. Simultaneously to gas exchange the produce temperature effecting enzymatic processes appears to be important. The objective of further investigations is to confirm the tendencies of the correlation between climatic impacts and quality changes.

5. REFERENCES

Ben-Yehoshua, S. (1987). Transpiration, water stress, and gas exchange. In: *Postharvest physiology of vegetables.* (Weichmann, J. , (Ed.)), 113-170. Marcel Decker, Inc., New York and Basel.

Dick, A. J. and J. M. Labavitch (1989). Cell wall metabolism in ripening fruits. *Plant physiol.* **89**, 1394-1400.

Kader, A. A. (1987). Respiration and gas exchange of vegetables. In: *Postharvest physiology of vegetables.* (Weichmann, J. , (Ed.)), 25-45. Marcel Decker, Inc., New York and Basel.

Kader, A. A. (1992). Postharvest technology of horticultural crops. *University of California, division of agriculture and natural resources, 2. edition, publication 3311.*

Kays, S. J. (1991). *Postharvest physiology and handling of perishable plant products.* Van Nostrand-Reinhold, New York.

Larcher, W. (1994). *Ökophysiologie der Pflanzen.* 5. Auflage, Eugen Ulmer, Stuttgart.

Linke, M. and M. Schreiner (1995). Keeping quality of lettuce (*Lactuca sativa* L. var. *capitata* L.) in postharvest - measurement and analysis of climatic factors and transpiration. *Gartenbauwiss.* (pressed).

Linke, M., M. Schreiner and M. Geyer (1994). Modelling climatic impacts on quality losses on perishable horticultural produce in postharvest period. *Abstracts of 2. IFAC/ISHS International Workshop „Mathematical and control applications in agriculture and horticulture".* Silsoe, UK, 6-3.

Monteith, J. L. (1975). *Principles of environmental physics.* Edward Arnold (Publishers) Limited, London.

Raschke, K. (1975). Stomatal actions. *Ann. Rev. Plant Physiol.* **26**, 309-340.

Shewfelt, R. L. and S. E. Prussia (1993). *Postharvest handling - a systems approach.* Academic Press, San Diego, California.

Sonego, L., R. Ben-Arie, J. Raynal and J. C. Pech (1995). Biochemical and physical evaluation of textural characteristics of nectarines exhibiting woolly break down: NMR imaging, X-ray computed tomography and pectin composition. *Postharvest Biol. Technol.* 5 (3), 187-198.

DEVELOPMENT OF A LOW COST VENTILATION RATE SENSOR FOR MICRO-ENVIRONMENTAL CONTROL IN STORAGE PLACES

D. Berckmans[1,2] , E. Vranken[1], V. Goedseels[1]

[1]Laboratory for Agricultural Buildings Research,
[2]National Fund for Scientific Research Belgium
Katholieke Universiteit Leuven, Kardinaal Mercierlaan 92, B-3001 Heverlee, Belgium

Abstract: A new turbine meter is developed to be used as a ventilation rate sensor in agricultural buildings and storage places. Starting from a previous sensor, which was introduced in 1983, several improvements were done to obtain a low cost air flow rate sensor with improved accuracy. The sensor is developed within a measuring range from 200 to 5000 m^3/h and this for pressure differences from 0 to 120 Pa, but the principle can be extended to higher ventilation rates and pressure differences. The accuracy of the current sensor is +/- 60 m3/h within the mentioned measuring range. This sensor can be integrated in the climate control equipment of a storage place to improve process control.

1. PROBLEM

1.1. The importance of the ventilation rate and the airflow pattern in storage places for agricultural products.

In literature, for some time already the ventilation rate through the building is considered to be one of the most important parameters in inside climate control in ventilated buildings and storage places. The identification of ventilation problems and early investigations of these provided a milestone in environmental control of agricultural structures (King, 1908; Bond, 1976). The influence of air flow rate on the values of physical micro-environmental parameters such as inside temperature, humidity and gas concentration has been described (Guss and Grout, 1973; Bruce, 1975; Clark and Cena, 1981). Carpenter (1974) explains the importance of the ventilation rate by its relationship with air temperature and with the air velocity within the structure. He explains the complexity of the role of the ventilation system by the interactions between the inside climate and on the other hand variables such as structure of the building, risk for disease, etc. The objective of the ventilation system is described as control of gas concentration, of air humidity and of inside temperature. Consequently the three quantitative criteria are known to determine the ventilation rate through the building and to provide the optimum conditions for the stored products by controlling the physical micro-environment. To achieve this, the two basic functions of a ventilation system are to provide effective control over the ventilation rate and efficient control over the airflow pattern within the ventilated structure. These should be attained simultaneously (Randall, 1981).
The air flow pattern in a ventilated structure forms the physical link between the control inputs (ventilation rate, cooling and heat supply) and the physical micro-environment around the biological material (ASHRAE, 1974). An adequate control of the ventilation rate through the building indeed has been proven to be decisive in establishing a suitable air flow pattern as well (Randall, 1975; Randall and Battams, 1979; Barber et al., 1982).
The influence of the ventilation rate and its control on energy use in modern storage places for agricultural products also has been described in literature (ASHRAE Handbook, 1977).
From literature it can be concluded that the ventilation rate trough a storage place is one of the

most important variables to be controlled.

1.2. *Control of the ventilation rate in agricultural structures*

Based on the effect of inside climate control on product quality in modern storage places and on the knowledge about the importance of the ventilation rate, research has been done on the efficiency of normal proportional air flow rate control. By statistical analysis of field data, analysis of control equipment in laboratory test installations and a theoretical approach it has been analysed why normal mechanical ventilation systems (proportional control) do not perform adequately in the field (Berckmans and Goedseels, 1986; Berckmans et al., 1988). Furthermore laboratory experiments show that the combination of an axial fan with the proportional ventilation rate controller cannot give effective control of ventilation in practice (Berckmans and Goedseels, 1986).
Consequently a new controller has been developed, in an attempt to achieve improved control of ventilation rate and hence more effective environmental control in agricultural buildings. The principle of this improved air flow rate controller, introduced in 1983, is that the signal of a low cost air flow rate sensor (price of about 130 U.S. dollar) is used to realise a feedback control (Berckmans and Goedseels, 1983; Berckmans and Goedseels, 1986; Berckmans et al., 1986).

2. OBJECTIVES

It is clear that the efficiency of this higher mentioned principle, as it is used today by different European producers of control equipment, depends on the accuracy of the air flow rate sensor that is integrated in the control system. The previous sensor which was developed for the introduction of this controller in 1983 had an accuracy limited to about 600 m^3/h over a measuring range of 500 to 5000 m^3/h.
The objective of this paper is to present a new low cost air flow rate sensor with improved accuracy to be used in modern agricultural buildings and storage places for fruits and vegetables. This means more specifically that the resulting air flow rate sensor :

- must be usable in the hard environment of a storage place with high temperature differences (from -10 °C to 55 °C and with a lot of moisture (60 to 95 % RH) and dust.
- must operate for a period of 10 years without any major maintenance,
- must permit measuring the ventilation rate in a chimney with a diameter of 0.5 m to 1.5 m and a tube length of at least 1 m and with turbulent air flow in it,
- has a small time constant (Max. of 25 sec.) to be usable as a control sensor,
- has a little weight (less than 2 kg) for a mounting in the chimney,
- has a measuring range from 200 to 20000 m^3/h with an acceptable accuracy (better than 100 m^3/h and low pressure influence (better than 100 m3/h from linearity)
- creates a minor pressure drop for the fan (better than 500 m3/h loss of maximum flow rate).
- must have a low price (less than 280 $) since it is the objective to integrate this sensor in the control equipment.

3. METHOD

3.1. *Position of the sensor*

In most of the modern storage places in Europe overpressure ventilation systems are used (e.g. grain drying). In spite of the fact that the building envelope has been improved during the last years and is carefully tight up against air infiltration, it still is difficult to realise a leakage-free construction in the field. To be sure of measuring the global ventilation rate through the building the air flow rate sensor is placed in the inlet section in case of an overpressure system (Figure 1), and in the exhaust chimney in case of an underpressure system.

3.2. *Principle of the sensor*

From the measurements of the characteristics of axial fans, it is known that a back flow effect can appear if the static pressure drop over the fan is high.
Because of the high pressure difference to overcome, the fan still keeps his rotational speed but takes the air back through its own impeller. The resulting total ventilation rate is decreasing while the fan still is using a high voltage and power input (Figure 2). In these conditions the air flow in the section at fan level is subject to important changes of air flow direction. Because of these effects on the turbulent flow in the ventilation duct it is concluded that the sensor should cover the whole duct section. Consequently a sensor that measures in one point of the section like a point in a pitot tube traverse is useless. Taking into account the required working environment of the sensor (see Objectives) and the condition of a low price (to become a sensor for control purposes) the choice was made for the principle of a free running turbine covering the whole ventilation duct section. In literature it is shown that a pressure correcting measurement can be realised with turbine meters of smaller diameters (Lee et al., 1981, 1982).

3.3. Test rig

In the higher mentioned position a flow rate sensor of the considered principle will be subject to changes in air flow rate and variations in pressure difference over the test chimney as well. Because of the principle of a free running turbine the measurements are focused on the relationship between rotational speed of the turbine, the measured air flow rate and the pressure difference over the test section. Consequently the procedures for fan testing with variation of pressure difference should be applied for testing this air flow rate sensor.
To measure the characteristics of the air flow rate sensor a test chimney is used with a diameter of 0.5 m and a length of 1 m (Figure 3). In this test chimney it was possible during the experiments to realise an air flow rate from 0 to 5000 m^3/h and this for pressure differences (underpressures in the test room) from 0 to 120 Pa. The test installation has been built according to the German Standards DIN 1952, the Belgian Standard NBN 688 and the British Standard BS 848. The principle of this test rig is that the air flow rate can be controlled using the fan of the test rig (nr. 30 in figure 3) while an orifice is used (with compensation for air temperature and air humidity) to measure the flow rate (nr. 27 in figure 3). As long as the second fan (nr. 23 in figure 3) is producing the same air flow rate as fan nr. 30, there is no pressure difference between the test room and the environment. In these conditions fan nr. 30 is taking care for the pressure losses through the system. From the moment where fan nr. 23 is producing a higher air flow rate than the one that is supplied by fan nr. 30, the underpressure in the test room is increasing. This makes it possible to change the pressure difference over the test section from 0 to 120 Pa and this for every air flow rate from 200 to 5000 m^3/h.
Since it takes a lot of time to realise constant steady state conditions in the installation and to do reliable measurements (because fluctuations of pressure etc.), the global test installation has been computerised (Figure 3) with the necessary statistical procedures in it. The orifice allows to measure the flow rate with an accuracy of 10 m^3/h in the range of 200 to 1500 m^3/h and an accuracy of 30 m^3/h in the range of 1500 to 5000 m^3/h. The rotational speed of the turbine is measured by using a fixed proximity switch with 4 indication points on the turbine. The system measures the rotational speed with an accuracy of +- 0.5 rev/min. The repeatability, to produce a certain flow rate, as calculated from 4 experiments was better than 3 %.

4. RESULTS

To realise the objectives, a number of improvements were done on the previous type of turbine. Each of these is described.

4.1. Reduction of the number of blades of the turbine meter.

At the time of the introduction of the previous type of turbine meter as a ventilation rate sensor for agricultural buildings and storage places an identical impeller as the one from the fan was used for practical reasons. This still is done by most of the European producers of this type of equipment and consequently most of the existing turbine meters for livestock buildings still have 6 or 8 blades on their impeller (Stylianos, 1989).
Since it can be assumed that a high number of blades does increase the resistance to the fluid flow, experiments were done to measure the air flow rate at 4 different levels of voltage to the fan and this for turbines with 8, 4 and 2 blades.

Table 1: Air flow rates (m^3/h) as measured by a 2-, 4- and 8-blade turbine with a same fan at different voltages to the fan and for different pressure differences (ΔP) to overcome.

ΔP (Pa)	0	19.6	39.2	58.9	78.5	98.1	117.7
(Volt)	turbine with 2-blade impeller						
100	1600 m^3/h						
140	3800	2700	1500				
180	5500	5200	4800	4200	3600	2000	
220	5500	5500	5500	4500	4000	3000	1500
(Volt)	turbine with 4-blade impeller						
100	1900						
140	3500	2500	2100				
180	5100	5000	4500	4000	3500	2100	2000
220	5500	5000	5000	4500	3900	2000	2000
(Volt)	turbine with 8-blade impeller						
100	1600						
140	3100	2300	1700				
180	4600	4500	4100	3600	3200	2000	1600
200	5000	4500	4500	4000	3500	2000	2000

From Table 1 it can be seen that the resulting air flow rates are higher (up to 900 m^3/h) when turbines with 2 blades are used instead of 8 blades. The pressure losses are minimised by using 2 blades. This is in agreement with the equation for the so called specific number of revolutions (Wislicenus G., 1965):

$$n_s = n \cdot Q / (\Delta P)^{3/4} \qquad (1)$$

The equation shows that for a constant value of n_s, the specific number of revolutions, all other variables (n, Q and ΔP) are determined and they will generate identical flow conditions in geometric identical machines. An increase of the specific number of revolutions only can be realised by an increase of the flow section or a reduction of the rotor-diameter.
Minimum pressure drop is achieved by maximising n_s and high values of the specific number are characteristic for axial designs (largest flow section for a given diameter). Moreover, the number of blades reduces as n_s increases. It can be concluded that, as opposed to current practices, it is desirable to use a turbine with only 2 blades on its impeller.

4.2. Position of the turbine meter.

It has already been explained that the ventilation rate sensor should be positioned in the neighbourhood of the fan (see Method 3.1).
For maintenance reasons the first sensor, as introduced in 1983, was positioned after the fan in an underpressure system. Since it can be expected that the fluid will have a more uniform flow pattern before passing through the fan, it is tested to determine whether the turbine should be positioned upstream or downstream with regard to the fan in the fluid flow. For both these positions of the turbine the air flow rate and the corresponding rotational speed of the turbine for different pressure differences over the test section were measured. From Figure 7 it can be concluded that the upstream position of the turbine with regard to the position of the fan does have an important positive influence on the characteristics of the turbine meter. Notwithstanding the great pressure influence and symptoms of back flow below an air flow rate of 2000 m^3/h, there is a good linearity for an air flow rate higher than 2000 m^3/h (Figure 5).

4.3. Design of the turbine impeller.

The impeller of the turbine has different components : the **boss**, the **bearings** and the **blade**.

- The **boss** normally should be as small as possible since, in the minimal section with diameter 0.5 m, a bigger boss decreases the usable area for fluid flow and increases the air velocity . It is known that the pressure losses in ducted flows are proportional to the square of this fluid velocity. But the minimization of the dimensions of the boss are limited by the fact that the positioning of the blades becomes impossible or to expensive on a too small boss. Moreover the use of too long blades should introduce turbulence. The boss used in this development has a diameter of 0.1 m.
- The **bearing** is a standard SKF enclosed bearing with low temperature Sodium-grease.
- The **length of the blade** is determinated by the difference between the tube diameter and the boss diameter. When using an inside tube diameter (chimney with an outer diameter of 0.50 m) of 0.45 m and with a tip clearance of 0.05 m, the length of the blade is 0.17 m.
- The **chord of the blade** is not commonly calculated in design procedures. Mostly the design is evaluated and improved by wind tunnel experiments.

Following considerations should be taken in account : the force exerted by the air on the turbine is proportional to the area of the blade. This prompts the use of a large chord. On the other hand a large blade creates more disturbances in the fluid flow. As a compromise, the chord was taken equal to 0.1 m in this development.
Calculation of the **blade angle** involves determination of the inlet angle β_1. The blade outlet angle β_2 can be determined on the basis of this inlet angle and the curvature of the blade (which is defined as β_2-β_1). Use has been made of the concept of velocity triangles (Figure 4). These vector diagrams, drawn for a random point of the stream path through the machine, make a study of the different velocity components possible :

- absolute flow velocity V (stationary frame of reference).
- relative flow velocity W (rotating frame of reference)
- peripheral velocity of the rotor blades U , being equal to the product of radius and angular velocity of the rotor.

The absolute velocity equals the sum of relative and peripheral velocity vectors.
The calculation of the **blade angle** has to be preceded by the determination of a 'design point', which is to be chosen with the specific application of the turbine meter in mind. In livestock buildings and storage places, the lower values of the ventilation rate are the most critical to control. A high rotational speed is necessary to minimise instabilities at these low air flow rates.

A blade geometry was calculated for a 500 m3/h-250 rev/min design point.

The **curvature of the blade** was expected to be small, since in this application, the turbine has to surmount merely friction losses (as it is not used for the transformation of kinetically to mechanical energy). This can be verified by calculation (Vranken et al., 1992) : the maximum curvature amounted to no more than 1°, and was in the actual design assumed to be zero.

4.4. Linearity of the relationship between flow rate and rotational speed of the turbine at different pressure differences.

To determine the steady state characteristics of the turbine meter its rotational speed was measured at different flow rates and at various pressure differences. Flow rates varied between 200 and 5000 m^3/h. For low flow rates up to 1600 m^3/h, the measuring interval was held on 200 m^3/h. For larger rates, the interval was expanded to 500 m^3/h. Pressure differences varied from 0 to 120 Pa, with a step of 20 Pa. The axial fan in use during these tests was a Woods-2100 fan. The complete range of flow rate from 200 to 5000 m^3/h was measured three times in identical circumstances. A linear regression analysis was carried out on the data. The R^2 was calculated as a measure for the goodness of fit. A good fit implies independence of pressure variations to a large extent. Calculation of the mean (MV) and maximum (MaxV) deviation from the regression-based predicted values establishes a criterion of accuracy.

Table 2: Variation of blade inlet angle β_1 with radius for the prototype; design/point : 500 m^3/h - 250 rev/min.

Rad (m)	Va (m/s)	V (m/s)	β_1 (°)
0,01	0,87	0,202	17°
0,02		0,524	
0,03		0,785	
0,04		1,047	
0,05		1,309	
0,06		1,571	
0,07		1,833	
0,08		2,094	
0,09		2,358	
0,010		2,620	
0,011		2,880	
0,012		3,142	
0,013		3,403	
0,014		3,665	
0,015		3,927	
0,016		4,184	
0,017		4,451	78°

Figure 6 shows the corresponding linear regression covering the complete flow rate interval from 0 to 5000 m^3/h. In figure 7 the result of a comparable analysis is shown for a turbine sensor of the previous type (diameter 0.45 m, pressure differences 0-120 Pa). Table 2 shows the comparison of the regression analyses on these results.

It is clear that there is an improvement in the linearity of the steady state characteristics of the turbine meter used as an air flow rate sensor. The corresponding improvement in accuracy is important since the mean deviation has been reduced from 488 m^3/h to 91 m^3/h which value is 1.8% of the full scale (5000 m^3/h). Since the maximum deviation is occurring for the very low air flow rates (see Figures 6 and 7), the reduction of this value might be of importance for automation of the set points of minimum ventilation rate.

4.5. Pressure drop caused by the freely rotating impeller of the turbine flow meter.

The air flow rate produced by an axial fan at a certain voltage supply is very sensitive to the static pressure to overcome by the fan as seen from the steady state characteristics of the fan. Consequently it is important to minimise the pressure drop caused by the air flow rate sensor positioned in front of the fan. Therefore the influence of a turbine meter, used as an air flow rate sensor, on the fan characteristics was measured. More specifically, since feedback control of the air flow rate will be used, it is important to know what maximum air flow rate can be produced by the fan.

Table 3: Results of the regression analyses of rotational speed of (the previous and the new) turbine meters as a function of the air flow rate (from 0 to 5000 m^3/h) and for pressure differences from 0 to 120 Pa.

	Previous Sensor	Prototype
R^2	0.672	0.992
Mean (m^3/h)	488	91
Max (m^3/h)	1267	346

As already indicated, the test rig can be used to determine fan characteristics. Measurement, at various flow rates, is required of the power needed to establish a certain pressure difference. The same pressure and flow rate intervals as in the foregoing were covered to realise a determination of the saturation values (the flow rates at maximum voltage supply of 220 V to the fan). The results are given in Table 3.

Following observation can be made : the introduction of the turbine meter in the measurement system has but a small effect on the realised maximum flow-rates. Maximum reduction of saturation values amounts to no more than 150 m^3/h.

4.6. Influence of the type of axial fan on the characteristics of the freely rotating turbine.

Table 4: Effect of the turbine meter to the maximum air flow rates produced by a Woods 2100 fan at a voltage of 220V.

ΔP (Pa)	with turbine meter (m^3/h)	fan without meter (m^3/h)
0	5550	5550
19.6	5300	5300
39.2	4950	5050
58.9	4550	4650
78.5	4050	4200
98.1	2450	2600
117.7	1950	2000

Since the efficiency of a turbine meter is influenced by the flow pattern as the air reaches the blades, it can be expected that the type of fan (different blades, etc.) does influence the characteristics of the turbine meter. Therefore the turbine meter was subjected to an identical series of tests, the only modification being the type of axial fan used. Instead of the Woods-2100, a Fancom 1445 fan, one of the more competitive types on the market, was chosen. These results will be compared with the corresponding measurement data under 4.5 for the Woods-fan. Figure 8 and Table 4 give the result of a linear regression analysis.

Table 5: Comparison of the R^2 and the mean deviation of the turbine meter characteristics if combined with two different types of fans for a flow rate range of 200 to 5000 m^3/h and pressure differences from 0 to 120 Pa.

	Woods-2100	FANCOM-1445
R^2	0.992	0.998
Mean	91 m3/h	58 m3/h
Reduction of max air flow	150 m3/h	150 m3/h

The whole flow rate range has been taken in consideration, pressure differences varying between 0 and 120 Pa. Compared to the Woods-2100, the mean deviation has been further reduced from 91 to 58 m^3/h which is 1.1% (instead of 1.8 % see 4.5) of the full scale (5000 m^3/h). In addition, it can be noticed from Table 5 that the reduction of saturation values for the Fancom-fan if combined with the turbine sensor is of an equally small measure as for the Woods. Maximum reduction amounts to 150 m^3/h.

4.7. Development of a prototype sensor for practical implementation in the field

The measurements described in the previous section are carried out on a first prototype sensor, made of wood. Since the manufacturing time of such a sensor is relatively high, the concept of a hand-made sensor can not be used for large production amounts and for a low cost sensor.

As a consequence a new prototype was developed for commercialisation purposes (Vranken et al., 1992, 1993a, 1993b, 1993c). The requirements for this sensor were even stronger than the one for the first prototype, as indicated in table 6.

Table 6: Technical requirements for a prototype air flow rate sensor (diameter 45 cm) suitable for commercialisation.

Accuracy	< 60 m3/h
Linearity	R2 > 0.99
Minimal rate	< 120 m3/h
Time constant`	< 5 sec
Max. static pressure drop	< 10 pa
Max. noise production	< 80 dBa

The main problem was the reduction of the time constant of the first prototype sensor from about 15 sec to less then 5 seconds. For this purpose the amount of material used for the blades of the propeller and the specific weight of the material was reduced by using plastics with a maximal thickness of 5 mm, and by reducing the blade chord from 100 to 75 mm.

A second adaptation was done at the level of the bearings, which resulted in a reduction of minimum range from about 200 to about 50 m3/h.

Finally a correction of the blade angle was necessary to reduce the maximal rotational speed and the noise production.

The technical results of the final sensor are given in table 7.

Table 7: Technical requirements for a prototype air flow rate sensor (diameter 45 cm) suitable for field use.

Accuracy	+/- 65 m3/h
Linearity	$R^2 = 0.998$
Minimal rate	18 m3/h
Time constant`	4 sec
Max. static pressure drop	10 pa
Max. noise production	81 dBa

In can be concluded that most of the requirements are fulfilled with this new prototype sensor.

5. CONCLUSIONS

The relationship between the rotational speed and the flow rate for the new developed turbine meter is highly linear ($R^2 = 0.995$). This implies a flow rate measurement of satisfactory accuracy, nearly independent of prevailing pressure differences (Mean deviation from regression-predicted values = 91 m3/h or 65 m3/h, depending on the type of axial fan in use). The comparable accuracy of the previous turbine sensors was of about 600 m^3/h.
It must be noted that this turbine meter can be used as a ventilation rate sensor in a chimney with a diameter of 0.5 m and a minimal length of 1 m and that this sensor then permits measuring the ventilation rate with an accuracy of 60 m^3/h in a range from 200 to 5000 m^3/h and this for pressure differences from 0 to 120 Pa. The sensor principle can be extended to higher diameters, ventilation rates and pressure differences.
The price of each prototype was around 220 U.S. dollar and should decrease for higher volume production.

The free running impeller of the flow rate sensor causes only limited pressure losses. Saturation values of the fan suffer a reduction of maximum 150 m^3/h.

The behaviour of the flow rate sensor is dependent of the type of axial fan in use. This is demonstrated by comparing the linearity of the relationship of rotational speed versus flow rate and this for the flow rate sensor in combination with two different axial fans (Woods-2100 and Fancom 1445, diameter 0.45 m). The flow rate sensor will only operate satisfactory when combined with an appropriate fan.

A second prototype sensor for practical field use has been developed with a an adapted blade geometry in order to reduce the time constant, the minimal range, the noise production and the production price. The accuracy of this sensor is even better (about 65 m3/h) than the first prototype. This sensor has been commercialised (Vranken et al.,1993)

6. NOTATION

g: acceleration due to gravity = 9.81 m/s^2
n: number of revolutions (1/s)
n_s: specific number of revolutions (see (1))
p: pressure (Pa)
ΔP: pressure difference (Pa)
r: root mean square of inner and outer radius of flow passages at turbine blades (m)
Rad.: radius of the blade (m)
Q: volume flow rate (m^3/s)
V: absolute velocity of fluid flow (m/s)
W: relative velocity of fluid flow (m/s)
U: peripheral velocity of rotor blade (m/s)

Greek symbols

ß. blade angle (grad).

Subscripts

a. axial component
θ. rotational component
1: inlet section
2: outlet section
in: outside
out: inside

7. REFERENCES

Albright, L.D. (1991) Environment Control for Animals and Plants, American Society of Agricultural Engineers, St. Joseph, Michigan, 453 pp.

Environmental control principles (1974) Supplement to the ASHRAE Handbook of fundamentals (1972). American Society of Heating, Refrigerating and Air-conditioning Engineers, New York, 253 pp.

ASHRAE Handbook - Fundamentals (1977) Environmental control for animals and Plants. American Society of Heating, Refrigerating and Air-conditioning Engineers, New York, 9.1 - 9.18 pp.

Barber, E.M., Sokhansanj, S., Lampman W.P. and Ogilvie, J.R. (1982) "Stability of airflow patterns in ventilated airspaces", American Society of Agricultural Engineers paper no. 82-4551, 10 pp.

Berckmans, D. and Goedseels, V. (1983) "Principles of the Enecom-R controller for livestock buildings", Internal Report in industrial contract, Katholieke Universiteit Leuven, 30 pp.

Berckmans, D. and Goedseels, V. (1986) "Development of new control techniques for the ventilation and heating of livestock buildings", Journal of Agricultural Engineering Research, 33, 1-12.

Berckmans, D., Goedseels V., and Geers, R. (1986) "Development of a controller for livestock buildings based on an air flow rate measurement device", Proceedings of the first European symposium on air condition and refrigeration, Brussels, 5-6 Nov., 61-68.

Berckmans, D., Vranken, E., Geers R. and Goedseels, V. (1988) "Efficiency of climate control equipment in pig houses", Farm Building Progress, 93, 15-22.

Berckmans, D., Vranken, E., Goedseels, V. (1992) "Nieuwe sensor voor meting van luchtdebieten in secties met turbulente stroming". Internal research publication Fancom-Katholieke Universiteit Leuven, 83 pp.

Berckmans, D., Goedseels V. and Van der Stuyft, E. (1989) "Influence of the climate controller on inside climate and on energy use in livestock buildings", Proceedings of the eleventh international congress on agricultural engineering, Dublin, 4-8 Sept.,1305-1311.

Bond, T.E. (1976) "Milestones in environmental control of animal production structures - and future outlook", American Society of Agricultural Engineers paper No. 76-4046

Bruce, J.M. (1975) "The open ridge as a ventilator in livestock buildings", Farm Buildings, R-D studies no. 7 and 8, November 1975, 8 pp.

Bruce, J.M. (1981) Ventilation and temperature control criteria for pigs, In : Environmental aspects of housing for animal production by Clark, J.A., Butterworths, London, 197-216.

BS 848 Part 1 (1980) Britisch Standards Institution. Fans for general purposes. Part 1 : Methods of testing performance, 80 pp.

Buffington D.E. and Skinner, T.C. (1976) "Energy analyses of buildings based on seasonal vs. design day weather data", ASAE-paper no. 82-4553, 14 pp.

Carpenter , G.A. (1974) Ventilation of buildings for intensively housed livestock, In : Heat loss from animals and man by Monteith, J.L. and Mount, L.E., Butterworths, London, 389-403.

Clark, J.A. and Cena, K. (1981) Monitoring the house environment, In: Environmental aspects of housing for animal production. Clark, J.A., Butterworths, London, 309-330.

DIN 1952 (1952) Deutsche Normen. Durchfluszmessung mit genomen Düsen, Blenden und Venturidüsen, 13 pp.

Guss, S.B. and Grout, A.R. (1973) "Effects of housing on the health of dairy cows", National Dairy Conference. ASME Publication SP-01-73, Febr. 1973, 120-127.

King, F.H. (1908) "Ventilation for dwelling, rural school and stables", Madison WI Published by the author.

Lee, W.F.Z., Fisher H.W. and Montag, D.R. (1981) Development of a self-checking gas turbinemeter, In : Flow, its measurement and control in science and industry, Vol. 2, by W.W. Durgin. Instrument. Society of America, pp. 453-469.

Lee, W.F.Z., Blakeshee D.C. and White, R.V. (1982) "A self-correcting and self-checking gas turbine meter", Journal of Fluids Engineering, Vol. 104, 143-149.

Loewer, O.J., Bridges, C.T., Bucklin,R.A. (1994) "On-Farm Drying and Storage systems", ASAE-publication 9, 560 pp.

NBN 688 NI (1966) "Belgische Norm. Debietmeting van fluïda door drukvalapparaten" - Algemeen, 32 pp.

Randall, J.M. (1975) "The prediction of airflow pattern in livestock buildings", Journal of Agricultural Engineering Research, 20, 199-215.

Randall, J.M. (1981) Ventilation system design, In : Environmental aspects of housing for animal production by Clark, J.A., Butterworths, London, 351-369.

Randall, J.M. and Battams, V.A. (1979) "Stability criteria for airflow patterns in livestock buildings", Journal of Agricultural Engineering Research, 24, 361-374.

Stylianos, L. (1989) "Onderlinge vergelijking van luchtdebietmeetsystemen voor agrarische toepassingen", Eindverhandeling voorgedragen tot het behalen van de graad van landbouwkundig ingenieur, Unpublished Thesis, Katholieke Universiteit Leuven, 133 pp.

Surbrook, T.C., Esmay M.L. and Bickert, W.G. (1979) "Control effectiveness and energy comparison for animal housing ventilation and heating systems", American Society of Agricultural Engineers paper no. 70-4526, 12 pp.

Turton, R.K. (1984) "Principles of Turbomachinery" E & F.N. Spon. London, 199 pp.

Vranken, E., Berckmans, D. Goedseels, V. (1992) "Optimalisatie Fancom-meetturbine". Internal research publication Fancom-Katholieke Universiteit Leuven, 56 pp.

Vranken, E., Freson, L. (1993) "Evaluatie en optimalisatie van een eerste commercieel prototype van een nauwkeurige luchtdebietsensor". Internal research publication Fancom-Katholieke Universiteit Leuven, 24 pp (Part I), 36 pp.(Part II).

Wislicenus, G. (1965) "Fluid mechanics of turbomachinery", Vol.I, Dover Publications, New York, 416 pp.

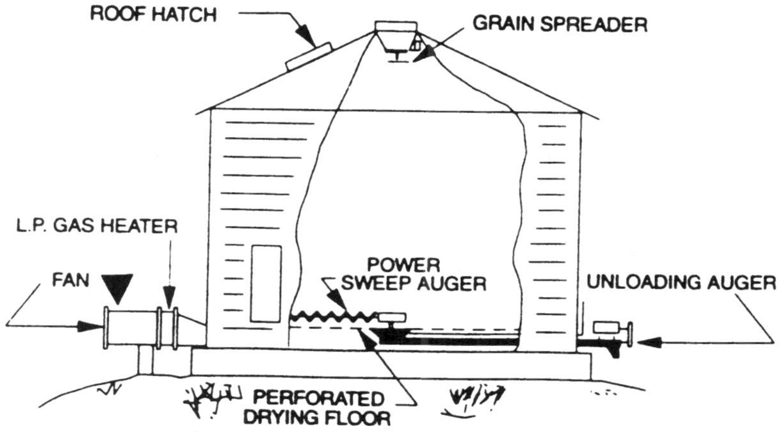

Fig. 1 Place where the flow rate sensor should be mounted because of possible leakages in the building envelope.

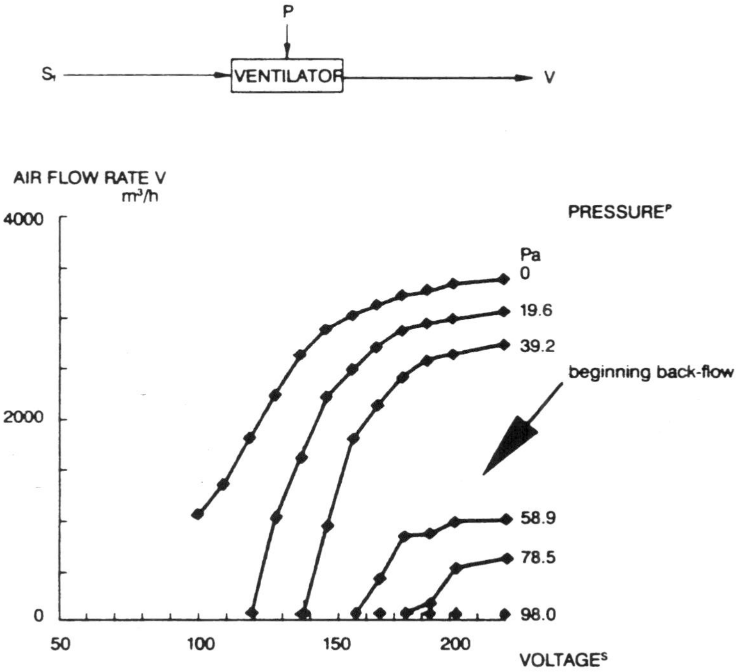

Fig. 2A Characteristics of an axial fan.

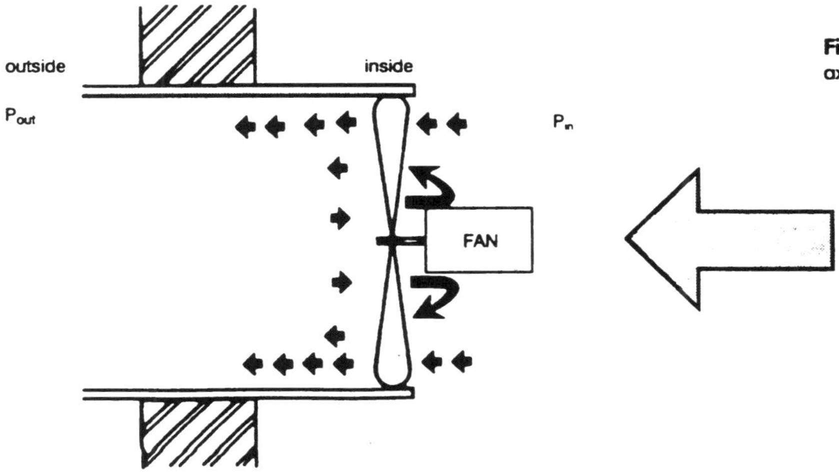

Fig. 2B Back flow effect of an axial fan. If $P_{in} << P_{out}$.

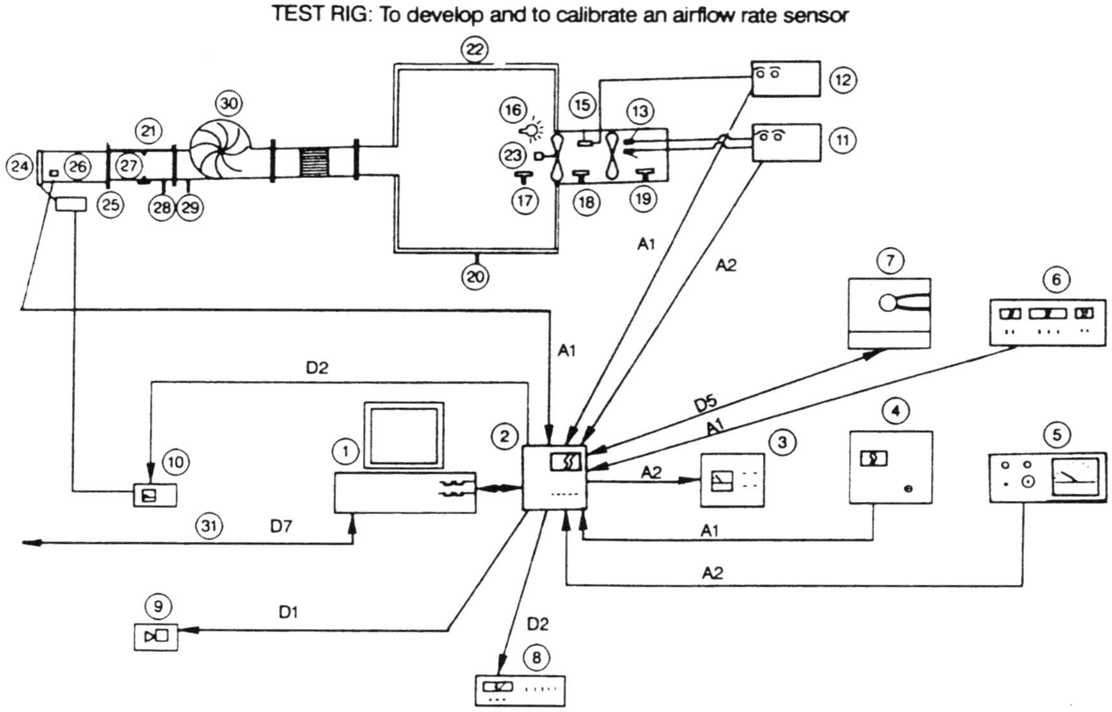

Fig. 3 Schedule of the test rig.

1. Minicomputer (Monitor, floppy disc, to control the measurement installation and to store measured data. **2.** Parallel-interface for digital and analogue signals. **3.** Linearised conversion 0-10 V_{DC} to 0-20 V_{AC}. **4.** Pressure difference sensor 0-26 Pa with digital and analogue output (Furness Control). **5.** Reference-pressure difference sensor 0-200 Pa (Furness Control FC0 60). **6.** Power and energy monitor 0-200 kW. **7.** Choice-switch to scan 10 channels. **8.** 3-channel voltmeter. **9.** Digital controlled acoustic signal. **10.** Controller for airflow control. **11.** Transducer for rotation speed and direction of the turbine. **12.** Transducer for the rotation speed of the fan. **13.** en. **14.** Proximity-switches. **15.** Photocell. **16.** Light source. **17, 18, 19, 20.** Positions for measuring the pressure difference (static and dynamic). **21.** Reference unit for airflow rate. **22.** Pressure chamber. **23.** Fan for testing. **24.** Diaphragm for airflow rate control. **25.** Engine for controlling the diaphragm. **26.** Position-sensor for the diaphragm. **27.** Orifice. **28, 29.** Pressure difference measurement. **30.** Second fan. **31.** Serial data transmission to the graphic workstation (Tektronix 4052A).

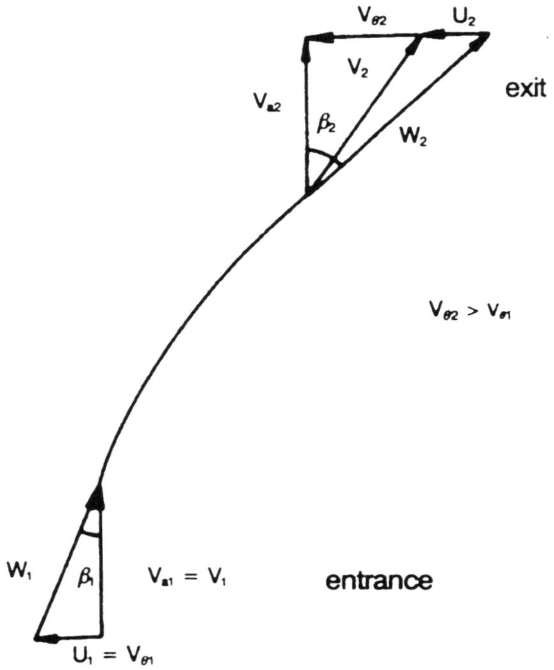

Fig. 4 Velocity triangles at inlet and outlet of a turbine blade.

Fig. 5a Rotational speed vs. flow rate for a 2-blade impeller, downstream position.

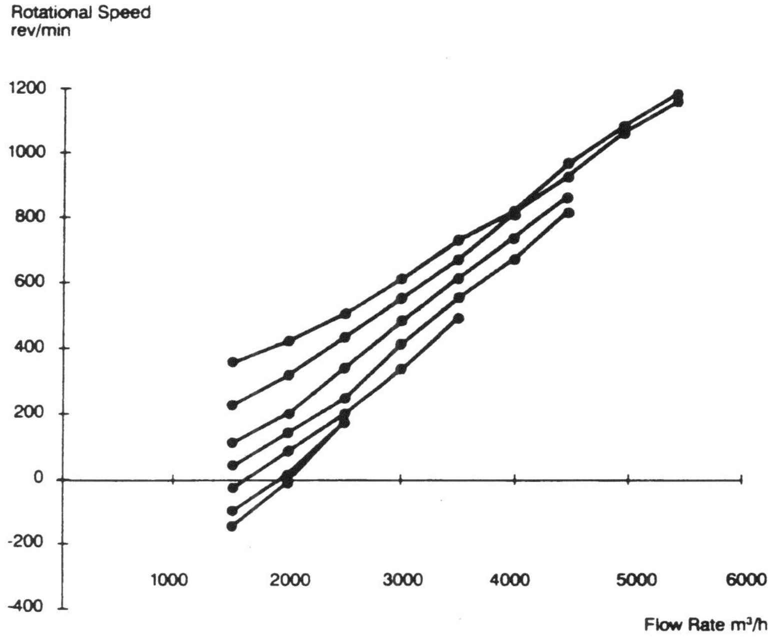

Fig. 5b Rotational speed vs. flow rate for a 2-blade impeller, upstream position.

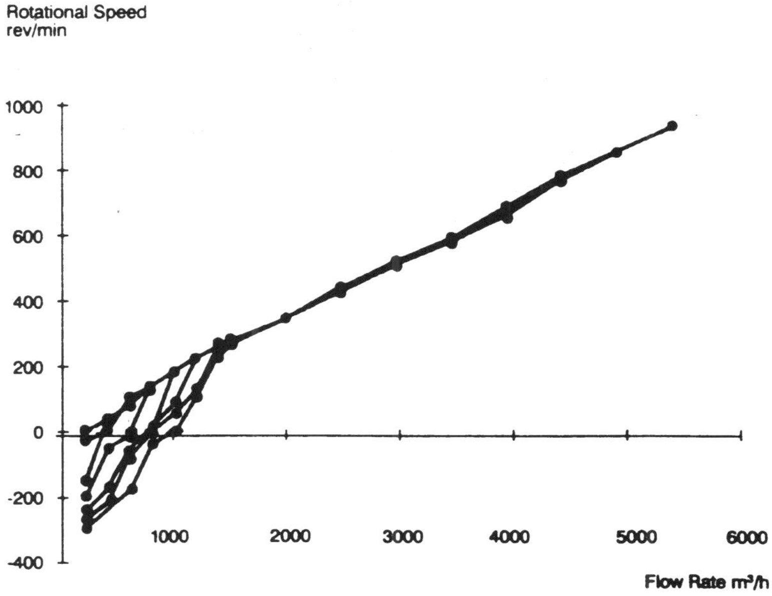

Fig. 5 Comparison of turbinemeter performance (old design) for upstream and downstream positioning (flow rate: 1500-5000 m^3/h; $\Delta p = 0\text{-}120$ Pa).

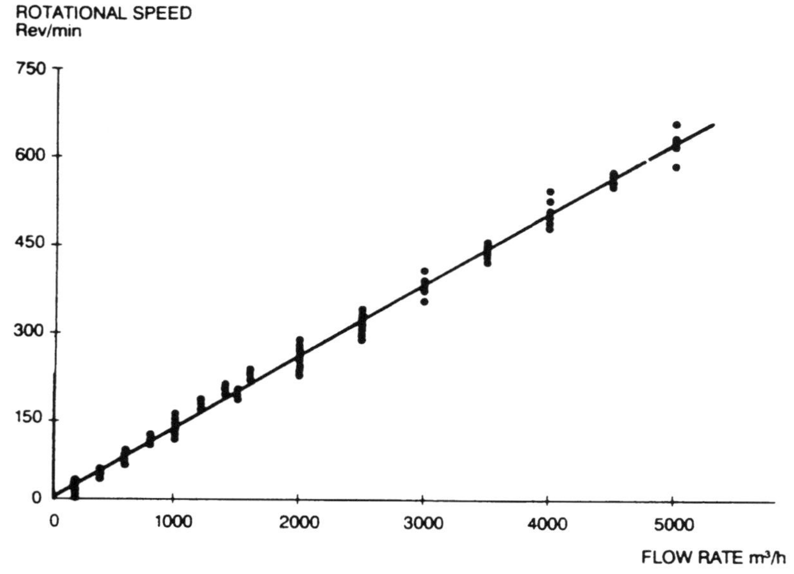

Fig. 6 Linear regression analysis. 2-blade impeller prototype (flow rate: 200-5000 m³/h, Δp: 0-120 Pa), upstream position, combined with Woods 2100-fan.

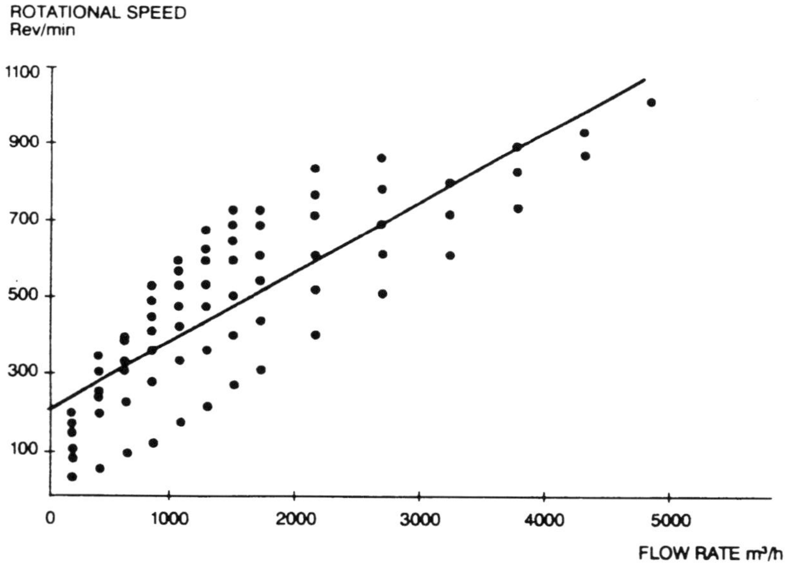

Fig. 7 Linear regression analysis. Turbinemeter old design (flow rate : 200-5000 m³/h, Δp-120 Pa), downstream position, combined with Woods 2100 – fan.

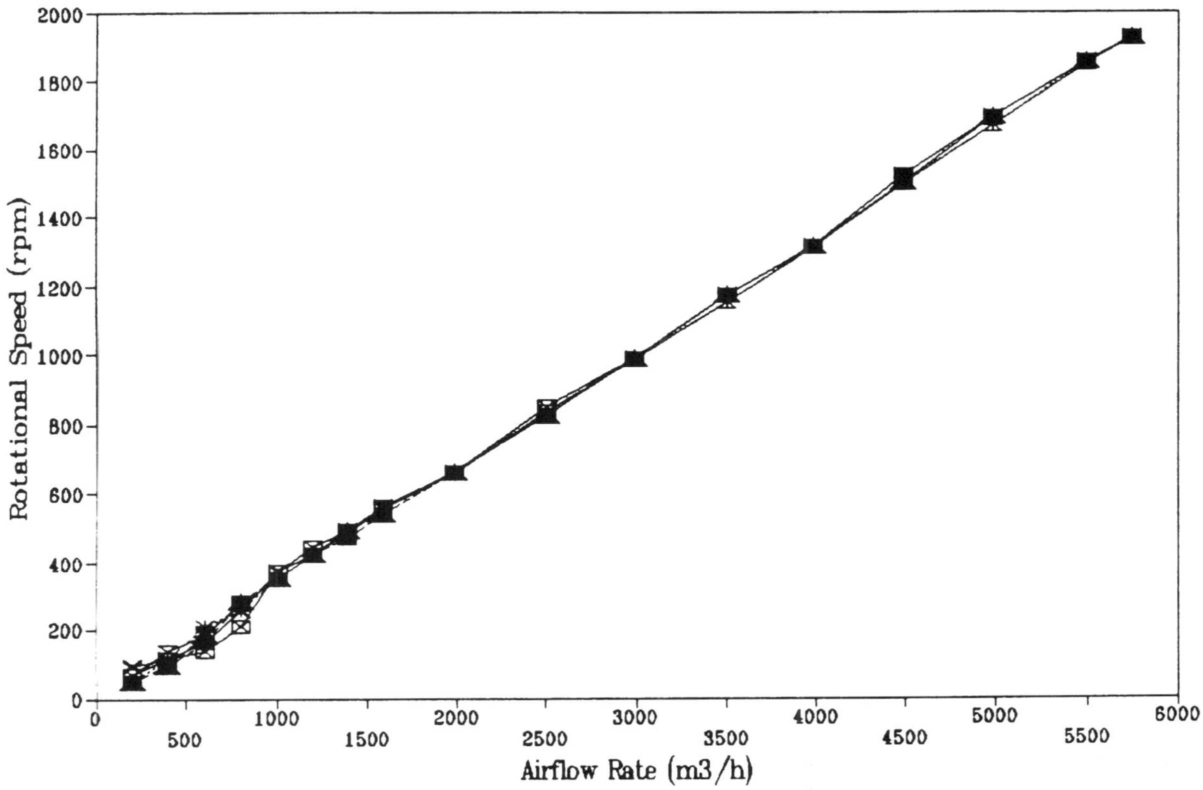

Fig. 8 Linear regression analysis, 2 blade impeller prototype (flow rate 200-6000 m3/h, Δp: 0-120pa), upstream position, combined with Fancom 1445-fan

CONTROLLING MOISTURE LOSS AS A TOOL TO REDUCE BRUISE SUSCEPTIBILITY

Verstreken E., García J.L.*, Ruiz-Altisent M.*, De Baerdemaeker J.

Department of Agroengineering and -Economics
Katholieke Universiteit Leuven
K. Mercierlaan 92
B-3001 Heverlee
Belgium

** Departamento de Ingenieria Rural*
Universidad Politecnica de Madrid
28040 Madrid
Spain

Abstract: A method to reduce the bruise susceptibility of apples by controlling the moisture loss of the fruit was evaluated. Previous research indicates that reduction of the relative humidity of the storage air leads to an immediate effect on the weight loss and on skin properties and to a lower bruise susceptibility of apples. The diffusion equation is used to determine the waterpotential profile inside the fruit during storage. Characteristics of the waterpotential distribution in the fruit are related to measured bruise volumes. The results indicate how this model can be used to control bruise susceptibility.

Keywords: damage, susceptibility, permeability, finite element method, simulation, quality control

1. INTRODUCTION

Literature and practice indicate that a certain amount of moisture loss is required to reduce the bruise susceptibility of fruits during handling. However the weight loss during storage should be kept as low as possible to avoid shriveling of the fruit and to avoid loss of product weight.

Mechanical damage of fruits is mainly caused by impact forces during harvest, handling and transport (García, et al., 1988), so the sensitivity of apples to damage by impact is emphasized in this work. Different factors which could influence the bruise susceptibility of apples are described by Johnson and Dover (1990). These factors are the rate of ripening, the mineral content, the harvest date and ripeness stage, the mechanical properties of the flesh and of the skin.

Bruise volumes appeared to be negatively correlated with the percentage of weight loss during storage (Johnson and Dover, 1990). This phenomenon is also described by other researchers (Garcìa, *et al.*, 1994; Horsfield, *et al.*, 1972), who state that the turgidity of the surface tissue (internal cellular stress) of fruits is affected by the relative humidity of the surrounding air.

The weight loss of fruits during storage mainly consists of the transpiration of water. The rate of transpiration is determined by the humidity and temperature of the air, the air circulation, stacking of the boxes and the ripeness stage of the fruit. This was also found in the (unpublished) experimental results obtained in 1994 in the framework of the EC-CAMAR project. McCarthy and Perez (1991) described the process of water loss of a drying apple slab with the diffusion equation of Fick and derived a diffusion coefficient for the flesh. Since the driving force for water transport is a difference in waterpotential, the distribution of waterpotential in the fruit will be simulated with the diffusion model in this work.

It is clear that an interaction exists between storage conditions, weight loss, ripeness stage, mechanical properties and bruise susceptibility. Reduction of the bruise susceptibility by controlled dehydration seems to be a promising application (Horsfield, *et al.*, 1972; Johnson and Dover, 1990), but the models required for prediction are not yet available.
In this report an approach for this problem is proposed. Simulations of the waterpotential distribution inside the fruit during storage under different conditions are related to bruise volumes.

2. MATERIALS AND METHODS

2.1. Experimental setup

Apples. A batch of 130 freshly harvested Delbare Estival apples was kept in storage for up to 12 days. The batch consisted of apples of 2 different harvest dates (10/08/94 and 17/08/94).
A similar experiment was conducted with 110 Golden Delicious apples.

Storage conditions. One group of Delbare Estival apples was kept at 2°C and 50 % RH and another group in perforated plastic bags to obtain ± 95 % RH. One batch of Golden Delicious apples was kept at 20°C and 40 % RH, while the other batch was kept at ± 95 % RH in perforated plastic bags.
After different storage periods, 10 apples of each group were removed from storage in both experiments and the weight loss, mechanical properties of flesh and skin and bruise susceptibility were determined

Mechanical properties of flesh and skin. The firmness of the fruit flesh (FF) was determined with a handhold penetrometer. The skin property measured was a modulus of elasticity determined from the maximum force/maximum deformation (SF/D) recorded at perforation of the skin with a 0.48 mm diameter pin.

Bruise volume measurements. A 52.1 g mass was dropped on the fruit from a constant height of 8 cm. The contact radius of the spherical head of the mass was 2 cm. After a couple of hours (2 to 4) the bruise depth B and diameter D were measured and the bruise volume BV was calculated (Chen and Sun, 1981) :

$$BV = \frac{\pi D^2 B}{6} \qquad (1)$$

Statistical analysis of the experimental results. An analysis of variance (SAS, SAS Institute Inc.) was conducted to evaluate the effect of storage time and relative humidity of the storage air on the evolution of the bruise susceptibility. A correlation analysis was performed on the whole experimental dataset to find interesting relationships.

2.2. Simulation of the water distribution

The driving force for water transport is not the difference in water concentration, but it is basically the difference between the waterpotential inside the apple and the waterpotential of the surrounding air. Compared with the heat transport : the diffusion of heat is caused by a difference in temperature (Luikov, 1966). The diffusion of water is caused by a difference in waterpotential and the water will flow from a higher to a lower potential (cf. electric current). This phenomena can be described with a diffusion equation, expressed in terms of waterpotentials (Luikov, 1966) :

$$\frac{\partial \psi_w}{\partial t} = D_p \frac{\partial^2 \psi_w}{\partial x^2} \qquad (2)$$

with ψ_w : waterpotential in apple
D_p : diffusion coefficient of waterpotential

The diffusion equation, used to describe the water transport in the fruit, is a partial differential equation. This is not easily solved for complex geometries. A numerical method like the finite element method can be used for this purpose. It is applied in this work with the software package ANSYS (Swanson Analysis Systems, Inc.). Solutions are obtained at discrete values of time and position in the object. A model of the object is entered by a series of interconnected elements and nodes. The geometry used in this work was a slice (1.5 cm thickness) taken from the equator part of the apple (fig. 7). This was done to reduce the number of nodes. The presence of a very thin (0.2 mm) but highly impermeable skin causes very high gradients near the apple surface. According to the objectives of this work, the highest interest is in the gradients near the surface. So a very fine mesh was required near the surface. The material properties are : the diffusivities of flesh and skin, $5.55.10^{-11}$ and $5.55.10^{-15}$ m^2/s respectively (Van Woensel, 1980). The initial condition is a uniformly distributed waterpotential in the fruit (-1.3 MPa, Van Woensel, 1980). The boundary condition is convective transport out of the object. The waterpotential of the air depends on the temperature and relative humidity and can be calculated from :

$$\psi_{wa} = \frac{RT}{\overline{V}_w} \ln \frac{RH}{100} \qquad (3)$$

The convection coefficient is calculated from dimensionless parameters correlated with experimental data for flow over a single sphere (Bejan, 1993) :

$$Sh = 0.3 + \frac{0.62 Re^{1/2} Sc^{1/3}}{\left(1 + (0.4 / Sc)^{2/3}\right)^{1/4}} \left[1 + \left(\frac{Re}{282000}\right)^{5/8}\right]^{4/5}$$

$$Sh = \frac{\overline{h_m} L}{D_{wa}} \; ; \; Re = \frac{vL}{\nu} \; ; \; Sc = \frac{\nu}{D_{wa}} \qquad (4\text{-}7)$$

for ReSc>0.

After simulating different storage periods under the imposed boundary conditions, the waterpotential distribution inside the fruit is obtained. Different characteristics of this profile are related to the measured bruise volumes.

The influence of the presence of the -very impermeable- apple skin on the profile of the waterpotential was investigated. This was done by giving the skin the same diffusivity as the flesh, thereby neglecting the presence of a skin.

The flesh and skin diffusivity were decreased with 25 %, which may occur in practice for apples of an earlier harvest date (smaller cells with thicker walls, more cell to cell contact due to less intercellular air spaces), in order to see the effect on the distribution of the waterpotential in the fruit.

3. RESULTS

3.1. Experimental results

Delbare Estival apples. The experimental results are shown in fig. 1 and 2, and tables 1 and 2. The data and the analysis of variance show that weight loss is instantaneously influenced by the relative humidity of the storage air (table 1 and 2). The bruise volumes for both harvest dates are higher after storage at 95 % RH than at 50 % RH, but the difference is not significant in the experimental data. There is a trend for a decreasing bruise volume after longer storage. This trend is significant at 50 % RH, but not significant at 95 %. The bruise volume was significantly higher for the later harvest date than for earlier harvest date. Flesh firmness was not affected by storage at different relative humidities for both harvest dates. The modulus of elasticity of the skin showed a more significant decrease at the lower humidity of the storage air. This is so for both harvest dates.

Golden Delicious apples. The experimental results are shown in fig. 3 and table 3. The same conclusions as for Delbare Estival apples can be made. Storage at different relative humidities had a significant effect on weight loss, modulus of elasticity of the skin and also bruise volume. No significant effect on the flesh firmness was observed.

The correlation analysis on the whole dataset indicates no significant correlation between bruise volume and flesh firmness, but bruise volume is significantly and strongly correlated with skin modulus of elasticity, weight loss (negative correlation) and mostly with the ratio skin modulus of elasticity/flesh firmness. A faster decrease in the modulus of elasticity of the skin compared to the decrease in flesh firmness (through storage in a dryer atmosphere) results in a faster decrease of the bruise volume. Apples of a later harvest date with a lower flesh firmness show higher bruise volumes. This confirms the results of Johnson and Dover (1990).

3.2. Simulation results

It was observed that in almost all cases the waterpotential gradient occurred within the apple skin (fig. 8). From fig. 4 and fig. 5 (for storage conditions in the Golden Delicious experiment) it can be seen that during the first few days a gradient is formed across the skin. During further storage the internal waterpotential will gradually (but very slowly) decrease. The same observations could be made for the conditions of the experiment with Delbare Estival apples.

The following parameters were extracted from the waterpotential profile :

- the waterpotential at the fruit surface (ψ_s)
- the waterpotential in the center of the skin (ψ_{sc})
- the waterpotential just beneath the skin (ψ_{bs})
- the waterpotential in the center of the fruit (ψ_c)
- the sum of the waterpotential values of all elements (sum(ψ))
- the sum of the waterpotential gradients of all elements (sum($\Delta\psi$))

From fig. 6 it is clear that the waterpotential just beneath the skin decreases fastest for the lower relative humidity. This decrease in waterpotential may be due to a lower turgor pressure in the superficial cell layers. This may also be the explanation for the observed faster decrease in bruise volume at the lower relative humidity of the storage air.In table 4 the correlations between these profile parameters and the experimentally determined bruise volumes are given. It should be mentioned that all the simulations were performed with the same values of skin and flesh diffusivity and for an initial homogeneous waterpotential. These results were compared with bruise volumes in apples of different variety and ripeness. One could expect different values of diffusivities for these fruits. Knowledge about the evolution of these values during ripening and the relationship with other measures of texture is necessary for further use of this model.

Neglecting the presence of a skin in a simulation, resulted in waterpotential profiles like fig. 9, where the gradient is spread out over a large part of the fruit flesh. It is clear that the skin is the limiting factor in the diffusion of water and that it is therefor essential to have a good idea of the evolution of the skin thickness and diffusivity during ripening.

The change in waterpotential values after decreasing the flesh and skin diffusivities with 25 % was higher for the more internal parts in the fruit and for the longer storage periods. The correlation coefficients were therefor best for these parameters (table 5). For the other parameters the correlation coefficients did not change compared to these obtained with the original data.

4. CONCLUSIONS

The measurement results on bruise volumes and storage conditions yield the following conclusions :

- Storage at different relative humidities of the storage air instantaneously resulted in significantly different weight losses and gradually in significantly different bruise volumes and elastic properties of the skin. Flesh firmness was not affected by the difference in relative humidity for the storage periods studied.
- Bruise susceptibility seemed to decrease during storage. An explanation for this behavior could be the decrease in modulus of elasticity of skin ánd flesh during ripening and also the cell wall degradation. This decrease in bruise susceptibility occurs faster at a lower relative humidity, probably caused by the faster decrease in modulus of elasticity of the skin. A lower relative humidity causes an increased water transpiration, which results in a lower turgidity of the superficial cell layers of the fruit and probably also in a lower modulus of elasticity of the skin.
- Bruise susceptibility was higher for the later harvest date, which can be explained by the larger cells and thinner cell walls which occur in apples of a later harvest date (Johnson and Dover, 1990).

It is possible to simulate the waterpotential distribution inside a fruit during storage under known conditions of temperature and relative humidity with a diffusion model. Parameters related to the change in waterpotential near the skin (waterpotential at the surface of, in or just beneath the skin) or parameters related to a global change in the waterpotential distribution (sum of waterpotential values or gradients of all elements) seem to be best correlated with the experimentally measured bruise volumes. It is however necessary to have a good estimate of the value of flesh and skin diffusivities, of the skin thickness, and how these are influenced by the ripeness stage. These parameters have a significant influence on the waterpotential profile. Especially the skin characteristics seem to be important, as was also observed in the experiments.

This research indicates the possibility to model the waterpotential distribution in fruit in order to give a prediction of the optimal storage conditions to reduce the bruise susceptibility. Information is required on the values of the skin and flesh diffusivities and skin thickness and how they are influenced by the ripeness stage, variety etc.

The next steps of the research will include :
- sensitivity analysis of the diffusion model to determine which parameters should be measured and with which accuracy.
- relationship between waterpotential and water concentration so that water concentration profiles can be obtained from the simulated waterpotential profiles and the weight loss can be calculated.
- experimental measurements of the required parameters and their change with ripeness stage.
- a control strategy to minimize bruise susceptibility.

LIST OF SYMBOLS

D_{wa}	: diffusion coefficient of water in air
$\overline{h_m}$	: average mass transfer coefficient
L	: diameter of sphere
R	: gasconstant
Re	: Reynolds number
RH	: air relative humidity
Sc	: Schmidt number
Sh	: Sherwood number
T	: air temperature
v	: air velocity
$\overline{V}_w$	: partial molar volume of the air
ψ_w	: waterpotential in apple
ψ_{wa}	: waterpotential of the air
ν	: kinematic viscosity

REFERENCES

Bejan, A. (1993). *Heat transfer,* Chap. 5. John Wiley & Sons, Inc., Singapore.

Chen, P.and Z. Sun (1981). Impact parameters related to postharvest bruising of apples. ASAE paper no. 81-3041.

García, C., M. Ruiz and P. Chen (1988). Impact parameters related to bruising in selected fruits. ASAE paper no. 88-6027.

García, J.L., M. Ruiz-Altizent and P. Barreiro (1994). Factors influencing apple and pear physical properties and bruise susceptibility. Report no. 94-G-062, presented at the AgEng meeting in Milano, Italy.

Horsfield, B.C., R.B. Fridley and L.L. Claypool (1972). Application of theory and elasticity to the design of fruit harvesting and handling equipment for minimum bruising. *Trans. of the ASAE,* **15(4)**, 746-750.

Johnson, D.S. and C.J. Dover (1990). Factors influencing the bruise susceptibility of Bramley's seeding apples. Seminoario Internacional sobre Danos por impacto en frutas y hortilizas, FIMA 90, Zaragoza, Spain.

Luikov, A.V. (1966). *Heat and mass transfer in capillary porous bodies,* Chap. 6. Pergamon Press, New York.

McCarthy, M.J. and E. Perez (1991). Model for transient moisture profiles of a drying apple slab using the data obtained with magnetic resonance imaging. *Biotechnol. Prog.*, 7, 540-543.

Van Woensel, G. (1980). *Waterpotentiaal veranderingen en mechanische eigenschappen van appels.* Licentiaatsverhandeling K.U.-Leuven.

TABLES AND FIGURES

Table 1. Delbare Estival apples harvested 10/08/94. Mean values (*) of weight loss (%), bruise volume (mm³), flesh firmness (N) and skin modulus of elasticity (N/mm) after different periods of storage.

Storage at 50 % RH

days	loss		BV		FF		SF/D	
0	0	A	170	A	60.4	A	5.5	A
5	0.55	B	163	A	57.2	A	5.1	AB
8	0.89	C	161	A	56.5	A	5.2	B
12	2.16	D	137	B	59.5	A	4.6	C

Storage at 95 % RH

days	loss		BV		FF		SF/D	
0	0	A	170	A	60.4	A	5.5	A
5	0.27	B	165	AB	57.8	A	5.7	AB
8	0.39	C	159	AB	56.6	A	5.8	AB
12	0.57	D	147	B	56	A	5.3	B

Table 2. Delbare Estival apples harvested 17/08/94. Mean values (*) of weight loss (%), bruise volume (mm³), flesh firmness (N) and skin modulus of elasticity (N/mm) after different periods of storage.

Storage at 50 % RH

days	loss		BV		FF		SF/D	
0	0.00	A	195	A	51.5	A	5.7	A
2	0.39	B	174	AB	48.4	A	5.2	B
5	1.49	C	168	CB	51.0	A	5.1	B
7	1.94	D	151	C	52.6	A	4.6	C

Storage at 95 % RH

days	loss		BV		FF		SF/D	
0	0 .00	A	195	A	51.5	A	5.7	A
2	0.10	B	184	A	47.5	A	5.7	A
5	0.22	C	178	A	47.7	A	5.4	AB
7	0.38	D	173	A	49.9	A	5.0	B

Table 3. Golden Delicious apples. Mean values (*) of weight loss (%) and bruise volume (mm³) after different periods of storage.

	40 % RH		95 % RH	
days	loss	BV	loss	BV
0	0.00	153.7 A	0.00	153.7 A
0.3	0.54	147.2 AB	0.02	152.4 A
1	0.62	133.4 ABC	0.06	162.1 A
4	2.05	132.7 ABC	0.70	154.6 A
5	2.40	121.4 BC	0.65	159.6 A
8	3.89	110.5 C	1.53	142.0 A

(*) Mean values with the same letter are not significantly different.

Table 4. Correlation analysis : simulated profile parameters versus measured bruise volumes.

	BV		BV
ψ_s	0.58	ψ_c	0.49
ψ_{sc}	0.56	sum(ψ)	0.6
ψ_{bs}	0.58	sum($\Delta\psi$)	-0.58

Table 5. Correlation analysis : adapted simulated profile parameters versus measured bruise volumes.

	BV		BV
ψ_s	0.58	ψ_c	0.58
ψ_{sc}	0.57	sum(ψ)	0.6
ψ_{bs}	0.61	sum($\Delta\psi$)	-0.58

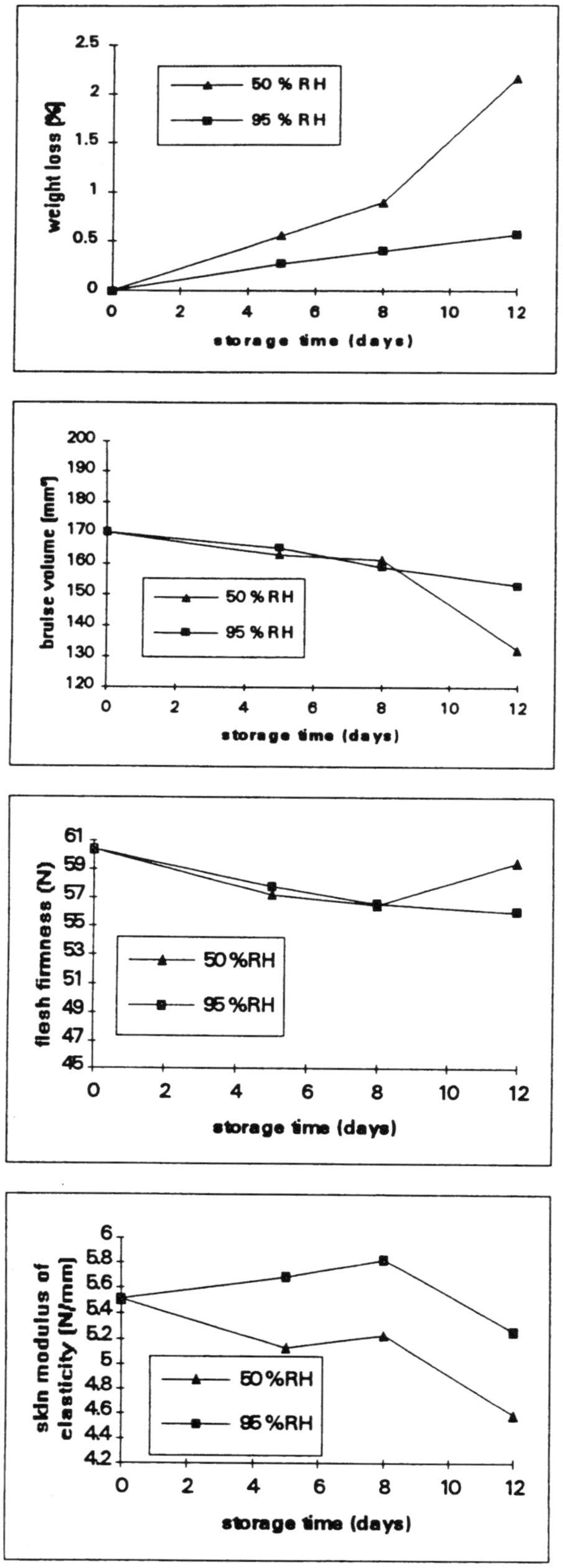

Fig. 1.a-d : results for Delbare Estival apples harvested 10/08/94 (°)

(°) a : weight loss (as % of original weight) after different periods of storage
b : bruise volume (mm³) after different periods of storage

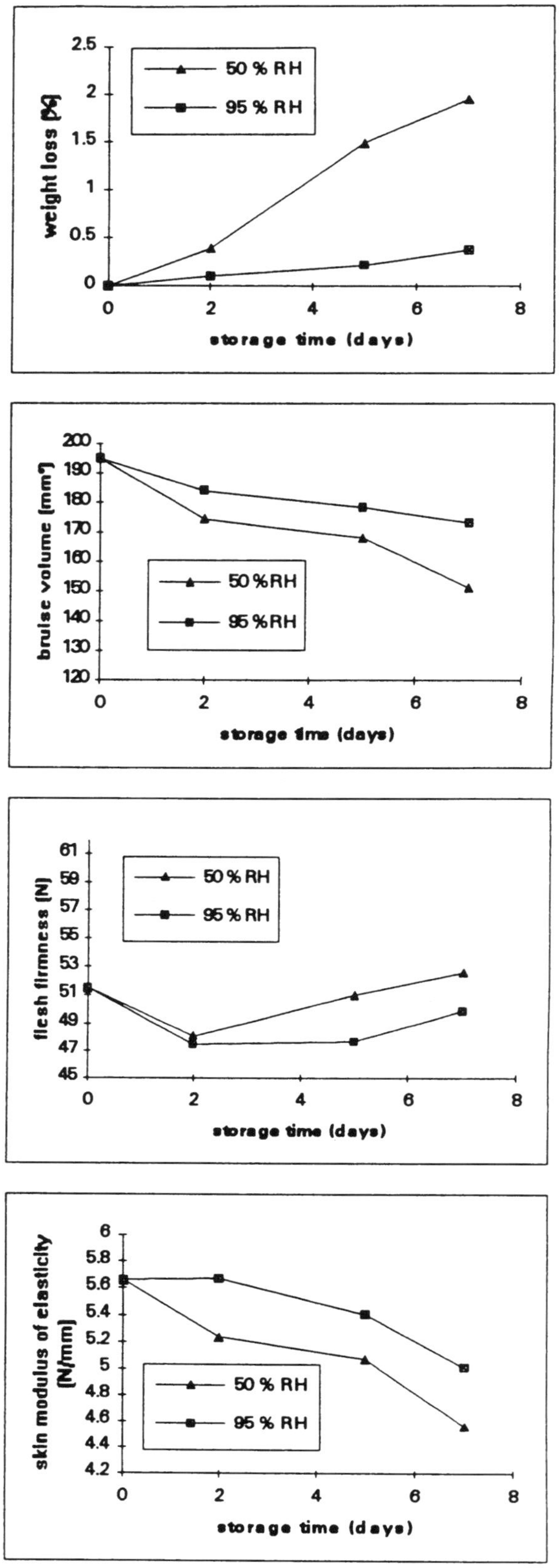

Fig. 2.a-d : results for Delbare Estival apples harvested 17/08/94 (°)

(°) c : flesh firmness (N) after different periods of storage
d : skin modulus of elasticity (N/mm) after different periods of storage

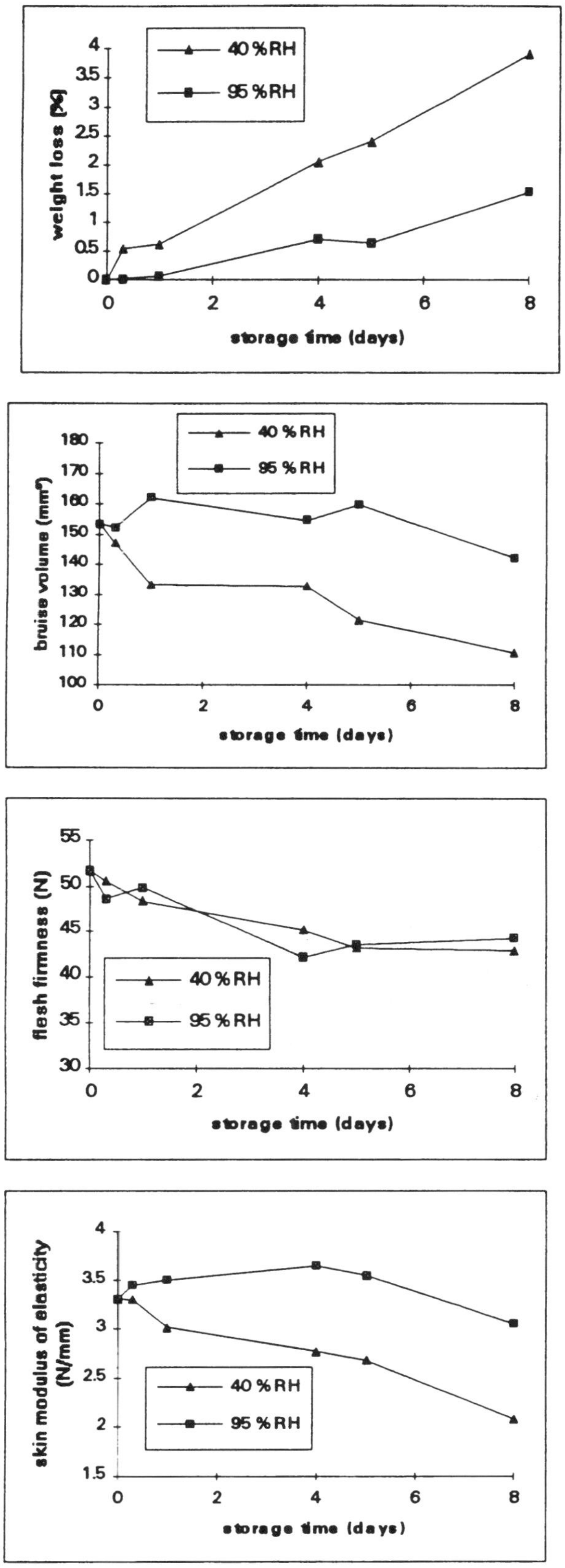

Fig. 3.a-d : results for Golden Delicious apples (°)

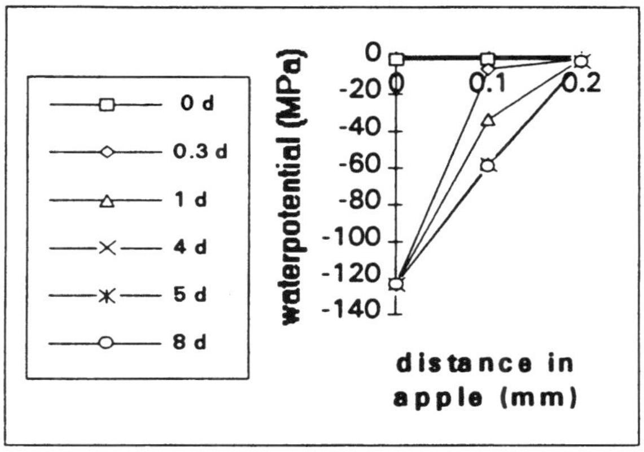

Fig. 4 : Evolution of profile of waterpotential in the apple skin during storage in 40 % RH (-124 MPa)

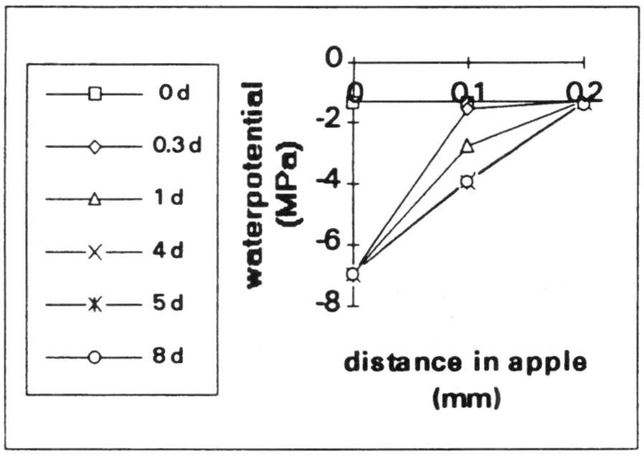

Fig. 5 : Evolution of profile of waterpotential in the apple skin during storage at 95 % RH (-6.9 MPa)

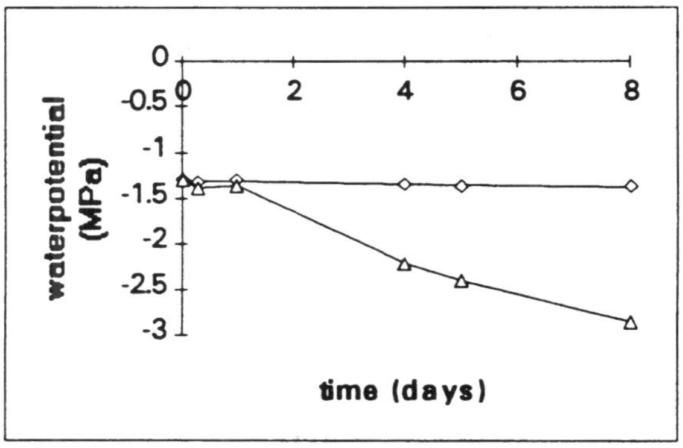

Fig. 6 : Evolution of the waterpotential just beneath the skin during storage at 40 (Δ) and 95 (◊) % RH for Golden Delicious apples

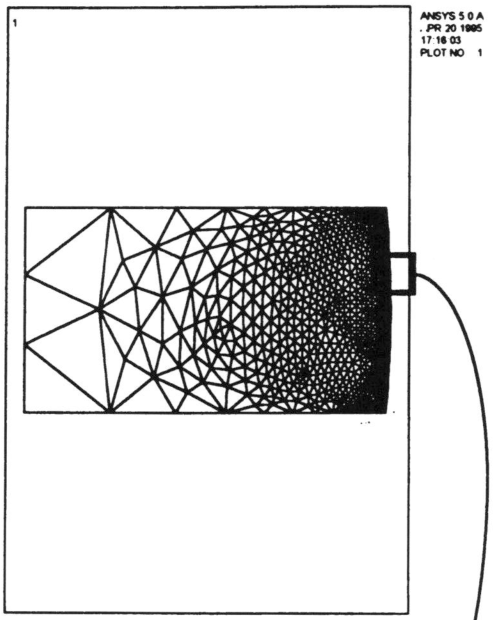

Fig. 7 : Geometry and mesh of the model of an apple slab, used for the simulations

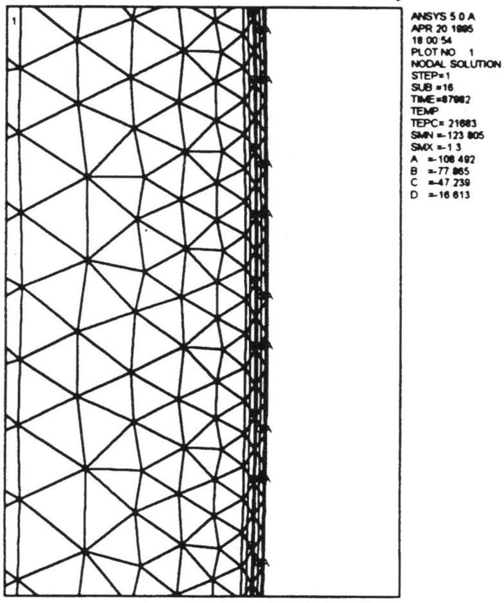

Fig. 8 : Simulation results for 1 day of storage at 40 % RH (-124 MPa) : zoom on the skin to view the contour display

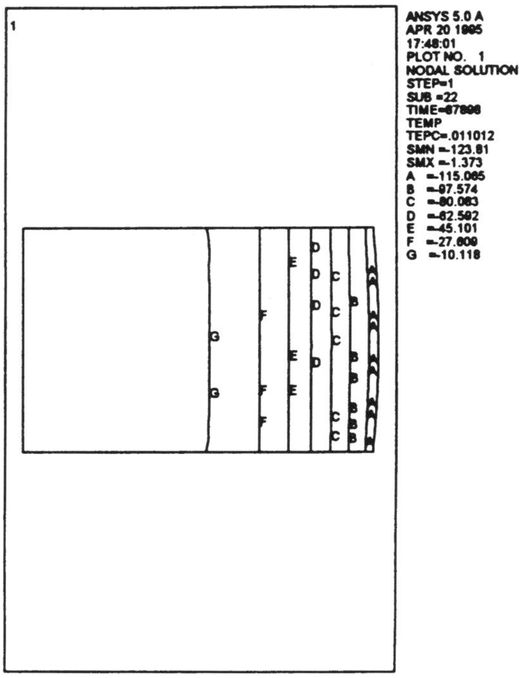

Fig. 9 : Simulation results for 1 day of storage at 40 % RH (-124 MPa) : simulation without the presence of a skin

APPLICATION OF PHYSICAL DATA TO QUALITY CONTROL OF MINIMALLY PROCESSED APPLES

Adam Kuczyński

Istitute of Agrophysics, Polish Academy of Sciences
Doświadczalna 4, 20-236 Lublin, Poland

Abstract: The objective was to develop a standard for the assessment and modelling of a difficult product, i.e. the fresh apple tissues viable in the course of handling and processing. Statistical distribution of tissue whiteness, firmness, soluble solids and susceptibility of apple slices to enzymatic browning was studied in relation to long cold storage of 33 cultivars. Analysis of wide range of parameters and their relation to slicing in air and under solution led to the conclusion that peelability and washability are responsible for the variation in browning that is important from the commercial point of view. The study emphasizes the need for a push towards more attractive apple cultivars, and that, requires quality control methods for their evaluation. Such products vary widely according to characteristics of unprocessed commodities and forms for consumption.

Keywords: optical properties, spectroscopy, agriculture, physiology, quality control.

1. INTRODUCTION

The volume of minimally processed fruits which are washed, peeled, sliced or grated and packed is increasing in consumer markets (Güntensperger, 1994). Since the plant tissue of such products is raw and respire after processing, minimally processed products - MPP are more perishable than unprocessed raw materials. MPP readily deteriorates in quality, especially in colour and texture, as a result of endogenous enzymes that, enhance respiration and microorganisms which lead to reduced shelf life.

Shelf life of minimally processed products has been extended by inhibition of metabolic reactions and aerobic spoilage caused by a wide variety of micro-organisms. Consumer markets demand less use of chemicals on such products. Hence, use of alternative and natural means is necessary.

Physiological, biochemical and microbiological storage stability of MPP may vary depending on a cultivar due to individual differences (Aubert *et al.*, 1992). Published information on apples is limited, but it is known that no single cultivar has been evaluated as most suitable for MPP (Kim *et al.*, 1993; Kuczyński, 1995).

Measurement of pigment content followed by enzymatic browning are analytical problems confronting the food technologist working with fruits slices. The method described below is one of the methods that can be practically applied in determining enzymatic browning. Cultivar differences are due mainly to the relative balance between hydroxycinnamide derivatives and flavan 3-ols in apple tissue (Oszmiański and Lee, 1990; Kuczyński *et al.* 1994). This can naturally be affected by many other factors when the fruit is harvested and during other operations involved in post-harvest storage and processing (Varoquaux and Wiley, 1994). In this sense white is the colour of freshness and cleanliness not only psychologically, but also as an indicator of freedom from contaminants (Saper and Douglas, 1987; Spanos and Wrolstad, 1992) and as such can be a measure of the quality of the product.

A number of scales are used specifically in the measurement of whites. Paper industry has been using blue reflectance as a its primary determinant of optical quality of pulp after bleaching, and textile industry has established a similar test to measure bleaching progress in cotton. To avoid further complications, of the measurements by choosing methods of assessment it is possible to used the absorbance at 440 nm (e.g. mercury arc lamps) which is adequate for monitoring processes such as browning (Cindy and Hicks, 1991; Kuczyński *et al.*, 1994). Parameter difference in absorbance at 440 nm using DA440 = A440(time) - A440(time=0min) has been introduced for the evaluation of pigmentation caused by enzymatic browning or estimation the effectiveness of inhibition (Kuczyński *et al.*, 1993; 1994).

The Hunter's L*a*b* method colour difference dE* give the best measurements for visual appearance of a product. This is another important dimension of colour measurement that is sophisticated and only little dependent on the biochemical composition of apples. For instance no difference occurred in colour (lightness L* value) of apple slices among the 12 cultivars just after cutting (Kim *et al.*; 1993).

Qualitative evaluation of fruit has been a matter of interest for researchers for many years. However, there is no clear definition of fruit quality. Many quality factors such a size, colour, flavor, firmness and taste are related to maturity. Since many quality-ripeness factors are related to physical properties of apples it is necessary to develop techniques to evaluate post-harvest ripeness based on physical properties. In the present work optical absorbance of the fresh product can be used as the universal parameter of this type.

2. MATERIAL AND METHOD

2.1 Apples and slices preparation.

Apples of 33 cultivars: *Sampion* (SAM), *Ozark Gold* (OZG), Elstar (ELS), *Fantazja* (FZJ), *Jonagold* (JOG), *Gala* (GAL), *Golden Delicious* (GDL), *Ligol* (LIG), *Jonathan* (JON), *Starking* (STK), *Red Fireside* (RFS), *Gloster* (GST), *Honeygold* (HGD), *Melrose* (MEL), *Macspur* (MPR), *Wealthy* (WHY), *Starkrimson* (STN), *Cortland* (CTR), *Prinz Albrecht von Preussen* (PAP), *James Grieve* (JGV), *Bancroft* (BKF), *Friedom* (FDM), *Beforest* (BFR), *Macoun* (MUN), *Prima* (PRI), *Billing Boscoop* (BIB), *McIntosh* (MCI), *Boiken* (BOI), *Florina* (FLO), *Idared* (IDR), *Liberty* (LBR), *Lobo* (LBO), *Spartan* (SPT) were grown at the same plantation at the Experimental Station Orchard Albigowa (Rzeszów voivodship) of the Skierniewice Research Institute of Pomology and Floriculture.

Apples were hand-picked on normal harvest dates in the region, estimated for these cultivar and were planed for cold storage. Fruits samples, representative for the varieties were selected and packed in carton boxes containing 80 apples, each cooled, and stored in conventional storage room (3-5°C and 90% RH) until analyzed. At 3 periods; early, medium and late storage life for market distribution of each cultivar, fruits were removed. Then 10 apple sample was conditioned at room temperature for 36 hours in order to obtain higher organoleptic quality.

Maturity from ripe to unripe was determined visually only ones, at the first term of cold storage. The basis for selection was skin colour variability in the box. This allowed for forming of 8 groups of 5 apples each ranging from ripe No. 1 to unripe No. 2 in the box. The fruit remained in cold storage in the same boxes. Five apples (marked by No. 1, 2, 3, 4 and 5), were selected randomly and processed as slices in air and the five apples under water, according to rigidly controlled technology (Kuczyński *et al.*, 1993).

Each sliced apple was immediately analyzed for light absorbance and then as follows: every 8 min for a total time of 32 min if cut in air, or every 16 min for 64 min period (6 samplings) if cut under water.

2.2 Browning, firmness and soluble solids measurements.

Diffuse reflectance at 440 nm was measured on slice surface by single beam spectrophotometer equipped with Ulbricht Sphere with 14 mm diameter viewing port. The results absorbance were collected by a microcomputer which also provided reliability and control of results by continual comparison of readings against an etalon: 57% reflectance at 440 nm and 800 mV data input. Slice surfaces were directly hand-held to the viewing port. A simple apparatus - i.e. Specol 10 ensured sphere protection against contamination and gave easy regeneration during a long process of investigations. The spectrophotometric method adopted for fresh slices was verified in earlier works (Kuczyński, *et al.*,1994).

The firmness of apple flesh was measured 32 mm or 64 mm after slicing by means of portable dynamometer equiped with a 7.9 mm Taylor puncturing plunger and was reported as maximum peak force - F(N). A slice was mounted on an angle support for tests performed at each slice without skin along the equator on fruit. Juice from slices of each apple was used for the soluble solids - SS(%) determination with an optical refractometer.

3. RESULTS

3.1 Maturity effect

Changes in time of absorbance parameter - A440 approximated since cutting by means of least squares method were presented on Fig. 1 (STATISTICA for Windows rel. 4.5) for fruit of different ripeness cut in air or under water. In general all cultivars showed a rapid increase in A440 values due to enzymatic browning (during the first 32 min if cut in air or 64 min if cut under water) but the tendency towards browning in some cultivars was slower than in others. After a time there were fewer changes and most cultivars maintained A440 values found in these periods throughout the rest of storage.

It is not always possible to see the influence of apple ripeness on graphs. Absorption values were calculated from the approximation model for individual cultivars: for time 0 min - A0, difference of absorbance after 32 min from cutting in air - DA32, and absorbance difference after 64 min from cutting under water - DA64. Only after a short storage time, changes are visible for some cultivars. Lack of uniformity of variances has been noticed in the studied cultivars absorbance parameters and for that reason the values of parameters A0, DA32, DA64 was normalized for the N(1; 0) distributions

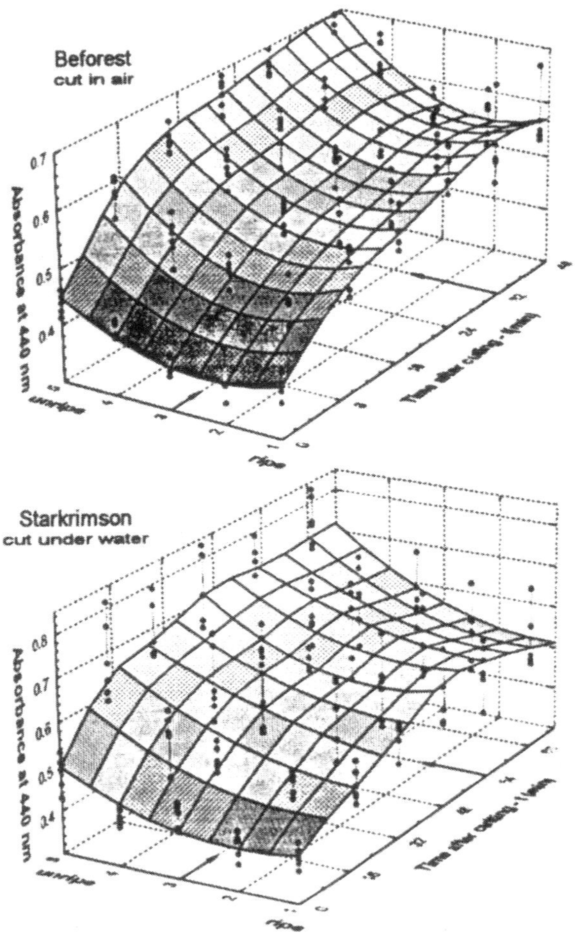

Fig. 1. The rate absorbance increase at 440nm of cut slices surface i.e. enzymatic browning and the effect of maturity of apples in either air (cultivar Beforest) or under water cutting (cultivar Starkrimson) slices procedure.

eliminating the effect of cultivars and periods. Figure 2 shows the result of analysis and mean values, standard errors for 5 classes of ripeness in standardized units; and Tab. 1 presents statistical parameters of the studied data distribution, standardized absorbance increase N(1) - N(5) and the real difference of absorbance A(1) - A(5) when going from class 1 to class 5.

Along the decrease of apple ripeness the increase of A0 tissue absorption and the increase of DA32 and DA64 parameters can be observed. Average changes for each

Table 1. Statistical parameters for the whole population of measurements and mean increase values at changing ripeness classes from 1 to 5 for normalized absorbance N and calculated absorbance A.

	Count	Average	Std. Dev.	Variability	del. N(0;1)	del. Abs.
A0	970	0.345	0.058	17%	0.44	0.03
dA32p	485	0.146	0.077	53%	0.65	0.05
dA64w	485	0.124	0.063	51%	0.34	0.02

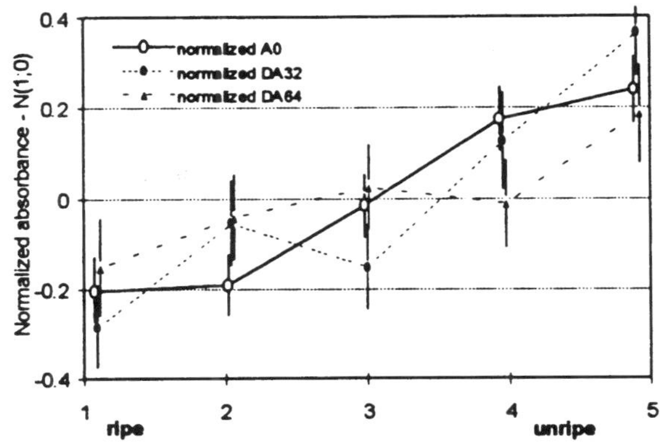

Fig. 2. The means and standard errors for five class of ripeness. The values were calculated from normalized distribution N(0;1) of original absorbance data A0, DA32 and DA64.

cultivar in the real absorbance caused by ripeness could be calculated from their standard deviation and normalized averages for ripeness. The chosen time of 32 min for air and 64 min for cutting under water formed two series of data for which both variance and means test proved no important differences. It led to the conclusion that in order to get the same absorption increase for slices browning while cutting under water the amount of time needed is two times longer than when cutting in air. At the same time the highest absorbance increase by 0.05 at lowering apple ripeness can be expected after cutting in air (Tab. 1).

3.2 Cultivars effect.

Cultivar differentiation was evaluated by means of A0, DA32 and DA64 parameters for median ripeness (class No. 3). The variability of absorbance parameters within a cultivar is considerable (Tab. 2). Coefficients of variation of the tested set ranged from 17 to 51 per cent. The cultivars presented on Fig. 3 were arranged according to the median of absorbance difference - DA64 of apples sliced in water. Based on absorbance of flesh just after cutting, cultivars could be divided into groups: ones with white A0 < 0.3 *Cortland, Spartan, Lobo, Boiken, Ligol, Macoun, McIntosh* and dark A0 > 0.4 such as *Gala, Honeygold, Jonagold, Starkrimson, Billing Boscop, Starking.*

Table 2. Statistical evaluation of the cultivar population by absorbance parameters -(ABS),firmness-F(N) and soluble solids -SS(%).

	Count	Average	Std. Dev.	Variability	Minimum	Maximum
A0	33	0.343	0.055	16%	0.21	0.44
dA32p	33	0.150	0.070	47%	0.01	0.28
dA64w	33	0.124	0.057	46%	0.02	0.21
F(N)	33	76.40	5.76	8%	66.8	90.8
SS(%)	33	12.4	1.2	10%	9.7	14.9

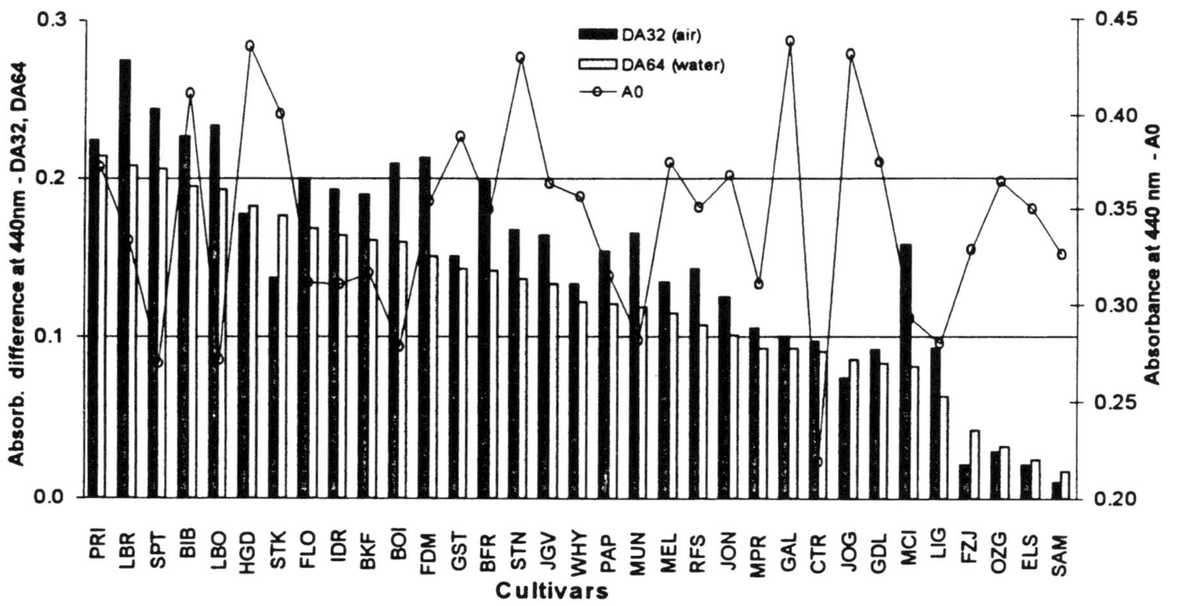

Fig. 3. Apple cultivars arranged according to increasing values of absorbance difference at 440nm after 64min - DA64 from slicing under water i.e. on the basis of progressing pigmentation from enzymatic browning of surface.

The values of absorbance difference - DA32 reached as high as 0.25 for the *Liberty, Spartan* cultivars. The increase of absorbance of the slices cut under water or in air was not statistically significant only for the such cultivars as: *Sampion, Elstar, Fantazja, Ozark Gold.* Correlations between cultivar evaluations by means of DA32 and DA64 has been significant - Spearman rank correlation r = 0.89, (Prob. <1%).

A significantly higher value of standard deviation has also been found for the evaluation of cultivars by means of DA32 parameter in relation both to A0 and DA64. It can be gathered from the above that despite a significant and considerably high correlation measurements of both DA32 and DA64 are taken by means of different technologies and for that reason are not in full agreement in their evaluation for the cultivars.

3.3 Firmness and soluble solids effect.

Firmness of apple slices - F represents a wide variation among cultivars ranging from 66.8N to 90.8N. Slices of cultivars : *Ligol, Florina, Red Fireside, Boiken, Liberty* were firmer (F > 84N) and *Prima, Lobo, Wealthy, Macspur, James Grieve* were less firm (F < 72N) among studied cultivars.

Soluble solids content -SS of the apple slices varied widely depending upon cultivar. The value ranged from low of <11.3% in *Wealthy, James Grieve, Prima, Friedom, Prinz Albrecht von Preussen* to high of >13.6% in *Billing Boscoop, Elstar, Golden Delicious, Red Fireside, Boiken.* However the variability of both parameters is very low when compared with the parameters of absorption (Tab. 2). It has been found that hardness and soluble solids evaluation for different cultivars are correlated r = 0.49, (Prob. =2.2%).

The significant correlations of hardness with absorbance difference DA32, r = 0.41, (Prob. =2.9%) and with DA64, r = 0.36, (Prob.=5.3%) tell us about the influence of hardness on the evaluation of browning. For DA32 regression coefficient is = 0.083 ± 0.03 and for dA64 regression coefficient is = 0.056 ± 0.027. On that basis the tendency of hard cultivars to brown more can be established.

A slight discrepancy in the evaluation of browning done by means DA32 parameter in relation to DA64 can explained by a found regression with the use soluble solids parameters:

$$dA64 = 0.727\ dA32 + 0.0013\ SS.$$

The above regression explains R2 = 99.2%, i.e. relation between both absorbance difference parameters in contrast to R2 = 83.1% without soluble solids.

4. DISCUSSION

The colour of apple slices is an important quality factor in marketing fresh fruits salad and for processing. It is determined by many factors: cultivars, maturity, method of processing, storage, use of browning inhibitors and possibly by other factors. Hence, there is much interest in identifying objective quality control procedures to provide as far as possible an objective measure of these factors.

Many methods to measure colour have been devised. The present study were conducted according to the best method to determine apple slices browning, and using the simple spectrophotometer to determine the range in fruits flesh whiteness of apple varieties for technology purposes. This methods is available for automatic control of produce quality.

Browning potential depends on the variety and was found to increase from ripe to unripe. The estimated relation of maturity to browning was described. The influence of cultivar hardness differences on slices browning was stated. The regression with soluble solids parameter proposed in that study was useful in describing the influence of washing process in the solutions.

Automatic control of optical properties influenced by enzymatic browning will play a significant role in mass production where is a chain from producer to customer. The present experiment can be properly validated for a real process even for a short time of slices storage. The 10 days browning with simple inhibitors have the same path of absorbance increase in time development (Kuczyński *et al.*, 1995). Knife cutting procedure in air or under solution is comparable with real technology. Newer carborundum abrasies, steam peeling or caustic acid, disturb cell walls and thus increase the change of microbial growth, hence they cannot be used for minimally processed refrigerated fruits (Varoquaux and Wiley, 1994).

The quantification method is new and was compared with tristimulus CIE L*a*b* colour description (Kuczyński *et al.*, 1993). Results obtained can be used in the determination of tolerance to quality changes in individual cultivars especially on the basis of natural variability of whiteness of fresh slices. They are naturally acceptable to the consumer.

Numerous features determine the attractive-ness of fresh and packed sliced apple. The instruments-type evaluation cannot provide for all the preferences, However the method is very accurate and may be used for solving also detailed cultivar problems of slices storage. This data may have significant contribution to the application of automatic control techniques.

5. ACKNOWLEDGEMENTS

Gratitude is expressed to Skierniewice Research Institute of Pomologie and Floriculture, for making valuable study material available and to Ing. F. Baran for help in preparing apple cultivars and determination of the periods of storage.

REFERENCES

Aubert, S., M.J. Amiot and J. Nicolas (1992). Browning criteria of apples. *Sci.Aliments.*, **12**, 625-647.

Cindy, B.S., T. and K. Hicks (1991). Sulfated polysaccharides inhibit browning of apple juice and diced apples. *J.Agric.Food Chem.*, **39**, 1719-1722.

Güntensperger, B. (1994). International Symposium on minimal processing of foods. *Trends in Food Sci. and Technologie*,5, 266-269.

Kim, D.M., N.L. Smith and C.Y. Lee (1993). Quality of minimally processed apple slices from selected cultivars. *J.Food Sci.*, **5**, 1115-1117,117.

Kuczyński, A. and P. Varoquaux (1992). Reflectance method for the study of initial colour and browning rate of white peach pulps. *Sci.Aliments*,12,213-221.

Kuczyński, A., P. Varoquaux and M. Souty (1993). Light reflectance method to control the enzymatic browning in 'Ready-to-eat' apple products. *Int. Agrophysics.*, **7**(1), 85-92.

Kuczyński, A., J. De Baerdemaeker and J. Oszmiański (1994). An optical reflectance method for studying the enzymatic browning reaction in apple. *Int. Agrophysics.*, **8**, 421-425.

Kuczyński, A. and P. Varoquaux (1995). A method for estimation of enzymatic browning and its suppression in apple slices. *Int.Agrophysics.*, **1**, in press.

Kuczyński, A. (1995). The effect of cultivar on apple slice whiteness and enzymatic browning. *Zemedelská Technika*, **43**(2), 51-54.

Oleszek, W., C.Y. Lee and K.R. Price (1989). Apple phenolics and their contribution to enzymatic browning reactions. *Acta Soc.Bot.Polonaise.*, **58**(2), 273- 283.

Oszmiański, J. and C.Y. Lee (1990) Enzymatic oxidative reaction of catechin and chlorogenic acid in a model system. *J.Agric.Food Chem.*, **38**, 1202-1204.

Sapers, G.M. and F.W. Douglas Jr (1987). Measurement of enzymatic browning at cut surfaces and in juice of raw apple and pear fruits. *J.Food Sci.*, **52**(5), 1258 -1262,1285.

Spanos, G.A. and R.E. Wrolstad (1992). Phenolics of apple, pear and white grape juices and their changes with processing and storage - A review. *J.Agric. Food Chem.* **40**, 1478-1487.

Varoquaux, P. and R.C. Wiley (1994). Biological and biochemical changes in minimally processed refrigerated fruits and vegetables. In: *Minimally processed refrigerated fruits and vegetables.* (R.C.Wiley, Ed.), Chap. 6 , pp. 226-269. Chapman & Hall, London.

OFF-LINE COLOUR MACHINE VISION FOR ANALYSING THE BISCUIT BAKING PROCESS

Thierry MOLL(1), Christophe GUIZARD(2), Gilles RABATEL(2)

(1) S.G.D.P. : 212 Avenue Jean Jaurès 75019 Paris
Tél : (33) 1 40 40 99 80 Fax: (33) 1 40 40 99 81
(2) CEMAGREF : BP 5095 34033 Montpellier Cedex 1
Tél : (33) 67 04 63 00 Fax: (33) 67 63 57 95

Abstract: A colour machine vision has been developed and used for off-line quality control of biscuits. One of the major objectives was to evaluate the feasibility of contact-less colour control during the biscuit baking process, including dimensional aspects. To confirm measurements, we have compared colorimeter, and CCD colour camera performance on 20 referenced colour samples and on a 300 biscuit's set. With this test, we intend to prove the efficiency of colour test with a camera. The results of colorimetric analysis are not strictly comparable with those of a laboratory colour system, but it is enough to classify biscuits.

Keywords: Food processing, image processing

1. INTRODUCTION

Consumers expect to find a constant product quality for a same biscuit brand, but it is not easy to obtain "identical" biscuits even during a short baking period. Baking is a complex process (Moore, 1991) where biochemical and physical transformations give its characteristics to the biscuit (taste, colour, cracking and organoleptic qualities). The conduct of oven relies on the knowledge of operators, each of them having his own way to conduct the baking process. The quality team controls products, in laboratory, every 15 minutes for the most frequent test. These controls are tedious because most of them are based on manual operations and suppose different measurement systems. Today, only a few factories are equipped with sensors and most of these are off-line (Guizard *et al*, 1992). To insure constant quality throughout the process and propose an appropriate diagnosis, it is necessary to collect valuable information. If tolerance margins, fixed for each product, are overstepped, the operator has to modify the settings of the oven by tuning the speed of the conveyor, the steam extraction and the temperature for each oven zone. However, due to the oven inertia and the delay induced by off-line controls, it will take several minutes to correct drifts. This observation currently leads research laboratories to investigate on an on-line baking regulation. This supposes that more precise and more frequent non destructive controls of the colour, the moisture and dimensions have to be done. In this context, various works (Mc Farlane, 1990), (Precetti, *et al*, 1991), have already been undertaken to control on-line baking indicators. As dedicated sensors evaluate each indicator, it increases the cost, the number of equipment and the complexity of data processing. These constraints have lead us to develop a tool based on the idea of a unique intelligent sensor (Robert, *et al*, 1993) without any contact. The diversity of shapes, decorations, product positioning on the conveyor band have directed us to use a sensor based on artificial vision.

2. PRESENTATION OF THE SYSTEM

2.1. Equipment

We used a standard image analysis equipment :

- A CCD colour camera 756 x 581 *(Model SONY XC 711P);*
- A 3.5mm focal lens;
- A 24 bits colour acquisition board *(Imaging Technology CFG);*
- A patented fluorescent lighting system [1];
- A laser diode structured light.

This off-line equipment has been integrated in a compact system (fig 1) to ease the operator's work. This system is piloted by :

- A colour image analysis software (OPTIMAS);
- A PC 486 33Mhz microcomputer.

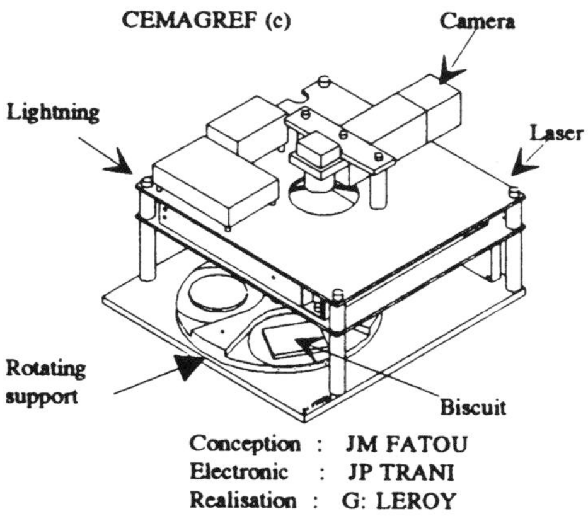

Fig. 1: Off line prototype

2.2. Calibration

A large part of the work has focused on a robust calibration of the system to insure a good reproducibility of measurements. The accuracy of our system has been evaluated according to two components : the repeatability and the reproducibility of measurements as describe in the AFNOR NF X 06-041 and ISO 5725 Norms. Series of measurements on a sample of 300 biscuits allowed to obtain repeatability and reproducibility indicators.

3. DIMENSIONAL CONTROL

We have chosen to measure three customary dimensions : the length and width for rectangular biscuits or the diameter for round ones, and the thickness. 2D vision principle allowed to obtain easily other morphological characteristics such as surface, perimeter, deformation or a precise cartography of the product. A laser light projection is used for the thickness control. This measurement is particularly important for operators because it guarantees a certain state of the baking, and the correct closing of the packaging. A vision based system allowed us to measure fairly easily the area and the diameter of round biscuits. Nevertheless, it was necessary to guarantee a perfect image segmentation to obtain a good reproducibility (biscuit separation from its support). In this case, the colour bringst very rich information to perform a reliable image segmentation. However, it's requires a good colour stability (cf. §4). The evaluation of the variation range and the reproducibility of the measurement has been made on a sample of 300 selected biscuits produce during a normal baking cycle. Results (fig 2) show that the reproducibility of the measurement is compatible with the observed variation of the diameter and the area. Similar results had been obtained during previous research (Krutz and Precetti, 1991).

	Area	Diameter
Variability	7.06%	5.75%
Reproducibility	1.12%	1.23%

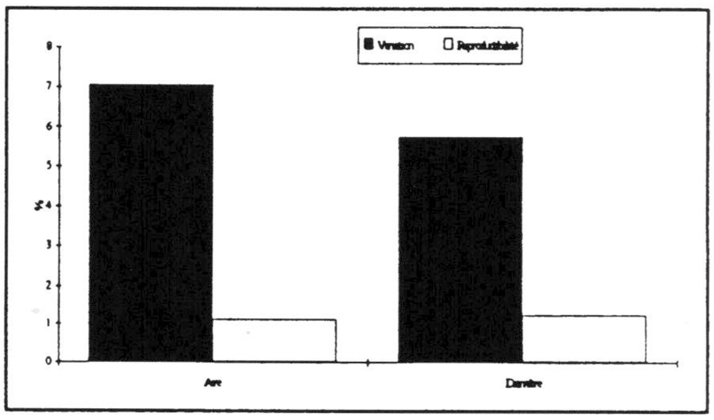

Fig. 2 : Dimensional measurement on a sample of 300 biscuits

3.1. Thickness measurement by laser projection

The biscuit thickness measurement is one of the main characteristics of the biscuits and it's critical for the packaging. Currently, the measurement is made off-line with calliper square help. For on-line solution ultrasonic sensors, have been proposed but they need a reference point which is difficult to obtain, because of undulations and deformations of the conveyor band. For this reason, before its integration into the off-line prototype, the thickness measurement has been tested on a "real" metallic conveyor band sample (fig 6).The thickness is measured by projection of a laser line directly on the biscuit surface. This laser beam forms an angle with the horizontal plan (fig 3) and the laser line splits into two parts : one on the biscuit surface and the other on the grid. This last is used as a reference with respect to the undulations of the conveyor band.

[1] CEMAGREF French patent N° 92 15666 *"Procédé et dispositif d'éclairage uniforme par conduction de lumière"*

We have to determine the distance **D** separating references and the trace on surface of the product and calculate the thickness 'e' (fig 4). With this method, the precision is only limited by the diffusion of the laser beam when the angle is too small.

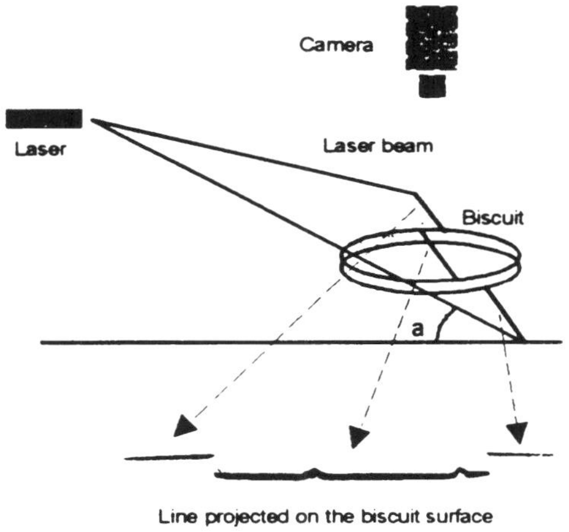

Fig 3: Measurement principle

Evaluated on eight samples (1.7 to 12 mm) for an angle of 15°, the results (fig 5) have showed a precision better than 50µm. This precision is compatible with the granularity of the biscuit surface (evaluated to 0.2mm). Holes and irregular texture on the biscuit surface disturbs the thickness measurement (fig 6). To determine the biscuit thickness, one solution has consisted in choosing the average thickness of the luminous trace on the product, after elimination of the holes. But when packaging tolerances are severe, the maximal thickness point will be more secure. The choice of the method is related to the type of product : these methods have been tested with a good reproducibility for products with strong relief (biscuit representing small animals).

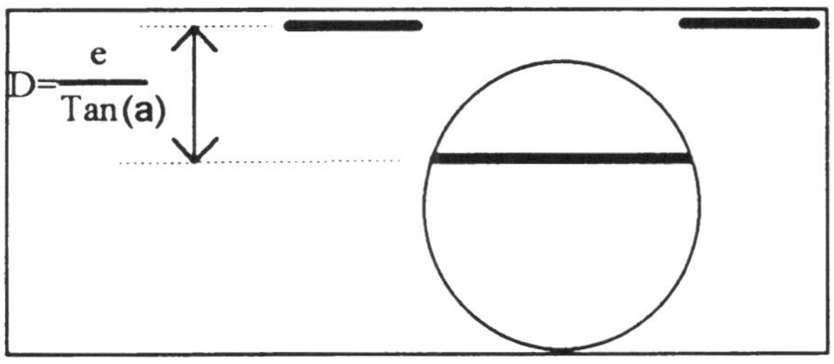

Fig. 4: Thickness measurement

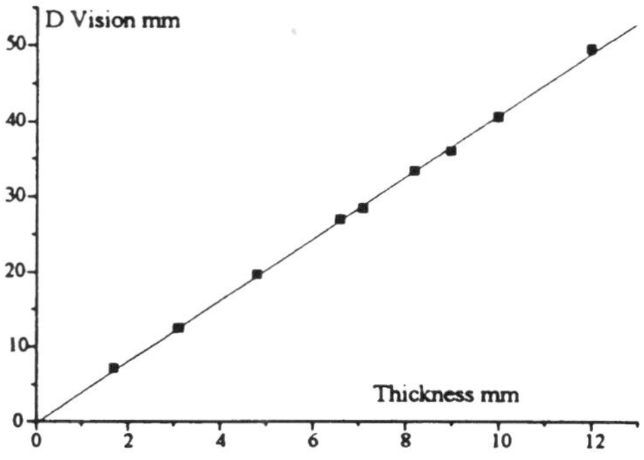

Fig. 5 : Tan (a) evaluation

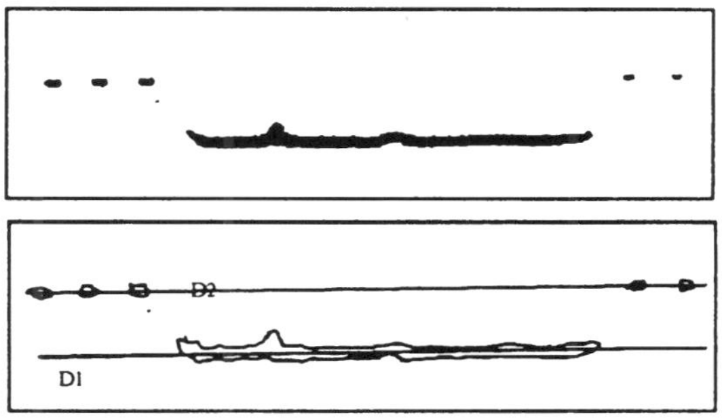

Fig. 6: Thickness measurement on conveyor grid.

3.2. Results

The stability of spatial calibration is efficient and the precision of the dimensional measurement satisfies manufacturer's expectations. So,it is possible to use off-line machine vision for the measurement of the diameter and the thickness of biscuits with a better accuracy than manual measurements, but the results are largely related to the colour stability.

4. COLOUR MEASUREMENT

The colour is certainly one of the most important criterion related to the product appearance. This is why the manufacturer evaluates regularly the colour of biscuits during the process using a colorimeter (Trouvé, 1991). In most cases, professionals use the "uniform" L*a*b* colour space (CIE 1976). For this reason we have tried to express results of the colour measurement in this normalised system. As we use a camera, a bi-dimensional information relative to the colour surface distribution can be obtained, which is impossible to obtain when using a standard colorimeter : colorimeter integrates the measure through a small surface with fixed size and at short distance from the product. Therefore, the heterogeneous surface of the biscuit (holes, inscriptions, drawing...) disturbs usual colorimeter measurement (Fig 7). Image processing allows to over step this constraint.

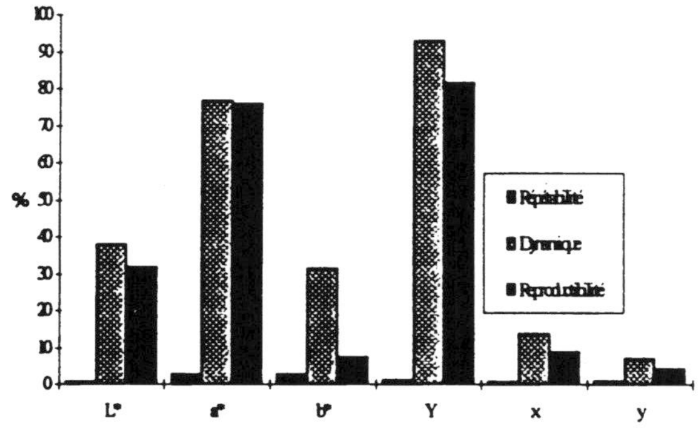

Fig 7 : Colour measurement with a colorimeter

Our objective is not to develop a tool that competes a colorimeter. We intended to be able to follow the drift of colour during the process and make measurements reproducible.

4.1. Colour calibration

The colour calibration is a major problem often met, but rarely solved by machine vision integrators. There are, along the acquisition and visualisation chain, many redundant adjustable settings that made the calibration empirical and the portability of the equipment not easy (Table 1).

	Possible changes
Lighting	Spectral Emission
Camera	Diaphragm and Gain
Acquisition board	3 R G B Separated inputs gain 3 R G B Separated inputs offset 3 R G B Separated outputs gain 3 R G B Separated outputs offset
Monitor Control	Contrast and Luminosity

Table 1 :Adjustable settings on a machine vision

Usually, operators use the monitor control as a reference to tune the system : they tune the camera diaphragm and the colour balance of the system in order to obtain a non saturated "good" image. It is difficult in these conditions to speak about reproducibility of any measurement. A colour calibration procedure has been developed to guarantee stability and reproducibility of the colour components.

The first work has consisted in analysing the lighting system. As we intended to compare the results of colour measurements to those of factory colorimeters, we had to be sure that our lighting system can be compared to a normalised illuminant.

For some technical reasons we have been brought to use compact fluorescent lamps as lighting source, even if they have not been designed for being used as reference lighting source for colour measurement (fig 8).

The method of lighting evaluation is based on the comparison of L*a*b* values obtained by computer simulation of the vision system : RGB components are calculated from random spectrum objects illuminated by normalised CIE D65 "Day Light" illuminant and our lamps, then L*a*b* components are calculated and represented graphically (fig 9 and 10).

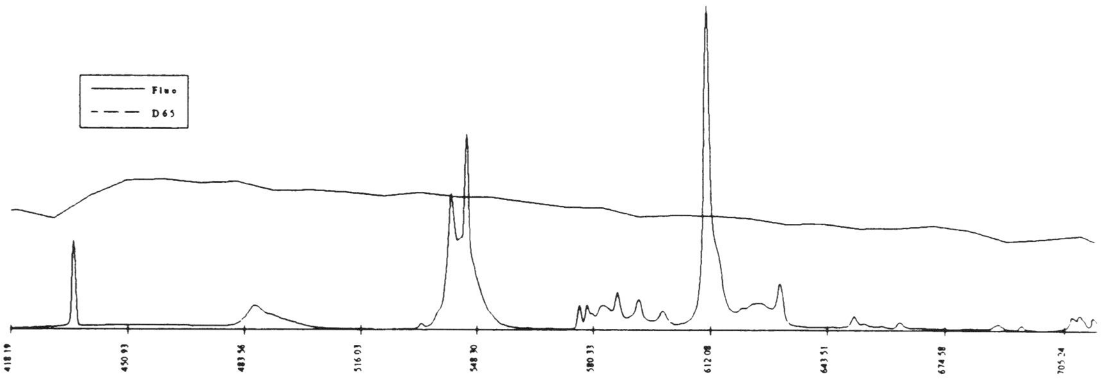

Fig 8: Fluorescent and D65 illuminants

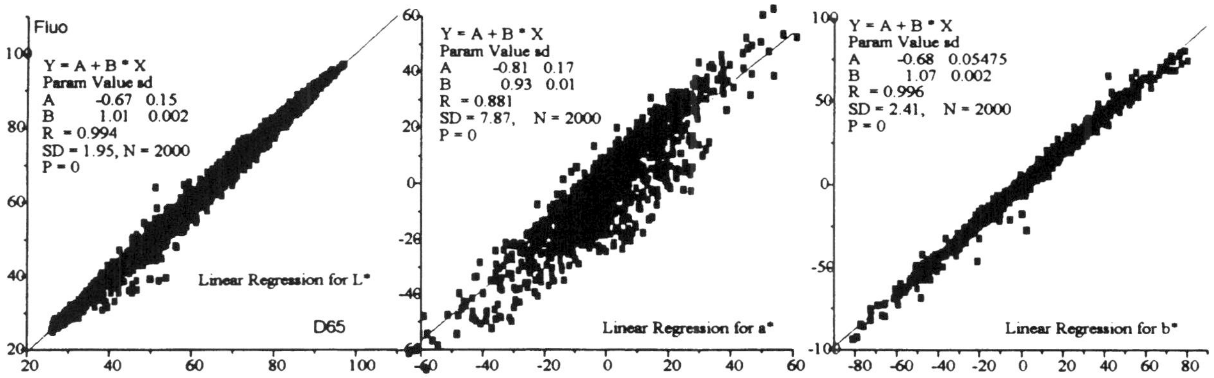

Fig 9 : L*a*b* fluo-compact and D65 illuminant comparison results (computer random spectrum)

Results obtained with random spectrum (Fig 9) can be considered as the worse possible results, because a real object has an absorption spectrum more continuous than those simulated and can often be approximated by a set of smooth functions like gaussians (Fig 10).
However, regression coefficients obtained show a good correlation except for a*. We have tested, successfully, the robustness of the method on real colour surfaces with different diaphragm adjustments.

Then, we have compared results obtained with a reference spectrocolorimeter (model ORIEL Multispec) to our fluo-compact lamp on real objects. A procedure of colour calibration has been developed by analysing a special hand-made 20 colours calibrated chart and by modifying automatically gain and offset of each of RGB channels during the acquisition, until the correct L*a*b* values are obtained.

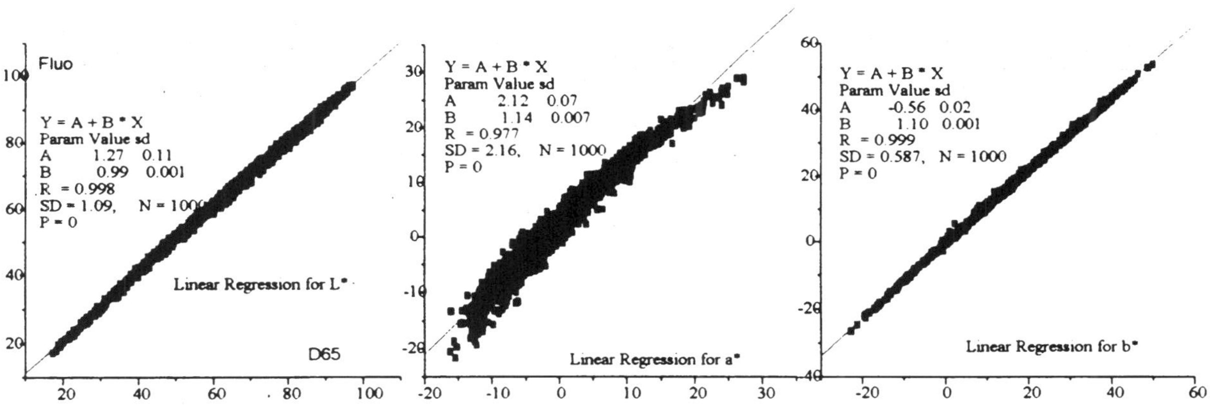

Fig 10 : L*a*b* fluo-compact and D65 illuminant comparison results (computer gaussian type spectrum)

For given diaphragm and shutter settings, values of gain and offset that minimised the global error constitute the optimal adjustment. The results of the correlation between colorimeter (Model MINOLTA CR 300) measurements and vision measurements in the L*a*b* system are presented here (fig 11).

The results are fairly good, but it is not possible to consider that our system can replace the colorimeter. The spectral measurement of RGB, XYZ or L*a*b* supposes that the object is illuminated by a lighting source as close as possible to the reference illuminant, and that RGB camera filters are as close as possible to the theoretical RGB filters normalised by the CIE. These ideal conditions are however difficult to achieve. A non continuous lighting source like fluorescent lamps cannot be compared to a normalised illuminant, and the CCD sensor use RGB filters without any guarantee of conformity with the theoretical filters. Therefore it is impossible to guarantee here to evaluate the exact CIE L*a*b* colour co-ordinate with our system. Nevertheless, despite the induced distortion of the L*a*b* colour space representation, the topological organisation is respected, in the case of materials presenting a continuous absorption spectrum like biological products.

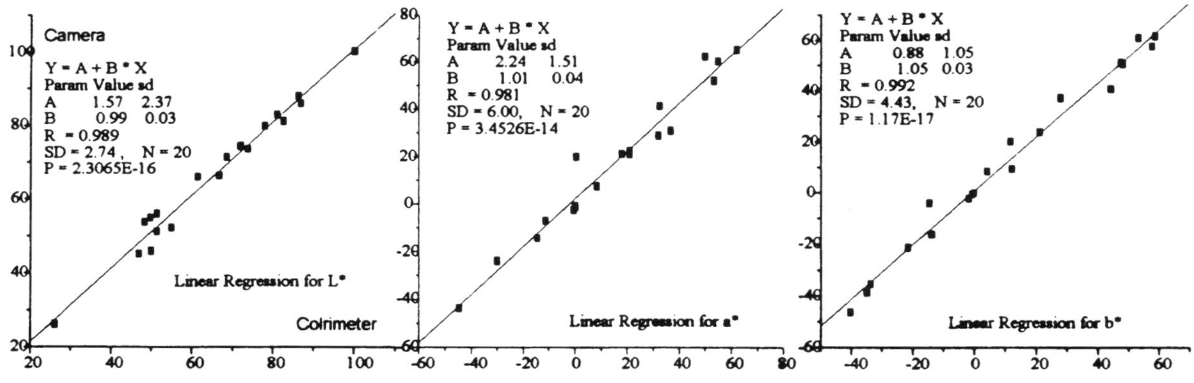

Fig 11 : Correlation between colorimeter and vision measurement of colour

4.2. Colour measurement on biscuits

During the first experiment we have measured the colour of the biscuit and evaluated the variation range and the reproducibility of L*a*b* values on a sample of 300 biscuits. The following results show that the vision is perfectly adapted to the relative colour measurement on biscuits. Therefore it is possible to follow the colour drift of biscuits during the production process. The analysed biscuit sample presents, for each component L*a*b*, a variation largely superior to its reproducibility error (Table 2).

	L*	a*	b*
Variation	28.8	12.9	16.4
Reproducibility	0.12	0.35	0.15

Table 2: Measurment variation and reproducibility error

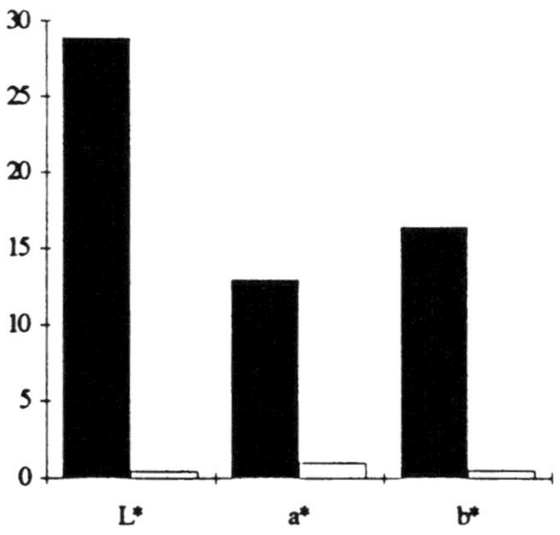

Fig 12: Colour measurement

5. CONCLUSION AND PERSPECTIVES

Spatial and colour calibration procedures are necessary to guarantee the stability and the reproducibility of the measurement system. Results have allowed to develop a system able to analyse geometric dimensions and the colour variation of biscuits during the baking process. Results shown here prove that it is possible to respect the specification of each manufacturer. The off-line prototype developed by this research prefigures the future on-line system and allows to consider in a close future a true automation of baking ovens.

6. REFERENCES

Guizard, C., V. Bellon, F. Sevila (1992) Vision *Artificielle dans les Industrie Agro-Alimentaires - Méthodes/Techniques/Choix*, Chp 1 and 2, DICOVA, ISBN 2-85362-305-X

Mc Farlane, I. (1990) - Simultaneous feed forward and feedback control applied to a biscuit backing oven.- *Automatic Control of Food Processes International Symposium II* - p 2.1

Krutz, G.W., Precetti C.J. (1991) - Automated cracker inspection using machine vision *ASAE* N° 91-7535.

Precetti C.J., M. Danzart M., G. Trystram., GW. Krutz (1991) - Accuracy of cracker size measurement using machine vision- *ASAE* N° 91-7535.

Moore, C.A. (1991) Automation in the Food Industry - Blackie.

Gonzalez R.C., P. Wintz (1987) - *Digital image processing* - Addison-Wesley Publishing Compagny - Second Edition.

Trouvé, A. (1991) - *La mesure de la couleur* - AFNOR CETIM.ISBN 2-12-450811-3

Robert, M., M. Marchandiaux, Porte M. (1993) - *Capteurs intelligents et méthodologie d'évaluation* - HERMES.

EVALUATION OF STARCH MODIFICATIONS FROM THERMO-MECHANICAL PROCESSES BY NMR

Armel Davenel, Philippe Marchal

CEMAGREF, 17 avenue de Cucillé, 34044 RENNES Cedex ,FRANCE

Abstract: Low resolution NMR technique was used to measure the effects of different process conditions on the transverse relaxation signal of different industrial products, sauces, cake batters and corn flakes. In sauces, the increase of the global relaxation time was related to cooking time and temperature. The cooking effect of batters was associated to a significant increase of the liquid-like starch protons and a transfer of sugared water in the starch partially gelatinized. The transverse relaxation behaviour of corn was very sensitive to the flaking process: a proton population tended to disappear with the increase of the flaking intensity.

Keywords: magnetic resonance, food processing, control equipment, relaxation analysis

1.INTRODUCTION

Food industry places increased emphasis on the development of sensors which can be used for monitoring processes. Low resolution NMR technique is well known as a rapid method to measure moisture and fat content of different kind of food products. Several previous works showed that ^{1}H NMR relaxation times of water and exchangeable starch protons in high moisture products were significantly affected by swelling and gelatinization mechanisms during heating (Lelievre and Mitchell,1975; Mora-Gutierrez, 1989). However very little information is available on the capability of this technique to follow starch modifications occurring during pure cooking or thermo-mechanical processes of complex and low or intermediary moisture food materials.

Leung, *et al.* (1979) observed two liquid components in the transverse relaxation of flour doughs. However, it would be hazardous to want interpret these variations of relaxation times in terms of water mobility. Proton relaxation is strongly affected by chemical exchange between water protons and exchangeable protons of macromolecules but also by cross relaxation. Oxygen-17 NMR would be the only NMR technique that reports purely on the behaviour of water (Belton,1990). Several publications report the water states by this technique in native starch products (Richardson, *et al.*, 1987), in starch/sucrose systems (Chinachoti and Strengle, 1990) and in starch gels (Lim, *et al.*, 1992).

However proton relaxation measurements in aqueous systems can potentially give relevant information about the state of biopolymers (Hills, 1992). More they are much more available to obtain rapid information in food plants. Using pulsed low resolution proton NMR (20MHz) Schierbaum (1992) found a remarkable linear relation between the solid-liquid ratio from NMR and X-ray crystallinity in thermally reversible maltodextrin gels. Teo (1992) applied the same method to study the starch retrogradation.

In this work, a similar method associating the [liquid/(liquid+solid)] ratio from free relaxation induction decay together with an analysis of the transverse relaxation behaviour of liquid protons from spin echoes sequence was used to study the effects of cooking time and temperature on cake batters and on sauces and the effect of the flaking intensity on corn flakes.

2.MATERIALS AND METHODS

2.1 Preparation of samples

Cake batters were mixed in a kitchen mixer and cooked in small baking cups with an electric oven at 220°C in the laboratory to test the effect of cooking time on the same standard industrial formula (table 1). For each baking of a formula, cups were sampled at 0, 2, 3, 4, 5.5, 9.5 and 13 mn during cooking.

Table 1 formula and chemical composition of cake batter

Ingredients (% wet basis)		Chemical composition (% wet basis)	
sugar	22.6	sucrose	22.6
flour	31.6	starch	19.6
eggs	27	protein	6.1
shortening	18	fat	18
baker powder	0.8	moisture	27.1

Complex industrial sauces were made from the same percentage of modified starch, mashed tomatoes and water mixed with about 30% sliced vegetables (carrot, pepper, on ion...). Sauces were cooked at 80°C and 90°C during 13, 25 and 40 mm. Sauces filtered before NMR measurements.

Corn flakes were realised with an industrial pilot in a pet food plant. Corn kernels first humidified at about 23% moisture content were softened in a steam chamber during 95 mm before flaking at five different spaces between metal rolls: 0.25, 0.4, 0.65, 0.85 and 1.1mm.

2.1 NMR measurements

The measurements in glass tubes of 10mm outer diameter were performed on a Bruker Minispec PC120 operating at 20 MHz for samples of sauces at 20°C, cake batters and centers of the cakes at 40°C to make sure that fat was totally in an amorphous state. Given their low specific weight, kernels and corn flakes were analysed at 40°C with a Bruker Minispec PC100 operating at 10 MHz and equipped with a 40 mm diameter probe.

Free relaxation decays were registered at 12 μs (FID12) after a 90° pulse with Minispec PC 120 (figure 1) and on the solid echo maximum (SEM) after a solid echo two pulse sequence (90°(0°)-τ-90°(90°); τ=15 μs) with Minispec PC100. Liquid-like free relaxation decays also were registered each millisecond between 50μs and 100μs after the beginning of the first pulse.

To reduce the effect of the heterogeneity of the magnetic field on the relaxation of the more mobil protons, true liquid relaxation signals were obtained by the 845 spin echoes amplitude maxima (0.2ms apart) of a spin echoes CPMG pulse sequence (figure 2). The delay between repetitions was 1s for cakes and corn products and 5s for sauces to allow the nuclear magnetization to return to its equilibrium value. To eliminate the effects of the variations of bulk density between the samples, all NMR intensities measured were divided by the FID12 value or, for corn products, by the SEM value, both approximatively proportional to the total number of protons.

Amplitude L of the liquid phase was estimated by the value of the free induction decay at 70 μs and then used to calculate a liquid/(solid+liquid) ratio (L/FID12 or L/SEM). True liquid relaxation curves M(t) can be often fitted by a sum of discrete exponential terms:

$$M(t) = \sum_{i=1}^{n} M_i(0).e^{-\frac{t}{T2_i}}$$

However the theory for the interpretation of proton relaxation in complex and water limited products is much less developed than for diluted ones. It should be hazardous to presume the number of exponentials in those systems. More in complex system like cakes the true relaxation spectrum F(T2) is more probably to be continuous, reflecting some heterogeneity in the distribution of water and fat.

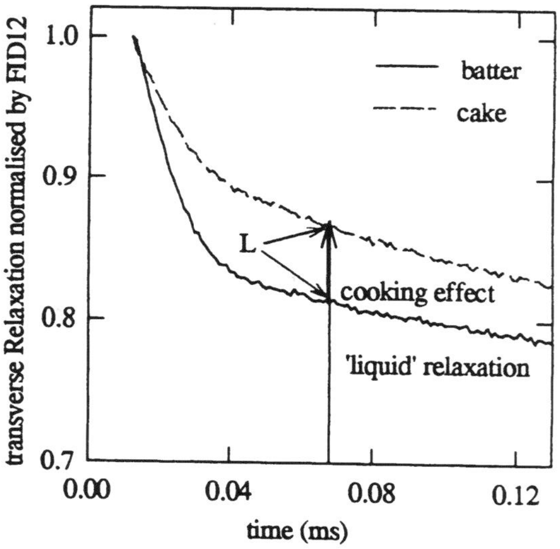

Fig. 1 Free relaxation decays of cake batter before and after cooking

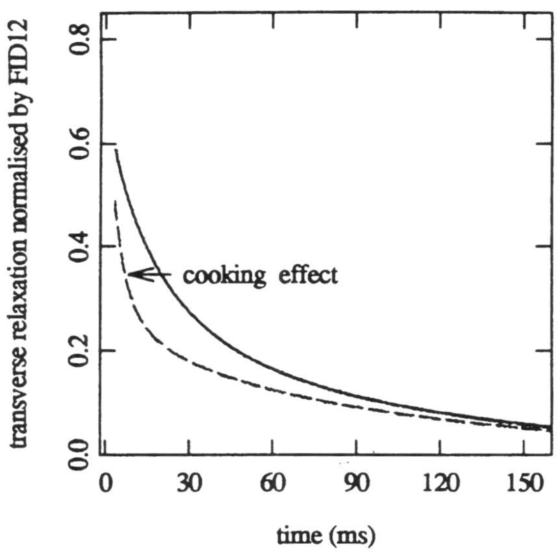

Fig. 2 True relaxation decays (CPMG curves) of cake batter before and after cooking

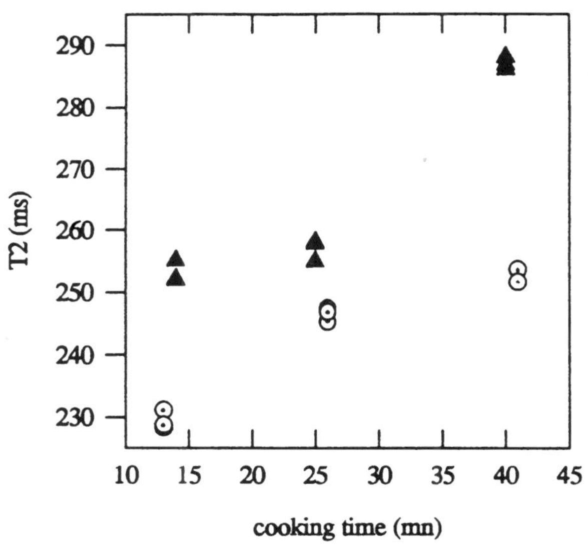

Fig. 3 Long transverse relaxation times of sauces at 80°C (O) and 90°C (▲)

We also used the same maximum entropy method (MEM) as Tellier, *et al.* (1993) applied to investigate the distribution of water during syneresis of curds. It allows no restrictive prior assumptions about the number of exponentials and introduces the least correlations in the reconstructed distribution from noisy data. Because MEM sofware used here only performed data sampled with constant spacing in the time domain, fittings were limited to spin echoes data.

3. RESULTS

Sauces. Moisture contents of all cooked sauces were 73 ± 1%. In its high moisture systems, no solid-like protons are detected and the liquid-like true relaxation curves can be explained by an isotropic three-state model with fast exchange between different proton species. With this model, one can expect to observe a single relaxation time which is based on a population-weighted average between "free" ($T2_f$), "hydration" ($T2_h$) water protons and exchangeable protons of macromolecules ($T2_{ex}$) expressed as:

$$\frac{1}{T2_{obs}} = \frac{P_f}{T2_f} + \frac{P_h}{T2_h} + \frac{P_{ex}}{T2_{ex}}$$

However, transverse relaxation curves of filtered sauces were better fitted by a sum of two exponential curves. The very small proton population observed (2.5% of total protons, T2=24-30 ms) could be assigned to non-exchangeable protons of macromolecules. Eventually, only the longer relaxation times of the abundant population time were usable to characterize the effects of the cooking parameters (figure 3).

Variance analysis showed that the observed transverse relaxation times were very significantly related to both time and temperature effects and were not affected by the slight variations of moisture content occurring during the experiments. When aqueous suspension of starch is heated, starch granules swell above the gelatinization temperature. This can result in an increase in the water hydration population and the relaxation time T2ex due to enhanced mobility of the polymer chains. These results suggest that NMR may be a relevant and rapid tool to control the cooking intensity of complex industrial sauces. Further investigations are underway to measure transverse relaxation time of the continuous phase in intact sauces with its all vegetable components.

Cakes. In intermediary moisture products, the molecular diffusion of water and dissolved sugar is strongly restricted. In complex systems like batter and cake, there are heterogeneities larger than the region sampled by any one water molecule in its own intrinsic relaxation time. Also multiexponential relaxation behaviour can be expected in its products. MEM allows to identify in batter three main populations (figure 4). The population POP1 with T2 less than 2ms could be assigned to water and starch exchangeable protons in starch granules characterized by a highly anisotropic mobility. The larger population POP3 (average T2, 20ms) counts about the half of the liquid-like population and would correspond to the larger part of sugared water outside the flour particles: its motion would be far less affected by starch or gluten interactions than the previous one. The third population POP4 with a large T2, about 100ms, is mainly due to fat protons, liquid-like to this temperature. Its amplitude decreased to 10 to 25 per cent after baking.

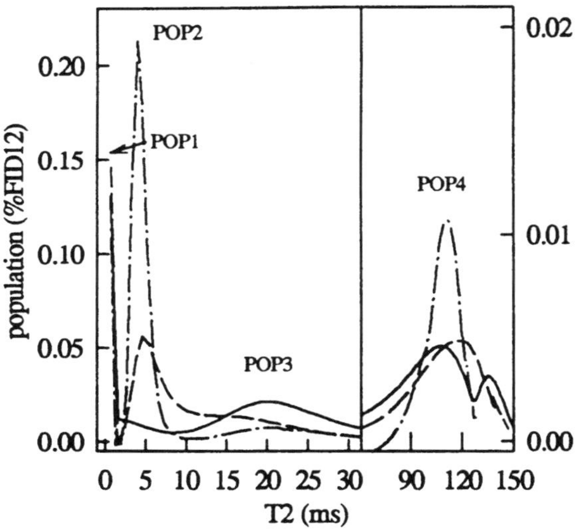

Fig. 4 Continous T2 spectra of batter (——), mid-cooked batter (- - -) and cake (— . —).

In batters, a part of this population POP4 could be allotted to water or sugared water not very affected by starch, for example residual droplets in fat emulsion.

In spite of moisture was sligthly reduced by about 1.5% during the about 5mn cooking phase, the L/FID12 ratio or total liquid protons POPT markedly increased about 8 per cent in the same time (figure 5). This behaviour could be attributed to a more or less important reducing of crystallinity related to the melting of sugar or starch. Previous work (Davenel and Marchal, 1995) showed that this increase was only correlated to the starch content. During melting a part of the starch protons would become liquid-like. During the cooking phase, the T2 distribution evolved and cake spectrum was characterised by a population POP2 (average T2, 5ms) situated between the two initial first populations.

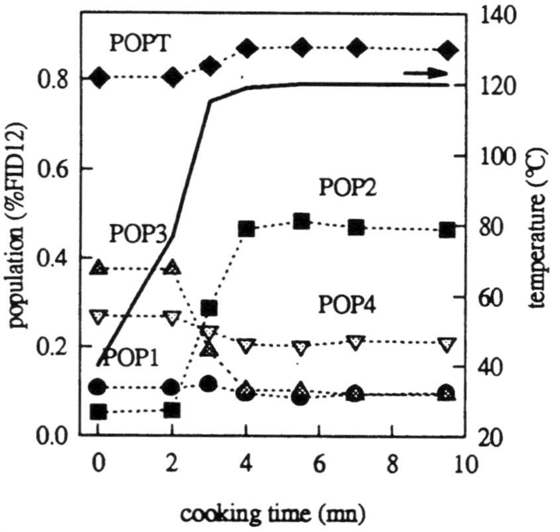

Fig. 5 Evolution of proton populations during batter cooking. Inside temperature of batter (——).

Only a small part of the population (T2, 20ms) subsisted in cake. The rise of the population POP2 in cakes is related to the swelling of the starch granules and this population would correspond to sugared water entrapped in gelatinized starch (Morra-Gutierrez and Baianu, 1989). Analysis of continuous T2 distribution of NMR relaxation signals is a very useful tool to have some comprehensive view of the modifications of relaxation times occurring during cooking of intermediary-moisture complex products.

Corn flakes. True relaxation decays of corn flakes sampled just after flaking and before drying are shown in figure 6. The moisture content of these samples was 25.2±.4%: the sharp evolution observed in the signal could be clearly assigned to the effect of the flaking intensity. After drying, the moisture content of flakes was between 12 and 15% wet basis.

Maximum entropy method split up signals into three proton populations. The first one, with relaxation time between 1 and 2 ms, was attributed to water protons in strong interaction with macromecules in starch granules. The population with the longer relaxation time about 100ms is attributed to very liquid fat protons in the literature. An intermediary population with a relaxation time T2int between 10-35ms tended to disappear with the reducing of the space between flaking rolls.

Variance analysis showed that the amplitude of this population is very significantly correlated to the space between flaking rolls (figure 7) whatever the moisture content of samples. Its amplitude did not decrease during the drying phase.This population could not be related to some water protons.

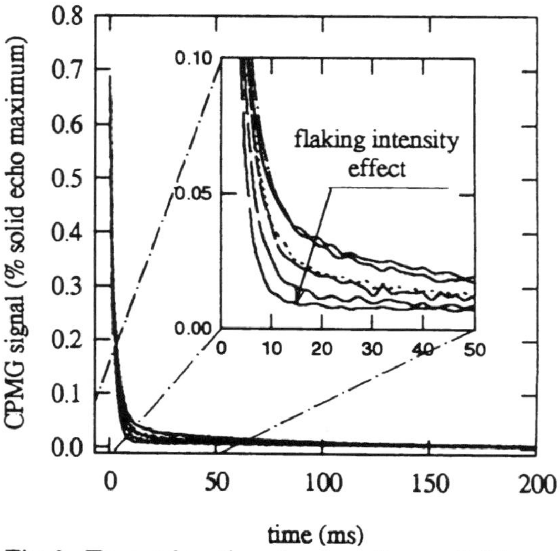

Fig.6 True relaxation decays of corn flaking with different intensity levels

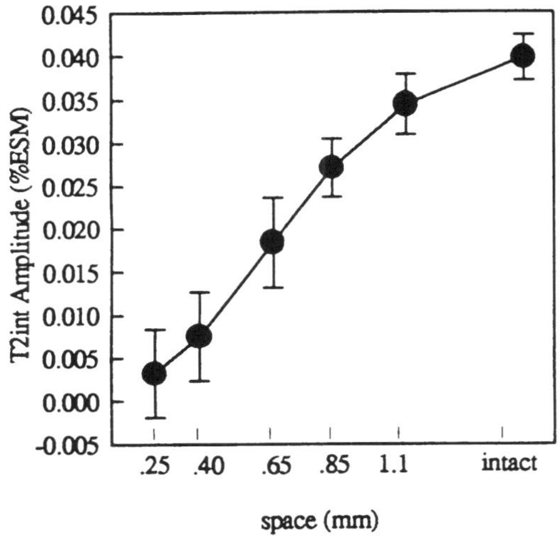

Fig.7 Amplitude of the intermediary population versus space between flaking rolls

Moisture content was strongly positively correlated to the amplitude of the first component (R2=.99%; SEC=.3% w. b.). The increase of the starch degradation with the flaking intensity was confirmed by measurements of enzymatic susceptibility to alpha-amylase by a kinetic method (Tollier and Guibot (1974). However , the evolution of the relaxation behaviour of corn flakes could not undoubtedly attributed to the starch modifications occurring during flaking process.

4. CONCLUSION

Transverse relaxation behaviour of different food products during thermal or thermo-mechanical processes was studied. Except for corn flakes, the NMR modifications observed were clearly related to some important starch modifications. These results show that low resolution NMR technique can be a relevant tool to know the state of starch in high or intermediary moisture foods cooking in different conditions.

5. REFERENCES

Belton, P.S. (1990). Can NMR give useful information about the state of water in foodstuffs? *J. Agric. Food Chem.*, **2**, 179-209.

Chinachotti, P. and T.R. Stengle (1990). Water mobility in starch/sucrose systems: an oxygen-17 NMR study. *J. Food Sci.*, **55**, 6, 1732-1734.

Davenel, A., P. Marchal P. and J.P. Guillement (1995). Rapid cooking control of cakes by low resolution NMR. In: *Magnetic Resonance in Food Science* (Webb, G.A., V.M. Gil, I. Delgafillo and P.S. Belton, Eds). In press. Royal Society of Chemistry, Cambridge.

Hills, B.P. (1992). The proton exchange croos-relaxation model of water relaxation in biopolymer systems. *Molec Phys.*, **76**, 3, 489-508.

Lelievre, J. and J.A. Mitchell (1975). A pulse NMR study of some aspects of starch gelatinization. *Starch*, **27**, 113-115.

Leung, H.K., J.A. Magnusson and B.L. Bruinsma (1979). Pulsed nuclear magnetic resonance study of water mobility in flour doughs. *J. Food Science*, **44**, 1408-1411.

Lim, H., C.S. Setser, J.V. Paukstelis and D. Sobczynska (1992). O-17 nuclear magnetic resonance studies on wheat starch-sucrose-water interactions with increasing temperature. Cereal Chemistry, 69, 4, 382-386.

Mora-Gutierrez, A. and I.C. Baianu (1989). 1H NMR relaxation and viscosity measurements on solutions and suspensions of carbohydrates and starch from corn: the investigation of carbohydrate hydration and stereochemical and aggregation effects in relation to 17O and 13C data for carbohydrate solutions. *J. agric. Food. Chem.*, **37**, 6, 1459-1465.

Richardson, S.J., I.C. Baianu and M.P. Steinberg (1987). Mobility of water in starch powders determination by nuclear magnetic resonance. *Starch*, **39**, 198-203.

Schierbaum, F., S. Radosta, W. Vorweg, V.P. Yuriev, E.E. Braudo and M.L. German (1992). Formation of thermally reversible maltodextrin gels as revealed by low resolution H-NMR. *Carbohydrate Polym.*, **18**, 155-163.

Tellier, C., F. Mariette, J.P. Guillement and P. Marchal (1993). Evolution of water proton nuclear magnetic relaxation during milk coagulation and syneresis: structural implications. *J. Agric. Food Chem.*, **41**, 2259-2265.

Teo, C.H. and C.C. Seow (1992). A pulsed NMR method for study of starch retrogradation. *Starch*, **44**, 288-292.

Tollier M.T. and Guilbot A. (1974). Le maïs-grain: préstockage, séchage et qualité. IV Charactéristiques de la fraction glucidique des échantillons de maïs-grain. *Ann. Zootech.*, **20**, 633-640.

SHAPE RECOGNITION ALGORITHM USING SIMULATED RETINAL FUNCTION FOR GRAFTING ROBOT SYSTEM

Y.Nishiura, H.Murase, N.Honami

Department of Regional Environmental Science, Faculty of Agriculture
University of Osaka Prefecture, 1-1 Gakuen-cho, Sakai, Osaka 593, Japan

ABSTRACT: Development of intelligent systems for bioproduction including a grafting system requires high level machine vision techniques. The required task is to extract useful information through the machine vision. It is of interest that introduction of more human like image processing including software and hardware aspects into the machine vision system is considered. In this study, a solid state receptor arranged over a polar coordinate system which serves as an artificial retina was fabricated. An algorithm which converts initial image data into a specific pattern containing useful features using 2 dimensional FFT and polar coordinate transformation was developed.

Keywords: Image sensors, Image recognition, Image processing, Robotics, Agriculture

1. INTRODUCTION

Automation and mechanization in many sectors of agriculture have been studied and realized. The automation and mechanization of fruit, vegetable and seedling production have been given priority. The development of a labor saving, automated, high quality seedling-production system has brought remarkable success. Such seedling production systems for fruits and vegetables are rapidly spreading among growers. The next task in developing automation and mechanization of fruit, vegetable and seedling production is to automate the management process for cultivation care for the plant growth.

Grafting is one of the techniques for improving the quality of seedlings. Grafting is a time and labor-consuming procedure in seedling production. Recently, development of an automatic grafting system has been initiated in Japan. A few automatic grafting systems have been already marketed. In this research work, a fully automatic grafting system was discussed including some aspects of a robotic system. Development of a preprocess handling systems for seedlings is crucial in a fully automatic grafting system. The preprocess handling system for seedlings involves development of a machine vision system for recognition of a seedling and development of an end effecter to cut seedlings to prepare a pair of scion and stock. Recognition of the shape of a seedling is an essential part in the handling process. There are various sensing techniques to recognize information from production objects or object information in relation to production methods. Pictorial instrumentation techniques, which are able to take information remotely and non-invasively, are some of the most effective techniques. The shape data have positional and size information for handling by a robot, and conditional information on the seedlings such as quality, age, health, and so on. In our past reports, a development of a vision system for preprocess handling of seedlings, especially a shape recognition algorithm, was discussed. The software consists of two parts. The first part of the program handles the extraction of a binary image of a seedling. The size measurement or positional detection(which we call shape recognition) is done in the second part. In that result, we found that our fabricated method was effective enough for measurement. However, especially, the qualitative information on seedlings is very important, as judged from the color and shape condition, the shape recognition has not been completely done.

In this report, we deal with the shape information as a pattern. The object of this study is the fabrication of a new method which can recognize a shape pattern. Such a recognition function in a machine is required for the development of automatic, intelligent and autonomic systems for bio-production management and harvesting. The recognition method in picture instrumentation is roughly classified into the following three types (Fig. 1).

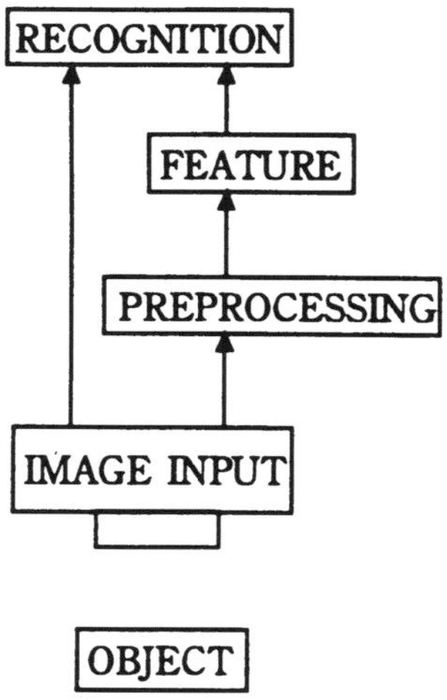

Fig. 1. Recognition method and process

1.1 Method of regarding judgment and recognition processing as important

Upon recognizing an object or object information, it is necessary and important to compare that with the knowledge data base learned in advance, to judge its appropriateness, or to construct the input/output structure of recognition system, after creating a rule on the basis of the knowledge data base. Then judgment and recognition are carried out. In the type in which judgment and recognition processing is regarded as important, when this input/output information is picture information and shape recognition of the object, the input pictorial data are dealt with directly as the object of judgment and recognition, without any selection processing of feature information during input of the pictorial information. To do this, a high level processing capacity for judgment and recognition is required, and it is necessary to learn well in advance like the human thought process. This has the advantage that a pictorial input apparatus can be simplified and has a high potential of ensuring arbitrariness. However, when the arbitrariness is put into practice, the information processing system inevitably reaches a high level and becomes complex in proportion to its extent. Therefore, the loads on hardware and software become significant. In other words, rough eyes and a wise brain should be created.

1.2 Method of regarding image acquiring process as important

On the contrary, the method of regarding an image acquiring process as important originates from the reason that a dull brain can see and understand well by having a good eye and simple judgment. Using this method, information for judgment and recognition is processed (for example, filtered or noise removed) and simplified in advance. Because the structure of the processing defines the quality of the data, it is important to extract useful information for this purpose from complex and chaotic information without removing the information necessary for judgment and recognition. In this case, the structure of the information is necessary to reach a definite decision, and its physical feature and regularity should be evaluated.

1.3 Compromise method

In this type, information is arranged by preprocessing to a certain extent. This will be the most general method. When a pictorial image is input, objective information must also be included. Upon extracting this information, techniques such as pictorial instrumentation for specific information or a feature extracting processing from pictorial information are necessary. In this process, the information is extracted through processing using various sensors and filters. Information used for the final judgment is summarized as binary information not exceeding four dimensions, and those information processing systems are constructed in a combination form of either parallel or serial. One of the simplest pictorial information processing systems for constructing this information can be taken as 2 dimensional multi value information in a plane (it may be a surface) shifted in time and depth, which constructs the shape information. For the shape information, information must be transformed into recognition in order to recognize the information. An algorithm is required in this case. The objective of this study is to transform the construction of pictorial data in order to briefly do shape recognition by pattern recognition in which a computer is useful. To practically do this, an algorithm for the processing system to transform the input pictorial data to pattern information including shape, is constructed using a 2 dimensional FFT and polar coordinate transformation.

2. EXAMPLES OF USING 2 DIMENSIONAL FFT

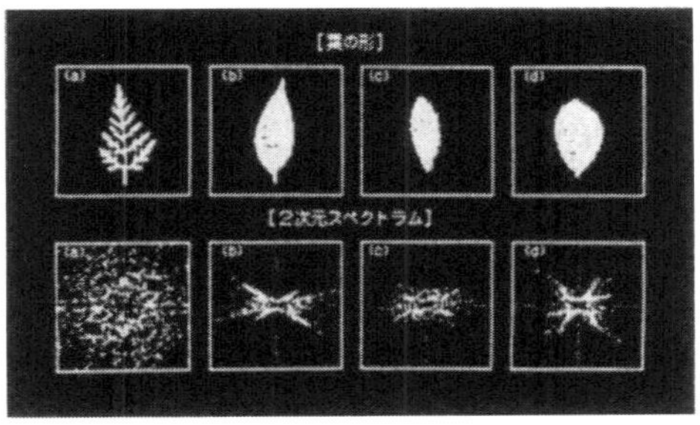

a. The shape of a leaf
(above : pattern, below : 2 dimensional spectrum)

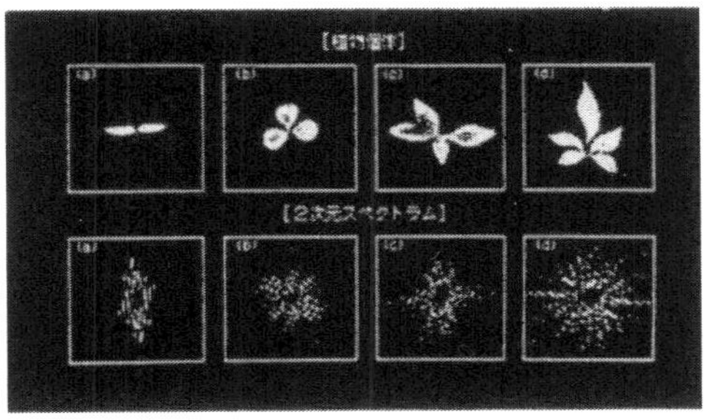

b. The number of leaves
(above : pattern, below : 2 dimensional spectrum)

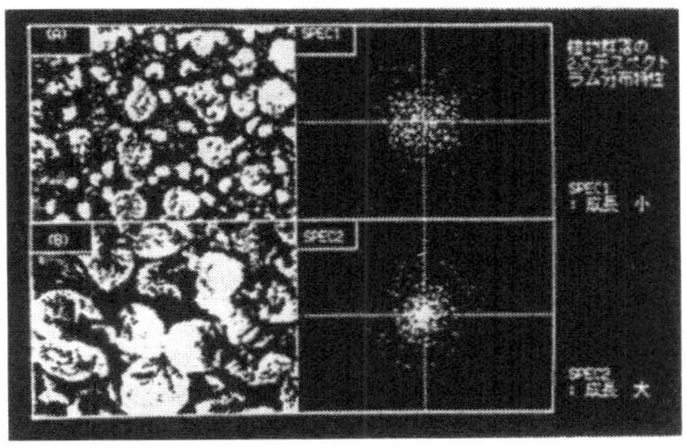

c. The plant density
(above : pattern, below : 2 dimensional spectrum)

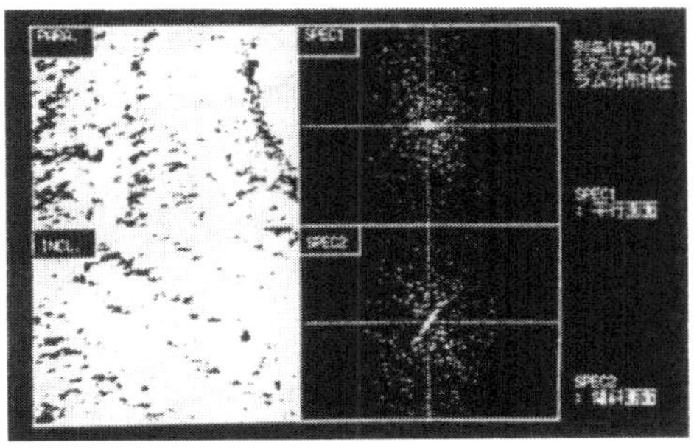

d. The direction of row
(above : pattern, below : 2 dimensional spectrum)

Fig. 2. The shape characteristics of plant and 2 dimensional spectrum pattern

Fig. 2(a) shows the shapes of a leaf from some kind of a plant and their 2 dimensional spectrum pattern. Then Fig. 2(b) shows the numbers of leaves and their 2 dimensional spectrum pattern. Fig. 2(c) shows the plant density and its 2 dimensional spectrum pattern. Fig. 2(d) shows the direction of a row and its 2 dimensional spectrum pattern. The characteristics of the shape and direction effect a 2 dimensional spectrum pattern. Those results give some rough information such as the direction and the edge of the shape, but it is difficult for the shape recognition to obtain all information from only the 2 dimensional spectrum pattern. It is necessary to process this data by any method. Then it is important for the shape recognition to standardize the shape pattern. Therefore, we try to do this by the polar coordinate method and 2 dimensional FFT transformation.

3. EXPERIMENTAL APPARATUS AND METHOD

3.1 Experimental apparatus

Fig. 3 shows the image acquisition equipment and processor. A monochromatic CCD camera with zoom and auto focus functions is used for the input equipment, and the image processing system is composed of an image input board, a controller (NEC, PC9801 VX) and a monitor. Quick BASIC is used for development of the processing program for the image input and image data.

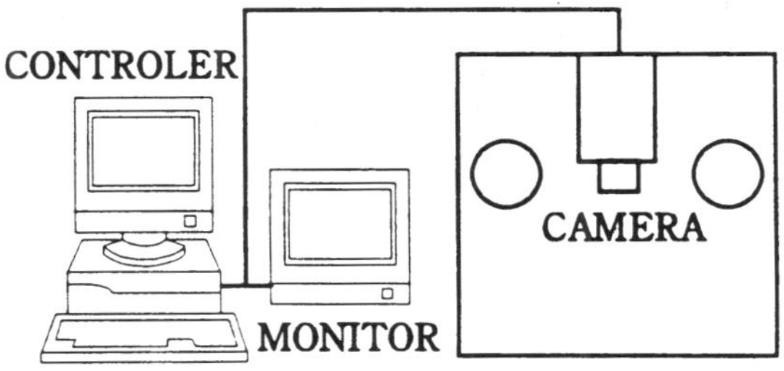

Fig. 3. Image acquisition system and processor

3.2 Experimental method

In order to simplify recognition, binary data transformed from multi value image data (integers from 0 to 255) are used. The pattern is constructed by enlarging the objective shape feature, 2 dimensional FFT calculation and polar coordinate transformation. Then it is processed into useful information for recognition by standardization of geometrical information such as size, position and direction. For multi value image data on various shapes, data with the same pattern and diverse sizes, position and direction are prepared and are processed by the following procedure. Multi value image data are also processed similar to the existing state.

Objects with diverse sizes, position and direction have to be judged the same in the shape recognition. Information is standardized and transformed into pattern information through the following processing (Fig. 4). First, with regard to an object' s size, as the number of pixels (resolving power) in the frame of the camera is defined, so the total pixel number (number of picture elements) in the objective pattern must be standardized in the binary state. Hence, the magnification of a camera lens is adjusted so as to have the same number as much as possible. The size in a frame is standardized by zooming. Next, in regard to position, in order to determine if it is the same no matter where it is, it is necessary to find the center of the object and transform it into relative coordinates. By taking notice of the 2 dimensional spatial frequency, it can be seen that the influence of position is canceled when the 2 dimensional FFT calculation is made, by considering the periodic continuity in the X- and Y-directions included in the pattern data. Finally,in regard to a direction, considering a polar coordinate having its center at the center of the figure (zero dimension in the case of FFT), spatial frequency, which has two dimensions of angle and diameter(radius) in the same way as in the X- and Y-directions, the influence of direction is canceled by the 2 dimensional FFT calculating. 2 dimensional information obtained by standardizing the information regarding size, position and direction in a certain way produces the pattern information of the figure.

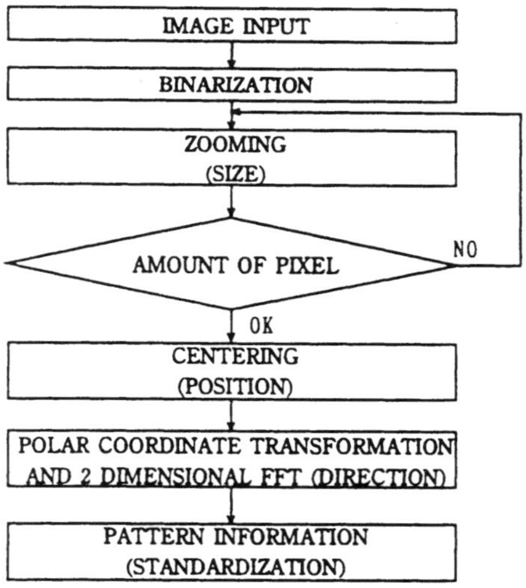

Fig. 4. Flow chart of shape recognition

4. EXPERIMENTAL RESULTS

4.1 Size (magnification, reduction)

Information about objects with the same shape and diverse size is transformed using the 2 dimensional FFT, the 2 dimensional spatial frequency spectrum is then found, and its results are summarized in a figure as shown in Fig. 5 (a:standard pattern, b: spectrum distribution of standard pattern, c: reduced pattern, d: spectrum distribution of reduced pattern). The figure has a zero dimension of spatial distribution at the center and shows a higher order towards both ends. For example, if the amount of data dealt with here is 16 x 16 (pixel), then up to $\pm$8th order can be found, and its spectrum pattern has features of point symmetry having its center at the origin. In order to define the feature of spectrum distribution, the direct current component is standardized to 255, and the data are standardized by allocating them proportionally to other values. In addition to the above, data have to be in the form of real/imaginary

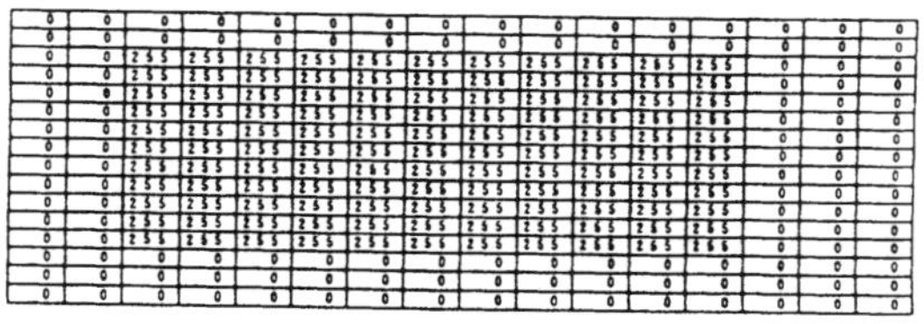

a. Standard pattern

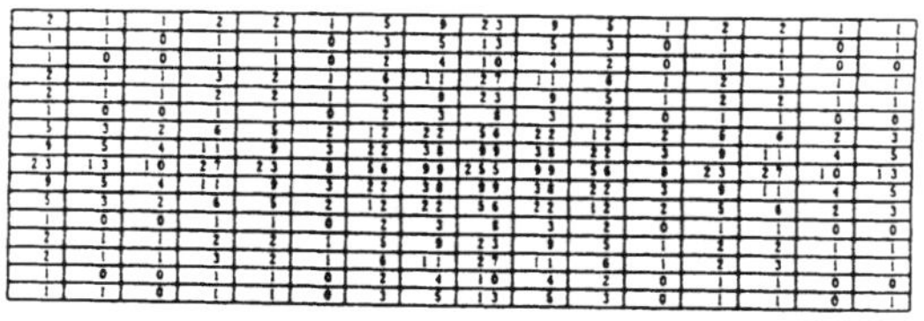

b. Spectrum pattern of standard

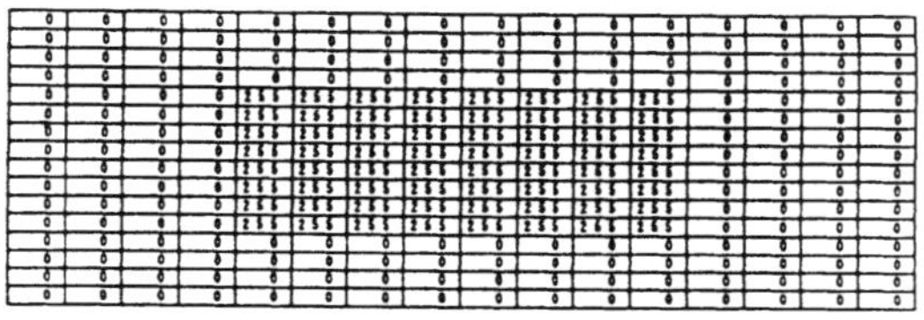

c. Small size pattern

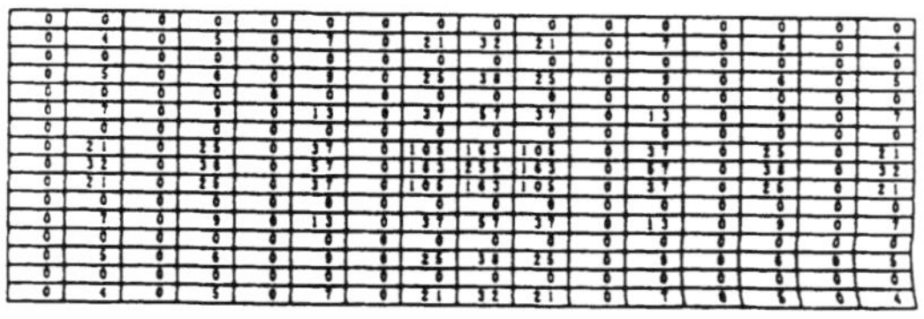

d. Spectrum pattern of small size

Fig. 5. Size of pattern and 2 dimensional spectrum pattern

numbers in order to be able to make the inverse transformation. If the data were made in the power spectrum, they could not be restored to their original form. The 2 dimensional spectral distribution figure is shown in the following using a similar method. Figure size is related to the spread of the high and low frequencies in the spatial frequency characteristics, and higher frequency components become richer. The 2 dimensional spectral pattern is apt to spread, with reduction in figure size, as shown in Fig. 5(d). The direct current component indicates the overall luminance sum total, and its value becomes larger with an increase in figure size.

4.2 Translation (position)

When the same shape pattern is translated and transformed by the 2 dimensional FFT, its 2 dimensional spatial frequency spectrum is found and summarized in a figure, which is shown in Fig. 6 (a: transformed translation figure, b: spectrum distribution of transformed translation figure). Considering the periodicity of the luminance value in the figure, the transformation of the pattern by translation in the X- or Y-direction is invariable regarding the 2 dimensional spatial frequency spectrum characteristics and remains unchanged with the standard pattern in Fig. 5.

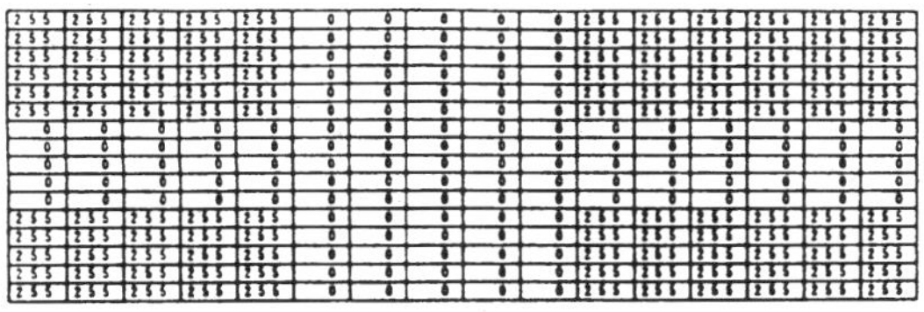

a. Pattern moved in parallel

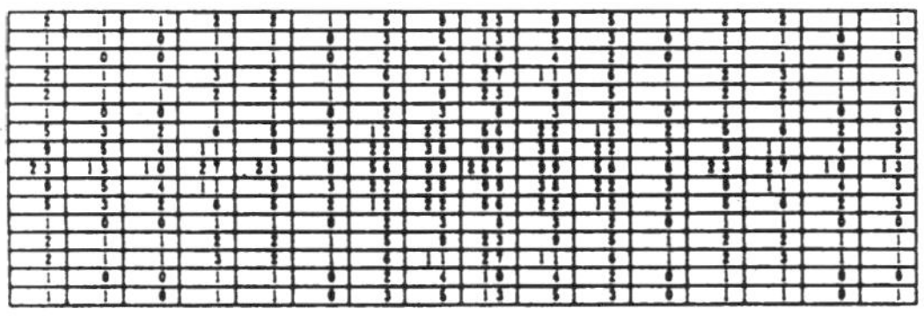

b. Spectrum pattern

Fig. 6. Pattern moved in parallel and 2 dimensional spectrum pattern

4.3 Rotation (direction)

When the same shape pattern with a diverse direction is transformed by the 2 dimensional FFT, its 2 dimensional spatial frequency spectrum is found and is made into a figure, which is shown in Fig. 7 (a: transformed figure by rotation, b: spectrum distribution of transformed figure by rotation).

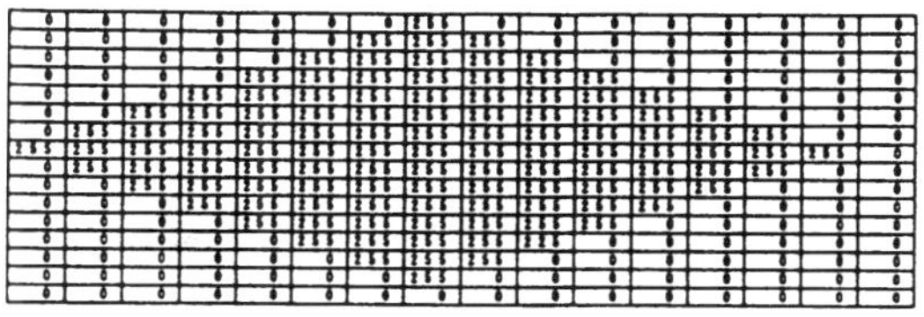

a. Rotated pattern

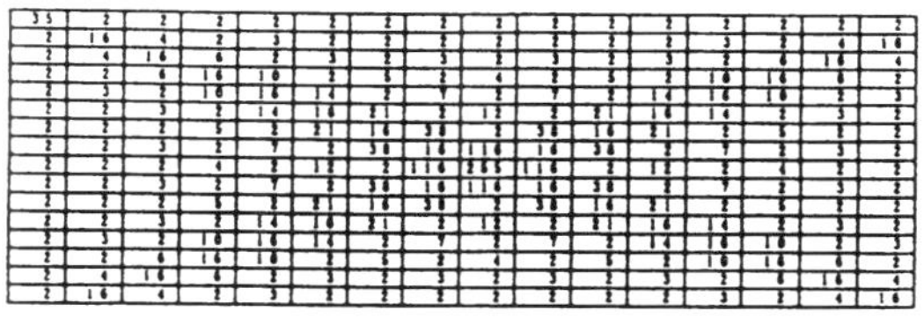

b. Spectrum pattern

Fig. 7. Rotated pattern and 2 dimensional spectrum pattern

The figure shows the rotated pattern of a standard shape at 45 degrees and the 2 dimensional spectrum distribution characteristics with a peak in the direction of the principal axis. Taking note of the information in the direction of the principal axis, it can be seen that it is similar to the state when a standard shape is rotated 45 degrees. However, its value is rather different, so that it is hard to recognize them as the same patterns. When the direction of the principal axis is provided, the 2 dimensional spectrum distribution is transformed into polar coordinates and is taken as a row of data related to the radial and circumferential directions. This is transformed by the 2 dimensional FFT again in order to standardize (make dimensionless) the direction information for the pattern. The results, as shown in Fig. 8, show nothing in common between the standard and the 45-degree rotated object:

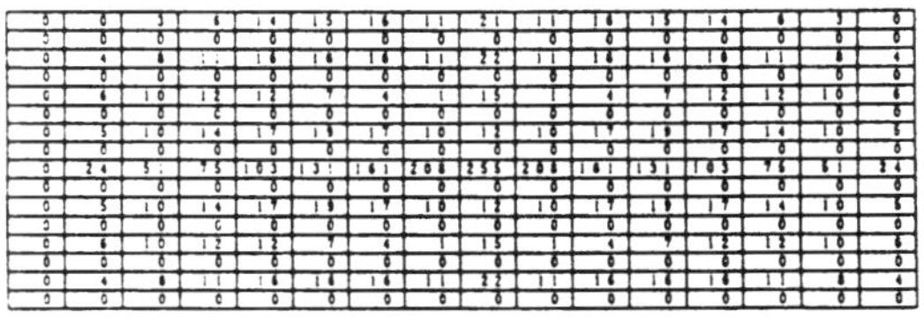

a. Standard pattern

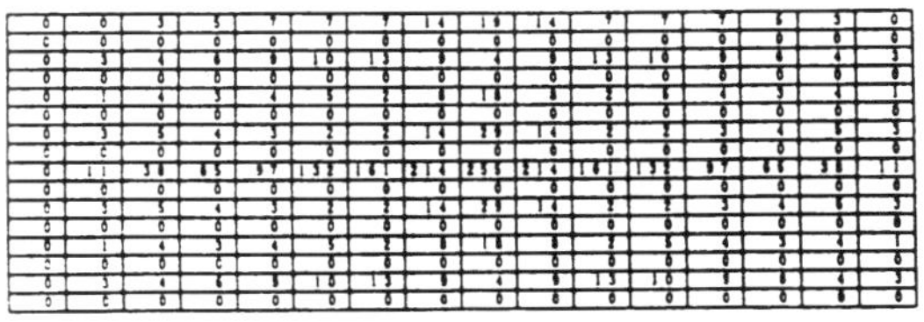

b. Rotated pattern

Fig. 8. Spectrum pattern after polar coordinate transformation

5. PICTURE ELEMENTS ARRANGED IN A POLAR COORDINATE SYSTEM

The gravity center of the figure is then found, and the picture element in the polar coordinate system is constructed with a center at the gravity center. By making the 2 dimensional FFT regarding the radial and circumferential directions after the transformation, both can be definitely recognized as about the same figure pattern(Fig. 9).

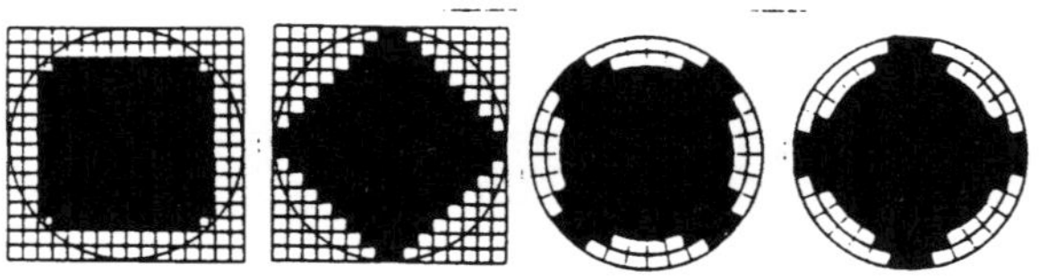

Fig. 9. Polar coordinate transformation about the original pattern

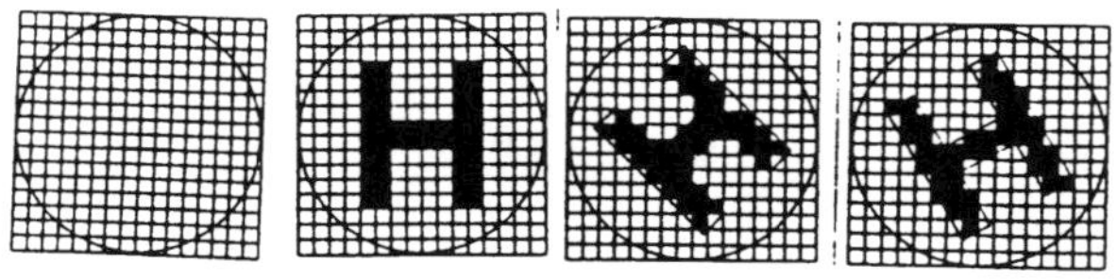

a. Rectangular coordinate system

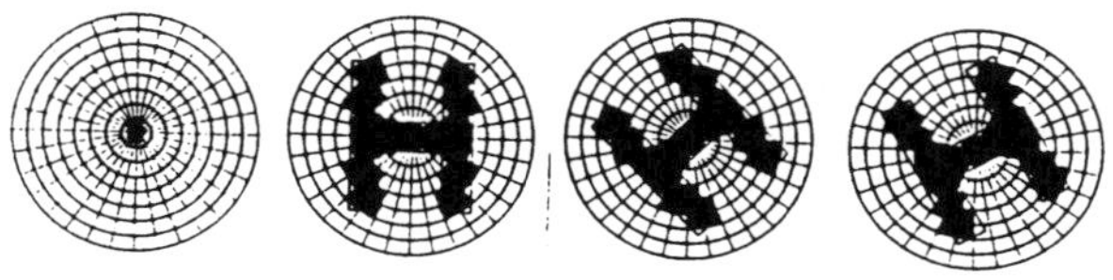

b. Polar coordinate system (Type A)

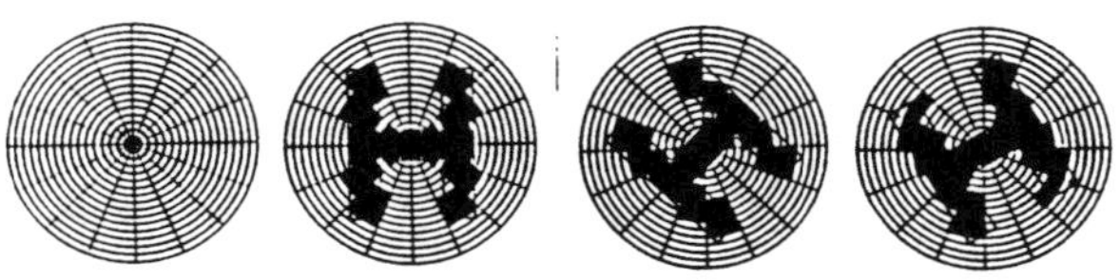

c. Polar coordinate system (Type B)

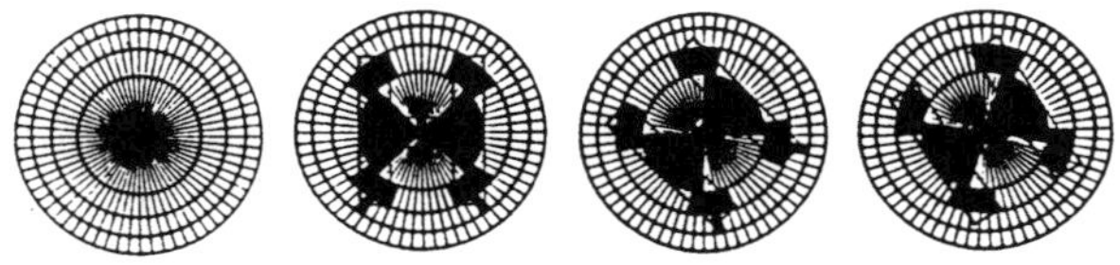

d. Polar coordinate system (Type C)

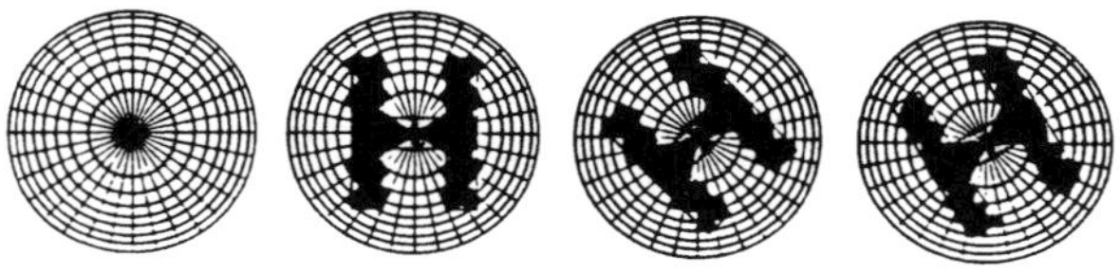

e. Polar coordinate system (Type D)

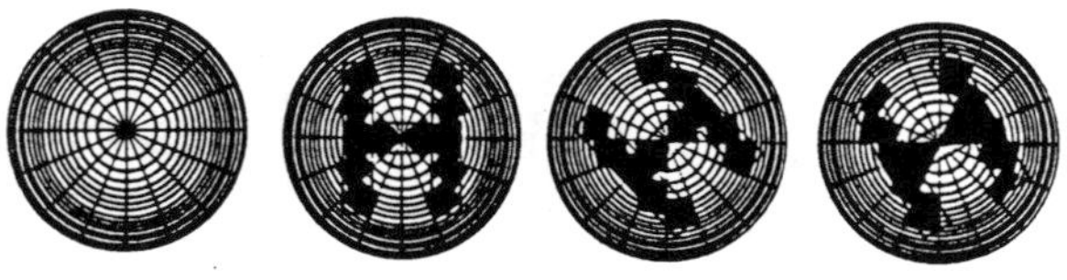

f. Polar coordinate system (Type E)

Fig.10. Several coordinate systems about an H

The shape model used here is a simple shape with the same shape and diverse direction. When a complex shape is used, higher resolution (number of pixels) is necessary from the point of view of precision of the polar coordinate transformation and the amount of available information, improvement in the information processing system and development of hardware for the polar coordinate picture element are required for the higher speed processing. Another important matter is the shape of one element in a picture. In this case, we used 8 pixels in the radial direction and 32 pixels in the circumferential direction, because the direction information was regarded as slightly more important. Moreover, in the radical direction the interval between cells was not equal in order to one picture element of the same size in the square. Fig.10 shows the results of an H pattern which is located at 0 degree, 30 degrees and 45 degrees. The A type and the B type have an equal interval between cells in the radical direction. The C type, the D type and the E type have equal squares at all cells. Then the ratio of the number of pixels in the radical direction to that in the circumferential direction is 4 to 64(the C type), 8 to 32(the A type and the D type), 16 to 16(the B type and the E type). Among all types, the D type can be most definitely recognized as about the same figure pattern. So that the number of pixels in the picture, the ratio of the number of pixels in the radical direction to that in the circumferential direction, and the interval between cells in the radical direction are very important parameters for the shape recognition. Then it is necessary to fit those parameters with the object, to clarify the characteristics of the object shape. Especially, it seems that human beings can process visual information if one changes the coordinate system on the retina flexibly accordingly to match the recognized object with this method, both characteristics, the continuity in the 2 dimensional FFT transformation and the sensibility of direction in the polar coordinate transformation, are utilized effectively.

6. CONCLUSION

Although it is becoming natural to construct picture elements using a rectangular coordinate system when recognition of the direction and shape of an object are the subject, the gravity center has to be focused. Thus it is newly proposed that picture elements be constructed using a polar coordinate system, which is an original idea. This construction of vision function is somewhat like the function of the human eye, and it is hoped that it can function like an artificial retina. We are now going to try to fabricate a flexible coordinate system using the FEM and the neural network process. This method offers some better results for shape recognition.

CONTROL OF THE STATE OF INTRACELLULAR WATER FOR HIGH QUALITY STORAGE

S. OSHITA*, Y. KAWAGOE*, Y. SEO* and A. HASHIMOTO**

* Dept. of Agric. Engineering, Fac. of Agriculture, Univ. of Tokyo, Yayoi 1-1-1, Bunkyo-ku, Tokyo 113, JAPAN
** Fac. of Bioresources, Mie Univ., Tsu-city, Mie 514, JAPAN

Abstract: When xenon gas dissolves into water, the formation of a hydrophobic hydration will occur and the water becomes structured. The creation of structured water was confirmed by proton N.M.R. measurement. The metabolism is expected to be suppressed if the intracellular water structured by being exposed to xenon under certain pressure and it was supported by the fact that germination of conidia of the barley powdery mildew fungus was suppressed. Suppression of respiratory metabolism was also observed for broccoli and eggplant when they were preserved under certain pressure of xenon.

Keywords: Agriculture, Bio control, Quality control, Non-polar gas, Water

1. INTRODUCTION

The suppression of metabolism plays an important role in keeping agricultural products fresh. The method of controlling a composition of surrounding gases and/or a temperature has been applied to reduce the respiratory metabolism of agricultural products and to maintain their freshness, but a new approach is still required to attain high quality preservation. Metabolism is based on intracellular chemical reactions which are, for the most part, enzyme mediated. As the diffusion of substrate regulates the rate of enzyme reaction, the decrease in the diffusion rate of substrate may lead to the suppression of metabolism. When nonpolar gas dissolves into water, the number of hydrogen-bonded water molecules increases by formation of a hydrophobic hydration and the water becomes structured. If the water in a cell becomes structured, the viscosity of water increases and the diffusion rate of a substrate decreases. Hence, a way for the suppression of metabolism can be thought to make intracellular water structured. Hydrophobic hydration with nonpolar gases and gas hydrates have been studied by many researchers, namely, Shinoda and Fujihira(1968), Parrish and Prausnitz(1972), Franks(1973), Garg et al.(1977), Davidson and Ripmeester(1978), and Arakawa et al.(1979), but little attention has been given to the application of the formation of structured water in the field of post-harvest technology.

The present study was designed, on the above-mentioned basis, to control the state of intracellular water by formation of a hydrophobic hydration in order to preserve agricultural products in high quality. The increase in the number of hydrogen-bonded water molecules caused by the dissolution of nonpolar gas was checked by the N.M.R. experiment. Then the method was applied to the preservation of agricultural products.

2. MATERIALS AND METHODS

2.1 Nonpolar gas and N.M.R. measurement

Among many nonpolar gases that create the formation of hydrophobic hydration to make the water structured, inert gases are considered to be most suitable since they can be expected not to cause biochemical changes in cells because of their low chemical reactivity. Xenon was selected as the inert gas in the present study because of the greater degree of structured water formation with xenon compared to other inert gases. In order to confirm the creation of water structured by the formation of hydrophobic hydration, a proton N.M.R. measurement was done. The distilled water and that subjected to xenon at partial pressure of 0.40 MPa were prepared. Both samples in the pressure N.M.R. tubes were kept at 279 K for 16 h and the spectra were recorded at the same temperature at 300 MHz. TMS [$(CH_3)_4Si$] was used as an external standard.

2.2 Markers for suppression of metabolism

In order to examine whether intracellular biochemical reactions are suppressed or not by applying xenon, conidia of the barley powdery mildew fungus (*Erysiphe graminis*) were used as markers for the examination of the suppression of metabolism. The suppression of metabolism, that is, biochemical reactions in conidia can be monitored by detecting the germination. The appearance of germ tube can easily be observed visually through the optical microscope and if the metabolism is suppressed, germ tubes will not appear. For this purpose, freshly harvested conidia were inoculated with a brush onto dissected coleoptile of barley (*Hordeum vulgare*). A single coleoptile cell is large enough with dimensions of around 100-150 μm and thin sections of coleoptile are easily prepared. Coleoptiles were excised from 9-day-old barley and single-cell layers of partially dissected coleoptiles were prepared for inoculation. Dissected coleoptiles, about 10 mm long and 3 mm wide, were composed of 1200 to 1500 cells. A coleoptile on which conidia were inoculated was placed in the cylindrical pressure container whose inside diameter was 20 mm and its height was 16 mm. The pressure container was equipped with viewing windows on the upper and bottom surface which make possible the observation of a coleoptile with conidia through an optical microscope. Xenon was superimposed on atmospheric pressure of air in the pressure container and xenon partial pressure was adjusted to a specified value in the range from 0 to 1 MPa. The pressure container with inoculated coleoptile was kept at 279 K for 7 days, then the appearance of germ tube was examined through the optical microscope at 400x magnification.

2.3 Preservation of agricultural products

Apparatus. A sample was placed under a given partial pressure of xenon in the pressure container. The inside diameter of the container was 130 mm and its height was 160 mm. The air-tight container having the same volume as the pressure container was prepared, as a control container, to keep a sample under atmospheric pressure of air without xenon. Containers were put in a constant temperature chamber to maintain those temperature to a desired value.

Sample and treatment. Broccoli and white eggplant were selected as samples. The reasons of sample selection were as follows: broccoli was used because of high respiratory activity and eggplant was chosen as a typical specimen of agricultural products which are easily damaged at low temperature of around 283 K (10 °C) or less by so called chilling injury.

Broccoli kept in a show case for several days after being harvested was cut lengthwise into halves. The one piece was placed in the pressure container and xenon was superimposed on atmospheric pressure of air to the initial partial pressure of 0.39 MPa. The other was placed in the control container. They were kept at 279 K for 16 days and the difference in color of cutting surface was evaluated by the use of color difference meter. Broccoli harvested at a farmer's field were also used to preserve under xenon and control condition to compare the quantity of CO_2 evolved from each sample. In this case, the initial partial pressure of xenon applied to the pressure container was 0.24 MPa. The control condition means the normal composition of air at atmospheric pressure. The temperature of both conditions was set to 275 K and preservation period was 6 days.

White eggplants harvested at the experimental farm of the University of Tokyo were placed in the pressure container and also in the control container. The environments in the pressure container contained xenon at initial partial pressure of 0.37 MPa and atmospheric pressure of air. The control container was filled with air at atmospheric pressure. They were kept at 288 K for 13 days and the concentration of CO_2 of each container was measured every day by the use of gas chromatography.

Assuming the ideal gas laws to apply, the mass of CO_2 is calculated from the equation below.

$$M_{CO_2} = 440 \times \frac{P_{total}\ V_f\ C}{R\,T} \qquad (1)$$

where

M_{CO_2}	:	mass of CO_2, mg
P_{total}	:	total pressure of mixed gas, Pa
V_f	:	volume of free space in container, m^3
C	:	volumetric concentration of CO_2, %
R	:	gas constant, $J\,K^{-1}\,mol^{-1}$
T	:	absolute temperature, K

3. RESULTS AND DISCUSSIONS

3.1 State of water

Fig.1 shows proton resonance line shapes for

distilled water and that subjected to xenon at partial pressure of 0.40 MPa (xenon solution). The upper spectrum for distilled water presented a narrow shape. In contrast to this, the lower spectrum for xenon solution gave a very wide shape having two peaks which was similar to the shape generally observed for solid sample. The difference in shape of spectra of both the distilled water and the xenon solution meant that the mechanism of relaxation of water molecules was different from each other. Adding to this feature, chemical shift was observed, that is, both of two peaks of the spectrum for xenon solution appeared in the lower side of magnetic field compared to the peak of distilled water. This can be interpreted that the introduction of a xenon solute gave rise to a change in water structure by the increase in the number of hydrogen-bonded water molecules. From the fact above, the creation of structured water was confirmed.

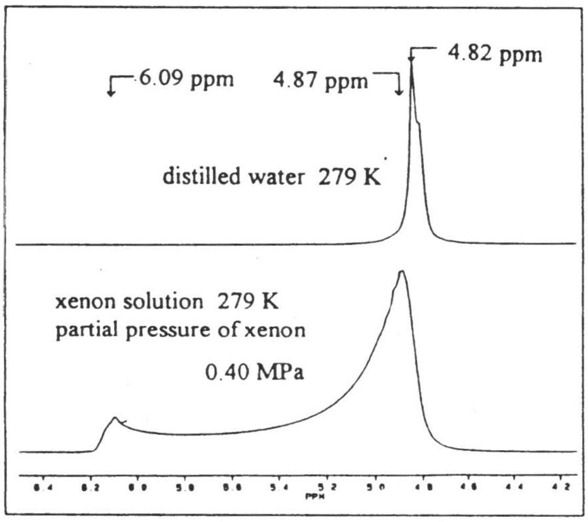

Fig.1 Proton N.M.R. spectra

3.2 *Suppression of metabolism*

Suppression of germination of conidia inoculated on coleoptiles was examined after 7-day preservation at 279 K under various xenon conditions. The suppression ratio was determined as the ratio of non-germinated viable conidia to the total number of viable conidia. This was considered as an index indicating suppression of metabolism, as mentioned above. Results were shown in Fig. 2. The higher partial pressure of xenon on the horizontal axis means the greater structured degree of intracellular water of conidia because the quantity of xenon dissolving into water increases as the partial pressure of xenon increases. The suppression of germination was 0.15 at the zero partial pressure of xenon, that is , at air normal air condition. This was not a singular phenomenon because 10 to 20 % of

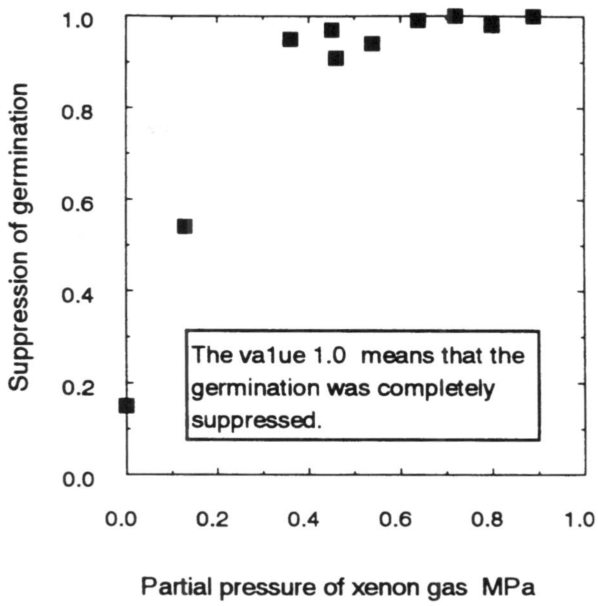

Fig.2 Suppression ratio of germination of powdery mildew fungi of barley

conidia will not germinate under normal air condition. The suppression of germination became higher as the partial pressure of xenon increased and it was beyond 0.9 in the range of partial pressure of xenon greater than 0.35 MPa. This indicated that the germination was suppressed for more than 90 % of viable conidia.

3.3 *Surface color and respiration of agricultural products*

There was no appreciable change in color of cutting surface of broccoli preserved at 279 K under xenon condition in the pressure container. Xenon dissolved into intracellular water in broccoli during 16-day preservation and the initial partial pressure of 0.39 MPa was decreased to 0.33 MPa at the end of the preservation procedure. In contrast to this, the change in surface color was observed for the broccoli preserved under normal composition of air in the control container. Color difference, ΔE, between cut surfaces of broccoli was measured to be 18.6 which meant distinct difference in color. The weight loss was less than or equal to 2 % for both broccoli because relative humidity in both container was kept to 100 %RH.

Table 1 shows the concentration of CO_2 and the total pressure of mixed gas for both the xenon condition in the pressure container and the control condition in the control container during the preservation of broccoli at 275 K for 6 days. The concentration of CO_2 in pressure container was always lower than that in control container. The observed Table 1 Concentration of CO_2 and total pressure of both xenon condition and control condition during the preservation at 275 K

Time, d	xenon condition P_{total}, MPa	C, %	control condition P_{total}, MPa	C, %
0.038	0.334	0.15	0.097	0.80
0.641	0.321	0.67	0.092	2.51
1.79	0.319	1.44	0.092	5.92
2.80	0.318	2.52	0.092	11.8
3.84	0.315	2.92	0.094	14.2
4.72	0.317	4.66	0.097	22.0
5.68	0.318	5.51	0.103	23.6

concentration of CO_2 soon became high for both containers because of those small volume, that is, 2124 ml. The cumulative quantities of CO_2 evolved from both broccoli were calculated from Table 1 by using equation (1) and results are shown in Fig. 3. The symbol of solid rectangle indicates the value for the broccoli in xenon condition whose partial pressure of xenon changed from 0.24 to 0.22 MPa because of the dissolution of xenon into broccoli during preservation. The symbol of open circle shows the value for the broccoli in control condition. After 2 days passed, the difference in cumulative quantity of CO_2 evolved from broccoli preserved in two experimental conditions became greater and it

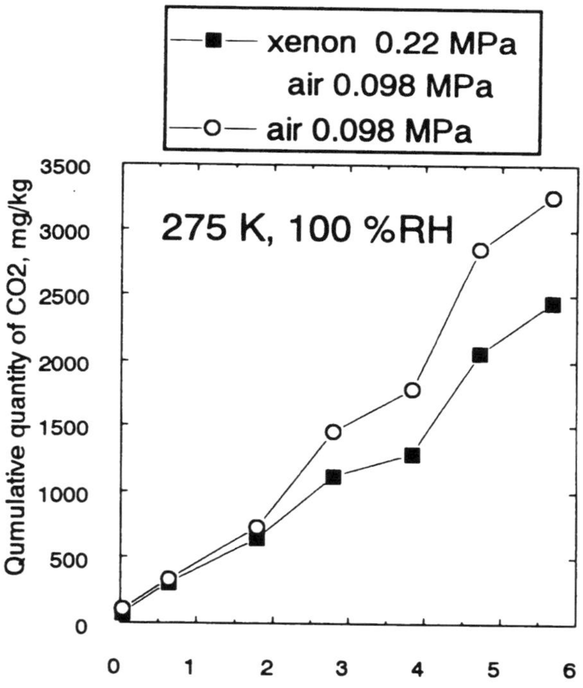

Fig.3 Qumulative quantity of CO2 evolved from broccoli

reached to 805 mg/kg at the end of preservation period. This clear difference supported that the respiratory metabolism was suppressed for the case of broccoli in xenon condition. The initial weights of broccoli in xenon and in control condition were 256.7 g and 266.8 g, and weight losses were 0.5 % and 1.2 % respectively.

Table 2 Concentration of CO_2 and total pressure of both xenon condition and control condition during the preservation of white eggplant at 288 K

Time, d	xenon condition P_{total}, MPa	C, %	control condition P_{total}, MPa	C, %
0.13	0.470	0.04	0.098	0.41
0.87	0.468	0.20	0.088	1.61
1.86	0.467	0.40	0.088	3.04
2.90	0.466	0.57	0.088	4.13
3.89	0.466	0.72	0.099	4.91
4.86	0.465	0.86	0.099	6.22
6.00	0.465	0.99	0.099	8.62
7.03	0.464	1.08	0.100	10.6
7.90	0.464	1.15	0.100	11.5
8.84	0.462	1.27	0.099	12.9
9.88	0.461	1.38	0.099	13.8
10.80	0.460	1.51	0.098	14.4
11.90	0.460	1.69	0.098	15.8
12.90	0.458	1.79	0.098	17.1

The preservation of white eggplant was conducted for 13 days. The temperature was adjusted to 288 K to avoid a chilling injury. The concentration of CO_2 and the total pressure of mixed gas for both pressure and control container are shown in Table 2. The concentration of CO_2 in the container of xenon condition showed always much lower than that in the container of control as same as the case of broccoli.

The cumulative quantity of CO_2 evolved from eggplants was calculated from the data on Table 2 by using equation (1) and was shown in Fig.4. The symbol of solid rectangle indicates the value for the eggplant in the container of xenon condition whose partial pressure of xenon became from 0.37 to 0.36 MPa during the preservation procedure because of the dissolution of xenon into the eggplant. The symbol of open circle shows the value for the white eggplant in control condition. The eggplant exposed to xenon condition evolved much lower quantity of CO_2 than the control over the preservation period and its difference between eggplants in two experimental conditions was up to 3129 mg/kg. The initial weights of eggplant in xenon and control condition were 86.2 g and 91.8 g, and weight losses were less than 1 % for both eggplants because of the same reason as the case of broccoli. Both eggplants after preservation were cut in halves in order to examine the inner color. The color changed from white to light brown only for the eggplant preserved in control condition, and the quality of eggplant in xenon condition was judged to be kept good from visual inspection. It was also confirmed from the results of eggplant that the respiratory metabolism was suppressed by creation of intracellular water structured.

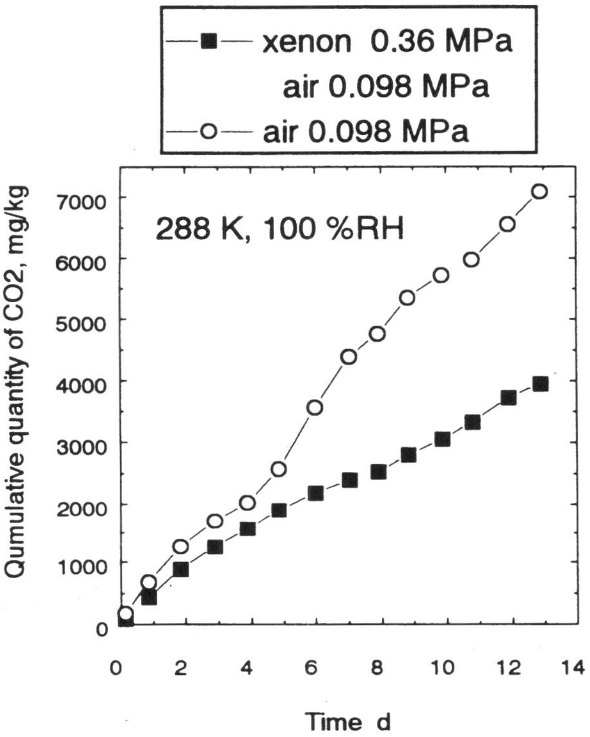

Fig.4 Qumulative quantity of CO2 evolved from white eggplant

4. CONCLUSIONS

Proton N.M.R. measurement indicated that the water became structured by formation of a hydrophobic hydration under given partial pressure of xenon. Xenon was applied at various pressures up to 1 MPa to conidia of the barley powdery mildew fungus inoculated on barley coleoptile cells which were used as a marker for suppression of metabolism. It was observed that the number of non-germinated viable conidia increased as the xenon pressure increased and complete suppression was attained in the range of high xenon pressure. The method was then applied to the preservation of broccoli and white eggplant. The broccoli cut in halves and preserved under partial pressure of xenon at 0.33 MPa at 279 K did not show appreciable change in color of cut surface, while the color change was observed for the broccoli preserved under control condition. For the broccoli preserved at 275 K, the cumulative quantity of CO_2 evolved from broccoli preserved at the xenon partial pressure of 0.22 MPa was lower than that of broccoli preserved under control condition and its difference became 805 mg/kg at the end of preservation period. The suppression of respiratory metabolism was examined also for eggplant. The cumulative quantity of CO_2 evolved from eggplant exposed to xenon at partial pressure of 0.36 MPa was much lower than that from eggplant kept in normal composition of air at atmospheric pressure at 288 K. After the preservation, the change in the inner color of eggplant was observed for only the eggplant in the control condition. The results suggested that the metabolism concerning to germination of conidia and the respiratory metabolism of agricultural products were suppressed in xenon condition.

The most appropriate preservation conditions of xenon pressure and temperature should be examined for each kind of agricultural product.

ACKNOWLEDGMENT

The authors express their appreciation for the financial support by The Iwatani Naoji Foundation's Research Grant and also for the financial and experimental support given by Takarazuka Research Center, SUMITOMO CHEMICAL CO., LTD..

REFERENCES

Arakawa, K., Tokiwano, K., Ohtomo, N. and H. Uedaira (1979). A note on the nature of ionic hydrations and hydrophobic interactions in aqueous solutions. Bulletin of The Chemical Society of Japan. 52(9),2483-2488.

Davidson, D. W. and J. A. Ripmeester (1978). Clathrate ices-Recent results. Journal of Glaciology. 21(85), 33-49.

Franks, F., Water-A COMPREHENSIVE TREATISE- (1973). , Vol.2, Chap.3, 115-234, Plenum Press, New York

Garg, S. K., Majid, Y. A., Ripmeester, J. A. and D. W. Davidson (1977). Reorientation and diffusion of water molecules in xenon hydrate and other structure. I. Clathrate hydrates, Molecular Physics, 33(3), 729-734.

Parrish, W.R. and J. M. Prausnitz (1972). Dissociation pressures of gas hydrates formed by gas mixtures, Ind. Eng. Chem. Process Des. Develop., 11(1), 26-35.

Shinoda, K. and M. Fujihira (1968). The analysis of the solubility of hydrocarbons in water, Bulletin of the chemical society of Japan, 41, 2612-2615.

F.T.-I.R. / ATR SPECTROMETER AS AT-LINE SUGAR SENSOR IN ENZYMATIC HYDROLYSIS

Véronique Bellon-Maurel*, Céline Vallat*, and Darrell Goffinet **

* *CEMAGREF, BP 5095, 34033 Montpellier Cedex 1, France*
***School of Agricultural Engineering, Purdue University, West Lafayette, IN 47907, USA*

Abstract: This paper discusses the use of Fourier Transform Infrared (FT-IR) spectroscopy coupled with an Attenuated Total Reflectance (ATR) accessory as applied to the quantification of individual sugar concentrations in mixtures extracted from starch hydrolysis (250-300 g/kg). Glucose and maltose are detected with the required precision, but not maltotriose or maltodextrines. The influence of external factors (proteins, salts, temperature) on the prediction is also evaluated. Proteins and salts, which occur in low concentrations in sugar solution have virtually no influence on the detection of sugar. Sample temperature does have an influence upon spectral measurements. In order not to affect the prediction ability, this influence is removed by the introduction of temperature variation in the model.

Keywords: Sensor, Spectrocopy, Carbohydrates

1. INTRODUCTION

The general objective of this study is to prove the feasibility of Fourier Transform Infrared Spectrometry as an at-line sensor to control the composition of sugar industry products. Earlier experiments have confirmed the suitability of FTIR/ ATR spectroscopy for monitoring sugar production from starch hydrolysis, especially when combined with multivariate regression methods (Principal Component Regression and Partial Least Square;, Bellon et Vallat, 1993; Cadet, et al, 1992). In order to develop an industrial analytical technique, it is still necessary to study several details:

1) Methodological Factors-- aspects occurring during and/or after measurement, including the model-making process and the repeatibility and reproducibility of results.

2) External Factors-- the factors surrounding the "unknown" liquid, namely the temperature of solution at the time of measurement and the existence of salts and proteins which may affect spectroscopic measurements.

2. MATERIALS AND METHODS

2.1 Material: Spectrometer and samples

A BRUKER IFS 25 FT-IR (Fourier Transform Infrared) spectrometer equipped with a SPECAC ZnSe ATR (Attenuated Total Reflectance) flat crystal accessory was used, with a glowbar source and a liquid nitrogen-cooled MCT detector. No filter was used.

Six samples were extracted from 2 hydrolysis tanks, under various hydrolysis conditions. They were analyzed by high-performance liquid chromatography (H.P.L.C.) to measure sugar contentrations. In order to provide a larger range of individual sugars, additional amounts of sugars were added to the mixture in various concentrations. Each original sample was partitioned into 4 subsamples, 3 of which were augmented with individual sugars. Total dry matter ranges from 200 to 300 g/Kg.

The interferograms were recorded with no optical filter, averaged on 100 scans and Fourier-transformed using the triangle apodization function on the 1300 to 850 cm^{-1} spectral range. A 4 cm^{-1} resolution (a total of 112 wavelengths per scan) was used.

2.2 Post-spectroscopic mathematical processing

The UNSCRAMBLER (CAMCO Inc, Norway) statistical software package was used to process all spectroscopic data. Partial Least Squares (PLS) Regression was used to correlate all data. The standard error of prediction (SEP) was used as an index of calibration performance. Data pre-treatment consists of centering and scaling values to unit variance.

Calibration: The total number of samples was 44, 30 of which were kept in the calibration set and 14 in the validation. The performance of the model was illustrated by the Standard Error of Prediction or SEP.

The precision of a calibration is twice the standard error of prediction. The precision-levels required by the industry are 8, 10, 5 and 5 g/Kg, respectively, for glucose, maltose, maltotriose and maltodextrines (also called DPn).

Repeatibility: Repeatibility is the tendency of a sensor to give the same response when used several times in the same experiment. The repeatibility index measured by both the difference between the maximum and minimum values of 5 measurements and the Standard Deviation (SD).

Reproducibility: Reproducibility is the tendency of a sensor to give the same response when used in the same experimental conditions, but not in the same experiment. Reproducibility was determined by recording the spectra of the same mixtures on different days and then predicting concentration. If the prediction showed no significant offset, the model was considered stable and reproducible.

2.3 Determining influence of external parameters.

Temperature: Temperature was controlled by the thermostatic ATR crystal and a regulated hot water bath. Spectra of 40 samples were taken at 3 different temperatures: 40,45 and 50°C.

Two different methods have been attempted to establish the calibration:

1) Calibration at 40°C on the 40 samples; predictions at 40, 45 and 50°C.

2) Calibration including all temperatures (that is to say, the same 14 samples at 3 different temperatures) said "universal calibration". Predictions were made at 40,45 and 50°C on the remaining samples.

If temperature has an influence, it is necessary either to determine how it alters the calibration line (by a model) or to integrate this influence into the calibration. The first case requires to model the influence of temperature on the spectra; in the second case, a "universal" calibration model integrates the temperature influence.

Proteins and salts: Proteins and salts are present during hydrolysis of starch at maximum concentrations respectively equal to 4g/Kg and 1g/Kg.

In order to simulate the observed protein concentration variations, proteins or salts were added to a fixed sugar mixture in different concentrations : 1, 2, 3 and 4 g/Kg for proteins and 0.3, 0.6, 0.9 and 1.3 g/Kg for salts. The influence of proteins -or salts- was estimated by calculation of the standard deviation (SD1) of the prediction values of the 5 samples (4 added samples plus original sample). It is compared to te repeatibility Standard Deviation (SD2)

3. RESULTS AND DISCUSSION

3.1 Evolution of sugar content

During hydrolysis, the concentrations of the individual sugars, i.e. glucose, maltose (DP2), maltotriose (DP3) and maltodextrines (DPn) evolve versus time (in hours) as shown in Figure 1. The points show the moments when samples were taken.

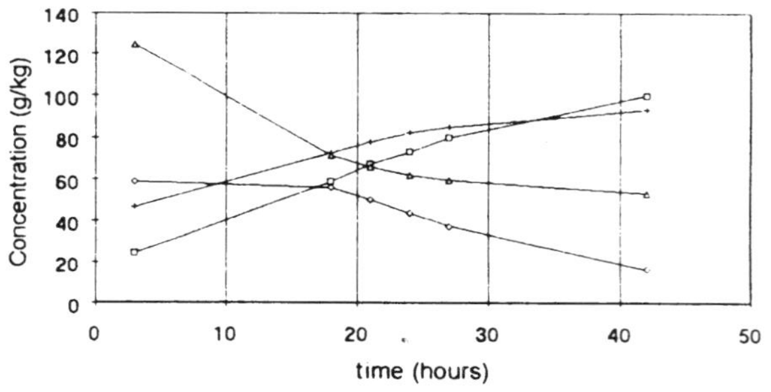

Figure 1: Evolution of the concentration of individual sugars during hydrolysis: □ glucose, + DP2,
◊ DP3, Δ DPn

3.2. Precision of the calibration model

The spectra of the different sugars are very similar as shown in Figure 2. Thus their processing requires very powerful signal processing techniques such as PLS.

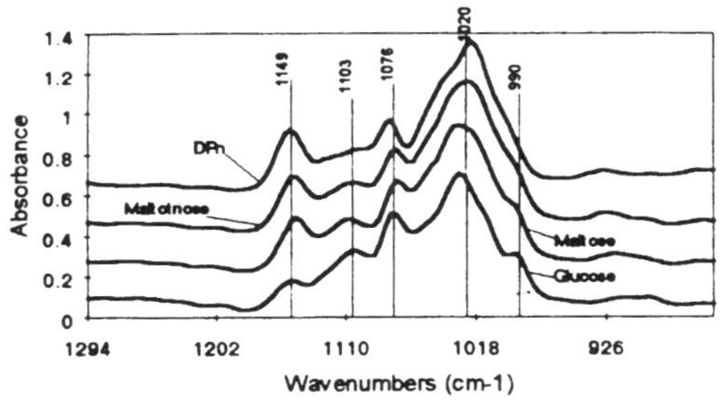

Figure 2: Spectra of individual sugars. Please note that, in fact, all spectra have the same base-line as that of glucose, and have been separated for ease of viewing.

PLS model has selected 7 factors which explain 91.7% of the variance in the chemical data. The standard errors of calibration and prediction are given in Table 1.

Table 1 Standard errors of calibration and prediction on real samples.

	n	Glucose	DP2	DP3	DPn
Range (g/Kg)		16-100	40-93	21-88	18-128
SEC (g/Kg)	30	3.2	3.9	4.8	4.2
SEP (g/Kg)	14	4.8	3.4	5.7	6

The predicted (Yp) versus real (Yr) values of sugar content are given in Figure 3. Predicted values are computed from the equation of calibration. Real values are measured by H.P.L.C. Glucose and maltose, which require a less severe precision than maltotriose and maltodextrines, meet the industry precision requirements.

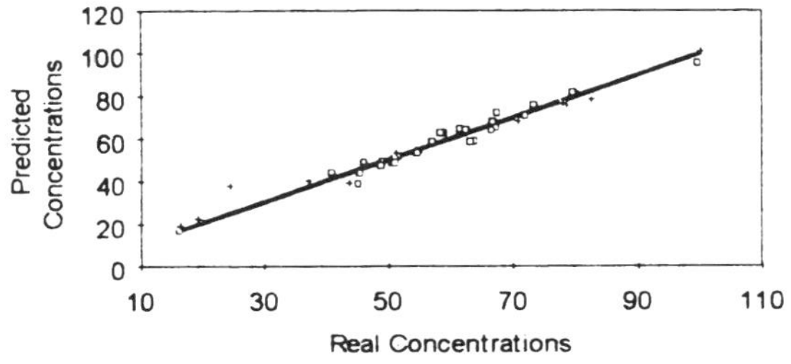

a) Glucose

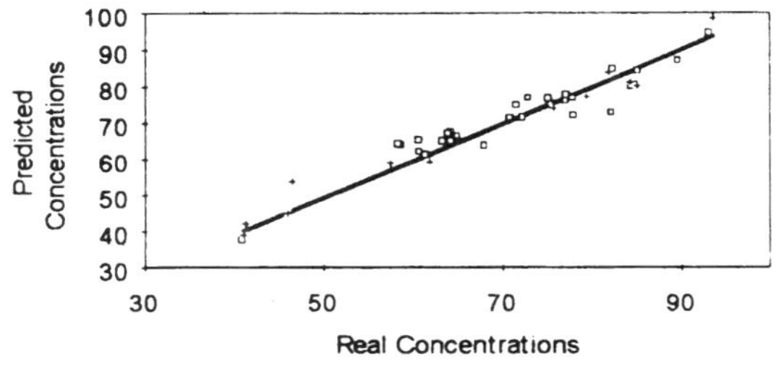

b) Maltose

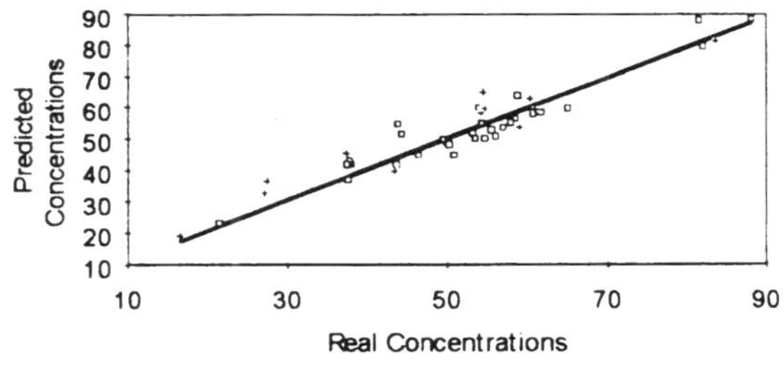

c) Maltotriose

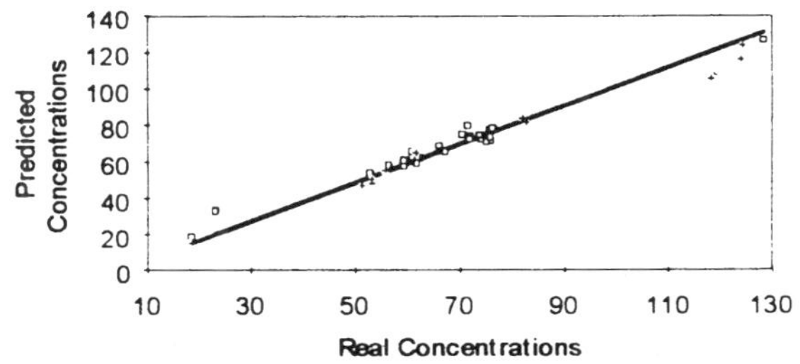

d) DPn

Figure 3: Regression curves of predicted versus real concentrations of sugars in a real mixture; □ Calibration, + Validation

3.3 Reproducibility and reproductibility

Repeatibility : The 2 repeatibility indices (Standard Deviation SD and difference between max and min values) are given in Table 2. The standard errors are small, in general less than ten times the prediction error (especially in the cases of glucose and maltose). This small standard error shows the good repeatibility of the measurements.

Table 2: Repeatibility indexes

	Glucose	DP2	DP3	DPn
Conc. (g/Kg)	99.48	93.03	21.37	52.78
Max.- Min.	1.17	0.85	1.36	1.73
SD	0.39	0.33	0.48	0.64

Reproducibility: To evaluate reproducibility, predictions on the same 40 samples have been made over a period of 5 days. As can be seen in Table 3, for each sugar, the SEP is very similar to the SEC, showing that no bias is introduced on the different days. More globally, the sum of error squares shows the agreement between SEC and SEP, and thus the good reproducibility, every external factor being kept equal.

Table 3: Comparison of Standard Error of Prediction (g/kg) obtained from the same model and the same samples on different days .

Day	SEC				
	Glucose	DP2	DP3	DPn	$R(S^2)$
1	2.617	3.287	*5.529*	*3.548*	7.799
2	2.419	3.034	*5.346*	*3.655*	7.550
3	3.596	3.301	*5.903*	*3.978*	8.631
4	3.844	3.782	*5.026*	*3.709*	8.252
5	2.765	3.519	*4.829*	*4.179*	7.798

(Results which are outside the proscribed limits of precision are in italics).

3.4. Influence of external factors

Temperature: Standard errors of calibration and prediction are given in Table 4 for the calibration at 40°C, and in Table 5 for the universal calibration. SEP1 and SEP2 refer to the calibration at 40°C and the universal calibration, respectively. The comparisons are made by the calculation of $R(S^2)$, I.e. the sum of the SEP squares of the different sugars, which allows a global estimation to find the best result, here corresponding to the lower value.

Table 4 : Standard Errors of calibration and prediction with calibration at 40°C.

T (°C)	Type	Standard Error Glucose	DP2	DP3	DPn	$R(S^2)$
40	Calib.	2.84	3.50	*3.83*	*3.66*	6.96
40	Pred.	3.80	3.53	*4.36*	*4.21*	7.98
45	Pred.	*5.13*	3.54	*6.20*	*5.18*	10.20
50	Pred.	*6.63*	4.30	*8.28*	*4.40*	12.26

In italics, figures which do not give a sufficient precision.

Table 5: Standard Errors of calibration and prediction with universal calibration.

T (°C)	Type	Standard Error Glucose	DP2	DP3	DPn	$R(S^2)$
All	Calib	2.951	2.50	*4.11*	*3.43*	6.61
40		2.662	3.63	*4.42*	*4.33*	7.65
45	Pred.	2.949	4.58	*5.30*	*4.40*	8.78
50		3.411	4.05	*4.55*	*3.92*	8.00

In italics, figures which do not give a sufficient precision.

The standard errors are much lower with the "universal calibration". These results show the influence of temperature. The way to bypass this influence is to establish a calibration including the temperature parameter, in which case a correction by standards is unneccessary and the results are improved; in this manner we detect glucose and maltose within the desired limits of precision. The temperature is not introduced as another variable of the model (in addition to intensities), but its variability is taken into account by building a calibration on spectra recorded at different temperatures.

Proteins: Table 6 gives the results of standard deviation obtained on samples with fixed sugar and variable added protein amounts (SD). This value is to be compared with the standard deviation obtained in repeatibility studies (SD repeat). There is no significative difference between both, and thus it can be concluded that the protein have no influence on the prediction.

Table 6 : Influence of proteins:comparison of standard deviations of samples with variable added protein amounts and of repeatitions.

Proteins added (g/Kg)	Glucose	DP2	DP3
0	92.83	69.28	48.31
1	90.57	67.91	49.72
2	91.35	67.56	50.65
3	93.22	72.32	44.01
4	93.97	72.09	43.86
SD	1.25	2.02	2.85
SD repeat.	1.02	2.21	2.19

This is explained by the fact that spectra of pure proteins (extracted from sugar mixtures at the end of the process) do not show any absorbance in the work range (only between 1680 and 1510 cm^{-1}, which is outside the work range).

Mineral salts: Table 7 provides the results obtained from measurements of samples with fixed sugar and variable salt concentrations. A repeatibility experiment was also performed on a single sample of salt solution (SD repeat.).

As in the case of proteins, the SDs of repeatibility for fixed salt solutions are very similar to those of the varying salt samples, with the exception of maltotriose. This again implies that the deviation in value is due solely to the error of repeatibility, and not the presence of salts. Moreover, the lack of influence can be attributed to the fact that salts (and proteins) occur in such small concentrations when compared to glucose, maltose, maltotriose, and maltodextrins. Therefore, variations in these low concentrations have no visible effect.

Table 7 : Influence of salts:comparison of standard deviations of samples with variable added salt amounts and of repeatitions.

Salts added g/Kg	Glucose	DP2	DP3
0	73.78	58.23	37.32
0.3	73.57	58.34	38.33
0.6	73.04	58.25	40.06
0.9	72.88	58.24	41.29
1.2	73.25	59.88	41.16
SD	0.33	0.65	1.57
SD repeat	0.37	0.44	0.54

4. CONCLUSION

Acceptable results were obtained for the detection of glucose and maltose in real solution extracted during hydrolysis: the standard errors of prediction respectively are 4.1, 3.4, 4.9 and 4.4 g/Kg for glucose, maltose, maltotriose and maltodextrines whereas the required precisions are respectively 8, 10, 5 and 5 g/Kg. Moreover, the repeatibility is satisfactory: the repeatibility error is generally not larger than 0.6 g/Kg and in average equal to 0.45 g/Kg, which gives a repeatibility precision better than 1g/Kg. The reproducibility of the method is also satisfactory with real samples. As the standard errors of prediction of measurement made on different days are equal to the standard errors of calibration, the time-stability of the model is shown. Moreover, no bias and offset adjustments are needed for each day, which is an important feature in automated industrial use. Therefore, the potential of this method for sugar analysis in industry is confirmed.

Among the external factors studied in this paper, temperature was found the most important: whereas reproducibility was good when the temperature was kept constant, a large shift was detected when the predictions were done at temperatures different from the calibration: whereas the calibration done at 40°C gives a global S.E.C. equal to 7 g/Kg, the predictions done at 40, 45 and 50°C give SEP respectively equal to 8, 10.1 and 12.2 g/Kg. To avoid the necessity of strictly regulating the temperature from hydrolysis tank to analysis cell (a costly and difficult alternative), a calibration which takes into account the possible fluctuations in temperature has been established. In this case, whereas the calibration (including 40, 45 and 50°C) gives a sum of squares of error equal to 6.6 g/Kg the predictions at 40, 45 and 50°C give sum of squares of error respectively equal to 8.7, 8 and 7.6 g/Kg which do not show any shift. Neither proteins nor salts, occurring in uncontrolled yet very small concentrations, have an appreciable effect upon the prediction of individual sugar concentrations. The standard deviation after addition of either salts or proteins was not significantly higher than the standard deviation of repetitions on the same sample. Therefore, they do not need to be taken into account.

The absence of non-correctible influences due to external factors facilitates the utility of the FT-IR / ATR method for the on-line analysis of individual sugars in concentrated liquid solution. Subsequent work consists of developing a program whose purpose is to coordinate and control all of the elements necessary to a continuous, real time, on-line application, including the ability to detect malfunctions.

ACKNOWLEDGEMENTS

We would like to acknowledge the contribution of Dr. Feinberg from CIQUAL (Paris), who allowed us to use the UNSCRAMBLER software, Dr Trystram from ENSIA (Massy) for his cooperation, Dr Mathaly, from ARD (Paris), for the analysis of the samples and Professeur Pourcin from the Université de Provence (Marseille) for his valuable advice.

REFERENCES

Bellon V. and Vallat C. (1993) Individual sugar content control by the use of F.T.-I.R. spectroscopy coupled with an A.T.R accessory. *International Conference I.C.E.F. 6, CHIBA (Japan), May 23-26.*

Cadet F., Bertrand D., Robert P., Maillot J., Dieudonné J., and Rouch C. (1991). Quantitative determination of sugar cane sucrose by multidimensional statistical analysis of their Mid-Infrared attenuated total reflectance spectra. *Applied Spectroscopy*, 45 (2),166-172.

NEAR INFRARED REFLECTANCE TESTING TO PREDICT SUCROSE AND MALIC ACID CONCENTRATIONS OF MANGOES

Hadi K. Purwadaria*, I Wayan Budiastra*, and Daniel Saputra**

** Dept. of Agricultural Engineering, IPB, P.O. Box 220, Bogor 16002, Indonesia*
*** Dept. of Agricultural Technology, Univ. of Sriwijaya, Palembang, Indonesia*

Abstract : Mangoes classification (*Mangifera indica*, var. *gedong*) into their tastes, sweet, sweet-sour and sour, commonly judged by aroma and experiences, is becoming increasingly difficult due to consumer demands on a prime quality guarantee. The near infrared diffuse reflectance system was developed and applied to 200 mango samples at the wavelength ranged from 1 400 to 1 975 nm. The Stepwise method was utilized in selecting the optimal NIR wavelengths for the sucrose and the malic acid of the samples measured by the HPLC analysis. The parameters were integrated to the calibration equations expressing the relationship between the NIR prediction and the measured sucrose and malic acid displaying agreeable correlation.

Keywords : mango, NIR, sucrose, malic acid, Stepwise method

1. INTRODUCTION

The production of fruits and vegetables in Indonesia increased by the average rate of 5.2 %/year from 4.96 million tons in 1987 to 6.26 million tons in 1991. However, Indonesia only shares less than one percent from the world market of fresh fruits and vegetables. Mango is one of the important export fruits with an increasing value from 579 thousand USD in 1990 to 857 thousand USD in 1992. Some importing countries like Singapore, Taiwan and Hongkong prefer sweet mangoes whilst others such as Japan and Europe favour sweet-sour ones. Recently, there is more and more demand from the consumers for a prime quality fruit with a preferred taste upon higher rewarding price.

The objective of this study is to develop a sortation system using the NIR (Near Infrared Reflectance) as a nondestructive method in classifying the mangoes based upon their taste sweet, sweet-sour and sour. The specific objectives of the first phase of the study, presented in this paper, are as follows 1) to assess the dominant individual sugar and acid of mangoes, 2) to assess the dominant sugar and acid concentration of mango at various levels of consumer preference, and 3) to predict the dominant sugar and acid concentration of mango by applying the NIR method.

2. EXPERIMENTAL APPARATUS AND PROCEDURE

2.1. *Experimental Apparatus*

The experimental apparatus for measuring the light reflectance in the near infrared region was set up as illustrated in Figure 1 following the system introduced by Ikeda *et al.* (1992). The apparatus consisted of a halogen lamp, chopper, filter, monochromator, integrating sphere, PbS sensor, lock in amplifier, A/D converter, D/O board, pulse motor controller and computer. The halogen lamp provided radiation in the wavelength of 700 - 2 500 nm. The filter was used to block a distinct wavelength region in order to eliminate the overlapping problem

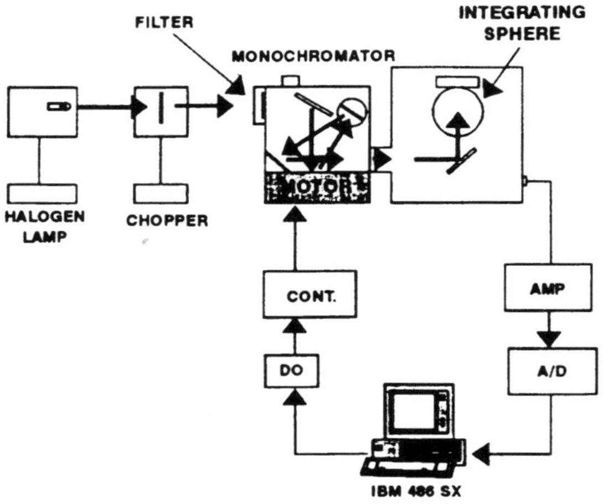

Figure 1. NIR apparatus system for the experiment

occurred at the grating monochromator. The scanning of the wavelength was driven by the stepping motor attached under the monochromator, in which its speed and revolution was controlled by the computer through the D/O board and the pulse motor controller. All of the optical system were manufactured by Shimadzu, Japan. The distance between the halogen lamp to the chopper was 60 mm and between the chopper to the monochromator was 30 mm. A computer program has been developed to drive the apparatus written in C language.

2.2. Material

Mangoes (*Mangifera indica*) var. *gedong* with various stages of maturity were harvested at three levels of maturity , 80, 90 and 100 days after fruiting from the orchard at Indramayu district, West Java. The mangoes were transported to Bogor in three hours, sorted, washed and stored at 10°C inside the cooling room.

2.3. Reflectance Measurement

The equatorial part of mango were cut in longitudinal direction from the skin to about 10 mm depth with a size of 30 mm x 10 mm x 10 mm. Prior to the measurement, the apparatus was warmed up for 30 minutes, setting up the entrance slit of monochromator in 500 nm, the amplifier gain in 100, the response time is 100 ms and the light intensity in 14 volt.

The sample was placed at a sample port, locating the skin part directly subject to light radiation. The calibration was done using $BaSO_4$ as the standard. The reflectance was calculated as the ratio of the reflectance of sample (volt) to that of standard (volt). The reflectance measurement was carried out in the wavelength of 1 400 - 1 975 nm.

2.4. Chromatographic Procedure

Right after the reflectance measurement, 20 - 40 g edible tissue was sliced out of the sample and blended for 2-3 minutes adding 100 ml of 100% ethanol. The extract was then filtered through Whatman No. 45 paper and washed with additional 80% ethanol (200 ml). The filtrate then reduced to a volume less than 25 ml by a rotary vacuum evaporator. The fruit concentrate was filled with distilled water up to 25 ml and filtered again through Whatman No. 42 paper and a Sep-Pak C_{18} cartridges. Two ml of the solution was injected to the HPLC apparatus. The carbohydrate column was used to analyze the sugar component of the sample, and the Bio-Rad reverse phase microguard column (ODS-10) was used for the acid analysis. The solution was eluted with 80% acetonitrile for sugar analysis and with phosphate buffer (pH 2.4) for malic acid each at the rate of 0.8 ml/min.

3. RESULTS AND DISCUSSION

Figure 2 illustrated examples of the light reflectance by the mangoes subjected to the wavelength of 1 400 - 1 975 nm. The light absorption by water, sugar and acid was indicated by the curve valley at the wavelength of 1 500, 1 800 and 1 900 nm, respectively.

The results of HPLC analysis for mangoes (Table 1) indicated that sucrose (6.18%) is the dominant component for individual sugar compared to fructose (1.3%) and glucose (0.58%). For individual acid, the malic acid was dominant (3.85%) compared to citric acid (1.56%), glycolic acid (0.17%) and oxalic acid (0.06%). Thus, the prediction by NIR was

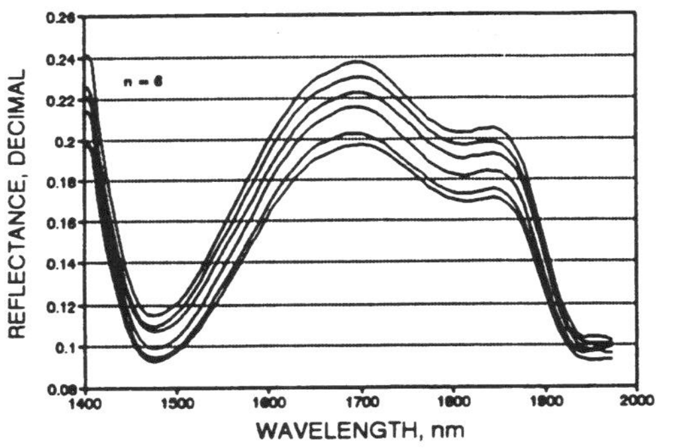

Figure 2. Mango reflectance under NIR system at various wavelength

carried out for the sucrose and the malic acid concentration.

The consumer organoleptic score for sweet-sour (4), sweet (3), sour (2) and bland (1) was assessed and related to the results of HPLC analysis for sucrose and malic acid (Figure 3). The individual sugar and acid concentration of the fruits harvested at 80 days after fruiting presented an erratic behaviour from sour, sweet and sweet-sour. However, at 90 days after fruiting the fruit indicated the trend to sweet-sour with the average of sucrose concentration was 6.2% and the malic acid was 3.6%. At 100 days after fruiting, the fruits turned to sweet with an increase of sucrose concentration to 6.8% and a decrease of malic acid to 2.4%. It is recommended that the mango is harvested 90 days after fruiting to obtain more sweet-sour fruits and 100 days after fruiting to obtain more sweet ones. The results of the image processing recording (Budiastra *et al.*, 1994) on the mango colour, weight, and projection area supported the recommendation for harvesting at 90 and 100 days after fruiting.

The prediction of sucrose and malic acid concentration by NIR method was carried out using multiple regression method (Ikeda *et al.*, 1992)

$$y = a_0 + a_1x_1 + a_2x_2 + \ldots\ldots\ldots\ldots + a_kx_k$$

where :

y	=	sucrose or malic acid concentrations
a_0	=	constant
$a_1, \ldots a_k$	=	regression coefficients
$x_1, \ldots x_k$	=	light reflectance at wavelength 1,.......k

The selection of an optimal set of wavelength was carried out by Stepwise method using the PROC STEPWISE of SAS (Statistical Analysis System) computer package program. Three optimal wavelengths, 1 333, 1 605, and 1 821 nm were selected for the sucrose calibration and a set of eight wavelengths 1 621, 1 629, 1 813, 1 821, 1 933, 1 941, 1 965 and 1 973 nm was for the malic acid calibration of harvested samples 90 days after fruiting (n = 200). The predicted sucrose and malic acid concentrations by the light reflectance using the selected optimal wavelength was calibrated to the results of HPLC analysis. The correlation coefficients (R) of the established calibration equation were 0.92 for sucrose and 0.98 for malic acid (Figure 4).

Table 1. Results of HPLC analysis on the individual sugars and acids of mangoes var. *gedong* (n = 25)

Days of Harvesting	Individual sugars, %			Individual acids, %			
	Fructose	Glucose	Sucrose	Oxalic acid	Glycolic acid	Malic acid	Citric Acid
80	1.48	0.76	5.97	0.06	0.15	5.30	2.15
90	1.31	0.58	6.09	0.07	0.37	3.47	1.48
100	1.10	0.38	6.48	0.05	0.00	2.78	1.04
Average	1.30	0.58	6.18	0.06	0.17	3.85	1.56

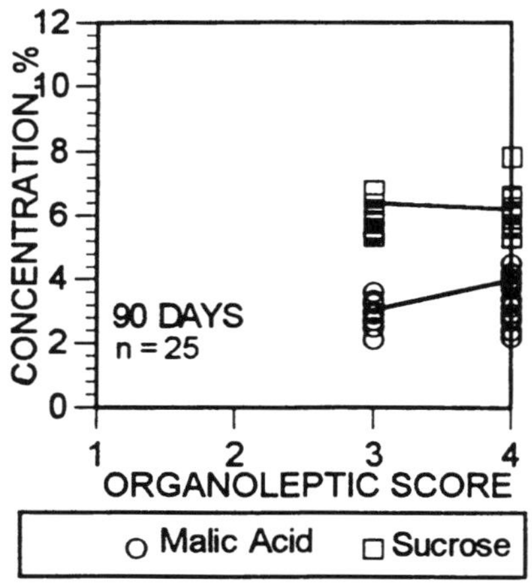

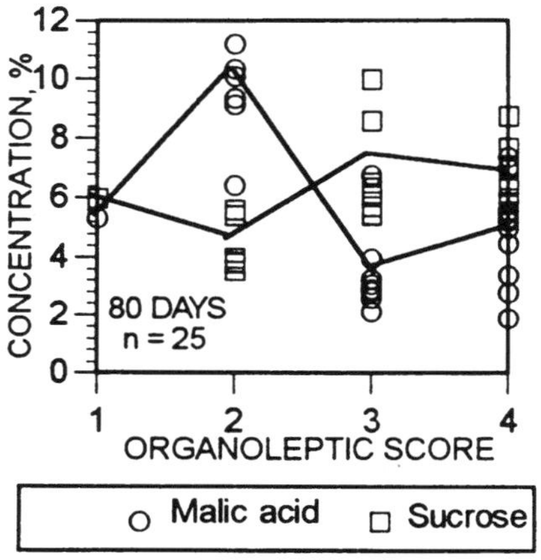

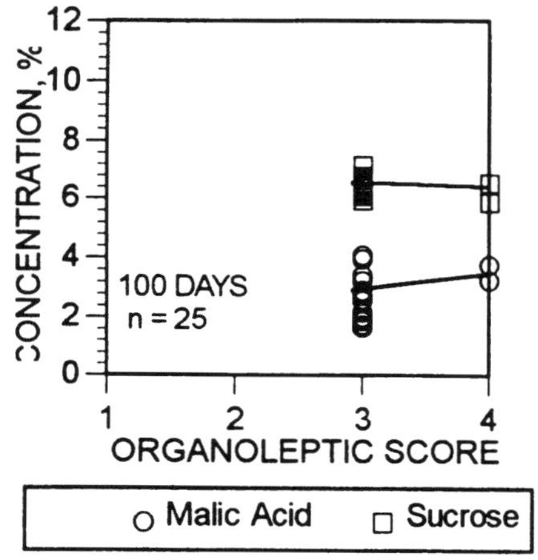

Explanation

Score 1 = Bland

Score 2 = Sour

Score 3 = Sweet

Score 4 = Sweetsour

Figure 3. Sucrose and malic acid concentration at various organoleptic score and days of harvesting

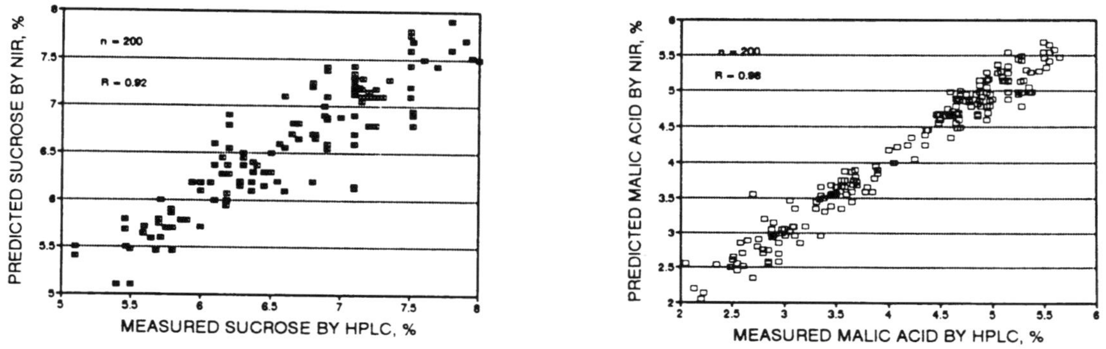

Figure 4. Predicted sucrose and malic acid by NIR method

4. CONCLUSIONS

1. The dominant individual sugar of mangoes var. *gedong* was sucrose with an average of 6.18%, and dominant acid was malic acid with an average of 3.85%.

2. The sweet-sour fruits indicated an average of 6.2% sucrose and 3.6% malic acid concentration at 90 day harvest after fruiting, and the sweet fruits contained 6.8% sucrose and 2.4% malic acid.

3. The selected NIR wavelengths to predict the fruit sucrose were 1533, 1605 and 1821 nm, and to predict the malic acid were 1 621, 1 629, 1 813, 1 821, 1 933, 1 941, 1 965, and 1 968 nm.

4. The NIR prediction on the sucrose and malic acid concentration of the fruits was found agreeable to the results of the HPLC analysis, respectively, with R=0.92 and R=0.98.

ACKNOWLEDGEMENT

This research is part of the works supported by the Centre of Research and Development for Scientific and Technology Project (PUSPIPTEK) and The National Research Council (DRN) of Indonesia through the Competitive Integrated Research 2 (RUT2) program. The authors are very grateful for the research grant and opportunity.

REFERENCES

Budiastra, I. W., D. Saputra, and H. K. Purwadaria (1995). Assessment of physical properties of mango by Image Processing. *Proceedings. Annual Meeting of Indonesian Society of Agricultural Engineers*, Bogor, 13-14 March, 1995.

Ikeda Y., I W Budiastra, T. Nishizu, and K. Ikeda (1992). On predicting concentrations of individual sugars and malic acid of the fruits by near infrared reflectance spectrometry. *Proceedings Advance on Agricultural Engineering and Technology*, II, Bogor, Indonesia, 12-15 October 1995.

Kawano S., H. Watanabe,and M. Iwamoto (1989). Determination of sugar content in intact peach by NIR spectroscopy (in Japanese). *Journal Japan Soc. Hort. Sci.* (**Suppl. 2**) 604-605.

Mouri K., and T. Iwao (1992). Studies on methods of quality evaluation of the fruits (in Japanese). *Proceedings 51$^{\underline{th}}$ Annual Meeting of JSAM*, Kyoto, Japan, 2-4 April 1992.

Saputra D., F.A. Payne, R.A. Lodder and S.A. Shearer (1992). Solutions of near infrared wavelength for monitoring milk coagulation using principal component analysis. *Transaction of the ASAE* **35(3)** : 1597-1605.

William, P., and K.H. Norris. 1987. *Near-infrared Technology in the Agriculture and Food Industry*. American Association of Cereal Chemist. Minnesota.

Yoshikawa T., K. Nagai, M. Sawa, and H. Tanaka. 1989. Non destructive method of quality evaluation of tomato (in Japanese). *Proc. JSAM Symp. on New Technology for Handling and Storage of Agric. Product*, Osaka, Japan, 15 October 1989 : 65-71.

Effects of storage conditions, fruit coatings and packaging on the marketable life of passionfruit.

By J K SSEMWANGA and A K THOMPSON
Silsoe College, Cranfield University, Silsoe, Bedford MK45 4DT. UK

Summary

Samples of passion fruit were coated with "Semperfresh", a commercially available fruit coating, at three concentrations or packed in polyethylene bags and stored at 7°C or 10°C and 60%, 83%, or 97% rh. Samples stored in polyethylene bags or those coated with 1 % "Semperfresh" had lower weight losses, less shriveling and were sweeter but were more moldy compared to control fruit. At 60% rh fruit coated with "Semperfresh" had lower weight losses than controls while those stored at 96% rh had similar weight losses. Moisture was lost mainly from the skin rather than the pulp, and shriveled samples had a similar flavour to that of smooth ones. Storage of fruit in polyethylene bags was shown to reduce moisture loss and shriveling and extend storage life to 28 days without adversely affecting the internal quality.

Key words: Passion fruit; "Semperfresh"; Polyethylene; Shriveling; Storage life; Flavour.

Introduction

The purple passion fruit, *Passiflora edulis* Sims forma *flavicarpa*, is produced in tropical and sub-tropical areas and exported to Europe by air, because of its high perishability, which results in a high price. Under ambient tropical conditions it rapidly looses moisture leading to excessive shriveling, and fermentation of the pulp can occur within 7-10 days of harvest (Pruthi, 1963). Like most exotic fruits it is largely an object of impulse buying so the shriveled appearance may deter many customers from purchasing it (Joy 1987). Marketable life, however, can be extended to 4 to 5 weeks by storage at 6.5°C to 7°C and 85%-90% rh. Lower temperatures may lead to disorders and predisposition to fungal attack while higher temperatures may lead to more moisture loss and fungal attack (Pruthi, 1963). Pruthi (1963) stored fruits in plastic bags for 4 weeks with low loss of moisture but did not determine whether off-flavours developed as has been reported in other crops (Hobson & Burton, 1989). A coating material called "Pro-long" was applied to passion fruit and found to lower the respiration rate, extend shelf life, reduce shriveling and preserve flavour during storage at 8°C, 16°C, or 24°C and 85% rh (Nyambati, 1984). "Pro-long" is an edible mixture of lipids and a polysaccharide, which is slowly permeable to carbon dioxide and oxygen (Kader *et al*, 1986). Currently the fruit is promoted and presented as a shriveled product with claims that shriveling signals the maximum development of sugar levels. Pruthi (1963), however, stated that although slightly shriveled fruits had a higher juice content they did not have better physico-chemical properties than smooth fruit.

This study investigated the possibility of extending the marketable life of fresh passion fruit while maintaining their quality. If this could be achieved it could provide the basic technical information needed in order to export the crop by sea rather than by air thus reducing its freight costs and market price.

Materials And Methods

Five trials were conducted on fruit which were already a deep purple colour. Fruit were airfreighted from Zimbabwe to UK for trials 1, 2, 4 and 5 and treatments were applied to fruit on arrival in U.K, between 6 to 7 days from harvest. For trial 3 the treatments were applied in Zimbabwe 2 days after harvest and then the fruit were airfreighted to U.K .A coating material known as "Semperfresh" was used in this study. "Semperfresh", which is similar to "Pro-long", is an edible mixture of sucrose esters of fatty acids and carboxymethyl cellulose, which is slowly permeable to carbon dioxide and oxygen. It was applied to fruit by dipping them in a solution for 10 seconds. 500 ppm benomyl was applied in the same manner. Packaging was in low density polyethylene bags of 25 microns thickness.

Experimental

Trials 1 and 2: a simple randomized block design with 6 treatments consisting of 4 "Semperfresh" levels (0, ½, 1 and 1½ %) + polyethylene bags + benomyl x 4 replicates stored at 10°C and 90 % rh for 15 days. Trial 3: a factorial split plot randomized block design with 3 levels of "Semperfresh" (0, 1 and 1½ %) + polyethylene bags x 2 relative humidities (83% and 99%) x 2 temperatures (7°C and 10°C) x 6 replications stored for 42 days. Trial 4: a factorial split plot randomized block design with 3 levels of "Semperfresh" (0, 1 and 1½ %) + polyethylene bags x 2 relative humidities (60% and 99%) x 2 temperatures (7°C and 10°C) x 6 replications stored for 28 days. Trial 5: a simple randomized experiment with 2 treatments, unwrapped and polyethylene bags, x 6 replicates stored for 42 days at 10°C and 60% rh.

Observations

Weight loss by weighing before and after storage and expressing the second measurement as a % of the first. Shriveling of the peel by scoring where 1 = none and 5 = maximum. Soluble solids by placing a droplet of juice in a refractometer (Atago model R11). Acidity by titration of the undiluted juice against standard sodium hydroxide solution. Firmness was measured by inserting a 2 mm probe through the intact fruit in an Instron Universal testing machine Model 1122, and measuring the fruit's resistance to penetration. Sensory evaluation by a triangular test with 12 panelists in trial 2. In trial 3 8 panelists were asked to taste the fruit and score it for sweetness and off-flavours on a 0 to 5 scale where 0 = no off flavours or low sweetness and 5 = maxima.

Results

Weight Loss and Shriveling

Weight loss increased with the duration of storage. Throughout the investigations fruit stored in polyethylene bags had significantly lower weight loss and less shriveling (P = 0.05) compared to control fruit (Table 1). Compared to controls, fruit coated with "Semperfresh" at the 1 % concentration had significantly less weight loss than the control in the fourth trial but the reduction at 1½ % was not significant. "Semperfresh"-treated fruit had a reduced rate of weight loss during the first two weeks of storage which remained constant until the fifth week when it increased slightly. "Semperfresh"-coated fruit lost significantly (P = 0.05) less weight at low rh, but at high rh the levels were comparable to controls (Table 2). Further, within each treatment there was less weight loss at high than at low rh.

Compared with the control there was no significant difference in the level of shriveling in samples coated with "Semperfresh" at any level applied. Temperature did not significantly (P = 0.05) affect the response of fruit to the various treatments after 28 days.

Table 1: *The effect of "Semperfresh", polyethylene packaging and benomyl (500 ppm) on weight loss of passion fruit in storage.*

Treatment	Trial 1, after 15 days (a)	Trial 1, after 15 days (b)	Trial 2, after 15 days (a)	Trial 2, after 15 days (b)	Trial 3, after 48 days (a)	Trial 3, after 48 days (b)	Trial 4, after 28 days (a)	Trial 4, after 28 days (b)
Semperfresh								
0.0%	2.789	6.30	2.944	9.10	2.843	7.47	2.944	9.99
0.5%	2.816	6.70	2.810	6.60	-	-	-	-
1.0%	2.829	7.00	2.828	7.20	2.795	6.80	2.843	7.81
1.5%	2.731	5.40	2.842	7.30	2.797	6.70	2.912	9.12
Polyethylene	2.410	1.10	2.451	1.60	2.420	1.25	2.414	1.19
Benomyl 500 ppm	2.841	7.20	2.945	9.30	-	-	-	-
LSD	0.106		0.179		0.068		0.051	
c.v.	2.6%		4.4%		4.4%		3.2%	

where: (a) = Transformed data (log [% weight loss+10])
(b) = Untransformed data (% weight loss)

Table 2: *The effect of "Semperfresh", polyethylene packaging and benomyl (500 ppm) on weight loss of passion fruit stored at 97% and 60% rh.*

Treatment	RH %	Week 1 (a)	Week 1 (b)	Week 2 (a)	Week 2 (b)	Week 3 (a)	Week 3 (b)	Week 4 (a)	Week 4 (b)
Semperfresh									
0.0%	97	2.42	1.23	2.54	2.77	2.59	3.35	2.64	3.98
	60	2.60	3.52	2.92	8.65	3.12	12.7	3.25	16.0
1.0%	97	2.40	1.07	2.51	2.29	2.55	2.76	2.58	3.27
	60	2.55	2.86	2.82	6.92	2.96	9.42	3.10	12.4
1.5%	97	2.41	1.14	2.53	2.65	2.57	3.14	2.65	4.15
	60	2.55	2.88	2.81	6.75	3.01	10.4	3.18	14.1
Polyethylene	97	2.35	0.44	2.39	0.99	2.41	1.11	2.42	1.27
	60	2.34	0.41	2.38	0.86	2.40	1.01	2.41	1.11
LSD_1		0.04		0.05		0.06		0.07	
LSD_2		0.04		0.06		0.07		0.07	
c.v.		2.2%		2.8%		3.1%		3.2%	

where: (a) = Transformed data (log[% weight loss+10])
(b) = Untransformed data (% weight loss)
LSD_1- for comparing any 2 means within the same week (columns)
LSD_2- for comparing means at the same level of rh within a week

Table 3: *The effect of "Semperfresh" and polyethylene bags on the level of Total Soluble Solids (TSS) on Passion Fruit stored for 42 days at two temperatures.*

Treatment	Main Effects		Interactions		
	log(%TSS+10)	% TSS	Temp. oC	log(%TSS+10)	% TSS
Semperfresh					
0.0%	3.07	11.71	10	3.097	12.15
			7	3.038	11.27
1.0%	3.18	14.17	10	3.174	13.97
			7	3.192	14.37
1.5%	3.19	14.27	10	3.174	13.92
			7	3.02	14.63
Polyethylene	3.12	12.74	10	3.157	13.52
			7	3.085	11.97
LSD	0.09		LSD_1	NS	
			LSD_2	NS	
c.v.	3.4%				

LSD_1- for comparing any two means
LSD_2- for comparing two means at the same temperature

TSS and Titrable Acidity

Coating with "Semperfresh" did not result in any significant effect on the level of soluble solids after 15 days' storage but after 42 days, samples treated with "Semperfresh" at both levels had a significantly higher percentage of soluble solids than controls (Table 3).

There was no difference in the level of soluble solids between treatments, and samples in polyethylene bags at 7°C had significantly higher titrable acidity than those held at 10°C (Table 4(a)). Temperature and rh did not modify the effect of "Semperfresh" on soluble solids levels. There was no significant difference between the soluble solids content of smooth (stage 1) and that of shriveled fruit (stage 4-5). The level of titrable acidity was not affected by coating with "Semperfresh". However, samples treated with "Semperfresh" at 1.0% and 1.5% concentrations and stored at 83% rh had significantly ($P = 0.05$) higher levels of titrable acidity than those stored at 99% rh (Table 4(a)).

After 42 days in storage samples coated with "Semperfresh" at 1.0% and 1.5% and held at 7°C had significantly higher ($P = 0.05$) soluble solids:acid ratios than control fruit at the same temperature while those in polyethylene bags did not. At 10°C there was no significant difference between treatments. Further, at high rh, samples coated with "Semperfresh" at both levels had higher soluble solids:acid ratios than control fruit, but not at low rh (Table 5(a)).

There was no significant effect ($P = 0.05$) of treatment with "Semperfresh" or polyethylene bags on peel firmness after 15 or 42 days' storage.

Flavour

After 15 days' storage there was no significant difference ($P = 0.05$) in flavour between "Semperfresh" coated or polyethylene packed samples and control fruit. Sensory evaluation for sweetness and off-flavours after 42 days did not show any significant difference ($P = 0.05$) between samples treated with

"Semperfresh", those stored in polyethylene bags or excessively shriveled (stage 4) and control fruit. However, in another batch of fruit 83% of the panelists preferred samples that had been sealed in polyethylene bags claiming that they had a less sharp flavour than controls. Juice and peel moisture content peel from shriveled samples (stages 4 - 5) had significantly lower (P = 0.05) moisture content than those from smooth fruit (stage 1) whereas there was no significant difference in the moisture contents of their respective juices (Table 6).

Table 4(a): *The effect of "Semperfresh" and polyethylene bags on the level of Titrable acidity of Passion Fruit stored for 42 days at different temperatures and relative humidity Units are (log[% acidity + 10]) w/v.*

Treatment	Temperature 10°C	Temperature 7°C	Relative Humidity High (99%)	Relative Humidity Low (83%)	Overall Means
Semperfresh					
0.0%	2.58	2.58	2.54	2.62	2.58
1.0%	2.60	2.54	2.48	2.66	2.57
1.5%	2.59	2.55	2.46	2.68	2.57
Polyethylene	2.56	2.70	2.66	2.60	2.63
LSD_1	0.09		0.09		NS
LSD_2	0.10		0.10		
c.v.	3.1%				

LSD_1- for comparing two means at different temperatures/rh
LSD_2- for comparing two means at the same temperature/rh

The corresponding untransformed data is presented in Table 4(b).

Table 4(b): *The effect of "Semperfresh" and polyethylene bags on the level of Titrable acidity of Passion Fruit stored for 42 days at different temperatures and relative humidity Units are (% acidity, w/v).*

Treatment	Temperature 10°C	Temperature 7°C	Relative Humidity High (99%)	Relative Humidity Low (83%)	Overall Means
Semperfresh					
0.0%	3.26	3.24	2.72	3.78	3.25
1.0%	3.54	2.85	2.03	4.36	3.19
1.5%	3.36	2.98	1.68	4.66	3.17
Polyethylene	2.97	4.89	4.31	3.55	3.93

Pathological attack and low temperature injury

After 15 days' storage at 10°C and 80 - 100% rh, neither fungicide treated nor control fruit had developed any mold. However, after 7 days at 25°C samples treated with "Semperfresh" were more moldy than the control or fungicide treated samples, those which had been treated with the 1½% "Semperfresh" concentration being more severely attacked than those with the ½% concentration. Some samples in plastic bags were found to have a pungent smell and disintegrating skin after 35 days. After 12 days' storage at 7°C, 25% of the samples treated with 1½% were moldy, as compared to 12.5% and 10% in the 1% "Semperfresh" and polyethylene respectively. The level of infection was more severe with the 1½% than the others. No statistical analysis was carried out on these results.

After 35 days low temperature injury in the form of blood-red patches on the peel was observed on some samples held at 7°C and the incidence seemed to be higher in the 1½% "Semperfresh" treatment than any other. No statistical analysis was done on this observation.

Table 5(a): *The effect of "Semperfresh" and polyethylene bags on the TSS:Acid ratio of Passion Fruit stored for 42 days at different temperatures and relative humidity Units are (log [TSS:Acid+10]).*

Treatment	Temperature		Relative Humidity		Overall Means
	10°C	7°C	High (99%)	Low (83%)	
Semperfresh					
0.0%	2.64	2.65	2.72	2.56	2.64
1.0%	2.65	2.88	2.94	2.60	2.77
1.5%	2.67	2.94	3.04	2.58	2.81
Polyethylene	2.71	2.53	2.58	2.66	2.62
LSD_1	0.16		0.16		0.10
LSD_2	0.15		0.15		
c.v.	4.6%				

LSD_1- for comparing two means at different temperatures/rh
LSD_2- for comparing two means at the same temperature/rh

The corresponding untransformed data is presented in Table 5(b)

Table 5(b): *The effect of "Semperfresh" and polyethylene bags on the TSS:Acid ratio of Passion Fruit stored for 42 days at different temperatures and relative humidity.*

Treatment	Temperature		Relative Humidity		Overall Means
	10°C	7°C	High (99%)	Low (83%)	
Semperfresh					
0.0%	4.04	4.32	5.39	2.97	4.18
1.0%	4.20	8.89	9.60	3.49	6.55
1.5%	4.53	10.78	12.14	3.17	7.65
Polyethylene	5.18	2.51	3.22	4.47	3.85

Table 6: *Comparison of juice and peel moisture contents between smooth and shriveled fruits (figures are mean of 5)*

	Initial weight (g)	Final weight (g)	Juice weight (g)	Juice m.c (%)	Peel weight (g)	Peel m.c (%)
Shriveled fruit	47.17	31.61	21.60	88.90	5.53	44.50
Smooth fruit	49.31	46.16	19.69	88.20	19.76	85.10

Discussion

The effect on moisture loss of packing fruit in polyethylene film bags agrees with results obtained by Pruthi (1963) for passion fruit and Ben-Arie (1984) with various fruits and vegetables. In the 4th trial fruit sealed in polyethylene bags for 42 days were preferred due to higher sugar levels which developed because a modified atmosphere leading to lowered respiration was created inside the bags. Off-flavours were not detected as has been reported for other crops stored in polyethylene bags (Hobson & Burton, 1989). A high relative humidity created inside the bags could have encouraged fungal attack.

The effect of "Semperfresh" in reducing weight loss agrees with the results obtained by Nyambati (1984) and Kader *et al* (1986) for plums, pears and nectarines. However, in most cases the effectiveness of "Semperfresh" reduced with increased concentration of the chemical, which contrasts with the reports from Nyambati (1984) and Kader *et al* (1986). Results obtained in this study are similar to those of Falana (1987) and Anyaoha (1988) for stored plantains and Thompson (1984) with "Pro-long" on courgettes. "Semperfresh" significantly reduced weight loss, but not at high rh. Anyaoha (1988) with plantains, Al-Zaemey (1988) and Salleh (1989) with bananas showed similar effects but found weight loss was in fact increased at high rh. Conversely, Cargnelli (1988), Flores (1988) with mango fruits, and Falana (1987) with plantains found "Semperfresh"-treated fruit had a reduced weight loss at high but not at low rh. Being hygroscopic (Semper Biotechnology 1990), "Semperfresh" probably absorbs moisture from the crop losing it to the atmosphere. Therefore at higher concentrations, it is less effective and, in some cases, increases weight loss. At low rh it would be expected to absorb more moisture from the crop and resulting in more weight loss than at high rh. This was not the case in this study, nor in those by Anyaoha (1988), Al-Zaemey (1988) and Salleh (1989). "Semperfresh" seems to have unpredictable effects on weight loss with different crops or even with different varieties of the same crop (Cargnelli, 1988).

Samples treated with "Semperfresh" had a reduced weight loss more at 10°C than at 7°C although not significantly so. These results agree with Nyambati (1984) who found "Pro-long" to reduce weight loss more at 24°C than at 16°C or 8°C. This could be due to two reasons. Firstly at low temperatures metabolism is so low that the fruit will lose little moisture with or without the coating. At higher temperatures the coating has more effect because of the high fruit metabolism. Secondly humidity is more likely to be low at higher temperatures which favours the action of the chemical. "Semperfresh" was not found to significantly affect the extent of shriveling unlike the findings of Nyambati (1984) who reported that "Pro-long" significantly reduced the degree of shriveling. Moisture loss was more rapid during the earlier stages of storage and therefore treatments aimed at preventing shriveling should be applied as early as possible after harvest. There was an indication that "Semperfresh" treatment of fruit resulted in more retention of soluble solids. Kader *et al* (1986) and Nyambati (1984) found similar results. This is, perhaps, due to lowered respiration rates in "Semperfresh"-treated samples. Samples treated with "Semperfresh" had higher sugar:acid ratio than control after 42 days but this difference was perhaps too small to be picked by the taste panel. The hypothesis that "Semperfresh" acts through lowering respiration rates, rather than retention of moisture is supported by the fact that, whereas "Semperfresh" failed to reduce weight loss significantly after 42 days, it retained significantly higher levels of sugars in the same trial. Furthermore, there was an indication that "Semperfresh"-treated fruit retained higher levels of soluble solids at 7°C than at 10°C which would support the hypothesis.

Samples treated with "Semperfresh" were attacked by mold. "Semperfresh" contains a carbohydrate-based medium (Semper Biotechnology, 1990), which could have provided nourishment to the fungus. This would explain why higher concentrations attracted more mold than lower ones. This, however, contradicts the findings of Al-Zaemey et al (1989) who found that "Semperfresh" treated bananas had a lower disease level than control fruit. The taste panel did not find any difference between shriveled (stage 4) and smooth stage 1) samples and tests with the refractometer did not yield any significant difference in sugar levels. Furthermore, results from the investigation with the moisture contents of the peel and the juice suggest that moisture loss was from the peel, rather than the juice. These results are in agreement with Pruthi (1963). The argument that shriveled fruits are sweeter because their juice has been concentrated through loss of moisture is mitigated by the findings of this work since, even after 42 days' storage (and at stages 4 - 5 shriveling), moisture was still being lost from the peel.

Symptoms of low temperature injury and the papery appearance observed in this work are similar to those described by Pruthi (1963). It is not clear why fruit treated with "Semperfresh" at 1.5% should have had a higher incidence of low temperature injury, contrary to claims by the manufacturers that "Semperfresh" reduced chill damage in tropical produce (Semper Biotechnology, 1990).

Acknowledgments

The authors wish to acknowledge the Food and Agriculture Organisation of the United Nations who provided the grant for this study and the staff of Semper Biotechnology Ltd for the assistance they rendered in applying treatments in Zimbabwe.

References

Al-Zaemey A B S (1988). An investigation into the effect of "Semperfresh" on the shelf-life and quality of banana fruits. MSc Thesis, Silsoe College, Cranfield Institute of Technology 29-47.

Al-Zaemey A B S, Falana I B and Thompson A K (1989) Effects of permeable fruit coatings on the storage life of plantain and bananas. Aspects of Applied Biology 20 73-82.

Anyaoha E T (1988). The interaction between coating plantain fruits with "Semperfresh" and humidity. MSc Thesis, Silsoe College, Cranfield Institute of Technology 43-45.

Ben-Arie R (1984). Developments in semi-permeable wrappings - the life of fresh fruits and vegetables. Eurofruit Congress **V**, 1-10.

Cargnelli S (1988). The effects of "Semperfresh" on the degradation of mango fruits. MSc Thesis, Silsoe College, Cranfield Institute of Technology 25-46.

Falana I B (1987). An investigation into the effects of a membranous food surface coating ""Semperfresh"" and plastic film bags on the storage life and quality attributes of plantains (Musa AAB). MSc Thesis, Silsoe College, Cranfield Institute of Technology 54-57.

Flores G M S A and Helens Sanchez T S U (1988). Uso de retardantes de maduracion en mangoes c.v. "Haden" producidos en Venezuela para exploracion. Fundacion servicio para el agricultor. Estacion experimental de Cogua, Venezuela. In Cargnelli, S. (1989) The effects of "Semperfresh" coatings on the degradation of mango fruits. MSc Thesis, Silsoe College, Cranfield Institute of Technology 25-46.

Hallam D and Molina J P (1988). The United Kingdom market for exotic fruit. Overseas Development Natural Resources Institute **Bulletin, 13,** 21-49.

Hobson G and Burton K S (1989). The application of plastic film technology to the preservation of fresh horticultural produce. Professional Horticulture, **3** 20-23.

ITC (International Trade Centre) Newsletter (1989). plus personal Communication with J. Henry, Senior Marketing Adviser, ITC, Geneva.

Joy C (1987). Selected European markets for specialty and tropical fruit and vegetables. Report of the Tropical Development Research Institute, **G 201**, 40-41.

Kader A A, Cantwell M and Mitchell F G (1986). Gas diffusion, water loss and quality of pears and stone fruits as affected by ""Semperfresh"" coating Progress Report, Dept. of Pomology, University of California, Davis, USA.

Nyambati M G (1984). The effect of permeable coatings on the storage of fruits. Prolong treatment on purple passion fruit Passiflora edulis Sims (Unpublished).

Pruthi J S (1963). Physiology, chemistry and technology of passion fruit. Advances in Food Research **12**, 203-282.

Salleh E V (1989). Rate of deterioration of Banana fruits as affected by coating with "Semperfresh" or storing in plastic film bags. MSc Thesis, Silsoe College, Cranfield Institute of Technology 69-72.

Semper Biotechnology Ltd (1990). "Semperfresh", a product overview, pp 1 - 11.

Thompson J A (1984). An investigation into the effects of a membranous surface coating and plastic film wrappings on the storage life of courgettes. MSc Thesis, Silsoe College. Cranfield Institute of Technology.

POSTHARVEST OPERATIONS AND QUALITY OF RAPESEED

Andrzej Stępniewski and BoguslawSzot

*Institute of Agrophysics, Polish Academy of Sciences,
P.O.Box 121, 20-236 Lublin, Poland*

Abstract: The investigations were made in order to find sources and reasons of quality losses in typical handling of rape. Samples were taken before and after succeeding operations to estimate the influence of each of them on physical state of seeds. The estimation was curried out with simultaneous registration of working parameters of machines. Evaluation of damage was made according to methods elaborated in the Institute of Agrophysics, which allow to measure the amount of micro and macro damage as well as the amount of foreign material. Moisture content of seed was also measure during the study.

The decrease of quality features of seed was noticed after each handling operation. However harvesting acted the most negatively and the level of damage noticed after harvesting decided about the amount of damaged seeds at the end of postharvest operations on rape. Drying caused the increase of macrodamage and decrease of microdamage. The amount of microdamage increased in transport. The moisture content after drying was correct according to standards.

Keywords: damage, quality, handling, transport, mechanical properties

1. INTRODUCTION

The great offer of rapeseed on the world market extorts better quality of seeds. Therefore rape producers are obligated to supply cleaner and undamaged seed to be equal to the demands of exporters. From the other hand the wide use of machinery in postharvest handling negatively influence physical state of seed (Stępniewski *et al.*, 1991). Seeds undergo various external loads while they are drying, cleaning, conveying and storing. Forces acting during these operations cause damage to seed, what results in worse quality of raw material for oil production. Szot (1987) reported decrease of basic strength features of seeds of new varieties of rapeseed. Genetic changes led to improvements of oil composition, but at the same time they also negatively influenced mechanical properties of seeds. New varieties were more susceptible to damage than old one. Szot and Stępniewski (1994) described mechanical properties of rapeseed in the aspect of it quality and quantity losses. Apart of harvesting and transport, drying played very important part in preparing seeds to storing and then to oil production. Stępniewski *et al.*, (1994) noticed decrease of technological value of rapeseed with the increase of drying temperature and initial moisture content. Stępniewski *et al.*, (1991) studied decreasing the quality of rapeseed during postharvest handling. However this results refered to specific circumstances - machines used were untipical, normaly used to handle beet pulp in sugar factory. Drum dryer as well as pneumatic transport were used and this played the most negative part in the whole postharvest process. They noticed also very negative influence of harvesting. Fiscus *et al.*, (1972) investigated physical damage of grain caused by various handling techniques. They studied the influence of various type of transport on the amount of damage seed.

The present study monitored changes in level of rapeseed damage from harvest till storing of seed before oil production. The described process was typical for polish circumstances and reffered to typical machines and apparatus used in Poland for preparation of seed to storage or production. The aim was also to find operations and/or machines acted the most negatively on quality features of rapeseed.

2. MATERIAL AND METHODS

Investigations were made during 1994 "harvest seasons" in BOLMAR factory - the biggest oil producer in south-east Poland. Rapeseed came from state and private farms and it was submitted right after harvesting. All other operations were made in BOLMAR factory and were under investigation. The basic samples (primary) represented physical state of rape after harvesting and were colected from trailors before unloading. The succeede operations of postharvest technological process of rapeseed are presented on the Figure 1. This process was divided into five stages: unloading, cleaning, storage of wet rapeseed, drying and storage of dry rapeseed. Between stages seeds were conveyed using various types of conveyors: screw, belt, Redler, chute and buckets elevators. Unloading was performed on tippler, seeds were poured into the charging hopper of the srew conveyor. It conveyed rape to the buckets elevator, which lifted it to the slide pipe and than to the sieve cleaner. But it was only rough separation of rapeseed from pieces of straw, pods, stones. Grains of cereals and weeds as well as other little pieces of foreign material weren't removed. Wet rape was stored in silo and during the same day it was dried. Vertical continuous drying stove was used, where hot air was a drying factor. The temperature of air at dryer's entrance was 110-130°C. Dry rapeseed was stored in silos till oil production.

Samples for investigations were colected before and after each operation. The first sample was taken before unloading and it represented seed after harvesting. The second was colected before cleaning and the next after storing of wet seed. The next two samples were taken before and after drying. The last sample was taken from silo of dry rape. Samples were taken every 15 minutes within 5 hours a day. Investigations were started at July 26-th. and finished at August 1-st. The total number of samples was 140. All samples were taken from a moving stream of rapeseed, considering accessibility of the sampling places. After colecting, samples were taken to the laboratory in order to check moisture content and their quality parameters. Moisture was checked by drying according to Polish Standards. Before studing quality features, all samples were kept open till seed achieve air-dry moisture content (5 - 7% w.b.).

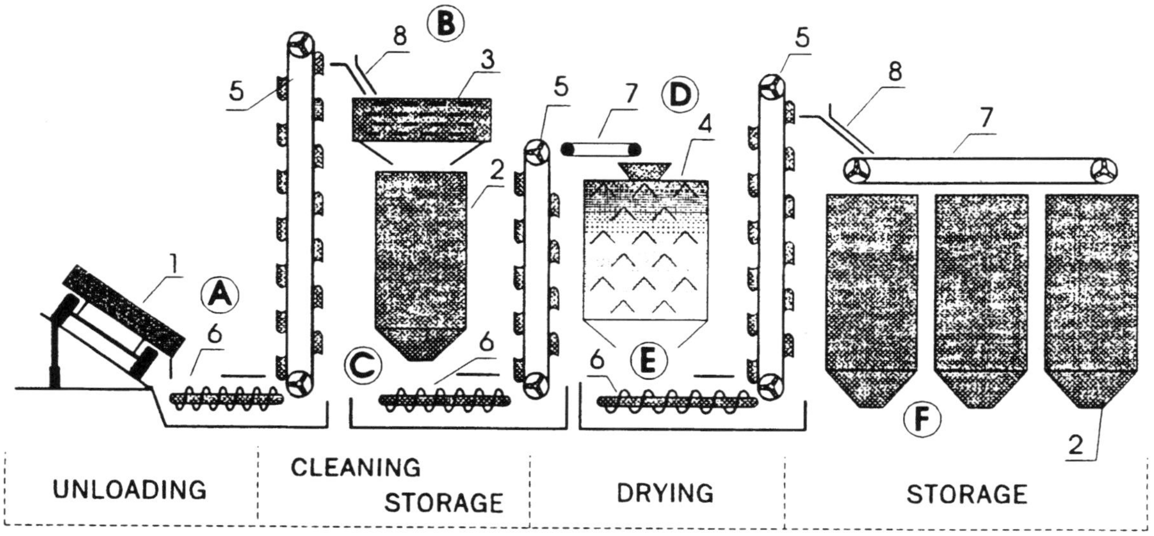

Fig.1. The scheme of typical postharvest technological process of rapeseed: A-F - sampling places, 1- tippler, 2- silo, 3- cleaner, 4- dryer, 5- bucket elevator, 6- screw conveyer, 7- belt conveyer, 8- chute

The following quality features were investigated: amount of macrodamage, amount of microdamage and amount of foreign material (pieces of straw, pods, grains of wheat and weeds etc.). Macrodamage and foreign material were expressed as a weight percentage of air dry matter. First foreign material was separated by hands from the 10g sample of rapeseed. Then damaged seeds - broken pieces of cotyledons and skins as well as seeds with visibly broken skin were separated. From the rest of seeds 10 subsamples of 100 seeds in each were put on the wet filter paper and kept for 2 hours. Seed swelled and microcracks of skin could be seen. The amount of microdamage was expressed in percents as number of seeds (from 100).

Analysis of variance (p=0.05) was made in order to estimate significance of differences.

3. RESULTS AND DISCUSSION

Moisture content of rape delivered to the factory was from 9 till 14%. Such a low moisture was observed because of very dry and hot weather before and during harvesting. In the most of cases seed required only sleight drying (drying up), which was conducted correctly and the final moisture content of seed was from 5 till 6%.

The amount of foreign material generally decreased during the whole investigated process, however statistical differences were insignificant (Fig. 2). Rape delivered to the factory after harvesting had 3.2% of foreign material in average, while at the end of the process - in the silo, there was 2.9% of this fraction. It should be stated that the biggest decrease of this fraction was noticed during cleaning (from 3.7 to 2.9% in average), but even this difference was insignificant. The amount of foreign material after cleaning was on the similar level till the last operation of the process.

The differences between macrodamage in succeeded operations were insignificant, but the tendency to the increase was observed (Fig. 3). An average amount of macrodamage after harvesting was 5.3% and it increased till 6.1% after drying. The slight increase of macrodamage level (0.8%) was noticed during cleaning and drying, while during transport this level decreased (about 0.3%).

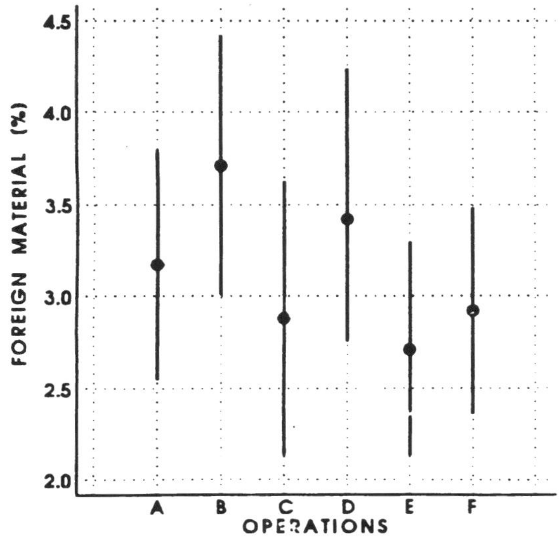

Fig. 2. Means and confidence intervals of the amount of foreign material in succeeded operations

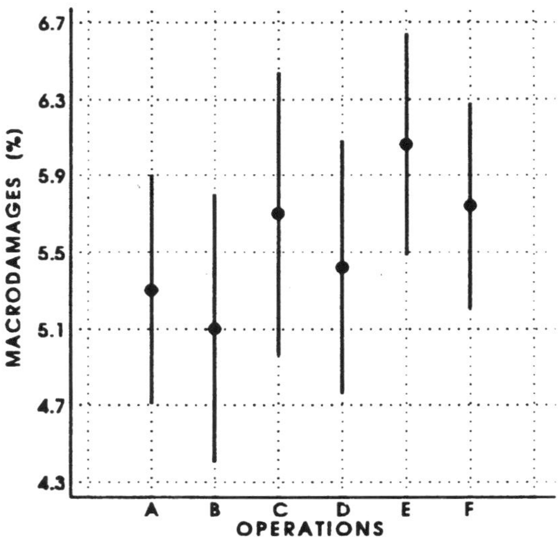

Fig. 3. Means and confidence intervals of the amount of macrodamage in succeeded operations

The level of microdamage increased at the begining and during drying rapidly decreased (Fig. 4). The smalest number of microdamage was noticed in samples after harvesting (2.2%) and they increased in each transport operation and during cleaning (insignificant difference) till 3.7% before drying. Then the difference was significant (1.3%) and the level of microdamage was in average 2.5% till storing in silo.

The decrease of the amount of microdamaged seeds during drying was connected with the increase of macrodamage occured in this process. It suggested, that microdamaged seeds had been broken while exposed on the higher temperature.

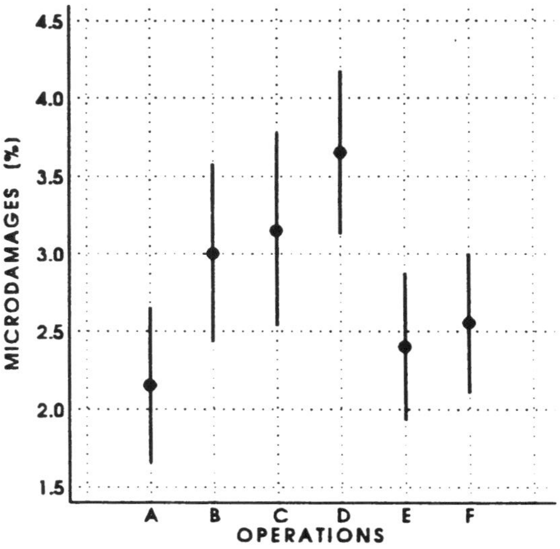

Fig. 4. Means and confidence intervals of the amount of microdamage in succeeded operations

Water and temperature stress in drying extended microdamage to dacrodamage.

Similar transport operations caused increase of microdamage and decrease of macrodamage. In spite of very differentiated samples, which were colected from various farms (different harvest conditions), the tendency to increase of macrodamage and decrease of microdamage could be observed during the investigated process.

The amount of foreign material was on similar, high level within all operations - what indicated insufficient cleaning. However unmistakable results could be achieved only on homogeneous lot of seed, what was impossible in practice (rape came from various farms). In the light of the above mentioned results, quality features of rapeseed after harvesting strongly influence final quality of this seeds. Each level of all investigated fractions was similar to the level after harvesting. The differences occured were statistically insignificant and indicated only a tendency. Postharvest operations on rapeseed hadn't any strong influence on it's quality. This opinion could be compared with the conclusions obtained few years ago by the same autors (Stępniewski *et al.* 1991). That time transport of dry seed influenced seed quality as strong as harvesting. It was pneumatic transport with quite high velocity, what resulted in very high level of macrodamage. In the present investigations this type of transport didn't occure, so the only cause of high amount of damaged seed was harvesting. Other operations didn't act so strongly.

4. CONCLUSIONS

The most significant influence on quality features of rapeseed had harvest operation. The amount of micro and macrodamage were on the similar level during all postharvest operations.

The increase of macrodamage and decrease of microdamage during drying was observed, what could be a result of extend of micro to macrodamage in consequence of water and temperature stress.

Transport operations gave the increase of the amount of microdamage.

Drying was conducted properly and the final moisture content of seed was correct according to the standards.

Cleaning was insufficient and the amount of foreign material was on similar level both after harvesting and at the end of postharvest operation in the silo.

5. ACKNOWLEDGEMENTS

The authors thank State Committee for Scientific Research for the sponsorships within grant :
No. 5 S306 001 07

6. REFERENCES

Fiscus, D.E., G.H. Forster and H.H. Kaufman (1972). Physical damge of grain caused by various handling techniques. *Transaction of ASAE* **14, 3**, 480-485.

Stępniewski, A., B. Szot. and R.L. Kushwaha (1991). Decreasing the quality of rapeseed during postharvest handling. *Proceedings of 8-th. International Rapeseed Congress*, 1251-1255.

Stępniewski, A., B. Szot., J. Fornal. and J..Sadowska (1994). Drying conditions and mechanical properties of rapeseed. *J. Food Physics*, **2**, 86-89.

Szot, B. (1987). An evaluation of the mechanical properties and susceptibility to damage of winter rape seeds. *Proceedings of the 7-th. International Rapeseed Congress*, 850-855.

Szot, B. and A. Stępniewski (1994). Studies on mechanical properties of rape in the aspect of its quantity and quality losses. *Zemedelska Technika*, in print.

Polish Standards PN-62/R-66163

Polish Standards PN-73/R-66147

Polish Standards PN-75/R-66145

EFFECTS OF CONTROLLED ATMOSPHERE STORAGE ON THE EXTENSION OF POSTHARVEST QUALITIES AND STORAGE LIFE OF TOMATOES

Ali Batu[1] and A. Keith Thompson[2]

1 Gazi Osman Pasa University, Agriculture Faculty, Food Science and Technology Department, 60100, Tokat, Turkiye.

2 Postharvest Technology Department, Silsoe College, Cranfield University, Silsoe, MK45 4DT Bedford, England.

Abstract: Tomatoes were harvested at the pink stage of maturity and stored at 13°C for 60 days in six different combinations of CO_2 all with 5.5 % O_2 compared to in 20.9 % O_2 with 0.3 % CO_2 (as control). The CO_2 levels affected the colour development of tomatoes, in that after 50 or 60 days storage there was no significant colour changes in those stored in 6.4 % and 9.1 % CO_2 with 5.5 O_2 but the tomatoes stored in the control, 0.2 % CO_2 and 3.2 % CO_2 all combined with 5.5 % O_2 had turned fully red. Colour values of fruit stored in 12.2 % and 15.6 % CO_2 with 5.5 % O_2 had actually decreased at that time. Fruit firmness, their total soluble solids content and the level of acceptability, assessed by panellist were highest for those stored in 9.1 % CO_2 with 5.5 % O_2 and those tomatoes had the best organoloptic properties and were still considered marketable after 60 days storage.

Keywords: Tomato, controlled atmosphere, storage life, colour, firmness, acidity, TSS

1. INTRODUCTION

After harvest, ripening continues and tomatoes can become over-ripe quickly depending on their temperature and harvest maturity. This results in loss of quality, restricted shelf life and, in some instances, wastage of fruit during distribution and marketing (Geeson et al., 1985). Fresh tomato quality is determined by appearance, colour, firmness and flavour (Risse et al., 1985). Delaying ripening and senescence, reducing the incidence and severity of certain physiological disorders are possible by keeping the fresh tomatoes in suitable storage environment.

Cooling is the most widely utilised technique to extend the life of perishable produce but controlled atmosphere storage (CAS) is a very important technique usually used in conjunction with refrigeration in maintaining quality and extending the post harvest life of fresh fruit and vegetable (Ben-Yehoshua et al., 1985; Peleg, 1985). Kader (1992) defined CAS as the modification of O_2, CO_2 and/or ethylene concentrations in the atmosphere surrounding the commodity to levels different from those in air and keeping constant of chosen concentrations of the gases. Generally reducing the O_2 and increasing CO_2 concentration of the storage atmosphere decreases the respiration rate of fruits and vegetables (Geeson et al., 1985). CAS is effectively used to extend the storage life of seasonal perishable products (Hardenburg et al., 1986). Preventation of ripening and associated changes in fruits is one of the main benefits of CAS (Kader, 1980). CAS reduce respiration rates as long as the levels of O_2 and CO_2 with in tolerated by commodity. Low O_2 and high CO_2 concentration can reduce the incidence and severity of certain physiological disorders such as those induced by ethylene and chilling injury of some commodities (Kader et al., 1989). The effects of this reducing O_2 and increased CO_2 on fruit varies with concentration, exposure time and variety but it generally retards any fungal infection and can inhibit the effects of ethylene in causing deterioration (Thompson, 1995).

Maturity at harvest is very important to composition and quality of tomatoes (Hobson, 1981; Yang et al., 1987; Ramana et al., 1987 and Gesson, 1989). This is especially a problem with tomatoes picked green since it is difficult to differentiate between mature and immature green fruits. Mature green and advanced mature green tomatoes will usually attain a much better flavour at the table ripe stage than those picked at the immature or partially mature stages (Grierson and Kader, 1986). The present investigation was, therefore, carried out to develop a modified storage atmosphere, using different levels of O_2 and CO_2, which would retard deterioration and extend the shelf life of tomatoes without detrimentally affecting their organoloptic qualities.

2. MATERIAL AND METHODS

Freshly harvested pink (first appearance of external pink red or tannis-yellow colour over 30 % but not more than 60 % red; Grierson and Kader, 1986) tomatoes (cv 'Criterium') were obtained from the greenhouse at Silsoe Research Institute.

The six tomatoes were placed in a plastic box (18 cm length, 9 cm depth and 11 cm width). Then 12 of those plastic boxes were each placed in an air tight 75 litre polyethylene container. The containers were connected to the gas distributor. The gas distributor was connected to a computer-controlled gas blender which was connected to cylinders of compressed oxygen, carbon dioxide and nitrogen. After sealing the containers, the O_2 content was rapidly reduced by flushing with the mixture of the desired concentration of gases. The appropriate atmospheres were obtained within a maximum of 24 hours after harvest. The output of the O_2 and CO_2 concentrations from the gas blander and the containers were monitored twice a day for the first week of the experiment then twice a week. The concentration of the gases used were 0.2 %, 3.2 %, 6.4 %, 9.1 %, 12.2 % and 15.6 % CO_2 all combined with 5.5 O_2 plus a control was contained 20.9 % O_2 with 0.3 % CO_2. The containers were stored in a temperature controlled room at 13°C (12.8-13.7°C) for 60 days. Each 10 days one plastic box was removed from each container. Skin colour, firmness, weight loss, titratable acidity and total soluble solids were measured.

Skin colour values were measured using a Minolta chromometer model CR 200 at three points on the circumference of the fruits. The instrument was calibrated against standard white colour plate (Y=93.9, x=0.313, y=0.321) (Anon, 1990).

A destructive deformation test was used to measure fruit firmness (Batu and Thompson, 1993), by applying a constant force using a 6 mm diameter round stainless steel probe with a flat end in an Instron Universal Testing Machine, model 1122 with cross-head and chart speeds of 20 mm minute^{-1}. The force required for the probe to penetrate the fruit and the deformation of the fruit up to the bioyield point were measured. The firmness was expressed as force/deformation in N mm $^{-1}$.

Soluble solids content was determined with an Atago digital refractometer model PR1. Acidity was determined by titrating juice to pH 8.1 with 0.3125 N NaOH using a Jenway digital pH meter model 3020.

The experiment was a complete randomised factorial design subject to analysis of variance. Least significant differences were calculate to compare means where appropriate.

3. RESULTS AND DISCUSSION

3.1. Fruit Colour

There were significant differences (p=0.05) between the treatments during storage period. Tomato colour changed more quicker when they were kept in the control (20.9 % O_2 and 0.3 % CO_2) and in 0.2 % CO_2 with 5.5 % O_2 than the other treatments. Those tomatoes reached their reddest colour over the first 20 and 30 days of storage respectively, then remained a similar colour. The colour of tomatoes stored in 3.2 % CO_2 and 6.4 % CO_2 both combined with 5.5 % O_2 also increased during the first 20 days of storage but levels were significantly lower than fruits stored in control and 0.2 % CO_2 with 5.5 % O_2. Then their colour also remained the similar pink stage of at harvest maturity. After 60 days of storage their skin colour were still at the light red stage.

The Minolta a*/b* values for fruits stored in 12.2 % CO_2 and 15.6 % CO_2 both combined with 5.5 % O_2 had actually decreased during the 60 days of storage. The decrease of colour values of fruits stored in 15.6 % CO_2 were higher than in 12.2 % CO_2. Although colour values of tomatoes stored in 9.1 % CO_2 with 5.5 % O_2 slightly increased during the first 20 days of storage. After 40 days of storage fruit colour remained the same and after 60 days of storage it was very similar to the initial fruit colour.

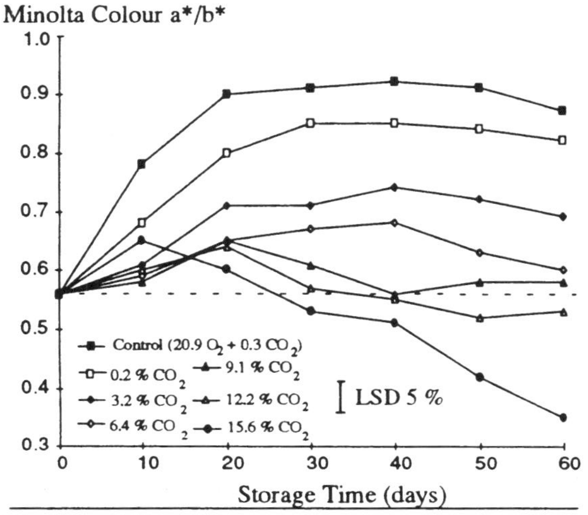

The dotted line represents to colour values at harvest Colour values were pink: 0.08-0.57; light red: 0.61-0.95; red: 0.95-1.22 (Batu and Thompson, 1994)

Figure 1. Changes of Minolta colour values of tomatoes harvested at the pink stage and stored in various controlled atmosphere environments at 13°C.

CO_2 has previously been shown to affect colour development of tomatoes. Buescher (1979) indicated that colour development was inhibited in tomatoes particularly when exposed to high levels (20 to 60 %) of CO_2 because the production of ethylene was suppressed (Kuba et al., 1989). Colour development of tomatoes were affected by the levels of CO_2. There was a good correlation between the CO_2 levels and red colour development of tomatoes. Yang et al., (1987) and Salunkhe and Wu (1973) showed that there was also a good correlation between red colour development and lycopene content of tomatoes. Slowing down or decreasing of red colour development of tomatoes was directly dependent on the level of ethylene required to stimulate ripening. In this experiment high level of CO_2 suppressed (data is not shown) the production of ethylene. Therefore red colour values of tomatoes were decreased in 12 % and 15 % CO_2 both combined with 5.5 O_2. It could be due to continuation of synthesis of β-carotene but not (or less) lycopene as mentioned by Buescher (1979). Another possibility was reported by Buescher (1979) that the retardation of lycopene development in high CO_2 exposed fruits may be explained on the basis that ethylene synthesis declined faster in the presence of high CO_2.

3.2. Fruit Firmness

All fruit softened during storage but some of those stored in CAS (0.2 %, 3.2 %, 6.4 % and 9.1 % CO_2 all combined with 5.5 % O_2) soften significantly ($p=0.05$) more slowly than those stored in 20.9 % O_2 with 0.3 % CO_2, 12.2 % CO_2 with 5.5 % O_2 and 15.6 % CO_2 with 5.5 % O_2. There was no significant ($p=0.05$) difference in firmness values between tomatoes stored in 20.9 % O_2 and 0.3 % CO_2 and in 15.6 % CO_2 with 5.5 % O_2 until the 40th day of storage. But after that time tomatoes stored in 15.6 % CO_2 with 5.5 % O_2 softened very rapidly (Figure 2).

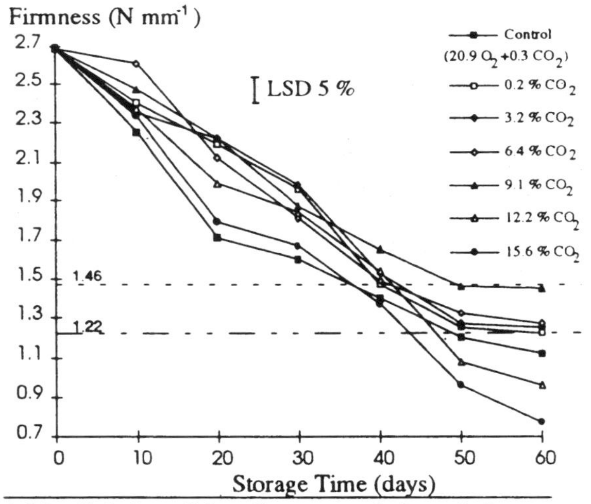

1.22; slightly soft but firm enough for salads, 1.46; very firm

Figure 2. Changes of firmness values of tomatoes harvested at the pink stage and stored in various controlled atmosphere environments at 13°C.

An arbitrary subjective estimate of tomato softness that would be the minimum acceptable on the market was established as between 1.46 (very firm) and 1.22 N mm^{-1} (slightly soft but acceptable). Tomatoes stored in 20.9 % O_2 with 0.3 % CO_2 and 15.6 % CO_2 with 5.5 % O_2 were shown to fall below this level after 35 days storage while all those stored in between 0.2 % and 12.2 % CO_2 with 5.5 %

O_2 were above this 1.45 N mm^{-1} level after 40 days of storage even fruits stored in 9.1 % CO_2 with 5.5 % O_2 remained around this level and were still firm after 60 days of storage. Fruits stored in 3.2 % CO_2 with 5.5 % O_2 and 6.4 % CO_2 with 5.5 % O_2 were judged to be still in a condition to used for salads even after 60 days of storage. We found that it would be possible to extend the keeping quality of tomatoes for at least 60 days when stored in 9.1 % CO_2 with 5.5 % O_2 at 13°C.

Of the quality factors investigated the most marked response was the maintenance of texture under elevated CO_2 concentrations (including 3.2 %, 6.4 % and 9.1 % all combined with 5.5 % O_2) but fruits stored at 9.1 % CO_2 with 5.5 % O_2 were found to be firmer. In this experiment fruits became softer in the 0.2 %, 3.2 % and 6.4 % CO_2 with 5.5 % O_2 because this amount of CO_2 was not enough to inhibit the production of ethylene which in turn inhibit the action of pectinestarase and polygalacturase enzymes which have been shown to be responsible for softening in tomatoes (Grierson and Tucker, 1983) Buescher (1979) also found that ethylene production of tomatoes in 5 % CO_2 was very similar to in air compared to ethylene production in 10 % and 20 % CO_2. But in our experiment increasing CO_2 concentration to 12 % and 15 % with 5.5 % O_2 gave the softer textured fruits because of CO_2 injuries. Beneficial effects on texture with 5 % CO_2 have been reported by many authors (Herner, 1987 and Kader, 1986) and Larsen and Watkins (1995) strawberries stored in 10 % CO_2 with 2 % O_2 at 20°C had a firm texture then those stored in air. Kader (1986) also reported that although elevated CO_2 atmospheres slowed the softening rate of tomatoes, the mechanism of controlled or modified atmosphere effects on texture of fresh fruits and vegetables is still not fully understood.

3.3. Total Soluble Solid (TSS)

TSS of tomatoes stored in 6.4 % CO_2 and 9.1 % CO_2 both combined with 5.5 % O_2 increased over 20 days of storage although there was a trend to decreasing TSS during storage after the 20th day, TSS values of those tomatoes were higher after 60 days of storage than their initial TSS values (Figure 3). Although there were a slight increase during the first 30 days of storage, after this time TSS values of tomatoes stored in 0.2 % and 3.2 % CO_2 with 5.5 % O_2 remained similar to their initial values. There was a trend to decreasing TSS values of 12.2 % CO_2 with 5.5 % O_2 and 15.6 % CO_2 with 5.5 % O_2 and control fruits with the trend higher in control fruit. Acidity levels have been shown to decrease during storage which would account for some of the losses in TSS but some of the losses would be accounted for in loss of sugars used up in the normal metabolism of the fruit. The TSS of tomatoes are made up largely of sugars with some organic acids and other chemicals (Hobson and Grierson, 1993).

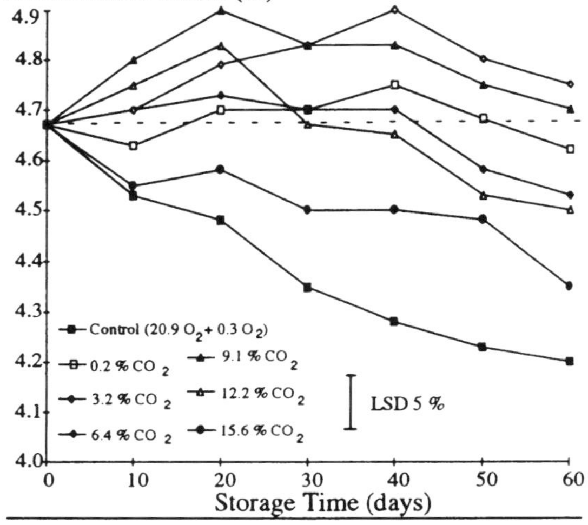

The dotted line represents to TSS at harvest

Figure 3. Changes of total soluble solids values of pink tomatoes stored in various controlled atmosphere environments at 13°C.

Conflicting results on changes in TSS and sugar levels have previously been reported in tomatoes. These included no changes in TSS (Esquerra and Bautista, 1990) and an increase in reducing sugar (Goodenough and Thomas, 1981). In tomatoes the rate of total sugar accumulation was considerably reduced by storage in 10 % CO_2 compared to in air (Burton, 1982b). However, low O_2 (1 %) storage was shown to inhibit sugar accumulation in tomatoes. The results of this experiment confirm this latter trend which might be related the suppression of the fruit's metabolism more affecting the breakdown of starch than the suppression in respiration rate which would use sugar as a substrate.

3.4. Titratable Acidity

Acidity of fruits in some treatments remained similar during the first 10 days storage period while it decreased in the rest. After 10 days the acidity levels generally linearly decreased (Figure 4). Although there were significant (p=0.05) differences in acidity values of tomatoes between treatments, there was no direct correlation with the concentrations of CO_2 levels. The acidity values of tomatoes stored in 12.2 CO_2 with and 15.6 % CO_2

both combined with 5.5 % O_2 were lower than other treatments during throughout storage, but after 30 days storage, control fruit were also lower. In comparison, acidity of tomatoes stored in 3.2 %, 6.4 % and 9.1 % CO_2 all combined with 5.5 % O_2 were higher than other treatments.

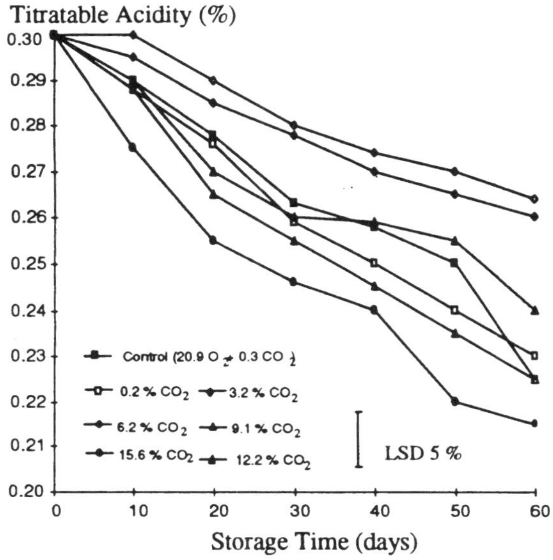

Figure 4. Changes of titratable acidity of tomatoes harvested at the pink stage of maturity and stored in various controlled atmosphere environments at 13°C.

It would seem logical that fruits stored in CAS with increased levels of CO_2 would be more acid because CO_2 is an acid gas which should be dissolved in the cell sap in proportion to its concentration in the surrounding atmosphere but our results disagreed with this. Goodenough and Thomas (1981) reported that acids increased steadily throughout maturation and ripening to a peak as the fruit changed colour and then declined somewhat. There is little information on changes of acidity during CAS of tomatoes. Parsons et al., (1970) found that titratable acidity increased with increasing CO_2 concentration from zero to 5 % CO_2 during CAS but there was no comment that levels of acidity of those tomatoes was acceptable by consumer, and there was also no information on comparison to control tomatoes at that time.

It is difficult to find enough information on the relation between acidity changes and CO_2 concentrations during CAS of tomatoes. But there is some conflicting reports on some other fruits. In recent review paper, Riquelme et al., (1994) reported that storage of strawberry in low O_2 and high CO_2 concentrations do not affect titratable acidity. It was also reported that in 60 % CO_2 there was no affect on titratable acidity of Valencia oranges, while storage of lemons under high CO_2 leads to accumulation of organic acids (Biale, 1960). Salunkhe and Wu (1974) reported that the titratable acidity of green beans increased during air storage, but decreased slightly in CAS. They also reported that titratable acidity of broccoli and asparagus decreased progressively with increasing concentration of CO_2 in the CAS.

4. CONCLUSION

Tomatoes stored in CAS had an extended marketable life This was shown in the delayed colour changes and delayed softening of the fruits. There were no detrimental physiological characteristics detected in the tomatoes (except in 12 % and 15 % CO_2 both combined with 5.5 O_2) even after 60 days storage in CAS, although there was some indication that prolonged storage in CAS can detrimentally affect their flavour. After 60 days of storage the lowest amount of infected fruits were 10 % in 9.1 % CO_2 with 5.5 % O_2 and those fruits were the most acceptable to consumers on the basis of the sensory evaluation.

REFERENCES

Anonymous (1990). Minolta, Precise Colour Communication. Colour Control From Feeling To Instrumentation. *Handbook*. Printed by Minolta Camera Co. Ltd. Japan.

Batu, A. and A.K. Thompson (1993). Effects of Cross-head Speed and Probe Diameter on Instrumental Measurement of Tomatoes Firmness. In *Proceedings of the International Conference for Agricultural Machinery and process Engineering*, pp. 1340-1345. Seoul, Korea.

Batu, A. and A.K. Thompson (1994). The Effects of Harvest Maturity, Temperature and Thickness of Modified Atmosphere Packaging Films on the Storage Life of Tomatoes. *Proceedings of the International Symposium on New Application of Refrigeration to Fruit and Vegetables Processing*. Istanbul. Turkey. (In press).

Ben-Yeshoshua, S. 1985. Individual Seal-packaging of Fruit and Vegetables in Plastic Film. A new Postharvest Technique. *HortScience*. **20**, 32-27.

Biale, J.B. (1960). Respiration of Fruits. In Handbook Der Plantepphysiologie. *Encyclopedia of Plant Physiology* (J. Wolf (ed)), pp. 536-592. Springer-Verleg. Berlin

Buescher, R.W. (1979). Influence of high temperature on physiological and compositional characteristics of tomato fruits. *Lebensm. WIC.*

Technology. **12**, 162-164.

Burton, W.G. (1982). Physiological Responses to Stress and Disease. In *Postharvest Physiology of Food Crops*. (W.G. Burton (Ed)). Chapter 9, pp. 199-217. Logman Group Ltd.

Esquerra, E.B. and O.K. Bautista (1990). Modified Atmosphere Storage and Transport of 'Improved Popa' Tomatoes. *ASEAN Food Journal*. **5**, 27-33

Geeson, J.D., K. M. Browne, K. Maddison, J. Sherperd and F. Guaraldi (1985). Modified Atmosphere Packaging to Extend the Shelf Life of Tomatoes. *Journal of Food Technology*, **20**, 339-349.

Geeson, J.D. (1989). Modified atmosphere packaging of fruit and vegetables. *Acta Horticulture*. **258**, 143-150.

Goodenough, P.W. (1991). Tomato, Cucumber and Gherkin. In *Food Flavours*. Part C. The Flavour off Fruits. (I.D. Mortan & A.J. Acleod (Eds.)), pp. 327-350. Elsevier Science. Publishing B.U.

Grierson, D. and G.A. Tucker (1983). Timing of Ethylene And Polygalacturonase Synthesis in Relation to the Control of Tomato Fruit Ripening *Planta*, **157**, 174-179.

Grierson, D. and A.A. Kader (1986). Fruit ripening and quality. In *The tomato crop*. (J.G. Atherton and Rudich J. (Eds)), pp. 241-280. Chapman and Hill Ltd. USA.

Hardenburg, R.E., A.E. Watada and C.I. Wang. (1986). The Commercial Storage of Fruits, Vegetables, and Florist and Nursery Stocks. US. Department of Agriculture. *Handbook*. 66.

Herner, R.C. (1987). High CO_2 Effects on Plant Organs. In *Postharvest Physiology of Vegetables*. (J. Weichmann (Ed)), pp. 239-253. Marcel Dekker Inc. New York.

Hobson, G.E. (1981) The Short Term Storage of Tomato Fruits. *Journal of Horticultural Science*. **56** , 363-368.

Hobson, G. and Grierson D. (1993) Tomato. In *Biochemistry of fruit ripening*. (G. Seymour, J. Taylor and G. Tucker (eds)), pp. 405-422. Chapman and Hall, Ltd. London.

Kader, A.A. 1980. Preventation of Ripening in Fruits by Using of Controlled Atmosphere Storage. *Food Technology:* **May**. 51-54.

Kader, A.A. (1986). Biochemical and Physiological Basis for Effects of Controlled and Modified Atmospheres on Fruits and Vegetables. *Food Technology*. **40**, 99-100, 102-104.

Kader ,A.A., D. Zagory and E.L. Kerbel. (1989) Modified Atmosphere Packaging of Fruit and Vegetables. *CRC Critical Reviews in Food Science and Nutrition* **28**,1-30.

Kader, A.A. (1992). Modified Atmospheres During Transport and Storage. In *Postharvest Technology of Horticultural Crops*. (Kader et al. (Eds)). Chapter 11, pp. 85-92. Agriculture and Natural Resources Publications. University of California. USA.

Kubo, Y., A. Inaba and R. Nakamura. (1989). Effects of High CO2 on Respiration in Various Horticultural Crops. *Journal of Japanese Society Horticultural Sciences*. **58**, 731-736.

Larsen, M., C.B. Watkins (1995). Firmness and Concentrations of Acetaldehyde, Ethylacetate and Ethanol in Strawberries Stored in Controlled and Modified Atmospheres. *Pastharvest Biology and Technology* **5**, 39-50.

Parsons, C.S., R.E. Anderson and R.W. Penney (1970). Storage of Mature Green Tomatoes in Controlled Atmospheres. *Journal of American Horticultural Science*. **95**, 791-796.

Peleg, K. (1985). Storage and Preservation Techniques. In *Produce Handling, Packaging and Distribution*. Chapter 3, p. 29-51. AVI Publishing Company Inc. Westport, Connecticut.

Ramana, S.V., Mohan-Kumar B.L. and Jayaraman K.S. (1987). Extension of Stored Life of Tomatoes Under Ambient Conditions by Continuous Flashing of Storage Atmosphere. *Indian Food Packer*. **41**, 24-29.

Riquelme, F., M.T. Pretel, G. Martinez, M. Serrano, A. Amoros and F. Romajoro. (1994). Packaging of Fruits and Vegetables: Recent Results. In *Food Packaging and Preservation*. (M. Mathlouthi (Ed)). Chapter 8, pp. 141-158. Blackie Academic and Professional. London.

Risse L.A., Miller W.R. and Ben-Yehoshua S. **(1985)** Weight Loss, Firmness, Colour and Decay Development of Individual Film Wrapped Tomatoes. *Tropical Science*. **25**, 117-121.

Salunke, D.K. and M.T. Wu. (1973). Effects of Low Oxygen Atmosphere Storage on Ripening and Associated Biochemical Changes of Tomato Fruits. *Journal of American Society of Horticultural Science*. **98**, 12-14.

Salunke, D.K. and M.T. Wu. (1974). Development in Technology of Storage and Handling of Fresh Fruits and Vegetables. In *Storage, Processing and Nutrition Quality of Fruits and Vegetables*. (D.K. Salunkhe (ed)), pp. 121-161. CRC Press.

Thompson, A.K. (1995). *Postharvest Technology of Fruit and Vegetables*. Logmans. London.

Yang, C.C., Brennan, P. Chinnan M.S. and Shewfelt. R.L. (1987). Characterisation of Tomatoes Ripening process As Influenced by Individual Seal-Packaging and Temperature. *Journal of Food Quality*. **10**, 21-33.

DEVELOPMENT OF AN ON-LINE DEVICE FOR FIRMNESS SENSING OF TOMATOES

Lesage P., Destain M.-F.

Department of Agricultural Engineering
Faculté des Sciences Agronomiques de Gembloux
5030 Gembloux (Belgium).

Astract: An on-line device for firmness sensing of tomatoes is presented. It consists in two small plungers (3 mm diameter) creating into the fruits a slight elastic peel deformation. The tomatoes are rolling on an endless conveyor and push on these plungers which are mounted at the extremities of two Cantilever beams. When a fruit passes over a plunger, the load applied on the beam produces a deflection. Both load and displacement are measured and the tomatoes firmness is computed in real time.

Keywords: Sensors, Fruits, Firmness, On-line device

1. INTRODUCTION

The firmness is a criterium often used to appreciate fruits quality as it is directly related to fruit development, ripening after maturity and ability to conservation. It is also related with the probability that fruits might bruise when subjected to different impacts.

"Firmness" in food materials corresponds to "hardness" in engineering materials (Sharma and Mohesin, 1970) and is measured by reference devices such as the EFFIGI probe and the INSTRON. The EFFIGI penetrometer consists in a 8 or 11.1 mm diameter plunger, manually operated. This device indicates the load applied to have peel collapse or tissue break (in the case where the peel has been removed). The INSTRON testing machine is a laboratory system providing the linear displacement of a 3 mm diameter probe at controlled speed. It is fitted with sensors of load and displacement, ensuring the ability of firmness computation. This device is generally used to indicate the collapse load, when testing fruits.

Other methods have been designed to realize non-destructive tests particularly suited for fruit testing. A firmness sorting system was developed which conveyed fruit horizontally at constant speed and caused them to impact on a rigid surface. Impact force characteristics were used to sort fruit into hard, firm and soft categories (Delwiche et al., 1989). A 3 mm diameter pin was used as a mechanical thumb to sense firmness of oranges and tomatoes and was included in devices to differentiate between firm and soft oranges, or hard-green, firm-red and soft-red tomatoes (Mizrach et al., 1992). Non-destructive techniques for measuring firmness using sonic or vibrational characteristics have been the subject of numerous investigations (De Baerdemaeker et al., 1982, Affeldt and Abbott, 1989, Abbott and Liljedahl, 1994; Vandewalle X. et al., 1994, Chen et al., 1994).

In previous studies, a non-destructive, manually operated firmness sensor was developed (Lesage et Destain, 1994). It was based on the creation of a slight elastic deformation of the tomato peel using a spring-loaded small plunger. The correlation between the firmness measured with this device and the collapse load obtained with the EFFIGI probe was good. Thanks to this device, it was shown that the firmness of a tomato is different according to the measurement location. The carpels, because of their cellular structure, contributes to the firmness, even for a low part, and hence enhances the global strenght of the fruit. Between the carpels, where the tomato medium consists in a viscous liquid, the mechanical resistance of the fruit is lower. Due to this heterogeneity, the firmess of a tomato vary around 12 % around the circumference.

On the other hand, in a lot of tomatoes picked at the same time, the firmess may vary as much as 25 %. The evolution of the fruits will be different according to that initial maturity state.

In this study, a small plunger is included in an on-line machine and some problems involved in this design are discussed.

2. SYSTEM DESIGN

2.1. Preliminary considerations

On basis of previous experimental work, the principle of the on-line firmness device is based on the slight progressive penetration of small plungers (3 mm diameter) into the peel fruits. The measurements are non-destructive provided that the plungers displacement remains lesser than 1.5 mm. According to the tomatoes firmness, this corresponds to forces of about 2 to 4 N. The desired minimum sorting rate is 5 fruit/s.

2.2. Mechanical design

The system is designed with two main endless belt conveyors, 5 cm wide, to carry the fruits at the specified speed. (Fig. 1). The conveyors are mounted in such a way that they have a special shape to hold the fruits laterally and to convey them well oriented in the sensing section. Furthermore, the two conveyors are spaced from 2 mm to position the plungers between them. Their lateral displacement is prevented since they bear a small rail guided on its course. Belt speed is controlled by a variable speed electric dc motor.

In the sensing section, tomatoes laying on the belts are maintained by an elastic device causing them to rotate while moving forward. This device consists in an endless belt having a speed higher than that of the support conveyors. The belt is free to move up and down, according to the fruit diameter (comprised between 57 and 67 mm). The force applied by the belt on the fruits is created by vertical springs. In that sensing section, the tomatoes push on to two small plungers of 3 mm diameter, sticked out of 1.5 mm in the conveyor axle. These values have been chosen to ensure that the peel deformation remains elastic and that no damage (mould, bruising marks, ...) appears several days after the tests. The two plungers are separated by about a half fruit circumference, in order to compute two times the firmness during the fruit course and to reproduce the trials of the manual EFFIGI testing, on both sides of the fruit.

Each of these small plungers is mounted at the extremity of a Cantilever beam. When a fruit passes over the plunger, the load applied on the beam produces a vertical deflection. The design equation of the beam is:

$$f = \frac{6\,F\,L^3}{E\,a\,e^3}$$

where

F	= applied load;
L	= length;
E	= modulus of elasticity;
a	= width;
e	= thickness;
f	= vertical deflection.

The stiffness of the Cantilever beam can be adjusted to a desired value located within the range of the stiffness values of the tested fruits. This is obtained by modifying its lenght.

2.3. Instrumentation and signal processing

The effective load and deflection of both Cantilever beams are measured with sensors.

Strain gages mounted on each beam (full bridge) provide out-of-balance voltages related with the load applied by the tomato on the plunger. The displacement of the beams extremity is measured by a linear variable differential transformer (LVDT).

The signals issued from the sensors are sent to a micrologger Campbell 21x and are stored in a PC, where the tomato firmness values are computed in real time by dividing the load by the corresponding displacement.

3. RESULTS AND DISCUSSION

The sensing machine is a suitable method for measuring on-line the tomatoes firmness. The sensitivity of the method is high and firmness data provide the basis for the decision of acceptance or rejection the fruit.

Nevertheless, before using this principle in completely reliable sorting machines, further tests have to be realized on larger scale. Particularly, a range of cut-off values has to be defined taking into account that there exists some variability in the tomatoes firmness, due to the fruits heterogeneity, as mentioned before.

The tomato skin is the most important component related to its mechanical strenght and its only protection against cracking damage. The creation of a slight elastic deformation of this peel with a measuring device is thus relevant. If the method is used to evaluate the firmness of rigid fruits (apples, for instance), further research need to be conducted to differentiate the biomechanical properties being sensed when using a destructive reference measurement (after removing the peel) from those being sensed with the non-destructive measurement (without removing the peel).

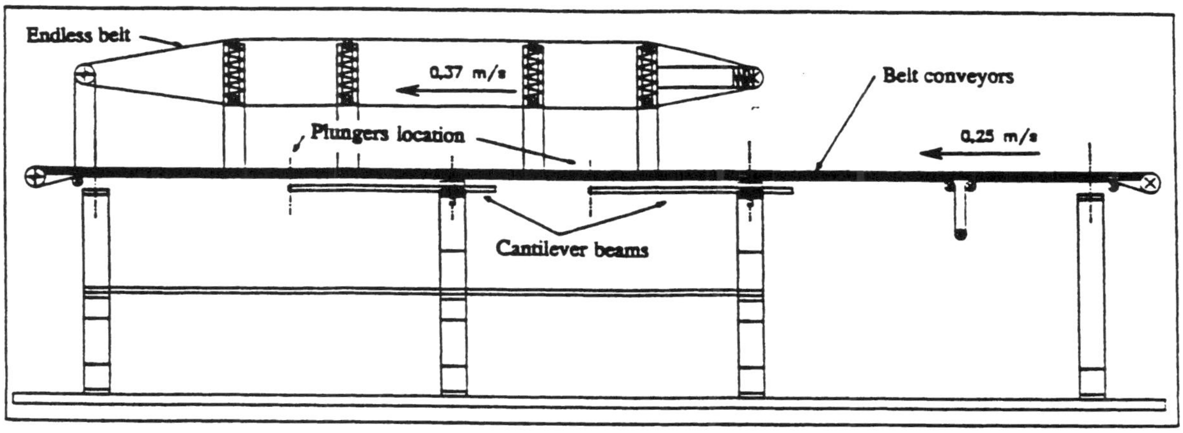

Fig. 1. Principle of the sensing machine.

4. CONCLUSION

An on-line tomato firmness sensing machine has been developed. It is based on the creation of a slight elastic deformation of the peel of the fruit. The load applied by the fruit and the resulting deformation are measured while the fruits are rolling on endless conveyors. These data provide the basis for the firmness computation.

ACKNOWLEGDMENTS

Funding has been provided by the "Ministère de la Région Wallonne, en application du Programme de Formation et d'Impulsion à la Recherche Scientifique et Technologique".

REFERENCES

Abbott, J.A. and Liljedahl, L.A. (1994). Relationship of sonic resonant frequency to compression tests and Magness-Taylor firmness of apples during refrigerated storage. Trans. of the ASAE. Vol. 37(4):1211-1215.

Affedt, H.A. and Abbott, J.A. (1989). Apple firmness and sensory quality using contact acoustic transmission. Land and water Use, Dood & Grace (eds), Balkema, Rotterdam, 2037-2045.

Chen, H., Duprat, F., Grotte, M., Loonis, D., Pietri, E.(1994). Influence of the inhomogeneity of the apple on the response spectra during nondestructive acoustic sensing of fruit firmness. Proceedings du Colloque "Plant Biomechanics", Montpellier.

De Baerdemaeker, J., Lemaitre, L., Meire, R. (1982). Quality detection by frequency-spectrum analysis of the fruit impact force. Trans. of the ASAE. Vol. 25(1):175-178.

Delwiche M.J., Tang S., Mehlschau J.J. (1989). An Impact Force Response Fruit Firmness Sorter. Trans. of the ASAE. Vol. 32(1), 321-326.

Lesage, P. and Destain, M.-F. (1994). Mise au point d'un dispositif de mesure non-destructif de la fermeté de fruits et légumes. Essais sur tomate. Proceedings du Colloque "Plant Biomechanics", Montpellier.

Mizrach, A., Nahir D.and Ronen B. (1992). Mechanical Thumb Sensor for Fruit and Vegetable Sorting. Trans. of the ASAE, Vol. 35(1), 247-250.

Sharma, M.G. and Mohesin N.N. (1970). Mechanics of deformation of a fruit subjected to hydrostatic presure. J. Agric. Engng. Res. 15 (1), 65-74.

Vandewalle, X., Langenakens, J., De Baerdemaeker, J. (1994). Influence of global shape and internal structure of the tomato on the reliability of firmness estimation by the acoustic impulse response technique. Proceedings du Colloque "Plant Biomechanics", Montpellier.

AUTHOR INDEX

Printed in Dunstable, United Kingdom

91898349R00185